INNOVATIVE FOOD SCIENCE AND EMERGING TECHNOLOGIES

The Science Behind Health

INNOVATIVE FOOD SCIENCE AND EMERGING TECHNOLOGIES

The Science Behind Health

Edited by

Rajakumari Rajendran
Anne George, MD
Nandakumar Kalarikkal, PhD
Sabu Thomas, PhD

Apple Academic Press Inc.
3333 Mistwell Crescent
Oakville, ON L6L 0A2
Canada

Apple Academic Press Inc.
9 Spinnaker Way
Waretown, NJ 08758
USA

First issued in paperback 2021

Exclusive worldwide distribution by CRC Press, a member of Taylor & Francis Group
No claim to original U.S. Government works
ISBN-13: 978-1-77463-150-8 (pbk)
ISBN-13: 978-1-77188-661-1 (hbk)

Library and Archives Canada Cataloguing in Publication

Innovative food science and emerging technologies / edited by Sabu Thomas, PhD, Rajakumari Rajendran, Anne George, MD, Nandakumar Kalarikkal, PhD.

Chapters stem from ICAFST--2015, the First International Conference on Advances in Food Science and Technology, held on November 20–22, 2015, in Kottayam, Kerala, India.
Includes bibliographical references and index.
Issued in print and electronic formats.
ISBN 978-1-77188-661-1 (hardcover).--ISBN 978-0-203-71140-8 (PDF)

1. Food industry and trade--Technological innovations--Congresses. I. Thomas, Sabu, editor II. Rajendran, Rajakumari, editor III. George, Anne, 1961-, editor IV. Kalarikkal, Nandakumar, editor V. International Conference on Advances in Food Science and Technology (2015 : Kottayam, India)

| TP370.I56 2018 | 664 | C2018-902504-2 | C2018-902505-0 |

CIP data on file with US Library of Congress

Apple Academic Press also publishes its books in a variety of electronic formats. Some content that appears in print may not be available in electronic format. For information about Apple Academic Press products, visit our website at **www.appleacademicpress.com** and the CRC Press website at **www.crcpress.com**

ABOUT THE EDITORS

Rajakumari Rajendran

Rajakumari Rajendran is a research fellow working at the International and Inter University Centre for Nanoscience and Nanotechnology, Mahatma Gandhi University, Kottayam, Kerala, India. She received a bachelor's degree in Pharmacy from the Tamilnadu Dr. MGR Medical University, India and a master's degree in Pharmacy (Pharmaceutics), Sastra University, India. She has industrial experience in dietary supplements and medicinal formulations and is currently engaged in developing dietary supplements at the International and Inter University Centre for Nanoscience and Nanotechnology, Mahatma Gandhi University, Kottayam, Kerala, India. She has awarded prestigious Inspire Fellowship (DST) and has research publications.

Anne George, MD

Anne George, MD, is Associate Professor at Government Medical College, Kottayam, Kerala, India. She did her MBBS (Bachelor of Medicine, Bachelor of Surgery) at Trivandrum Medical College, University of Kerala, India. She acquired a DGO (Diploma in Obstetrics and Gynecology) from the University of Vienna, Austria; Diploma Acupuncture from the University of Vienna; and an MD from Kottayam Medical College, Mahatma Gandhi University, Kerala, India. She has organized several international conferences, is a fellow of the American Medical Society, and is a member of many international organizations. She has five publications to her name and has presented 25 papers.

Nandakumar Kalarikkal, PhD

Nandakumar Kalarikkal, PhD, is Associate Professor and Head of Advanced Materials Laboratory, School of Pure and Applied Physics, Mahatma Gandhi University, Kottayam, India. He is also the Joint Director of the International and Inter University Centre for Nanoscience and Nanotechnology at the same university. He is a fellow of many professional bodies. He has authored many professional journal articles and has co-edited several books. He is actively involved in research, and his group works on the synthesis, characterization, and applications of various nanomaterials, ion irradiation effects on various novel materials, and phase transitions.

Sabu Thomas, PhD

Sabu Thomas, PhD, is the Pro-Vice Chancellor of Mahatma Gandhi University and Founding Director of the International and Inter University Center for Nanoscience and Nanotechnology, Mahatma Gandhi University, Kottayam, Kerala, India. He is also a full professor of polymer science and engineering at the School of Chemical Sciences of the same university. He is a fellow of many professional bodies. Professor Thomas has (co-)authored many papers in international peer-reviewed journals in the area of polymer science and nanotechnology. He has organized several international conferences. Professor Thomas's research group has specialized in many areas of polymers, which includes polymer blends, fiber-filled polymer composites, particulate-filled polymer composites and their morphological characterization, ageing and degradation, pervaporation phenomena, sorption and diffusion, interpenetrating polymer systems, recyclability and reuse of waste plastics and rubbers, elastomeric crosslinking, dual porous nanocomposite scaffolds for tissue engineering, etc. Professor Thomas's research group has extensive exchange programs with different industries and research and academic institutions all over the world and is performing world-class collaborative research in various fields. Professor Thomas's Center is equipped with various sophisticated instruments and has established state-of-the-art experimental facilities, which cater to the needs of researchers within the country and abroad.

Professor Thomas has published over 750 peer-reviewed research papers, reviews, and book chapters and has a citation count of 31,574. The H index of Prof. Thomas is 81, and he has six patents to his credit. He has delivered over 300 plenary, inaugural, and invited lectures at national/international meetings over 30 countries. He is a reviewer for many international journals. He has received MRSI, CRSI, nanotech medals for his outstanding work in nanotechnology. Recently Prof. Thomas has been conferred an Honoris Causa (DSc) by the University of South Brittany, France, and University Lorraine, Nancy, France.

CONTENTS

List of Contributors .... *xi*

List of Abbreviations .... *xv*

Preface .... *xxi*

PART I: Health Benefits and Industrial Perspectives of Superfoods, Seafoods, and Potential Foods 1

1. **Lion's Mane: The Medicinal Mushroom That Offers New Hope for Peripheral Neuropathy** 3
    Kah-Hui Wong and Syntyche Ling-Sing Seow

2. **Understanding Probiotics: A Review** 25
    Mradula Gupta, Somesh Sharma, and Nitin Kothari

3. **Aquatic Enzymes and Their Applications in the Seafood Industry** 43
    Parvathy U., Jeyakumari A., George Ninan, L. N. Murthy, S. Visnuvinayagam, and C. N. Ravishankar

4. **Utilization and Value Addition of Culinary Banana: The Potential Food for Health** 69
    Prerna Khawas and Sankar Chandra Deka

5. **Modification of Culinary Banana Resistant Starch and Its Application** 107
    Prerna Khawas and Sankar Chandra Deka

PART II: Potential Source of Proteins, Amino Acids, and Oils, and Safety of Food and Food Products 157

6. **Gelatin from Cold Water Fish Species and Their Functional Properties** 159
    Nor Fazliyana Mohtar and Conrad O. Perera

7. **Fish Processing Waste Protein Hydrolysate: A Rich Source of Bioactive Peptides** 199
    Girija G. Phadke, L. Narasimha Murthy, Asif U. Pagarkar, Faisal R. Sofi, and K. K. Sabha Nissar

8. Effect of Oregano Essential Oil on the Stability of
    Microencapsulated Fish Oil 215
    Jeyakumari A., Zynudheen A. A., Binsi P. K., Parvathy U., and Ravishankar C. N.

9. Safety Considerations of Raw Meat: An Indian Perspective.................225
    Vishva Maru and Vaijayanti V. Ranade

10. Acrylamide in Food Products: Occurrence, Toxicity,
    Detoxification, and Determination 249
    Kalyani Y. Gaikwad and K. A. Athmaselvi

PART III: Innovative Methods in Processing and Preservation
    of Food and Bioactive Components...287

11. Improving Quality of Ready-to-Eat, Minimally Processed
    Produce Using Non-Thermal Postharvest Technologies.........................289
    Rohanie Maharaj

12. Radiation Processing: An Emerging Post Harvest Preservation
    Method for Improving Food Safety and Quality 317
    Sumit Gupta and Prasad S. Variyar

13. Foods Preserved with Hurdle Technology 351
    Anurag Singh, Ashutosh Upadhyay, and Ankur Ojha

14. Radio Frequency Applications in Food Processing...........................375
    V. K. Shiby, Aisha Tabassum, and M. C. Pandey

15. Microencapsulation of Bioactive Food Ingredients: Methods,
    Applications, and Controlled Release Mechanism—A Review 399
    Jeyakumari A., Parvathy U., Zynudheen A. A., L. Narasimha Murthy,
    S. Visnuvinayagam, and Ravishankar C. N.

16. Electrospinning as a Novel Delivery Vehicle for Bioactive
    Compounds in Food..425
    Rajakumari Rajendran, Apparao Gudimalla, Raghavendra Mishra, Shivanshi Bajpai,
    Nandakumar Kalarikkal, and Sabu Thomas

17. An Underutilized Novel Xeric Crop: Kair (Capparis decidua) 455
    Yamini Chaturvedi and Ranjana Nagar

18. Value Addition of Underutilized Crops of India by Extrusion
    Cooking Technology...471
    Duyi Samyor, Amit Baran Das, and Sankar Chandra Deka

19. Fungal Retting Technology of Jute...493
    Avijit Das

20. **Composting of Cashew Apple Waste Residue for Cultivating Paddy Straw Mushroom**..511

Meghana V. Desai, David G. Gomes, Babu V. Vakil, and Vaijayanti V. Ranade

PART IV: Development, Optimization, Characterization, and Applications of Food Products 521

21. **Quality Characteristics of Custard Made from Composite Flour of Germinated Finger Millet, Rice, and Soybean**..........................523

Charanjit S. Riar and Surwase S. Bhaskarroa

22. **Kinnow Peel—Rice-Based Expanded Snacks: Investigating Extrudate Characteristics and Optimizing Process Conditions**..............................541

Himanshu Prabhakar, Shruti Sharma, Savita Sharma, and Baljit Singh

23. **Development and Partial Characterization of Biodegradable Film from Composite of Lotus Rhizome Starch, Whey Protein Concentrate, and Psyllium Husk**..561

Sakshi Sukhija, Sukhcharn Singh, and Charanjit S. Riar

24. **Green Leaf Protein Concentrate and Its Application in Extruded Food Products** 575

Jayabrata Saha and Sankar Chandra Deka

25. **Development of Spray-Dried Honey Powder with Vitamin C and Antioxidant Properties Using Maltodextrin as a Carrier**................595

Yogita Suhag, Gulzar Ahmad Nayik, and Vikas Nanda

Index..609

LIST OF CONTRIBUTORS

K. A. Athmaselvi
Department of Food Process Engineering, School of Bioengineering, SRM Institute of Science and Technology, Chennai, Tamil Nadu, India. E-mail: athmaphd@gmail.com

Shivanshi Bajpai
Yuveraj Dutta PG College, Lakhimpur, India

Binsi P. K.
ICAR—Central Institute of Fisheries Technology, Kochi 682029, India

Surwase S. Bhaskarroa
Department of Food Engineering and Technology, SLIET, Longowal 148106, Sangrur, Punjab, India

Dr. Yamini Chaturvedi
Department of H.Sc., Govt. of Rajasthan, Jaipur, Rajasthan, India

Amit Baran Das
Department of Food Engineering and Technology, Tezpur University, Napaam 784028, Sonitpur, Assam, India

Avijit Das
Division of Quality Evaluation and Improvement, National Institute of Research on Jute and Allied Fiber Technology, Kolkata 700040, India. E-mail: avijitcrri@gmail.com

Sankar Chandra Deka
Department of Food Engineering and Technology, Tezpur University, Napaam, Tezpur 784028, Sonitpur, Assam, India. E-mail: sankar@tezu.ernet.in

Meghana V. Desai
Department of Microbiology, Guru Nanak Khalsa College of Arts, Science and Commerce (Affiliated to University of Mumbai), Nathalal Parekh Marg, Matunga, Mumbai 400019, Maharashtra, India

Kalyani Y. Gaikwad
Department of Food Process Engineering, School of Bioengineering, Food Safety and Quality Management, SRM Institute of Science and Technology, Chennai, Tamil Nadu, India. E-mail: Kalyani.ga@gmail.com

David G. Gomes
Department of Microbiology, Guru Nanak Khalsa College of Arts, Science and Commerce (Affiliated to University of Mumbai), Nathalal Parekh Marg, Matunga, Mumbai 400019, Maharashtra, India

Apparao Gudimalla
Department of Nanotechnology, Acharya Nagarjuna University, Guntur 22510, Andhra Pradesh, India International and Inter University Centre for Nanoscience and Nanotechnology, Mahatma Gandhi University, Kottayam, Kerala, India

Mradula Gupta
School of Bioengineering and Food Technology, Shoolini University of Biotechnology and Management Sciences, Solan, Himachal Pradesh, India. E-mail: mridulagupta1988@gmail.com

Sumit Gupta
Food Technology Division, Bhabha Atomic Research Centre, Mumbai 400085, India

Jeyakumari A.
Mumbai Research Centre of ICAR—Central Institute of Fisheries Technology, Vashi, Navi Mumbai 400703, India. E-mail: jeya131@gmail.com

Nandakumar Kalarikal
International and Inter University Centre for Nanoscience and Nanotechnology, Mahatma Gandhi University, Kottayam, Kerala, India
School of Pure and Applied Physics, Mahatma Gandhi University, Kottayam, Kerala, India.
E-mail: nkkalarikkal@mgu.ac.in

Prerna Khawas
Department of Food Engineering and Technology, Tezpur University, Napaam, Tezpur 784028, Sonitpur, Assam, India

Nitin Kothari
Department of Pharmacology, Pacific Medical College, Udaipur, Rajasthan, India

Rohanie Maharaj
Biosciences, Agriculture and Food Technologies (BAFT), EC, IAF Campus, The University of Trinidad and Tobago, Caroni North Bank Road, Centeno, Piarco, Trinidad & Tobago. E-mail: rohanie.maharaj@utt.edu.ttrad

Vishva Maru
Department of Microbiology, Guru Nanak Khalsa College of Arts, Science and Commerce (Affiliated to University of Mumbai), Nathalal Parekh Marg, Matunga, Mumbai 400019, Maharashtra, India

Raghavendra Mishra
International and Inter University Center for Nanoscience and Nanotechnology, Mahatma Gandhi University, Kottayam, Kerala, India

Nor Fazliyana Mohtar
School of Fisheries and Aquaculture Sciences, Universiti Malaysia Terengganu, Kuala Terengganu 21030, Malaysia

L. Narasimha Murthy
Mumbai Research Centre of CIFT—Central Institute of Fisheries Technology, Vashi 400703, Navi Mumbai, Maharashtra, India

Ranjana Nagar
Department of H.Sc., University of Rajasthan, Jaipur, Rajasthan, India

Vikas Nanda
Department of Food Engineering and Technology, Sant Longowal Institute of Engineering and Technology—Deemed University, Longowal 148106, Sangrur, Punjab, India

Gulzar Ahmad Nayik
Department of Food Engineering and Technology, Sant Longowal Institute of Engineering and Technology—Deemed University, Longowal 148106, Sangrur, Punjab, India

George Ninan
ICAR—Central Institute of Fisheries Technology, Kochi 682029, Kerala, India

K. K. Sabha Nissar
Department of Fishery Biology, Faculty of Fisheries, Sher-e-Kashmir University of Agricultural Science and Technology of Kashmir, Jammu and Kashmir 190006, India

Ankur Ojha
Department of Food Science and Technology, National Institute of Food Technology, Entrepreneurship and Management, Sonepat, India

Asif U. Pagarkar
Marine Biological Research Station, Near Murugwada, Pethkilla, Ratnagiri 415612, Maharashtra, India

M. C. Pandey
Defence Food Research Laboratory, Siddharthanagar, Mysore 570011, India

Parvathy U.
Mumbai Research Centre of ICAR—Central Institute of Fisheries Technology, Vashi, Navi Mumbai 400703, India. E-mail: p.pillai2012@gmail.com

Conrad O. Perera
School of Chemical Sciences, Food Science Programme, The University of Auckland, Private Bag 92019, Auckland 1142, New Zealand. E-mail: conradperera@gmail.com

Girija G. Phadke
Mumbai Research Centre of CIFT—Central Institute of Fisheries Technology, CIDCO Admin Building I, Sector I, Vashi 400703, Navi Mumbai, Maharashtra, India. E-mail: girija_cof@yahoo.com

Himanshu Prabhakar
Punjab Agricultural University, Ludhiana, Punjab, India. E-mail: himanshup825@gmail.com

Rajakumari Rajendran
International and Inter University Center for Nanoscience and Nanotechnology, Mahatma Gandhi University, Kottayam, Kerala, India

Vaijayanti V. Ranade
Department of Microbiology, Guru Nanak Khalsa College of Arts, Science and Commerce (Affiliated to University of Mumbai), Nathalal Parekh Marg, Matunga, Mumbai 400019, Maharashtra, India. E-mail: vaijayantiranade@gmail.com

C. N. Ravishankar
ICAR—Central Institute of Fisheries Technology, Kochi 682029, India

Charanjit S. Riar
Department of Food Engineering and Technology, Sant Longowal Institute of Engineering and Technology—Deemed University, Longowal 148106, Sangrur, Punjab, India. E-mail: charanjitriar@yahoo.com

Jayabrata Saha
Department of Food Engineering and Technology, Tezpur University, Napaam, Tezpur 784028, Assam, India

Duyi Samyor
Department of Food Engineering and Technology, Tezpur University, Napaam 784028, Sonitpur, Assam, India

Syntyche Ling-Sing Seow
Mushroom Research Centre, Institute of Biological Sciences, Faculty of Science, University of Malaya, Kuala Lumpur 50603, Malaysia

Savita Sharma
Punjab Agricultural University, Ludhiana, Punjab, India

Shruti Sharma
Punjab Agricultural University, Ludhiana, Punjab, India

Somesh Sharma
School of Bioengineering and Food Technology, Shoolini University of Biotechnology and Management Sciences, Solan, Himachal Pradesh, India

V. K. Shiby
Defence Food Research Laboratory, Siddharthanagar, Mysore 570011. India. E-mail: shibyk@gmail.com

Anurag Singh
Department of Food Science and Technology, National Institute of Food Technology, Entrepreneurship and Management, Sonepat, India. E-mail: anurag.niftem@gmail.com

Baljit Singh
Punjab Agricultural University, Ludhiana, Punjab, India

Sukhcharn Singh
Department of Food Engineering and Technology, Sant Longowal Institute of Engineering and Technology—Deemed University, Longowal 148106, Sangrur, Punjab, India

Faisal R. Sofi
Department of Fish Processing Technology, College of Fishery Science, Sri Venkateshwara Veterinary University, Muthukur524344, Andhra Pradesh, India. E-mail: dr.sofi786@gmail.com
Division of Post-Harvest Technology (Fisheries), Faculty of Fisheries Sher-e-Kashmir University of Agricultural and Technology of Kashmir, Jammu and Kashmir 190006, India

Yogita Suhag
Department of Food Engineering and Technology, Sant Longowal Institute of Engineering and Technology—Deemed University, Longowal 148106, Sangrur, Punjab, India. E-mail:er.yogita18@gmail.com

Sakshi Sukhija
Department of Food Engineering and Technology, Sant Longowal Institute of Engineering and Technology—Deemed University, Longowal 148106, Sangrur, Punjab, India. E-mail: dietsakshi.2007@gmail.com

Aisha Tabassum
Defence Food Research Laboratory, Siddharthanagar, Mysore 570011, India

Sabu Thomas
School of Chemical Sciences, Mahatma Gandhi University, Kottayam, Kerala, India
International and Inter University Centre for Nanoscience and Nanotechnology, Mahatma Gandhi University, Kottayam, Kerala, India. E-mail: sabuthomas@mgu.ac.in

Ashutosh Upadhyay
Department of Food Science and Technology, National Institute of Food Technology, Entrepreneurship and Management, Sonepat, India

Babu V. Vakil
Department of Microbiology, Guru Nanak Khalsa College of Arts, Science and Commerce (Affiliated to University of Mumbai), Nathalal Parekh Marg, Matunga, Mumbai 400019, Maharashtra, India

Prasad S. Variyar
Food Technology Division, Bhabha Atomic Research Centre, Mumbai 400085, India. E-mail: prasadpsv@rediffmail.com

S. Visnuvinayagam
Mumbai Research Centre of ICAR—Central Institute of Fisheries Technology, Vashi, Navi Mumbai 400703, India

Kah-Hui Wong
Department of Anatomy, Faculty of Medicine, University of Malaya, Kuala Lumpur 50603, Malaysia
Mushroom Research Centre, Institute of Biological Sciences, Faculty of Science, University of Malaya, Kuala Lumpur 50603, Malaysia. E-mail: wkahhui@um.edu.my

Zynudheen A. A.
ICAR—Central Institute of Fisheries Technology, Kochi 682029, India

LIST OF ABBREVIATIONS

3-MCPD	3-monochloropropandiol
5′-GMP	guanosine-5′-monophosphate
5′-IMP	inosine-5′-monophosphate
AAC	ascorbic acid content
ACEI	ACE inhibitory
AMG	amyloglucosidase
ANNs	artificial neural networks
AOA	antioxidant activity
APEDA	Agriculture and Processed Export Development Authority
ASE	accelerated solvent extraction
BD	bulk density
BHA	butylated hydroxyanisole
BHT	butylated hydroxytoluene
BOD	biological oxygen demand
BP	blood pressure
BSA	bovine serum albumin
BSE	bovine spongiform encephalopathy
BV	breakdown viscosity
CAGR	compound annual growth rate
CCFS	Central Committee for Food Standards
CCRD	central composite rotatable design
CFB	Cytophaga–Flexibacter–Bacteroidetes
CMC	carboxymethyl cellulose
CNP	cellulose nanopaper
CO_2	carbon dioxide
COD	chemical oxygen demand
CSLM	confocal laser scanning microscope
CSTEE	Scientific Committee on Toxicity and the Environment
CV	coefficient of variation
DEAE cellulose	diethylaminoethyl cellulose
DEEC	direct expansion extrusion cooking
DEFT	direct epifluorescent filter techniques
DHA	docosahexaenoic acid
DNA	deoxyribonucleic acid

EC number	enzyme commission number
ECM	extracellular matrix
EE	encapsulation efficiency
ELISA	enzyme-linked immune assay
EMC	enzyme modified cheese
EOs	essential oils
EPA	eicosapentaenoic acid
EPR	electron paramagnetic resonance
ER	expansion ratio
ESR	electron spin resonance
EW	electrolyzed water
FACIT	fibril-associated collagens with interrupted triple helices
FAO	Food and Agricultural Organization
FDA	Food and Drug Administration
FFAs	free fatty acids
FPH	fish protein hydrolysates
FV	final viscosity
GA	genetic algorithm
GC	gas chromatography
GI	gastro intestinal
GMP	good manufacturing practices
GOPOD	glucose oxidase/peroxidase reagent
GRAS	generally recognized as safe
HACCP	hazard analysis and critical control points
HDP	hydrodynamic pressure processing
HI	hydrolysis index
HMF	hydroxymethylfurfural
HPBCD	hydroxypropyl betacyclodextrin
HPLC	high-performance liquid chromatography
HT	hurdle technology
HTST	high-temperature short time
HV	hold viscosity
IARC	International Agency for Research on Cancer
ICGFI	International Consultative Group on Food Irradiation
IFIP	International Project in Field of Food Irradiation
IFPA	International Fresh-cut Produce Association
IM	intermediate moisture
IR	infrared drying
IRR	infrared drying and radiation

IUBMB	International Union of Biochemistry and Molecular Biology
IVSD	in vitro starch digestibility
KFD	Kyasanur forest disease
KNOS	kinin-nitric oxide system
LaL	lambda lysozyme
LDPE	low-density polyethylene pouches
LED	light-emitting-diodes
LOD	level of detection
LPC	leaf protein concentrate
LSD	least significant difference
MAP	modified atmosphere packed
MCPD	monochloropropanediol
MFJ	model fruit juice
MMP	matrix metalloproteinase
MNV-1	murine norovirus-1
MOFPI	Ministry of Food Processing Industries
MR	Maillard reaction
MRM	multiple reaction monitoring mode
MS	mass spectrometry
MSG	monosodium glutamate
MTO	microencapsulated tuna oil
MW+G	microwave-grill combo
NEPS	neutral endopeptidase system
NIR	near-infrared
NIRJAFT	National Institute of Research on Jute and Allied Fibre Technology
NLS	native lotus rhizome starch
NMPPB	national meat and poultry processing board
NTOH	nucleation temperatures
	ohmic heating
ORP	oxidation-reduction potential
OSA	octenyl succinic anhydride
PAGE	polyacrylamide gel electrophoresis
PAH	polycyclic aromatic hydrocarbons
PAL	phenylalanine ammonia lyase
PCL	polycaprolactone
PDA	potato dextrose agar
PDCAAS	protein digestibility corrected amino acid score
PE	polyethylene
PEF	pulsed electric field

PEO	polyethylene oxide
PER	protein efficiency ratio
PET	polyethylene terephthalate
PFEC	pellet-to-flaking extrusion cooking
PFOA	perfluorooctanoic acid
PH	psyllium husk
PL	pulsed light
PLA	polylactic acid
PME	pectin methylesterase
PP	polypropylene
PPO	polyphenol oxidase
PSL	photostimulated luminescence
PT	pasting temperature
PUFA	polyunsaturated fatty acid
PV	peak viscosity
PVA	polyvinyl alcohol
PVC	polyvinyl chloride
RAS	renin-angiotensin system
RCS	rennin-chymase system
RDS	readily digestible starch
RF	radio frequency
RH	relative humidity
ROS	reactive oxygen species
RS	resistant starch
RSM	response surface methodology
RTE	ready-to-eat
SDS	slowly digestible starch
SEM	scanning electron microscopy
SFE	supercritical fluid extraction
SME	specific mechanical energy
SmF	solid submerged
SPI	soy macromolecule isolate
SSF	solid state
SSPS	soybean soluble polysaccharide
SV	setback viscosity
TAGs	triacylglycerides
TBA	thiobarbituric acid
TBARS	thiobarbituric acid reactive substances
TBHQ	tertiary-butyl-hydroquinone
TD	tray drying

TG	thermogravimetric
TLC	thin-layer chromatography
TPC	total phenolic content
TT	thermal treatment
TTA	total titratable acidity
TVs	tyrosine values
UAE	ultrasonic-assisted extraction
USAEC	United States Atomic Energy Commission
VP	vacuum packaging
WAC	water binding capacities
WAI	water absorption index
WHO	World Health Organization
WPC	whey protein concentrate
WVP	water vapor permeability
WVTR	water vapor transmission rate

PREFACE

Food can be defined as the nutritive material taken by an organism for growth, work, and for maintaining the vital processes. Food sustains life, and it is viewed as a pure source of nutrition. The study of food science is used to modify and improve the food in our society. The knowledge of food science and nutrition helps to improve the health and quality of life and is certainly something everyone is passionate about. Food science and technology is the field devoted to the application of basic sciences and engineering to study the nature of foods, deterioration, the principles of food processing, and the improvement of foods for the consuming public. The field of food technology is growing rapidly, and its development is making tremendous impact on life sciences and public health. The importance and significance can be gauged by the fact that it has made huge advancements over the course of time and is continuing to influence various sector.

ICAFST—2015, the First International Conference on Advances in Food Science and Technology, was held on November 20–22, 2015, in Kottayam, Kerala, India. It was jointly organized by Ayurveda-und Venen-Klinik, the Institute for Holistic Medical Sciences (IHMS), and Institute of Macromolecular Science and Engineering (IMSE).

The new emerging field of food science and technology has put forward many challenging opportunities in the use of these smart healthy foods. Food technologists show great interest in new strategies regarding food chemistry, a main branch of food technology. Recently food engineering has gained much interest, and due to its unique properties is receiving global attention nowadays. The growing interest among the industrialists and researchers in the field of food technology is the driving force behind this book.

The conference emphasized various aspects of preparation, characterization, morphology, properties, recyclability and advances, and challenging opportunities of food science. This book addresses a lot of information in the field of advances in food science and technology and processing. During the three-day conference, distinguished scientists specializing in various disciplines discussed basic studies, applications, recent advances, difficulties, and breakthroughs in the field of food science and technology and also addressed topics of novel issues. It included discussions on formulations, manufacturing techniques, biodegradable flexible packaging, packaged foods, beverages, fruits and vegetable processing, fisheries, milk, and milk products, frozen food and thermo processing, grain processing, meat and poultry processing, rheological characteristics foods, heat exchangers in the food industry, emerging technology for food processing, food and health, nature cures, food supplements, and spice and spice processing. Additionally, there was a poster session with more than 40 posters to encourage budding researchers in this field.

The book, *Innovative Food Science and Emerging Technologies,* is a collection of chapters from the delegates who presented their papers during the conference. This book includes a wide variety of topics in food science, manufacturing process, formulations, and their applications. We appreciate the efforts and enthusiasm of the contributing authors and acknowledge those who were prepared to contribute but were unable to do so at the time. The guest editors are Rajakumari Rajendran, Research Scholar (Inspire Fellow), International and Inter University Centre for Nanoscience and Nanotechnology, Mahatma Gandhi University, Kottayam, India; Dr. Anne George, MBBS, DGO, MD, MAMS, Institute of Holistic Medicine Sciences, Kottayam, India; Dr. Nandakumar Kalarikkal, School of Pure and Applied Physics, Mahatma Gandhi University, Kottayam, India; and Prof. Sabu Thomas, Director, International and Inter University Centre for Nanoscience and Nanotechnology, Professor, School of Chemical Sciences, Mahatma Gandhi University, Kottayam, India.

This book is intended to serve an important reference resource on recent research accomplishments in the area food science. In this book, we have given special emphasis on the new trends and developments in the field of food science that will be very helpful to food technology students and food scientists. We would like to thank all who kindly contributed chapters to this book. We are also very thankful to Apple Academic Press, for their kind help and assistance in preparation and publication of this book.

—Prof. Sabu Thomas

PART I

Health Benefits and Industrial Perspectives of Superfoods, Seafoods, and Potential Foods

CHAPTER 1

LION'S MANE: THE MEDICINAL MUSHROOM THAT OFFERS NEW HOPE FOR PERIPHERAL NEUROPATHY

KAH-HUI WONG[1,2*] and SYNTYCHE LING-SING SEOW[2]

[1]*Department of Anatomy, Faculty of Medicine, University of Malaya, Kuala Lumpur 50603, Malaysia*

[2]*Mushroom Research Centre, Institute of Biological Sciences, Faculty of Science, University of Malaya, Kuala Lumpur 50603, Malaysia*

Corresponding author. E-mail: wkahhui@um.edu.my

CONTENTS

Abstract 4
1.1 Introduction..4
1.2 Medicinal Properties of *H. erinaceus* 5
1.3 Neuroregenerative Potential of *H. erinaceus*...................................11
1.4 Conclusion 21
Acknowledgments...21
Keywords 21
References...21

ABSTRACT

Peripheral nerves are complex structures that can be found throughout the body reaching almost all tissues and organs to provide motor and/or sensory innervation. They have the unique capability to regenerate after injury. Insights into the regeneration of peripheral nerves after injury may have implications for neurodegenerative diseases of the nervous system. We investigated the ability of *Hericium erinaceus* fresh fruit bodies in the treatment of nerve injury following peroneal nerve crush in Sprague-Dawley rats by daily oral administration. The activities of water-soluble polysaccharide were compared to activities exhibited by mecobalamin (vitamin B12) that has been widely used in the treatment of peripheral nerve disorders. Hotplate test revealed an acceleration of sensory recovery of hind limbs in the treated groups compared to negative controls. Immunofluorescence staining of peroneal nerves with anti-rat endothelial cells antigen-1 that recognized endothelial cells showed that polysaccharides effectively restore the integrity of the blood–nerve barrier after crush injury. After nerve injury, the dorsal root ganglion (DRG) neurons become hyperexcitable and generate neuropathic pain. Activation of Akt and p38 mitogen-activated protein kinase (MAPK) signaling pathway mediates inflammatory pain and facilitates neuronal survival and outgrowth after crush injury. DRG neurons ipsilateral to the crush injury in rats of treated groups expressed higher immunoreactivity for Akt and p38 MAPK compared to negative controls, which indicates enhanced regenerative capacity of sensory neurons. Our findings suggest that *H. erinaceus* is capable of promoting peripheral nerve repair after injury.

1.1 INTRODUCTION

Hericium erinaceus (Bull.: Fr.) Pers. (Hericiaceae, higher Basidiomycetes, also known as Lion's Mane, Monkey's Head, Hedgehog Mushroom, Satyr's Beard, Pom Pom Blanc, Igelstachelbart, and Yamabushitake) is one of the edible and medicinal mushrooms distributed in Asia, Europe, and North America. It is a saprophytic inhabitant on dead trunks of hardwoods, including oak, walnut, beech, maple, sycamore, elm, and other broadleaf trees. It is found most frequently on logs or stumps and is one of the wood-destroying fungi that cause white rot. The first report of cultivation was published in China in 1988 (Suzuki & Mizuno, 1997). This mushroom can be grown on artificial log using bottles and polypropylene bags, making it possible to constantly market this mushroom all year round (Mizuno, 1999).

H. erinaceus has been well known for hundreds of years and treasured in traditional Chinese and Japanese cookery and herbal medicine. In China, it is called Houtou, as its fruit bodies look like the head of a baby monkey, and Shishigashira (Lion's Head). It is one of the famous four dishes in China (the other three are sea cucumber, bear palm, and bird's nest) (Mizuno, 1999). In Japan, it is called Yamabushitake because it resembles the ornamental cloth worn by Yamabushi-Buddhist monks practicing asceticism in the mountains. It is also called Jokotake (funnel-like), Usagitake (rabbit-like), and Harisen-bontake (porcupine fish-like) according to its appearance. Japanese scientists have studied and confirmed the biological activities of *H. erinaceus* as a highly prized medicinal mushroom (Mizuno, 1999).

1.2 MEDICINAL PROPERTIES OF *H. erinaceus*

The health benefits of *H. erinaceus* as a curative for problems of the digestive tract such as stomach and duodenal ulcers are widely known among Chinese doctors. The effectiveness of *H. erinaceus* tablets in the treatment of ulcers, inflammations, and tumors of the alimentary canal was proven in the clinical trial subjects of Shanghai Third People's Hospital (Chen, 1992). Ingestion of this mushroom was reported to have a remarkable effect in extending the life of cancer patients. Pills were used in the treatment of gastric and esophageal carcinoma (Ying et al., 1987). Further, sandwich biscuits supplemented with the fruit bodies were used in the prevention and treatment of nutritional anemia of preschool children (Liu et al., 1992).

Chemotherapeutic resistance to drugs is a major obstacle to the successful treatment of human hepatocellular carcinoma. Lee and Hong (2010) demonstrated that purified components of *H. erinaceus* act as enhancers to sensitize doxorubicin (Dox)-mediated apoptotic signaling, and this sensitization can be achieved by reducing Cellular FLICE-inhibitory protein (c-FLIP) expression via Jun N-terminal kinases (JNK) activation and enhancing intracellular Dox accumulation via the inhibition of nuclear factor-kappa beta (NF-kB) activity. These findings suggest that *H. erinaceus* in combination with Dox serves as an effective tool for treating drug-resistant human hepatocellular carcinoma.

Methicillin-resistant *Staphylococcus aureus* (MRSA) is currently one of the most prevalent pathogens in nosocomial infections. Erinacines A, B (Kawagishi, 2005), and K (Kawagishi et al., 2006) were isolated as anti-MRSA compounds from the mycelium. A clinical trial in Japan showed that MRSA in some patients disappeared after they consumed the mushroom.

The hot water extract of dried fruit bodies is used as a health drink. It has been a practice to extract the mushroom with water or pickle it in brewed wine. A sports drink named Houtou was employed in the 11th Asia Sport Festival (1990) in China (Mizuno, 1999). Alcohol-based extracts prepared from pure mycelium cultured on organic brown rice are produced by Fungi Perfecti Co. (Olympia, WA, USA) and tablets of polysaccharide are manu-factured by Shanghai Baixin Edible and Medicinal Fungi of the Edible Fungi Institute (Shanghai, China). In Malaysia, capsules containing 100% pure powder of *H. erinaceus* are manufactured by Reishilab (Selangor, Malaysia) and marketed under ANI *Hericium* 450 mg, and an essence is produced by Vita Agrotech (Selangor, Malaysia) by combining three types of mushrooms, namely, *Agaricus brasiliensis*, *H. erinaceus*, and *Auricularia auricula-judae*.

Kawagishi et al. (2008) reported that the most promising activity of *H. erinaceus* is the stimulation of nerve growth factor (NGF) synthesis by heri-cenones from fruit bodies and by erinacines from mycelium. Phenol deriva-tives identified as hericenones C, D, E (Kawagishi et al., 1991), F, G, and H (Kawagishi et al., 1993) were isolated from fruit bodies of *H. erinacoeus*. Diterpene-xyloside possessing cyathan skeletons derivatives identified as erinacines A, B, C (Kawagishi et al., 1994), D (Kawagishi et al., 1996a), E, F, G (Kawagishi et al., 1996b), H, I (Lee et al., 2000), P (Kenmoku et al., 2000), Q (Kenmoku et al., 2000), J, K (Kawagishi et al., 2006), and R (Ma et al., 2008) as well as erinacol (Kenmoku et al., 2004) were isolated from mycelium of *H. erinaceus*. These compounds accelerated the synthesis of NGF and are observed to stimulate neurons to regrow. Shimbo et al. (2005) showed that erinacine A increased catecholamine and NGF content in the central nervous system (CNS) of rats. Rats treated with this compound had increased levels of both noradrenaline and homovanillic acid in the locus coeruleus at 4 weeks of age and increased levels of NGF in both LC and hippocampus at 5 weeks of age.

A double-blind trial was performed on 50 to 80-year-old Japanese men and women diagnosed with mild cognitive impairment in order to examine the efficacy of oral administration of *H. erinaceus* in improving cogni-tive functioning by using a cognitive function scale based on the Revised Hasegawa Dementia Scale (HDS-R) (Mori et al., 2009). The subjects of the *H. erinaceus* group took four 250 mg tablets containing 96% of dry powder three times a day for 16 weeks. Cognitive function scale scores increased with the duration of intake. Laboratory tests showed no adverse effect of *H. erinaceus* and it was effective in improving mild cognitive impairment (Mori et al., 2009).

Nagano et al. (2010) investigated the clinical effects of *H. erinaceus* on menopause, depression, sleep quality, and indefinite complaints, by means of a questionnaire investigation. They detected a difference between groups using the Kupperman Menopausal Index (KMI), the Center for Epidemiologic Studies Depression Scale (CES-D), the Pittsburgh Sleep Quality Index (PSQI), and the Indefinite Complaints Index (ICI). Their results showed that *H. erinaceus* intake may reduce depression and anxiety. It may also be relevant to frustration and palpitation because *H. erinaceus* intake lowers scores for the terms frustrating and palpitation. For this reason, they suggested a mechanism that is different from the NGF-enhancing action of *H. erinaceus*. Mori et al. (2011) examined the effects of *H. erinaceus* on amyloid $\beta(25-35)$ peptide-induced learning and memory deficits in mice. *H. erinaceus* prevented impairments of spatial short-term and visual recognition memory induced by intracerebroventricular administration of amyloid $\beta(25-35)$ peptide and may be useful in the prevention of cognitive dysfunction.

Neurotrophic effects of *H. erinaceus* dried fruit bodies on the neurons of hippocampal slices in rats have been studied (Grygansky et al., 2001; Moldavan et al., 1999). The mushroom exerted neurotrophic action or excitation of hippocampal neurons at concentrations that did not affect the growth of the neurons in vitro or evoke toxic effects that cause damage to the neurons. Hippocampal neurons as part of the limbic system are closely related to the regulation of motivation-emotional responses, memory, and other mental functions. A characteristic feature of this structure is an extreme sensitivity to slight changes in the intercellular liquid composition, which is considerably greater than that of neocortex and cerebellar neurons (Artemenko et al., 1983). The extract also promoted normal development of cultivated cerebellar neurons and demonstrated a regulatory effect on the process of myelin genesis in vitro after myelin damage (Kolotushkina et al., 2003). The myelin sheath is an important component of neurons that is involved in the transmission of nerve messages. Injury of myelin's compact structure leads to an impairment and severe illness of the nerve system.

Since 2000, *H. erinaceus* has been successfully domesticated via adaptation to tropical climate in Malaysia. This mushroom grows and produces fruit bodies in lowlands of tropical temperature (Fig. 1.1). It has been shown that cultivation conditions did not affect selected bioactive properties of *H. erinaceus* grown in tropical Malaysia (Wong et al., 2009a).

The mushroom has been extensively tested in in vitro trials in our laboratories as the neurite outgrowth stimulator in the cultured cells of the neural hybrid clone NG108-15 (Wong et al., 2007; Lai et al., 2013). This hybrid neuronal cell was developed in 1971 by Bernd Hamprecht (American Type

Culture Collection, 2004) by fusing mouse N18TG2 neuroblastoma cells with rat C6-BU-1 glioma cells in the presence of inactivated Sendai virus.

FIGURE 1.1   **(See color insert.)** *H. erinaceus* fresh fruit bodies grown in tropical climate of Malaysia.

Proper neuronal migration and establishment of circuitry are key processes for nervous system functioning. During development, neurons extend numerous processes, also termed neurites, which differentiate into dendrites and axons. The inability of damaged nerve fibers to regenerate is an active process under the control of molecules that inhibit and repulse growing neurites. Therefore, major efforts in nervous system drug discovery research are focused on the identification of compounds that affect the growth of neurites.

Figure 1.2 shows the effects of aqueous extracts at 0.2% (v/v) on neurite outgrowth of NG108-15 cells after 72 h of incubation. Obvious enhancement of neurite outgrowth was observed in cells treated with positive control (NGF 20 ng/mL) and extracts of fresh fruit bodies, freeze-dried fruit bodies, and mycelium. Cells exhibited neurite networks with an exuberant outgrowth of long, diverse, beaded, multipolar, and fine-meshed branching neurites. However, neurites were not induced by cells not treated with extract (negative control) and cells treated with extracts of oven-dried fruit bodies. In

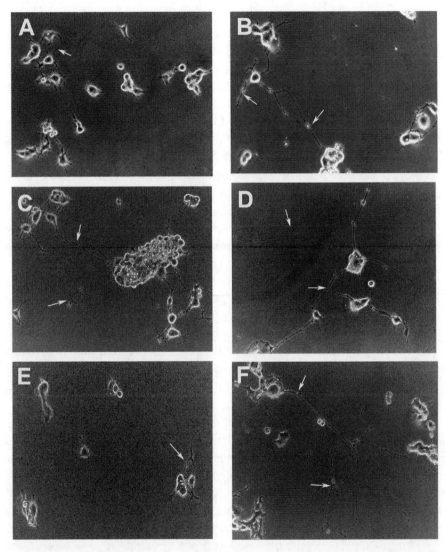

FIGURE 1.2 (See color insert.) The morphology of neural hybrid clone NG108-15 cells treated with 0.2% (v/v) aqueous extracts of fruit bodies and mycelium of *H. erinaceus* after 72 h of incubation at 37 ± 2°C in a 5% CO_2 humidified incubator ×100. NG108-15 cells without extract or treated with NGF (20 ng/mL) were used as negative and positive controls, respectively. Arrows indicate neurites. (A) Negative control without extract. (B) Positive control: NGF (20 ng/mL). (C) Fresh fruit bodies aqueous extract 0.2% (v/v). (D) Freeze-dried fruit bodies aqueous extract 0.2% (v/v). (E) Oven-dried fruit bodies aqueous extract 0.2% (v/v). (F) Mycelial aqueous extract 0.2% (v/v). (A) and (E): Cells attached strongly and exhibited short cellular extensions, but these have not elongated adequately to be scored as neurites. (B–D) and (F): Cells show an exuberant outgrowth of long, diverse, beaded, and branching neurites.

negative control or cells treated with extracts of oven-dried fruit bodies, they attached strongly and exhibited short cellular extensions, but these had not elongated adequately to be scored as neurites. There was very little neurite networks formation. Based on the scoring of cells that possessed neurites, it was observed that aqueous extract of fresh fruit bodies showed the most potent stimulation of neurite outgrowth in vitro. Therefore, we further explored the ability of a water-soluble polysaccharide from the aqueous extract of fresh fruit bodies in the enhancement of peripheral nerve regeneration after crush injury in vivo (Wong et al., 2009b, 2011, 2012, 2016).

Peripheral nerve injury is a serious health concern for society, which always results in restricted activity or life-long disability. Peripheral nerve injuries are encountered in clinical practice due to accidental trauma, acute compression, or surgery. Traffic crashes usually induce traumatic nerve injury resulting in the disruption of intraneural circulation. This condition induces demyelination, remyelination, axonal degeneration, axonal regeneration, focal, multifocal, or diffuse nerve fiber loss, and endoneural edema (Lundborg, 1988). After injury, free radicals are elevated, producing more tissue damage and retarding the recovery process.

Although microsurgical treatments for nerve injuries have been improved over the past decades, the outcome of peripheral nerve injury repair remains unsatisfactory. Peripheral nervous system has an intrinsic ability for repair and regeneration, therefore, making it a well-accepted neuroregeneration research model.

Studies of neurotrophic factors are aimed at finding new and more effective treatments for nerve disorders. These substances, which are produced naturally by the body, protect neurons from injury and encourage their survival. Neurotrophic factors also maintain normal function in mature neurons and stimulate axonal regeneration. The effects of these powerful chemicals on the peripheral nervous system may eventually lead to treatments that can reverse nerve damage and cure peripheral nerve disorders.

Drug therapy is commonly used to promote axonal regeneration in the treatment of nerve injuries (Schumacher et al., 2007) including crush injury, transected injury, or large nerve gap. Therefore, searching for effective drugs, especially those of natural origin, has gained extensive attention. Immunosuppressant and anti-inflammatory drugs may accelerate the rate of nerve regeneration following injury. However, they are associated with severe side effects such as high blood pressure, kidney problems, and liver disorders (Wierzba et al., 1984). Therefore, it is important to search for natural substances and possible new drug treatments that could affect nerve regeneration.

We investigated the neuroregenerative properties of polysaccharide from locally cultivated *H. erinaceus* on sensory functional recovery following crush injury to the peroneal nerve by behavioral experiment as assessed by hot plate test. Restoration of blood–nerve barrier (BNB), expression of protein kinase B (Akt), and p38 MAPK in the dorsal root ganglia (DRG) were examined by immunofluorescence techniques. Further, motor endplates and nerve fibers were evaluated in the extensor digitorum longus (EDL) muscles by combined silver-cholinesterase staining.

1.3   NEUROREGENERATIVE POTENTIAL OF *H. erinaceus*

The rat experiment was approved by the Institutional Animal Care and Use Committee, Faculty of Medicine, University of Malaya. Crush injury to the peroneal nerve (Fig. 1.3) results in foot drop (Fig. 1.4). It can be defined as a significant weakness of ankle and toe dorsiflexion. The foot and ankle dorsi-flexors include the tibialis anterior, extensor hallucis longus, and extensor digitorum longus. These muscles help the body clear the foot during swing phase and control plantar flexion of the foot on heel strike. The rats tend to drag the dorsum of their foot until reinnervation of axons into these muscles.

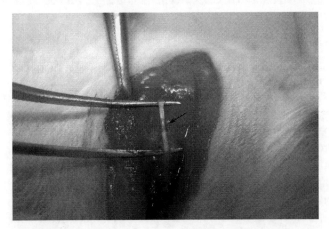

FIGURE 1.3   (See color insert.) A crush injury was created using a fine watchmaker forceps no. 4 for 10 s on the peroneal nerve at 10 mm from extensor digitorum longus muscle and complete crush was confirmed by the presence of a translucent band across the nerve as indicated by an arrow. All operations were performed on right limb and the left limb served as an uninjured control. (Reprinted from International Journal of Medicinal Mushrooms, Vol 14, Issue 5, K.-H. Wong, M. Naidu, P. David, R. Bakar & V. Sabaratna, Neuroregenerative Potential of Lion's Mane Mushroom, Hericium erinaceus (Bull.: Fr.) Pers. (Higher Basidiomycetes), in the Treatment of Peripheral Nerve Injury (Review), pp. 427-446, © 2012, with permission from Begell House, Inc.)

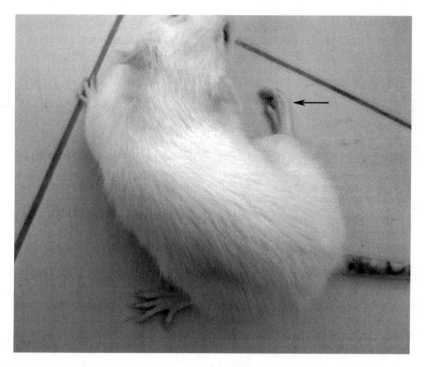

FIGURE 1.4   (See color insert.) Gait changes associated with peroneal nerve injury–joint contracture, making measurement impossible because the rat walks on the dorsum of the affected foot. Arrow indicates the operated limb.

Negative control group received daily oral administration of distilled water (10 mL/kg body weight per day), experimental group received polysaccharide of *H. erinaceus* (10 mL/kg body weight/day) and positive control group received mecobalamin (Natural Factors® batch 020564, 130 µg/kg body weight/day) for 14 days to function as pre-treatment before injury. Mushrooms have always been prepared for medicinal use by hot water extraction as in brewing of teas or decoctions in traditional Chinese medicine for prevention of oxidative stress-related diseases. In our study, pre-treatment with polysaccharide for 14 days was employed to build up strength and immune system before peripheral nerve injury.

The hot plate test is commonly used for evaluating thermal pain sensitivity. During the experiment, the rats were introduced into an open-ended cylindrical space with a floor consisting of a heated plate (Panlab, S. L. Spain). The plate heated to 50°C produces two behavioral components that can be measured in terms of their reaction times, namely paw licking and jumping (Fig. 1.5).

FIGURE 1.5   (See color insert.) Hot plate apparatus comprises of an open-ended cylindrical space with a floor of heated plate. Skin of the plantar surface of the foot was stimulated by gently placing the rat on the heated surface. Thermal nociception was evaluated by observing the withdrawal reflex latency (WRL) of the operated limb in response to heat stimulation (as indicated by an arrow) by a built-in *timer* activated by an external foot switch. (Reprinted from Wong, Kah-Hui et al . Restoration of sensory dysfunction following peripheral nerve injury by the polysaccharide from culinary and medicinal mushroom, Hericium erinaceus (Bull.: Fr.) Pers. through its neuroregenerative action. **Food Sci. Technol (Campinas)**, Campinas , v. 35, n. 4, p. 712-721, Dec. 2015 .   Available from http://www.scielo.br/scielo.php?script=sci_arttext&pid=S0101-20612015000400712&lng=en&nrm=iso   https://creativecommons.org/licenses/by/4.0/deed.en)

Polysaccharide from *H. erinaceus* could attenuate thermal hyperalgesia induced by peroneal nerve injury and enhanced the reflex response of the operated limb after 8 days of injury and these significant changes were observed subsequently during the remainder of the time period (Fig. 1.6). This effect is due to faithful reinnervation of sensory receptors by their original axons. Even when good motor recovery occurs, sensory deficits, particularly in proprioception, may impair the functional outcome. Sensory recovery may continue for a longer period of time than motor recovery as the axons of sensory neurons provide cutaneous sensation to a larger area than it had prior to the injury (Harden et al., 2012).

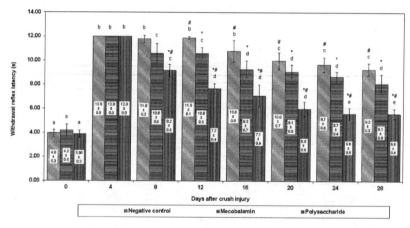

FIGURE 1.6 Mean WRL for right hind limb induced by thermal stimulation. Values in seconds (s) were obtained by performing WRL test to evaluate the nociceptive function. Each bar represents WRL from six animals per group. Asterisks (*) indicate significant differences ($p < 0.05$, DMRT) in values for different groups compared to the negative control within a same experimental day. Hash signs (#) indicate significant differences ($p < 0.05$, DMRT) in values for different groups compared to mecobalamin within a same experimental day. The same alphabet indicates no significant differences and different alphabets indicate significant differences ($p < 0.05$, ANOVA) between different experimental days in the same group. (Reprinted from Wong, Kah-Hui et al . Restoration of sensory dysfunction following peripheral nerve injury by the polysaccharide from culinary and medicinal mushroom, Hericium erinaceus (Bull.: Fr.) Pers. through its neuroregenerative action. **Food Sci. Technol (Campinas)**, Campinas , v. 35, n. 4, p. 712-721, Dec. 2015 . Available from http://www. scielo.br/scielo.php?script=sci_arttext&pid=S0101-20612015000400712&lng=en&nrm=iso https://creativecommons.org/licenses/by/4.0/deed.en)

Figure 1.6 presents the data for the WRL during the healing period of 4 weeks. As expected, during the first week following peroneal nerve crush, rats were unable to respond to the hot stimulus. This indicates complete loss of thermal and nociceptive sensitivity at the sole of the operated foot. Signs of recovery of foot's withdrawal response began at week 2 after crush injury. Thereafter, the WRL steadily improved during the 4-week recovery time. However, the normal value of 4 s was unachievable (Hu et al., 1997).

Negative control rats showed no signs of recovery in WRL response during the healing period of 4 weeks. The WRL at week 4 for the polysaccharide group was almost in the normal value range of 5.6 ± 0.4 s. When the mean WRLs were compared at each time interval, the mean WRLs of the polysaccharide-treated rats were significantly less than the mecobalamin group from day 12 onwards except day 24 ($p < 0.05$). Therefore, WRL in rats treated with polysaccharide recovered faster and better compared to those receiving mecobalamin from day 12 onwards ($p < 0.05$). There was no significant difference in mean WRLs between rats in mecobalamin and negative control groups after 4 and 8 days of injury.

Rat endothelial cell antigen (RECA-1)-positive cells delineated the tube-shaped blood vessels in contralateral uninjured nerve (Fig. 1.7A). Immunoreactivity for RECA-1 decreased in negative control group after crush injury (Fig. 1.7B). However, there was a bright immunofluorescence for the antigen in the regenerating axons distal to the injury site in the treated groups (Fig. 1.7C,D). Blood vessels regained their tube-shape in the treated groups at the same time point. Expression of RECA-1 occurred in regenerating axons and upregulated by polysaccharide from *H. erinaceus* after crush injury. Figure 1.7E shows the intensity measurement of RECA-1 immunoreactivity in the peroneal nerve axon. In injured nerve, the intensity followed the order polysaccharide (20.71 ± 2.9) > mecobalamin (13.55 ± 2.2) > negative control (9.23 ± 0.6). Intensity of RECA-1 was significantly higher ($p < 0.05$) in the polysaccharide and mecobalamin groups compared to negative control group.

Injury to neurons results in complex sequences of molecular responses that play an important role in the successful regenerative response and the eventual recovery of sensory function. Peripheral sensitization causes activation of Akt and p38 MAPK in small DRG neurons. These contributed to pain hypersensitivity found at the site of tissue damage and inflammation. In general, DRG neurons can be divided into large (>1200 μm^2), medium (600–1200 μm^2), and small (<600 μm^2) neurons. Small neurons respond to thermal, mechanical, and chemical nociceptive stimulations, whereas large neurons transmit touch and proprioceptive sensations (Wen et al., 2009). Akt and p38 MAPK did not co-localize with NF-200 in large DRG neurons. It is possible that polysaccharide from *H. erinaceus* could trigger the expression of protein kinases that regulate nociceptive function and inflammation during nerve regeneration.

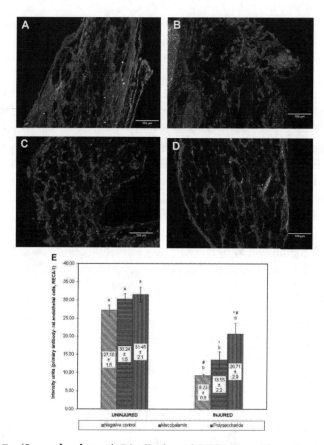

FIGURE 1.7 (See color insert.) Distribution of RECA-1-positive microvessels in the peroneal nerve. Fluorescence imaging staining for RECA-1 as green fluorescent lines. DNA was stained as blue fluorescence. (A) Uninjured nerve (contralateral side). (B) Distal to the injury site of injured nerve in negative control group. (C) Distal to the injury site of injured nerve in positive control group. (D) Distal to the injury site of injured nerve in polysaccharide group. Scale bar = 100 μm. (E) RECA-1 levels in the peroneal nerve as measured by intensity of immunoreactivity. Each bar represents 24 sections of DRG from 6 animals per group. Asterisks (*) indicate significant differences ($p < 0.05$, DMRT) in values for different groups compared to the negative control within a same uninjured/injured category. Hash signs (#) indicate significant differences ($p < 0.05$, DMRT) in values for different groups compared to mecobalamin within a same uninjured/injured category. The same alphabet indicates no significant differences and different alphabets indicate significant differences ($p < 0.05$, ANOVA) between different categories in the same group. (Reprinted from Wong, Kah-Hui et al . Restoration of sensory dysfunction following peripheral nerve injury by the polysaccharide from culinary and medicinal mushroom, Hericium erinaceus (Bull.: Fr.) Pers. through its neuroregenerative action. **Food Sci. Technol (Campinas)**, Campinas , v. 35, n. 4, p. 712-721, Dec. 2015 . Available from http://www.scielo.br/scielo.php?script=sci_ arttext&pid=S0101-20612015000400712&lng=en&nrm=iso https://creativecommons.org/ licenses/by/4.0/deed.en)

Figure 1.8 shows the in vivo expression of Akt within the DRG after crush injury. Double-labeled immunofluorescence demonstrated no co-localization of Akt and NF-200 in large DRG neurons. The population of large neurons is defined by the expression of NF-200 as green fluorescence. Akt staining as orange fluorescence was targeted to small neurons. Immunoreactivity for Akt was not detected in small neurons of contralateral DRG from uninjured nerve (Fig. 1.8A). Immunoreactivity of ipsilateral DRG in negative control (Fig. 1.8B) and mecobalamin group (Fig. 1.8C) were less than that observed in polysaccharide group (Fig. 1.8D). There was a bright immunofluorescence for Akt in small neurons of ipsilateral DRG from injured nerve in polysaccharide group. Intensity measurement of Akt immunoreactivity in DRG was shown in Figure 1.8E. In ipsilateral DRG, the intensity recorded followed the order polysaccharide (31.67 ± 1.2) > mecobalamin (30.76 ± 0.9) > negative control (23.41 ± 0.4). The significant increase in intensity was detected in polysaccharide group compared to negative control and mecobalamin groups ($p < 0.05$, $n = 6$).

Figure 1.9 shows the expression of p38 MAPK within the DRG after crush injury. Double staining of p38 MAPK with NF-200 showed no colocalization in the DRG. Immunoreactivity for p38 MAPK was not detected in neurons of contralateral DRG (Fig. 1.9A). In ipsilateral DRG, the intensity recorded followed the order polysaccharide (34.65 ± 0.2) > mecobalamin (29.81 ± 0.4) > negative control (22.31 ± 0.6) (Fig. 1.9E). Although all small neurons were stained brightly with p38 MAPK (Fig. 1.9B–D), higher level of the antigen was detected in ipsilateral DRG of rats treated with polysaccharide compared to negative control and mecobalamin ($p < 0.05$, $n = 6$).

Akt promotes cell survival by inhibiting apoptosis through phosphorylation and inactivation of several targets, including Bad, forkhead transcription factors, c-Raf, and caspase-9 (Franke et al., 1997). p38 MAPK is regulated by stress-inducing signals such as osmotic stress, UV radiation, pro-inflammatory cytokines and certain toxins (Okada et al., 2007). Activation of MAPK pathway is essential for neurite outgrowth, regeneration, synaptic plasticity, and memory functions in mature neurons (Sweatt, 2001).

Axonal reinnervation of the EDL muscle was enhanced in polysaccharide group than in negative control group after 14 days of crush injury as demonstrated by combined silver-cholinesterase staining (Fig. 1.10). An important difference between muscles of the groups occurs with respect to the regenerating axons. Extensor digitorum longus muscles of negative control group contained a mixture of degenerating and regenerating axons, and migration of macrophages to remove degenerated myelin and axon fragments, a process called Wallerian degeneration. Functional connection

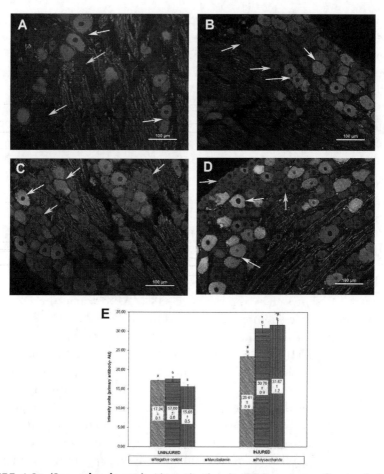

FIGURE 1.8 (See color insert.) Akt activation in DRG neurons after crush injury. Double immunofluorescence staining between Akt (orange) and NF-200 (green) in contralateral and ipsilateral DRG. DNA was stained as blue fluorescence. (A) DRG from uninjured nerve (contralateral side). (B) DRG from injured nerve in negative control group. (C) DRG from injured nerve in positive control group. (D) DRG from injured nerve in polysaccharide group. Akt did not co-localize with large neurons. White arrows indicate large neurons, whereas yellow arrows indicate small neurons. Scale bar = 100 μm. (E) Akt levels in the DRG as measured by intensity of immunoreactivity. Each bar represents 24 sections of DRG from six animals per group. Means with different letters in uninjured/injured category are significantly different ($p < 0.05$). (Reprinted from Wong, Kah-Hui et al . Restoration of sensory dysfunction following peripheral nerve injury by the polysaccharide from culinary and medicinal mushroom, Hericium erinaceus (Bull.: Fr.) Pers. through its neuroregenerative action. **Food Sci. Technol (Campinas)**, Campinas , v. 35, n. 4, p. 712-721, Dec. 2015 . Available from http://www.scielo.br/scielo.php?script=sci_arttext&pid=S0101-20612015000400712&lng=en&nrm=iso https://creativecommons.org/licenses/by/4.0/deed.en)

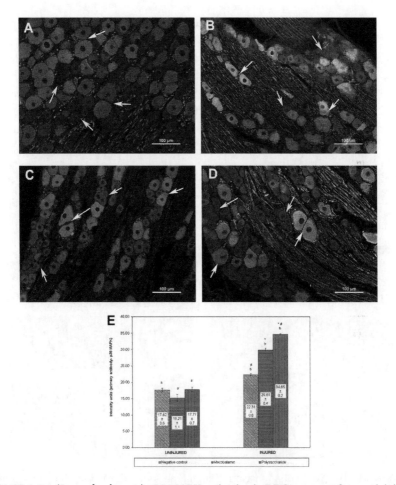

FIGURE 1.9 (See color insert.) p38 MAPK activation in DRG neurons after crush injury. Double immunofluorescence staining between p38 MAPK (orange) and NF-200 (green) in contralateral and ipsilateral DRG. DNA was stained as blue fluorescence. (A) DRG from uninjured nerve (contralateral side). (B) DRG from injured nerve in negative control group. (C) DRG from injured nerve in positive control group. (D) DRG from injured nerve in polysaccharide group. p38 MAPK did not co-localize with large neurons. White arrows indicate large neurons, whereas yellow arrows indicate small neurons. Scale bar = 100 μm. (E) p38 MAPK levels in the DRG as measured by intensity of immunoreactivity. Each bar represents 24 sections of DRG from six animals per group. Means with different letters in uninjured/injured category are significantly different ($p < 0.05$). (Reprinted from Wong, Kah-Hui et al . Restoration of sensory dysfunction following peripheral nerve injury by the polysaccharide from culinary and medicinal mushroom, Hericium erinaceus (Bull.: Fr.) Pers. through its neuroregenerative action. **Food Sci. Technol (Campinas)**, Campinas , v. 35, n. 4, p. 712-721, Dec. 2015 . Available from http://www.scielo.br/scielo.php?script=sci_ arttext&pid=S0101-20612015000400712&lng=en&nrm=iso https://creativecommons.org/ licenses/by/4.0/deed.en)

between motor neuron and EDL muscle fibers has not re-established at this stage. In rats treated with mecobalamin and polysaccharide, loose axon bundles indicate that regeneration process is ongoing. Axon bundles are more compact and regeneration process is more advanced after 14 days of crush injury compared to 7 days after injury. In treated groups, high density of regenerating axons reinnervating motor endplates can be observed. This indicates reestablishment of connection between motor neuron and EDL muscle fibers, leading to functional recovery.

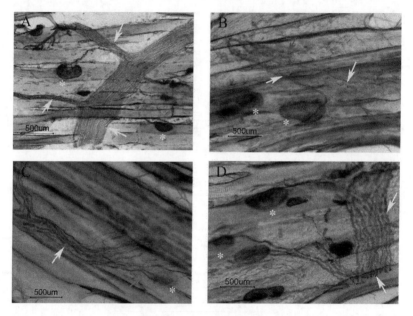

FIGURE 1.10 (See color insert.) The morphology of silver-cholinesterase-stained longitudinal section of EDL muscle in normal unoperated limb and operated limb after 14 days of peroneal nerve crush injury. Yellow arrows indicate the axons. Violet arrows indicate the degenerating axons. *Red arrows* indicate polyneuronal innervation. Asterisks indicate the motor endplates. Scale bar = 500 μm. (A) Normal unoperated limb. Axon bundles are clear and compact. (B) Operated limb in negative control group—distilled water (10 mL/kg body weight/day). Wallerian degeneration can be detected. Degenerated axons are being phagocytosed by the cooperative action of denervated Schwann cells and infiltrating macrophages. (C) Operated limb in positive control group—mecobalamin (130 μg/kg body weight/day). Loose axon bundles indicate regeneration process is ongoing. (D) Operated limb in polysaccharide group—*H. erinaceus* fresh fruit bodies (10 mL/kg body weight/day). Axon bundles are more compact and regeneration process is more advanced compared to positive control group. (Reprinted from International Journal of Medicinal Mushrooms, Vol 14, Issue 5, K.-H. Wong, M. Naidu, P. David, R. Bakar & V. Sabaratna, Neuroregenerative Potential of Lion's Mane Mushroom, Hericium erinaceus (Bull.: Fr.) Pers. (Higher Basidiomycetes), in the Treatment of Peripheral Nerve Injury (Review), pp. 427–446, © 2012, with permission from Begell House, Inc.)

1.4 CONCLUSION

We demonstrated a neuroregenerative role of a water-soluble polysaccharide from *H. erinaceus* fresh fruit bodies in the peripheral nervous system. Given the complexity of nerve *regeneration, f*urther studies are needed to fully elucidate molecular mechanisms that *H. erinaceus* utilizes to promote growth and regeneration of axons.

ACKNOWLEDGMENTS

This work was supported by University of Malaya RG268-13AFR and Ministry of Higher Education Malaysia through High Impact Research Grant F000002-21001.

KEYWORDS

- *Hericium erinaceus*
- polysaccharides
- peripheral nerve injury and regeneration
- sensory functional recovery
- blood–nerve barrier
- signaling pathways

REFERENCES

American Type Culture Collection. ATCC® Product Information Sheet for HB-12317. Cell Line Designation: NG108-15 [108CC15] ATCC® Catalog No. HB-12317. 2004.

Artemenko, D. P.; Gerasimov, B. D. The Blocked Effect of Adenosine and ATP on the Synaptic Transmitting in the Isolated Rat Brain Slices. *Neurophysiology* **1983,** *15,* 639–647.

Chen, G. L. *Studies on the Cultivation and Medicinal Efficacy of Hericium erinaceus;* The Edible Fungus Research Institute of Shanghai Academy of Agricultural Science: Shanghai, China, 1992.

Franke, T. F.; Kaplan, D. R.; Cantley, L. C. PI3K: Downstream AKTion Blocks Apoptosis. *Cell* **1997,** *88,* 435–437.

Grygansky, A. P.; Moldavan, M. G.; Kolotushkina, O.; Kirchhoff, B.; Skibo, G. G. *Hericium erinaceus* (Bull.: Fr.) Pers. Extract Effect on Nerve Cells. *Int. J. Med. Mushrooms* **2001,** *3,* 152.

Harden, R. N.; Richardson, K.; Shufelt, M.; Revivo, G. Neuropathic Pain: An Interdisciplinary Approach. In *Neuropathic Pain: Mechanisms, Diagnosis and Treatment;* Simpson, D. M., McArthur, J. C., Dworkin, R. H., Eds.; Oxford University Press: New York, NY, 2012; pp 472–484.

Hu, D.; Hu, R.; Berde, C. B. Neurologic Evaluation of Infant and Adult Rats before and after Sciatic Nerve Blockade. *Anesthesiology* **1997,** *86,* 957–965.

Kawagishi, H. Anti-MRSA Compounds from *Hericium erinaceus* (Bull.: Fr.) Pers. *Int. J. Med. Mushrooms* **2005,** *7,* 350.

Kawagishi, H.; Ando, M.; Sakamoto, H.; Yoshida, S.; Ojima, F.; Ishiguro, Y.; Ukai, N.; Furukawa, S. Hericenones C, D, and E, Stimulators of Nerve Growth Factor (NGF)-Synthesis, from Mushroom *Hericium erinaceum. Tetrahedron Lett.* **1991,** *32,* 4561–4564.

Kawagishi, H.; Ando, M.; Shinba, K.; Sakamoto, H.; Yoshida, S.; Ojima, F.; Ishiguro, Y.; Ukai, N.; Furukawa, S. Chromans, Hericenones F, G, and H from the Mushroom *Hericium erinaceum. Phytochemistry* **1993,** *32,* 175–178.

Kawagishi, H.; Masui, A.; Tokuyama, S.; Nakamura, T. Erinacines J and K from the Mycelia of *Hericium erinaceum. Tetrahedron* **2006,** *62,* 8463–8466.

Kawagishi, H.; Shimada, A.; Shirai, R.; Okamoto, K.; Ojima, F.; Sakamoto, H.; Ishiguro, Y.; Furukawa, S. Erinacines A, B, and C, Strong Stimulators of Nerve Growth Factor (NGF)-Synthesis, from the Mycelia of *Hericium erinaceum. Tetrahedron Lett.* **1994,** *35,* 1569–1572.

Kawagishi, H.; Shimada, A.; Shizuki, K.; Mori, H.; Okamoto, K.; Sakamoto, H.; Furukawa, S. Erinacine D, a Stimulator of NGF-Synthesis, from the Mycelia of *Hericium erinaceum. Heterocycl. Commun.* **1996a,** *2,* 51–54.

Kawagishi, H.; Shimada, A.; Hosokawa, S.; Mori, H.; Sakamoto, H.; Ishiguro, Y.; Sakemi, S.; Bordner, J.; Kojima, N.; Furukawa, S. Erinacines E, F, and G, Stimulators of Nerve Growth Factor (NGF)-synthesis, from the Mycelia of *Hericium erinaceum. Tetrahedron Lett.* **1996b,** *37,* 7399–7402.

Kawagishi, H.; Zhuang, C.; Yunoki, R. Compounds for Dementia from *Hericium erinaceum. Drugs Future* **2008,** *33,* 149–155.

Kenmoku, H.; Sassa, T.; Kato, N. Isolation of Erinacine P, a Parental Metabolite of Cyathane-xylosides, from *Hericium erinaceum* and Its Biomimetic Conversion into Erinacines A and B. *Tetrahedron Lett.* **2000,** *41,* 4389–4393.

Kenmoku, H.; Tanaka, K.; Okada, K.; Kato, N.; Sassa, T. Erinacol (cyatha-3,12-dien-14b-ol) and 11-O-acetylcyathin A3, New Cyathane Metabolites from an Erinacine Q Producing *Hericium erinaceum. Biosci. Biotechnol. Biochem.* **2004,** *68,* 1786–1789.

Kolotushkina, E. V.; Moldavan, M. G.; Voronin, K. Y.; Skibo, G. G. The Influence of *Hericium erinaceus* Extract on Myelination Process In Vitro. *Fiziol. Zh.* **2003,** *49,* 38–45.

Lai, P. L.; Naidu, M.; Sabaratnam, V.; Wong, K. H.; David, P.; Kuppusamy, U. R.; Abdullah, N.; Malek, S. R. Neurotrophic Properties of *Hericium erinaceus* (Bull.: Fr.) Pers. Grown in Tropical Climate of Malaysia. *Int. J. Med. Mushrooms* **2013,** *15,* 539–554.

Lee, J. S.; Hong, E. K. *Hericium erinaceus* Enhances Doxorubicin-induced Apoptosis in Human Hepatocellular Carcinoma Cells. *Cancer Lett.* **2010,** *297,* 144–154.

Lee, E. W.; Shizuki, K.; Hosokawa, S.; Suzuki, M.; Suganuma, H.; Inakuma, T.; Li, J.; Ohnishi-Kameyama, M.; Nagata, Y.; Furukawa, S.; Kawagishi, H. Two Novel Diterpenoids, Erinacines H and I from the Mycelia of *Hericium erinaceum. Biosci. Biotechnol. Biochem.* **2000,** *64,* 2402–2405.

Liu, X.; Lu, Y.; Xu, X.; Suo, Y. A Study on Preparation of Sandwich Biscuits with Hedgehog Fungus and Therapeutic Approach for Nutritional Anemia of Preschool Children. *Acta Nutrimenta Sinica.* **1992,** *14,* 109–111.

Lundborg, G. Intraneural Microcirculation. *Orthop. Clin. North Am.* **1988,** *19,* 1–12.

Ma, B. J.; Zhou, Y.; Li, L. Z.; Li, H. M.; Gao, Z. M.; Ruan, Y. A New Cyathane-xyloside from the Mycelia of *Hericium erinaceum. Z. Naturforsch.* **2008,** *63b,* 1241–1242.

Mizuno, T. Bioactive Substances in *Hericium erinaceus* (Bull.: Fr.) Pers. (Yamabushitake), and its Medicinal Utilization. *Int. J. Med. Mushrooms* **1999,** *1,* 105–119.

Moldavan, M. G.; Grygansky, A. P.; Kirchoff, B. In *Hericium erinaceus (Bull.: Fr.) Pers. Extracts Effect on the Neurons Impulse Activity in Stratum Pyramidale of Zone CA1 Hippocampal Slices in Rats.* Proceedings of the Third International Conference on Mushroom Biology & Mushroom Products, Sydney, Australia, October 12–16, 1999.

Mori, K.; Inatomi, S.; Ouchi, K.; Azumi, Y.; Tuchida, T. Improving Effects of the Mushroom Yamabushitake *(Hericium erinaceus)* on Mild Cognitive Impairment: A Double-blind Placebo-controlled Clinical Trial. *Phytother. Res.* **2009,** *23,* 367–372.

Mori, K.; Obara, Y.; Moriya, T.; Inatomi, S.; Nakahata, N. Effects of *Hericium erinaceus* on Amyloid β(25-35) Peptide-induced Learning and Memory Deficits in Mice. *Biomed. Res.* **2011,** *32,* 67–72.

Nagano, M.; Shimizu, K.; Kondo, R.; Hayashi, C.; Sato, D.; Kitagawa, K.; Ohnuki, K. Reduction of Depression and Anxiety by 4 Weeks *Hericium erinaceus* Intake. *Biomed. Res.* **2010,** *31,* 231–237.

Okada, Y.; Ueshin, Y.; Isotani, A.; Saito-Fujita, T.; Nakashima, H.; Kimura, K.; Mizoguchi, A.; Oh-hora, M.; Mori, Y.; Ogata, M.; Oshima, R. G.; Okabe, M.; Ikawa, M. Complementation of Placental Defects and Embryonic Lethality by Trophoblast-specific Lentiviral Gene Transfer. *Nat. Biotechnol.* **2007,** *25,* 233–237.

Schumacher, M.; Guennoun, R.; Stein, D. G.; De Nicola, A. F. Progesterone: Therapeutic Opportunities for Neuroprotection and Myelin Repair. *Pharmacol. Ther.* **2007,** *116,* 77–106.

Shimbo, M.; Kawagishi, K.; Yokogoshi, H. Erinacine A Increases Catecholamine and Nerve Growth Factor Content in the Central Nervous System of Rat. *Nutr. Res.* **2005,** *25,* 617–623.

Sweatt, J. D. The Neuronal MAP Kinase Cascade: A Biochemical Signal Integration System Subserving Synaptic Plasticity and Memory. *J. Neurochem.* **2001,** *76,* 1–10.

Suzuki, C.; Mizuno, T. XI. Cultivation of Yamabushitake (*Hericium erinaceum*). *Food Rev. Int.* **1997,** *13,* 419–421.

Wen, Y. R.; Suter, M. R.; Ji, R. R.; Yeh, G. C.; Wu, Y. S.; Wang, K. C.; Kohno, T.; Sun, W. Z.; Wang, C. C. Activation of p38 Mitogen-activated Protein Kinase in Spinal Microglia Contributes to Incision-induced Mechanical Allodynia. *Anesthesiology* **2009,** *110,* 155–165.

Wierzba, K.; Wańkowicz, B.; Piekarczyk, A.; Danysz, A. Cytostatics and Immunosuppressive Drugs. *Side Eff. Drugs Annual.* **1984,** *8,* 395–425.

Wong, K. H.; Sabaratnam, V.; Abdullah, N.; Naidu, M.; Keynes, R. Activity of Aqueous Extracts of Lion's Mane Mushroom *Hericium erinaceus* (Bull.: Fr.) Pers. (Aphyllophoromycetideae) on the Neural Cell Line NG108-15. *Int. J. Med. Mushrooms* **2007,** *9,* 57–65.

Wong, K. H.; Sabaratnam, V.; Abdullah, N.; Kuppusamy, U. R.; Naidu, M. Effects of Cultivation Techniques and Processing on Antimicrobial and Antioxidant Activities of *Hericium erinaceus* (Bull.: Fr.) Pers. Extracts. *Food Technol. Biotechnol.* **2009a,** *47,* 47–55.

Wong, K. H.; Naidu, M.; David, P.; Abdulla, M. A.; Abdullah, N.; Kuppusamy, U. R.; Sabaratnam, V. Functional Recovery Enhancement Following Injury to Rodent Peroneal Nerve by Lion's Mane Mushroom, *Hericium erinaceus* (Bull.: Fr.) Pers. (Aphyllophoromycetideae). *Int. J. Med. Mushrooms* **2009b,** *11,* 225–236.

Wong, K. H.; Naidu, M.; David, P.; Abdulla, M. A.; Abdullah, N.; Kuppusamy, U. R.; Sabaratnam, V. Peripheral Nerve Regeneration Following Crush Injury to Rat Peroneal

Nerve by Aqueous Extract of Medicinal Mushroom *Hericium erinaceus* (Bull.: Fr) Pers. (Aphyllophoromycetideae). *Evid. Based Complement. Alternat. Med.* **2011,** *2011,* 580752. DOI:10.1093/ecam/neq062

Wong, K. H.; Naidu, M.; David, P.; Bakar, R.; Sabaratnam V. Neuroregenerative Potential of Lion's Mane Mushroom, *Hericium erinaceus* (Bull.: Fr.) Pers. (Higher Basidiomycetes), in the Treatment of Peripheral Nerve Injury (Review). *Int. J. Med. Mushrooms* **2012,** *14,* 427–446.

Wong, K. H.; Kanagasabapathy, G.; Naidu, M.; David, P.; Sabaratnam, V. *Hericium erinaceus* (Bull.: Fr.) Pers., a Medicinal Mushroom, Activates Peripheral Nerve Regeneration. *Chin. J. Integr. Med.* **2016,** *22,* 759–767. DOI: 10.1007/s11655-014-1624-2

Ying, J. Z.; Mao, X. L.; Ma, Q. H.; Zong, Y. C.; Wen, H. A. *Icons of Medicinal Fungi from China* (Translation Xu YH). Science Press: Beijing, China, 1987.

CHAPTER 2

UNDERSTANDING PROBIOTICS: A REVIEW

MRADULA GUPTA[1*], SOMESH SHARMA[1], and NITIN KOTHARI[2]

[1]School of Bioengineering and Food Technology, Shoolini University of Biotechnology and Management Sciences, Solan, Himachal Pradesh, India

[2]Department of Pharmacology, Pacific Medical College, Udaipur, Rajasthan, India

*Corresponding author. E-mail: mridulagupta1988@gmail.com

CONTENTS

Abstract 26
2.1   Introduction 26
2.2   The History and the Definition of Probiotics 27
2.3   Mechanism of Probiotics 33
2.4   Future Advances 36
Keywords 38
References 39

ABSTRACT

Probiotics are the preparation of live microorganisms which beneficially affect the host by improving the properties of indigenous microbes. Probiotics are used to improve intestinal health and stimulate the immune system. Due to various health benefits to the host, probiotics are increasingly gaining importance and are a subject of extensive research. This review focuses on the history, concept, and criteria for the selection of probiotics. The health-related effects as well as the types of probiotics currently used are well explained. Commercially, the probiotics have been incorporated in various products such as dairy-based ones including fermented milk, ice cream, buttermilk, milk powder, and yogurts and non-dairy based such as soy-based products, cereals, and a variety of juices as appropriate means of probiotic delivery to the consumer. Desirable probiotic properties such as acid tolerance, bile tolerance, and antimicrobial activity are discussed further. Many mechanisms from the studies are trying to explain how probiotics could protect the host from the intestinal disorders. These mechanisms are briefly listed in the review. Summarily, in this review, the well detailed history of probiotics, their applications in the food as well as health area and future prospects of probiotic products and processes are discussed.

2.1 INTRODUCTION

The word "probiotic" comes from Greek language "pro bios" which means "for life" opposed to "antibiotics" which means "against life." These microorganisms are helpful in maintaining intestinal microbial balance and play a beneficial role in health. The probiotic microorganisms are mostly of the strains of the genera *Lactobacillus* and *Bifidobacterium*, but strains of *Bacillus*, *Pediococcus*, and some yeasts have also been found as suitable candidates. Different enzymes, vitamins, capsules or tablets, and some fermented foods are probiotic products which contain microorganism which beneficially affects host health. They can contain one or several species of probiotic bacteria. Most of the products of human consumption are produced in fermented milk or given in powders or tablets.

Human intestinal tract is an enormously complex ecosystem that includes both facultative anaerobic and aerobic microorganisms (Naidu et al., 1999). It is estimated that about 300–400 different cultivable species belonging to more than 190 genera are present in the colon of healthy adults. Among the known colonic microbial flora only a few major groups (main flora, according to Gedek (1993) dominate at levels around 10^{10}–10^{11}/g, all of which are strict anaerobes, such as *Bacteroids*, *Eubacteria*, *Bifidobacteria*, and *Peptostreptococci*. Facultative aerobes are considered to the subdominant flora, constituting *Enterobacteriaceae*, *lactobacilli*, and *streptococci*. Minor groups of pathogenic and opportunistic organisms, the so-called residual flora according to Gedek (1993), are always present in low numbers. There are lots of factors that may disturb the balance of the gut microflora such as age, diet, environment, stress, and medication (Fuller, 2007). The lifestyle is changing and it is difficult to have a healthy intestine and the balance of the bacteria must be maintained. Several factors make the host susceptible to illnesses. That is why, probiotics are suggested as food to provide the balance of intestinal flora (Holzapfel et al., 1998).

TABLE 2.1 Microorganisms Applied in Probiotic Products.

Lactobacillus species	*Bifidobacterium* species	Others
L. acidophilus	B. bifidum	Enterococcus faecalis
L. rhamnosus	B. animalis	Enterococcus faecium
L. gasseri	B. breve	Streptococcus salivarussubsp. Thermophilus
L. casei	B. infantis	Lactococcuslactissubsp. lactis
L. reuteri	B. longum	Lactococcuslactissubsp. cremoris
L. delbrueckii subsp. Bulgaricus	B. lactis	Propionibacterium freudenreichii
L. crispatus	B. adolascentis	Pediococcusacidilactici
L. plantarum	–	Saccharomyces boulardii
L. salivarus	–	Leuoconostocmesenteroides
L. johnsonii	–	–
L. gallinarum	–	–
L. plantarum	–	–
L. fermentum	–	–
L. helveticus	–	–

2.2   THE HISTORY AND THE DEFINITION OF PROBIOTICS

Foods are no longer considered by consumers only in terms of taste and immediate nutritional needs, but also in terms of their ability to provide specific health benefits beyond their basic nutritional value. Currently, the largest segment of the functional food market is provided by the foods targeted toward improving the balance and activity of the intestinal micro-flora (Saarela et al., 2002). Consumption of foods containing live bacteria is the oldest and still most widely used way to increase the numbers of advantageous bacteria in the intestinal tract. Such bacteria are called "probiotics" and have been predominantly selected from the genera *Lactobacillus* and *Bifidobacterium*, both of which have been extensively studied and established as valuable native inhabitants of the gastro intestinal (GI) tract (Fuller, 1989; Salminen et al., 1998; Capela et al., 2006). Various microorganisms, particularly species of *Lactobacillus* and *Streptococcus*, have traditionally been used in fermented dairy products to promote human health as well as food functionality and flavor.

Metchnikoff first introduced the probiotic concept in 1908, which observed the long life of Bulgarian peasants, who consumed fermented milk foods. He suggested that lactobacilli might counteract the putrefactive effects of gastrointestinal metabolism. In this century, which is elapsed since Metchnikoff's research, scientists and consumers have accepted the probiotic concept throughout the world (Fuller, 1992). The longevity of Caucasians was related to the high intake of fermented milk products Metchnikoff (1907), as elucidated in his bestselling book *The Prolongation of Life*. Lactic acid bacteria (LAB) belongs to a group of Gram-positive, non-sporulating, non-respiring cocci, or rods, which produce lactic acid as a major metabolic end product during the fermentation of carbohydrates (Salminen et al., 1998). In the world, the concept of providing functional foods including beneficial components rather than removing potentially harmful components is gaining ground in recent years. It may consider a functional food with the special property of containing live, beneficial microorganisms. Functional foods and nutraceutical can prevent and treat diseases. Yogurt and other fermented milk containing probiotics may be considered the first functional foods. The increasing cost of healthcare, the steady increase in life expectancy and the desire of the elderly for improved quality of their lives are driving factors for research and development in the area of functional foods. Although the concept of functional foods was introduced long ago with Hippocrates and his motto "Let food be your medicine," fairly recently the body of evidence started to support the hypothesis that diet may play an important role in modulation of important physiological functions in the body. Among a number of functional compounds recognized so far, bioactive components from fermented foods and probiotics certainly take the center stage due to their long tradition of safe use and established and postulated beneficial effects. The first clinical trials in the 1930s focused on the effect of probiotics on constipation and research has steadily increased since then. Today, probiotics are available in a variety of food products and supplements. Food products containing probiotics are mostly dairy products due to the historical association of lactic acid bacteria with fermented milk. The fermentation of dairy foods presents one of the oldest methods of long-term food preservation. The origin of fermented milk can be traced back long before the Phoenician era and placed in the Middle East. Traditional Egyptian fermented milk products, Laban Rayeb and Laban Khad, were consumed as early as 7000 BC. Their tradition claims that even Abraham owed his longevity to the consumption of cultured milk (Kosikowski & Mistry, 1997). Initially, established in the middle and far east of Asia, the tradition of fermenting milk was spread throughout the east Europe and Russia by the Tartars, Huns, and Mongols during their conquests. As a consequence, a wide range of fermented dairy products still exists in these regions and some popular products such as yogurt and kefir are claimed to originate from the Balkans and Eastern Europe (Azizpour et al., 2009).

The term "probiotic" firstly used in 1965 by Lilly and Stillwell to describe substances which stimulate the growth of other microorganisms. After this year, the word "probiotic" was used in different meaning according to its mechanism and the effects on human health. The meaning was improved to the closest one we use today by Parker in 1974. Parker defined "probiotic" as "substances and organisms which contribute to intestinal microbial balance." In 1989, the meaning use today was improved by Fuller. Thus, probiotic is a live microbial supplement which affects host's health positively by improving its intestinal microbial balance. Then this definition was broadened by Havenaar and Huis in't Veld (1992) including mono or mixed culture of live microorganisms which applied for animal and man (Çakır 2003; Guarner et al., 2005; Sanders 2003).

In the following years, many researchers studied probiotics and defined them in several ways. They are listed as follows.

1. "Living microorganisms, which upon ingestion in certain numbers, exert health benefits beyond inherent basic nutrition" by Schaafsma (1996).
2. "A microbial dietary adjuvant that beneficially affects the host physiology by modulating mucosal and systemic immunity, as well as improving nutritional and microbial balance in the intestinal tract" by Naidu et al. (1999).
3. "A live microbial food ingredient that is beneficial to health" by Salminen et al. (1998).

4. "A preparation of or a product containing viable, defined micro-organisms in sufficient numbers, which alter the microflora (by implantation or colonization) in a compartment of the host and by that exert beneficial health effects in this host" by Schrezenmeir and de Vrese (2001).
5. "Live microorganisms which when administered in adequate amounts confer a health benefit on the host" is accepted by FAO/WHO (report in October 2001) (Guarner et al., 2005; Sanders, 2003; Klaenhammer, 2000).

2.2.1   THE EFFECTS OF PROBIOTICS ON HEALTH

There are lots of studies on the health benefits of fermented foods and probiotics. These health-related effects can be considered as shown below (Çakır, 2003; Scherezenmeir & De Vrese, 2001; Dunne et al., 2001; Dugas et al., 1999)

- Improving immune system.
- Managing lactose intolerance.
- Prevention of colon cancer.
- Reduction of cholesterol and triacylglycerol plasma concentrations (weak evidence).
- Lowering blood pressure.
- Reducing inflammation.
- Reduction of allergic symptoms.
- Beneficial effects on mineral metabolism, particularly bone density and stability.
- Reduction of *Helicobacter pylori* infection.
- Antimicrobial effect.

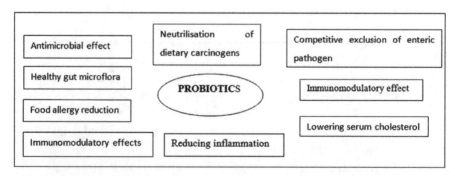

FIGURE 2.1   Beneficial effects of probiotics.

2.2.2   LACTOSE INTOLERANCE

Most of the humans commonly non-Caucasians become lactose intolerant after weaning. Due to the lack of essential enzyme β-galactosidase, lactose intolerant people are unable to metabolize lactose. When these people consume milk or lactose-containing products, symptoms including abdominal pain, bloating, flatulence, cramping, and diarrhea occur. If lactose passes from the small intestine, it is converted to gas and acid in the large intestine by the colonic microflora. Also, the presence of breath hydrogen is a signal for lactose maldigestion. Studies have shown that the addition of certain starter cultures to milk products, allows the lactose intolerant people to consume those products without the usual rise of breath hydrogen or associated symptoms (Fooks et al., 1999; Quewand & Salminen, 1998; Lin et al., 1991). The beneficial effects of probiotics on lactose intolerance are explained by two ways: (1) To lower lactose concentration in the fermented foods due to the high lactase activity of bacterial preparations used in the production. (2) Increased lactase active lactase enzymes entering the small intestine with the fermented product or with the viable probiotic bacteria and reducing the lactose levels (Salminen et al., 2004). Lactose intolerant people can easily consume yogurt because yogurt has bacterial β-galactosidase enzyme as compared to milk in which lactose is converted into lactic acid. Furthermore, the LAB which is used to produce yogurt, *Lactobacillus bulgaricus* and *Streptococcus thermophilus* are not resistant to gastric acidity. Hence, the products with probiotic bacteria are more efficient for lactose intolerant humans. The major factor which improves the digestibility by the hydrolysis of lactose is the bacterial enzyme β-galactosidase. Another factor is the slower gastric emptying of semi-solid milk products such as yogurt. So the β-galactosidase activity of probiotic strains and other lactic acid bacteria used in dairy products is really important. β-galactosidase activity within probiotics varies in a huge range. It has to be considered that both the enzyme activity of probiotic strain and the activity left in the final product have beneficial role in those who are lactose intolerant (Salminen et al., 2004).

2.2.3   DESIRABLE PROBIOTIC PROPERTIES

In order to be able to exert its beneficial effects, a successful potential probiotic strain is expected to have a number of desirable properties. Some of them will be discussed in more detail. A potential probiotic strain does need to fulfill all such selection criteria (Quwehand et al., 1999).

2.2.3.1   ACID AND BILE TOLERANCE

Bacteria used as probiotic strains are joined with the food system in a journey to the lower intestinal tract via the mouth. In this food system, probiotic bacteria should be resistant to the enzymes like lysozyme in the oral cavity. Then, the journey of food continues in the stomach and enters the upper intestinal tract which contains bile. In this stage, strains should have the ability to resist the digestion process. It is reported that it takes 3 h for the first entrants to get released from the stomach. To be able to be probiotic, strains need to be resistant to the stressful conditions of the stomach (pH 1.5–3.0) and the upper intestine which contains bile (Chou & Weimer, 1999; Çakır, 2003). Bile salts are synthesized in the liver from cholesterol and sent to the gallbladder and secreted into the duodenum in the conjugated form (500–700 mL/day). In the large intestine, these acids suffer some chemical modifications (deconjugation, dehydroxylation, dehydrogenation, and deglucuronidation) due to the microbial activity. Conjugated and deconjugated bile acids show antimicrobial activity especially in *Escherichia coli* subspecies, *Klebsiella* spp., and *Enterococcus* spp. in vitro. The deconjugated acid forms are more effective on Gram-positive bacteria (Dunne et al., 1999; Çakır, 2003; Chou & Weimer, 1999).

2.2.3.2   ANTIMICROBIAL ACTIVITY

Antimicrobial activity is one of the most important selection criteria for probiotics. Antimicrobial activity targets the enteric undesirables and pathogens (Klaenhammer & Kullen, 1999). Antimicrobial effects of lactic acid bacteria are formed by producing some substances such as organic acids (lactic, acetic, and propionic acids), carbon dioxide, hydrogen peroxide, diacetyl, low molecular weight antimicrobial substances, and bacteriocins (Ouwehand & Vesterlund, 2004; Çakır, 2003). Here are some examples: *Lactobacillusreuterii*, which is a member of normal microflora of human and many other animals, produce a low molecular weight antimicrobial substance reuterin; subspecies of *Lactococcus lactis* produce a class I bacteriocin, nisin A; *Enterococcus faecalis* DS16 produces a class I bacteriocin cytolysin; *Lactobacillus plantarum* produces a class II bacteriocin plantaricin S; *Lactobacillus acidophilus* produces a class III bacteriocins acidophilucin A (Ouwehand & Vesterlund, 2004). Production of bacteriocins is highly affected by the factors of the species of microorganisms, ingredients, and pH of the medium, incubation temperature, and time. Nisin, produced by *L. lactis* subsp. lactis is a well-known bacteriocin and it is allowed to be used in certain food preparations as a preservative (Çakır, 2003).

2.3 MECHANISM OF PROBIOTICS

Probiotic microorganisms are considered to support the host health. However, the support mechanisms have not been explained (Holzapfel et al., 1998). There are studies on how probiotics work. So, many mechanisms from these studies are trying to explain how probiotics could protect the host from the intestinal disorders. These mechanisms are listed below (Rolfe et al., 2000; Çakır et al., 2003; Salminen et al., 1999; Castagliuolo et al., 1999):

1. Production of inhibitory substances: Production of some organic acids, hydrogen peroxide, and bacteriocins which are inhibitory to both Gram-positive and Gram-negative bacteria.
2. Blocking of adhesion sites: Probiotics and pathogenic bacteria are in a competition. Probiotics inhibit the pathogens by adhering to the intestinal epithelial surfaces by blocking the adhesion sites.
3. Competition for nutrients: Despite the lack of studies in vivo, probiotics inhibit the pathogens by consuming the nutrients which pathogens need.
4. Stimulating of immunity: Stimulating of specific and non-specific immunity may be one possible mechanism of probiotics to protect the host from intestinal diseases.

This mechanism is not well documented, but it is thought that specific cell wall components or cell layers may act as adjuvants and increase humoral immune response. Some other possible mechanisms are suppression of toxin production, reduction of gut pH, and attenuation of virulence (Fooks et al., 1999).

2.3.1 PROBIOTICS AND FOOD PRODUCTS

There is a wide range of varieties of probiotic food products. The main products existing in the market are dairy-based ones including fermented milks, cheese, ice cream, buttermilk, milk powder, and yogurts, the latter accounting for the largest share of sales (Stanton et al., 2001). Major companies engaged in offering probiotic food products in the Indian market includes Mother

Dairy, Amul, Danone Yakult, Nestle, Tablets India, Dr. Reddy Laboratories, Unique Biotech, Zeus Biotech, and others. Mother Dairy has the dominant position in the Indian probiotic functional food and beverage market, followed by Amul. Tablets India, on the other hand, has established itself as a major brand in probiotic drug and dietary supplement market.

Non-dairy food applications include soy-based products, nutrition bars, cereals, and a variety of juices as appropriate means of probiotic delivery to the consumer (Ewe et al., 2010; Sheehan et al., 2007). The factors that must be addressed in evaluating the effectiveness of the incorporation of the probiotic strains into such products are, besides safety, the compatibility of the product with the microorganism and the maintenance of its viability through food processing, packaging, and storage conditions. The product's pH for instance is a significant factor determining the incorporated probiotic's survival and growth, and this is one of the reasons why soft cheeses seem to have a number of advantages over yoghurt as delivery systems for viable probiotics to the gastrointestinal tract (Medina et al., 1994; Gardiner et al., 1998; Kehagias et al., 2006). Current technological innovations provide ways to overcome probiotic stability and viability issues offering new options for their incorporation in new media and subsequent satisfaction of the increasing consumer demand.

2.3.2   PRODUCT CONSIDERATIONS

Prebiotics are non-digestible food ingredient that promotes the growth of beneficial microorganisms in the intestines. Probiotics and prebiotics are marketed as health, or functional foods whereby they are ingested for their advantages in the digestive tract and/or systemic areas such as the liver, vagina, or bloodstream. Unlike new drugs or pharmaceuticals, which are screened intensively for safety and effectiveness, probiotics and prebiotics are less rigorously assessed. It is therefore relatively easy to launch a new product, and legislation against such products is loose. Nevertheless, consumers should be provided with an accurate assessment of physiological, microbial, and safety aspects. Several criteria for the appropriate use of probiotics and prebiotics exist and may be summarized as follows. They should exert a proven beneficial effect on the consumer, preferably with a mechanistic explanation of how this occurred.

1. Be non-pathogenic, non-toxic, and free of adverse side effects;
2. Maintain stability in the product;

3. Contain a large number of viable cells (for probiotics);
4. Survive well in the GI tract (the best products should be resistant to gastric acid, small gut secretions and have a good ability to influence bacteria already in the gut);
5. Have good sensory and mouthfeel properties;
6. Preferably be isolated from the same species as the intended use; and have accurate product labeling and content.

Much effort has been concentrated on identifying probiotic bacteria and characterizing their beneficial credentials. It is generally considered that probiotic bacteria must possess certain properties. The probiotic must survive passage through the upper regions of the GI tract, and persist in the colon. There must be no adverse host response to the bacterium, its components, or metabolic end products. The probiotic should be antagonistic to pathogenic organisms in the gut, and must be genetically stable. Chosen microorganisms must be capable to industrial processes and remain viable in the final food product for the successful introduction of the probiotic concept into the food market. Determination of mechanisms involved in probiotic function, such as the production of antimicrobials, mucosal adhesions, and organic acids are enabled due to advances in the genetics of probiotic strains usually LAB or yeasts. However, this also opens the research area for the possibility of modifying existing strains to increase survival and efficacy in the human GI tract.

2.3.3   PROBIOTICS AS COMMERCIAL PRODUCTS

Probiotic foods are a group of functional foods with growing market shares and large commercial interest Shah (2000). Probiotics have been used for centuries in fermented dairy products. However, the potential applications of probiotics in non-dairy food products have not received formal recognition. Recently, interest to plant-based food products and beverages of probiotics has increased and the selection of new probiotic strains and the development of new application have gained much importance. The uses of probiotics have been very beneficial to the human health to play a key role in normal digestive processes and in maintaining the animal's health. However, a number of uncertainties concerning technological, microbiological, and regulatory aspects exist (Saarela et al., 2000). The presence of probiotics in commercial food products has been claimed for certain health benefits. The probiotics are actively used in the food industry and have been incorporated in various food products such as in fermented dairy foods, non-dairy products, plant-based products, functional foods, etc. This has led to industries focusing on different applications of probiotics in food products and creating a new generation of "probiotic health" foods. Among probiotics carrier food products, dairy drinks were the first commercialized products that are still consumed in larger quantities than other probiotic beverages.

Functional dairy beverages can be grouped into two categories: fortified dairy beverages (including probiotics, prebiotics, fibers, polyphenols, peptides, sterol, stanols, minerals, vitamins, and fish oil), and whey-based beverages (Saxelin et al., 2010). Vegetables and fruits are reported to contain a wide variety of antioxidant components, including phytochemicals. Phytochemicals, such as phenolic compounds, are considered beneficial for human health, decreasing the risk of degenerative diseases by reduction of oxidative stress and inhibition of macromolecular oxidation. There is a genuine interest in the development of non-dairy based functional beverages with probiotics because they serve as a healthy alternative for dairy probiotics, are cholesterol free and also favor consumption by lactose intolerant consumers.

The expectations of probiotic bacteria have become the most demanding for any bacterial group. Probiotics have become a very important element to everyday health food products, and their global market is estimated above US$ 28.8 billion by 2015 (Tilley et al., 2014). Nowadays, consumers are very concerned of chemical preservatives and processed foods, even though it provides a degree of safety and food diversity never seen before. However, consumers accept easily LAB as a natural way to preserve food and promote their health. In the last decade, the interest in bacteriocins produced by LAB has increased dramatically (Tripathi & Giri, 2014).

2.4 FUTURE ADVANCES

Microencapsulation technologies have been developed to protect the bacteria from damage caused by external environment. By the introduction of a drinking straw delivery system containing a dry form of the probiotic bacterium, beverage manufacturers can now provide it to the consumer. In addition, viable spores of a spore-forming probiotic are available in the market offering advantages during processing. In the same time, the potential of the production of antibiotics—substances with antimicrobial properties—by bifidobacteria is being explored in order to be applied in the food arena (Pszczola et al., 2012; O'Sullivan et al., 2012).

2.4.1   BIOTECHNOLOGY AND PROBIOTICS

With the revolution in sequencing and bioinformatics technologies well under way, it is timely and realistic to launch genome-sequencing projects for representative probiotic microorganisms. The rapidly increasing number of published lactic acid bacterial genome sequences will enable utilizing this sequence information in the studies related to probiotic technology. If genome sequence information is available for the probiotic species of interest, this can be utilized, for example, to study the gene expression (transcription) profile of the strain during fermenter growth. This will enable better control and optimization of the growth than is currently possible. Transcription profiling during various production steps will allow following important genes for probiotic survival during processing (e.g., stress and acid tolerance genes) and identifying novel genes important for the technological functionality of probiotics (Matias et al., 2014). Increasing knowledge of genes important for the technological functionality and rapid development of the toolboxes for the genetic manipulation of *Lactobacillus* and *Bifidobacterium* species will in the future enable tailoring the technological properties of probiotic strains. However, before wide application of tailored strains in probiotic food products, safety issues are of utmost importance and have to be seriously considered for each modified strain. In addition to dairy products, fruit juices have been shown to be suitable carriers for probiotics. The limiting factor for many of the probiotic strains is the low pH of the juices. There is growing interest in consumers toward healthier foods and probiotic fruit, vegetable juices, and cereal-based beverages can serve as a good option. But, the application of probiotic cultures in non-dairy products and environments represents a great challenge and needs to be researched at the industrial level for commercial production of these healthy products. Development of novel, economical, and technological matrices is a dire necessity to bring the non-dairy probiotic foods on par with the demand they have to their nature of healthy alternatives to dairy probiotic foods. Although there is a great potential for the use of fruit juices as probiotic products, very few reports on their preparation and production are available. Growing awareness among people and newer advances in the medical field has brought them into the *limelight* (Gupta & Sharma, 2016). Hence, there is a scope for further research in this area. While developing, functional properties, stability, sensory acceptance, especially related to taste, appeal, and price are to be kept in mind, as these factors play a major role in their successful commercialization. Care should be taken while selecting the probiotics to avoid removal of micronutrients from the product or to produce biogenic amines. As all cultures or strains may not have probiotic properties, selection of strain(s) with potential probiotic properties plays a major role in the success of the non-milk probiotic products. Technological issues that can affect the survival of probiotic cultures throughout the production process and during storage should also be addressed while formulating new probiotic products. Functional properties are extremely important to get a competitive advantage in the world market. Hence, care should be taken while confirming the functional attributes of starters before incorporating in the product. In conclusion, research on non-dairy probiotic products can be widened to better understanding and exploiting the benefits of non-dairy probiotic products for the mankind. Use of probiotics in combination with non-dairy probiotic products can also be attempted to produce synbiotic products.

Traditional cell culture methods, as well as the alternative techniques (direct imaging and visual enumeration, nucleic acid-based enumeration methods, and flow cytometry and cell sorting), offer advantages and limitations for enumerating probiotic microorganisms. The new methods and techniques show considerable promise for quantifying live microorganisms in different metabolic states. But, the probiotic efficacy cannot be predicted solely on the basis of viable cells. Very few microorganisms have been subjected to thorough in vitro studies confirming their specific health-promoting activity, and even fewer have been subsequently subjected to and passed the appropriate human trials. Additionally, probiotics can be dangerous, as they have been linked to an increase in mortality rate if administered to severely immunocompromised patients. Subsequent studies are needed to evaluate the health-promoting activity of probiotic bacteria (Céspedes et al., 2013).

KEYWORDS

- probiotics
- microflora
- microorganisms
- β-galactosidase
- lactose intolerance

REFERENCES

Azizpour, K.; Bahrambeygi, S.; Mahmoodpour, S.; Azizpour, A.; Mahmoodpour, S. History and Basic of Probiotics. *Res. J. Biol. Sci.* **2009,** *4* (4), 409–426.

Çakır, İ. Determination of Some Probiotic Properties on Lactobacilli and Bifidobacteria. Ph.D. Thesis, Ankara University, 2003.

Capela, P.; Hay, T. K. C.; Shah, N. P. Effect of Cryoprotectants, Prebiotics and Microencapsulation on Survival of Probiotic Organisms in Yoghurt and Freeze-dried Yoghurt. *Food Res. Int.* **2006,** *39* (2), 203–211.

Castagliuolo, I.; Riegler, M. F.; Valenick, L.; LaMont, J. T.; Pothoulakis, C. *Saccharomyces Boulardii* Protease Inhibits the Effects of Clostridium Difficile Toxins A and B in Human Colonic Mucosa. *Infect. Immun.* **1999,** *67* (1), 302–307.

Céspedes, M.; Cardenas, P.; Staffolani, M.; Ciappini, M. C.; Vinderola, G. Performance in Nondairy Drinks of Probiotic L. Casei Strains Usually Employed in Dairy Products. *J. Food Sci.* **2013,** *78* (5), M756–M762.

Chou, L. S.; Weimer, B. Isolation and Characterization of Acid-and Bile-tolerant Isolates from Strains of *Lactobacillus Acidophilus*. *J. Dairy Sci.* **1999,** *82* (1), 23–31.

Dugas, B.; Merccnier, A.; Wijnkoop, I. L.; Arnaud, C.; Dugas, N.; Postaire, E. Immunity and Probiotics. *Immunol. Today* **1999,** *20* (9), 387–390.

Dunne, C.; Murphy, L.; Flynn, S.; O'Mahony, L.; O'Halloran, S.; Feeney, M.; Morrissey, D., et al. Probiotics: From Myth to Reality. Demonstration of Functionality in Animal Models of Disease and in Human Clinical Trials. In *Lactic Acid Bacteria: Genetics, Metabolism and Applications;* Siezen, R. J., Kok, J., Abee, T., Schaafsma, G., Eds.; Springer: Netherlands, 1999; pp 279–292.

Dunne, C.; O'Mahony, L.; Murphy, L.; Thornton, G.; Morrissey, D.; O'Halloran, S.; Feeney, M.; Fitzgerald, G.; Daly, C.; Kiely, B.; O'Sullivan, G. C.; Shanahan, F.; Collins, J. K. In Vitro Selection Criteria for Probiotic Bacteria of Human Origin: Correlation with in Vivo Findings. *Am. J. Clin. Nutr.* **2001,** *73* (2), 386s–392s.

Ewe, J. A.; Wan-Abdullah, W. N.; Liong, M. T. Viability and Growth Characteristics of Lactobacillus in Soymilk Supplemented with B-vitamins. *Int. J. Food Sci. Nutr.* **2010,** *61* (1), 87–107.

Fooks, L. J.; Fuller, R.; Gibson, G. N. Prebiotics, Probiotics and Human Gut Microbiology. *Int. Dairy J.* **1999,** *9* (1), 53–61.

Fuller, R. History and Development of Probiotics. 2007. http://www.albertcllasic.net/probiotics.php (accessed June 3, 2007).

Fuller, R. History and Development of Probiotics. In *Probiotics*. Fuller, R., Ed.; The Scientific Basis, Chapman and Hall: London, 1992; 1–8.

Fuller, R. Probiotics in Man and Animals. *J. Appl. Bacteriol.* **1989,** *66,* 365–378.

Gardiner, G.; Ross, R. P.; Collins, J. K.; Fitzgerald, G.; Stanton, C. Development of a Probiotic Cheddar Cheese Containing Human-derived Lactobacillus Paracasei Strains. *Appl. Environ. Microbiol.* **1998,** *64* (6), 2192–2199.

Gedek, B.; Enders, C.; Ahrens, F.; Roques, C. The Effect of Saccharomyces Cerevisiae (BIOSAF Sc 47) on Ruminal Flora and Rumen Fermentation Pattern in Dairy Cows. *Ann Zootech.* **1993,** *42,* 175.

Guarner, F.; Perdigon, G.; Corthier, G.; Salminen, S.; Koletzko, B.; Morelli, L. Should Yoghurt Cultures be Considered Probiotic? *Br. J. Nutr.* **2005,** *93* (6) 783–786.

Gupta, M.; Sharma, S. Probiotics in Limelight. *J. Innov. Biol.* **2016,** *3* (1), 276–280.

Havenaar, R.; Huis in't Veld, J. H. Probiotics: A General View. *Lactic Acid Bacteria.* **1992,** *1,* 151–170.

Holzapfel, W.; Petra H.; Snel, J.; Schillinger, U.; Huis in't Veld, J. H. Overview of Gut Flora and Probiotics. *Int. J. Food Microbiol.* **1998,** *41* (2), 85–101.

Kehagias, C.; Koulouris, S.; Arkoudelos, J. S.; Samona, A. Viability and Biochemical Activity of Bifidobacteria in Association with Yoghurt Starter Cultures in Bifidus Milk and Bio-yoghurt during Storage at 4°C. *Egypt. J. Dairy Sci.* **2006,** *34* (2), 151.

Klaenhammer, T. R. Probiotic Bacteria: Today and Tomorrow. *J. Nutr.* **2000,** *130* (2), 415S–416S.

Klaenhammer, T. R.; Kullen, M. J. Selection and Design of Probiotics. *Int. J. Food Microbiol.* **1999,** *50* (1), 45–57.

Kosikowski, F. V.; Mistry, V. V. Process Cheese and Related Products. *Cheese and Fermented Milk Foods;* FV Kosikowski LLC: Westport, CT, 1997; Vol. 1, pp 328–352.

Lin, M. Y.; Savaiano, D.; Harlander, S. Influence of Nonfermented Dairy Products Containing Bacterial Starter Cultures on Lactose Maldigestion in Humans. *J. Dairy Sci.* **1991,** *74* (1), 87–95.

Matias, N. S.; Bedani, R.; Castro, I. A.; Saad, S. M. I. A Probiotic Soy-based Innovative Product as an Alternative to Petit-suisse Cheese. *LWT Food Sci. Technol.* **2014,** *59* (1), 411–417.

Medina, L. M.; Jordano, R. Survival of Constitutive Microflora in Commercially Fermented Milk Containing Bifidobacteria during Refrigerated Storage. *J. Food Prot.* **1994,** *57* (8), 731–733.

Metchnikoff, E. *The Prolongation of Life: Optimistic Studies;* Chalmers Mitchell, P., Ed.; Heinemann: London, 1907.

Naidu, A. S.; Bidlack, W. R.; Clemens, R. A. Probiotic Spectra of Lactic Acid Bacteria (LAB). *Crit. Rev. Food Sci. Nutr.* **1999,** *39* (1), 13–126.

O'Sullivan, D. J. Exploring the Potential to Utilize Antibiotic Producing Bifidobacteria to Create Dairy Ingredients with Increased Broadspectrum Antimicrobial Functionalities Yields Encouraging Results. *Food Technol.* **2012,** *66* (6), 45–50.

Ouwehand, A. C.; Vesterlund, S. A. T. U. Antimicrobial Components from Lactic Acid Bacteria. *Food Sci. Technol.* **2004,** *139,* 375–396.

Parker, R. B. Probiotics, the Other Half of the Ntibiotic Story. *Anim. Nutr. Health.* **1974,** *29,* 4–8.

Pszczola, D. E. What Makes a Winning Ingredient? Which Ingredients are Game-changers? Some of the Factors that Help Determine a "Winning" Ingredient Development will be Looked at in This Review of Ingredient Launches from the 2012 IFT Annual Meeting & Food Expo. *Food Technol.* **2012,** *6* (8), 58.

Quewand, A. C.; Salminen, S. J. The Health Effects of Cultured Milk Products with Viable and Non-viable Bacteria. *Int. Dairy J.* **1998,** *8* (9), 749–758.

Quwehand, A. C.; Kirjavainen, P. V.; Shortt, C.; Salminen, S. Probiotics: Mechanisms and Established Effects. *Int. Dairy J.* **1999,** *9* (1), 43–52.

Rolfe, R. D. The Role of Probiotic Cultures in the Control of Gastrointestinal Health. *J. Nutr.* **2000,** *130* (2), 396S–402S.

Saarela, M.; Lähteenmäki, L.; Crittenden, R.; Salminen, S.; Mattila-Sandholm, T. Gut Bacteria and Health Foods—The European Perspective. *Int. J. Food Microbiol.* **2002,** *78* (1), 99–117.

Saarela, M.; Mogensen, G.; Fondén, R.; Mättö, J.; Sandholm, T. M. Probiotic Bacteria: Safety, Functional and Technological Properties. *J. Biotechnol.* **2000,** *84* (3), 197–215.

Salminen, S.; Ouwehand, A.; Benno, Y.; Lee, Y. K. Probiotics: How Should They be Defined? *Trends Food Sci. Technol.* **1999,** *10,* 107–110.

Salminen, S.; Bouley, C.; Boutron, M. C.; Cummings, J. H.; Franck, A.; Gibson, G. R.; Isolauri, E.; Moreau, M. C.; Roberfroid, M.; Rowland, I. Functional Food Science and Gastrointestinal Physiology and Function. *Br. J. Nutr.* **1998,** *80* (1), S147–S171.

Salminen, S.; Gorbach, S.; YuanKun, L.; Benno, Y.; von Wright, A.; Ouwehand, A. Human Studies on Probiotics: What is Scientifically Proven Today? *Lactic Acid Bacteria: Microbiology and Functional Aspects,* 3rd ed.; CRC Press: Boca Raton, FL, 2004; pp 515–530.

Sanders, M. E. Probiotics: Considerations for Human Health. *Nutr. Rev.* **2003,** *61* (3), 91–99.

Saxelin, M.; Lassig, A.; Karjalainen, H.; Tynkkynen, S.; Surakka, A.; Vapaatalo, H.; Järvenpää, S.; Korpela, R.; Mutanen, M.; Hatakka, K. Persistence of Probiotic Strains in the Gastrointestinal Tract When Administered as Capsules, Yoghurt, or Cheese. *Int. J. Food Microbiol.* **2010,** *144* (2), 293–300.

Schaafsma, G. State of Art Concerning Probiotic Strains in Milk Products. *Int. Dairy Fed. Nutr. News Lett.* **1996,** *5,* 23–24.

Schrezenmeir, J.; de Vrese, M. Probiotics, Prebiotics, and Synbiotics—Approaching a Definition. *Am. J. Clin. Nutr.* **2001,** *73* (2), 361s–364s.

Shah, N. P. Probiotic Bacteria: Selective Enumeration and Survival in Dairy Foods. *J. Dairy Sci.* **2000,** *83*(4), 894–907.

Sheehan, V. M.; Ross, P.; Fitzgerald, G. F. Assessing the Acid Tolerance and the Technological Robustness of Probiotic Cultures for Fortification in Fruit Juices. *Innov. Food Sci. Emerg. Technol.* **2007,** *8* (2), 279–284.

Stanton, C.; Gardiner, G.; Meehan, H.; Collins, K.; Fitzgerald, G.; Lynch, P. B.; Ross, R. P. Market Potential for Probiotics. *Am. J. Clin. Nutr.* **2001,** *73* (2), 476s–483s.

Tilley, L.; Keppens, K.; Kushiro, A.; Takada, T.; Sakai, T.; Vaneechoutte, M.; Degeest, B. A Probiotic Fermented Milk Drink Containing Lactobacillus Casei Strain Shirota Improves Stool Consistency of Subjects with Hard Stools. *Int. J. Probiotics Prebiotics* **2014,** *9* (1–2), 23.

Tripathi, M. K.; Giri, S. K. Probiotic Functional Foods: Survival of Probiotics during Processing and Storage. *J. Funct. Foods* **2014,** *9,* 225–241.

CHAPTER 3

AQUATIC ENZYMES AND THEIR APPLICATIONS IN THE SEAFOOD INDUSTRY

PARVATHY U.[1*], JEYAKUMARI A.[1], GEORGE NINAN[2],
L. N. MURTHY[1], S. VISNUVINAYAGAM[1], and C. N. RAVISHANKAR[2]

[1]*Mumbai Research Centre of ICAR—Central Institute of Fisheries Technology, Vashi, Navi Mumbai 400703, India*

[2]*ICAR—Central Institute of Fisheries Technology, Kochi 682029, Kerala, India*

Corresponding author. E-mail: p.pillai2012@gmail.com

CONTENTS

Abstract 44
3.1   Introduction 44
3.2   Specificity of Enzymes 45
3.3   Classification of Enzymes 46
3.4   Important Enzymes from Aquatic Sources 47
3.5   Recovery and Purification of Enzymes 51
3.6   Seafood Applications of Aquatic Enzymes 54
3.7   Future and Prospects 62
Keywords 63
References 63

ABSTRACT

Advancements in recombinant technology and protein engineering have resulted in enzymes finding wide applicability in various food, industrial, and therapeutical purposes. Compared to chemical techniques, they offer the advantage of substrate specificity and elevated activity facilitating better control of the production processes. Recent researches have revealed novel properties for enzymes from aquatic sources with relatively more stability and activity compared to plant or animal derived enzymes. Initially, the commercial use of enzymes in the food industry was limited confining to some fermentation and curing alone mainly based on endogenous fish proteases. But recently, they have emerged to play an important role in modern food industries to produce novel and diversified range of products. More extensive research needs to be carried out to identify the most specific and promising enzymes and to determine optimal conditions for their application. Such an approach would facilitate commercialization possibilities of enzymes at a much lower cost.

3.1   INTRODUCTION

Enzymes are used in a variety of industrial processes to create an array of foods (Shahidi & Kamil, 2001), cosmetics (Guerard et al., 2010), medicinal products (Chung et al., 2004), in pulp, and paper industry (Bajpal, 1999), leather industry (Senthilvelan et al., 2012), synthesis of surfactants (Sarney & Vulfson, 1995), in textile industry (Araujo et al., 2008), etc. They offer advantages over chemical techniques including substrate specificity and elevated activity allowing better control of the production processes. Recent advances in recombinant technology and protein engineering have resulted in enzymes being evolved as an important molecule widely applicable in various industrial and therapeutical purposes (Gurung et al., 2013). Studies on the enzymes and their existence started long back, and some earlier studies were performed by Jon Jakob Berzelius, a Swedish chemist in 1835 who termed their chemical action as catalytic wherein the phenomenon of catalysis makes possible biochemical reactions necessary for all life processes. In 1877, Wilhelm Friedrich Kühne, a physiology professor from the University of Heidelberg first used the term enzyme, which means "in leaven," derived from a Greek word. The first enzyme "urease" was obtained in pure form from the "jack bean" by James B. Sumner of Cornell University in 1926 which paved way toward the basis of enzyme technology and this work earned him the 1947 Nobel Prize.

All known enzymes are proteins mainly globular where the tertiary structure has given the molecule a generally rounded, ball shape. They are high molecular weight compounds ranging from 10,000 to 2,000,000 primarily made up of amino acid chains linked together by peptide bonds. Most enzymes can be denatured under adverse conditions such as changes in temperature, pH, presence of solvents, salts, and other reagents, etc., which disrupt the three-dimensional structure of these proteins. Enzymes require the presence of other compounds referred to as cofactors for activation of their catalytic activity. This entire active complex, that is, apoenzyme (protein portion) together with the cofactor is referred to as the holoenzyme.

According to Holum (1968), the cofactor may be:

1. A coenzyme—a non-protein organic substance, usually vitamins or made from vitamins which is dialyzable, thermostable, and is temporarily and loosely attached to the protein part.
2. A prosthetic group—an organic substance which is dialyzable, thermostable, and is firmly and permanently attached to the protein or apoenzyme portion.
3. A metal-ion-activator—positively charged metal ions which temporarily bind to the active site of the enzyme, giving an intense positive charge to the enzyme's protein. These include K^+, Fe^{2+}, Fe^{3+}, Cu^{2+}, Co^{2+}, Zn^{2+}, Mn^{2+}, Mg^{2+}, Ca^{2+}, and Mo^{3+}.

3.2 SPECIFICITY OF ENZYMES

Enzymes are natural biocatalysts gifted with high catalytic power, noteworthy substrate specificity, and ability to work under mild reaction conditions (Karan et al., 2012). They accelerate the reactions by providing an alternative reaction pathway of lower activation energy. Without enzymes, these reactions take place at a rate far too slow for the rate of metabolism. Enzymes take part in the reaction by altering the rate of reaction but do not undergo permanent changes and so remain unchanged at the end of the reaction. Most chemical catalysts are usually not very selective and they catalyze a wide range of reactions. In contrast, enzymes are usually highly selective, catalyzing specific reactions only. This specificity is due to the shapes of the enzyme molecules and is one of the properties of enzymes that make them important as research and diagnostic tools. Enzyme specificity can be categorized into:

1. Absolute specificity—the enzyme will catalyze only one particular reaction.
2. Group specificity—the enzyme will act only on molecules that have specific functional groups, such as amino, phosphate, and methyl groups.
3. Linkage specificity—the enzyme will act on a particular type of chemical bond regardless of the molecular structure.
4. Stereochemical specificity—the enzyme will act on a particular steric or optical isomer.

3.3 CLASSIFICATION OF ENZYMES

Enzymes are classified based on the chemical reactions they catalyze (Table 3.1) and they are allotted Enzyme Commission number (EC number) (Webb, 1992). In enzyme nomenclature system, every EC number is associated with a recommended name for the respective enzyme with most enzyme names ending in "ase," except in the case of some originally studied enzymes such as pepsin, rennin, and trypsin. The first EC report (1961), proposed a system for classification of enzymes that also serves as a basis for assigning code numbers to them and the current edition in 1992, published by the International Union of Biochemistry and Molecular Biology, contains 3196 different enzymes.

TABLE 3.1 Classification of Enzymes Based on Type of Reactions (Webb, 1992).

Enzyme commission number and class of enzyme	Type of reaction
EC 1-Oxidoreductases	Catalyzes oxidation reactions involving the transfer of electrons from one molecule to another. Typical, for example, dehydrogenases
EC 2-Transferases	This class of enzymes catalyzes the transfer of groups of atoms from one molecule to another. For example, Aminotransferases or transaminases
EC 3-Hydrolases	This group catalyzes hydrolysis resulting in the cleavage of substrates by water. In general, larger molecules are broken down to smaller fragments by hydrolases
EC 4-Lyases	Reaction involves the addition of groups to double bonds or the formation of double bonds through the removal of groups. For example, Pectate lyases
EC 5-Isomerases	Isomerases catalyze the transfer of groups from one position to another in the same molecule
EC 6-Ligases	Ligases join molecules together with covalent bonds with the input of energy in the form of cofactors, such as ATP

3.4 IMPORTANT ENZYMES FROM AQUATIC SOURCES

Aquatic environment offers enormous potential as a source of enzymes with unique physical, chemical, and catalytic properties. More research has revealed the presence of enzymes from marine organisms with novel properties possessing relatively more stability as well as activity compared to plant or animal-derived enzymes (Bull et al., 2000; Kin, 2006). This is on account of the complexity of the marine environment that provides the microbial population with unique genetic structures and life habitats compared to microorganisms in the terrestrial environment (Stach et al., 2003) leading to boosted marine microbial enzyme technology with valuable products as pharmaceuticals, food additives, and fine chemicals (David & Michael, 1999). Hence, there is considerable interest in the utilization of these marine microorganisms as a new promising source of enzymes with wide application potentials, whose immense genetic and biochemical diversity is still in its infant stage (Ghosh et al., 2005).

3.4.1 PROTEASE

About 60% of all industrial enzymes are proteases, widely used in detergent industry, leather industry, pharmaceutical industry, and food industry (Kumar et al., 2004; Zhang et al., 2005). Proteases (protein hydrolases) catalyze amide (peptide) bond hydrolysis in protein or peptide substrates. Based on their site of action on the substrate, they are classified as exo or endopeptidases. Exopeptidases cleave the peptide bond proximal to the amino or carboxy termini of the substrate. Based on the site of action at the N or C terminus, they are classified as aminopeptidases and carboxypeptidases. Endopeptidases cleave peptide bonds distant from the termini of the substrate. Based on the functional group present at the active site, endopeptidases are further classified into four prominent groups, that is, serine proteases, aspartic proteases, cysteine proteases, and metalloproteases. According to Svendsen (1976), the catalysis by the major class of proteases, the serine proteases occurs as three consecutive reactions: Formation of the Michaelis complex between the original peptide chain (the substrate) and the enzyme; Cleavage of the peptide bond to liberate one of the two resulting peptides; a nucleophilic attack on the remains of the complex to split off the other peptide and to reconstitute the free enzyme.

Studies on proteolytic activity of enzymes present in fish viscera began in the 19th century and was apparently first observed by Stirling in 1884

(Gildberg, 1988). In 1972, Nobou Kato isolated a new type of alkaline protease from marine Psychrobacter, and since then quite a few proteases have been continually obtained from marine microorganisms. Proteolytic enzymes that have been investigated in fish and aquatic invertebrates include gastric proteinases, intestinal proteinases, muscle proteases, and hepatopancreas proteinases (Shahidi & Kamil, 2001). There is considerable demand for these enzymes with right combination of properties for specific application in the industry (Haard & Simpson, 2012). Several proteases are used in food industry for the modification of proteins, recovery of protein from bones and fish waste, meat tenderization, etc., papain for chill proofing and haze removal in beverages; neutral proteases in baking industry for dough conditioning; alkaline proteases in detergent industry for stain removal; fungal proteases, chymosin, other proteases in dairy industry for the replacement of calf rennet, whey protein processing, and production of enzyme-modified cheese (EMC), etc. (Kumar et al., 2008).

3.4.2   LIPASE

Lipases are involved in diverse biological processes ranging from routine metabolism of dietary triglycerides to cell signaling and inflammation (Tjoelker et al., 1995; Spiegel et al., 1996). They play an important role in post-mortem quality deterioration in seafoods during handling, chilled, and frozen storage. Lipases catalyze the breakdown of fats and oils with subsequent release of free fatty acids, diacylglycerols, monoglycerols, and glycerol. Besides this, they are also efficient in various reactions such as esterification, transesterification, and aminolysis (Babu et al., 2008). This provides an opportunity for the fats and oils industry to produce new types of triacylglycerols, esters, and fatty acids and to improve the quality of the existing products produced by employing conventional technologies. While comparing with the traditional methods for fish degreasing, namely, expression method, extraction method, and alkaline processing, using lipases has incomparable advantages. Lipases act under mild conditions and are highly stable in organic solvents, show broad substrate specificity thus, increasing the applicability of lipases in the fish processing field (Kojima & Shimizu, 2003; Aryee et al., 2007). With the exploitation of marine resources, the pelagic fishes have become the primary target of resourceful and potential lipases. Nayak et al (2003) have observed fish intestine and liver as a potential source of lipases. Different fish species such as sardine (Smichi et al., 2010), grey mullet (Aryee et al., 2007), and carp (Gorgun & Akpinar, 2012), etc., have been exploited for these enzymes.

3.4.3   CHITINASE AND CHITOSANASE

Chitin is an abundant biopolymer with non-toxic properties and is quite abundant in the crust of insects and crustaceans. It is made up of a linear chain of acetylglucosamine groups, and chitosan is obtained by deacetylation of chitin. Enzymes with chitinolytic activity include "true" chitinases and lysozymes. Chitinase is an endo-b-N-acetylglucosaminidase which gives random hydrolysis of b-1,4-linkages in poly- and oligo-saccharides of N-acetylglucosamine. Lysozymes have a similar distribution in nature to chitinases and are generally defined as N-acetylmuramyl hydrolases. Among the natural chitinous resources, fishery wastes (shrimp/crab shells and squid pens) are especially rich (Liang et al., 2014). Marine zooplankton is also sources of untapped chitin, which could be a rich source of carbon and energy for growth and reproduction of chitin-degrading microorganisms. Investigations had been carried out for the isolation of chitinase from different species of marine bacteria, seaweeds, fishes, etc. (Osama & Koga, 1995; Koga et al., 1999).

3.4.4   ALGINATE LYASES AND AGARASE

The brown algae are one of the largest marine biomass resources. Alginate has a wide range of applications with the degraded low-molecular fragment showing more potential. Alginate lyases, characterized as either mannuronate or guluronate lyases are a complex copolymer of α-l-guluronate and its C5 epimer β-d-mannuronate (Kim et al., 2011). They have been isolated from a wide range of organisms including algae, marine invertebrates, and marine and terrestrial microorganisms (Wong et al., 2000). Discovering and characterization of alginate lyases will enhance and expand the use of these enzymes to engineer novel alginate polymers for applications in various industrial, agricultural, and medical fields.

Agar is a highly heterogeneous polysaccharide having wide range of food applications and cosmetic applications. Agarases are characterized as α-agarases and β-agarases according to the cleavage pattern and the basic units are agarobiose and neoagarobiose, respectively. Agarases have potential application in degrading the cell wall of red algae for extraction of labile substances with biological activities, such as unsaturated fatty acids, vitamins, and carotenoids from algae. Agarases have been isolated from many sources, including seawater, marine sediments, marine algae, marine mollusks, fresh water, and soil as well as engineered microorganisms (Fu

& Kim, 2010). They find wide applications in food industry, cosmetics, and medical fields producing oligosaccharides with remarkable activities favoring human health. They are also used as a tool enzyme for biological, physiological, and cytological studies (Wang et al., 2004).

3.4.5 CARRAGEENASES

Carrageenases are glycoside hydrolases that specifically degrade carrageenan, a highly anionic polysaccharide found in the cell wall of many red algal species. Carrageenases are also known to be produced by marine invertebrates feeding on carrageenophytes. Carrageenan and carrageenin are mainly used in food industry as a coagulant, adhesive, stabilizer, and emulsifier. In addition, it has also been widely applied in the pharmaceutical and cosmetics industries. The oligosaccharides obtained from carrageenan degradation show a variety of specific physiological activities, such as antiviral, anti-tumor, anti-coagulation, immunoregulation, anti-angiogenesis, etc. (Yuan et al., 2005; 2006; Keisuke et al., 2004).

3.4.6 COLLAGENASE AND COLLAGENOLYTIC ENZYMES

Collagen is the protein found in connective tissues of animals and is the specific collagenase substrate. Collagenase enzyme can be isolated from microbial cells, animal tissues, and digestive organs of different fish and invertebrates and finds application in fur and hide tanning to help ensure a uniform dying of leathers (Kanth et al., 2008). However, the most common application of these enzymes appears to be in the pharmaceutical industry. The enzymes that are capable of degrading collagen (including cathepsin and elastase) are known generally as collagenolytic enzymes. All other collagenolytic enzymes are members of the matrix metalloproteinase (MMP) family and act at a neutral pH (Visse & Nagase, 2003).

3.4.7 EXTREMOZYMES

Extremophilic microorganisms can be screened from marine environments on account of their extremely complex conditions and they may have some specific physiological principles, which can produce unique biocatalysts that function under extreme conditions such as extremely high or low temperature, high pressure, acidic or alkaline conditions, etc. compared to those prevailing in various industrial enzymes (Niehaus et al., 1999). A basic understanding of the stability and function of extremozymes under extreme conditions sets the way for significant innovations in biotechnology (Karan et al., 2012).

3.5 RECOVERY AND PURIFICATION OF ENZYMES

Enzyme purification involves removal of other proteins as well as non-proteins present in the preparation to achieve pure biological catalyst. This assists in understanding their mechanism and developing them for application in various fields. Improvements in materials, use of computerized instruments and in vivo tagging has made protein separation and purification more controllable and predictable (Janson, 2011). Purification of an enzyme protein is generally a multi-step process and their purification strategy involves exploitation of their biophysical and biochemical characteristics such as relative concentration in the source, amino acid composition, sequence, subunit structure, size, shape, net charge, isoelectric point, solubility, heat-stability, hydrophobicity, ligand/metal-binding properties, and post-translational modifications. Based on their properties, a combination of various methods can be used for separation of enzymes with objectives like high recovery, highly purified enzyme protein, reproducibility of the methods, economical use of the chemicals (reagents), and shorter time for complete purification (Kumar, 2008).

The efficiency of enzyme purification can be monitored by several steps of which spectrophotometric or colorimetric method for enzymatic activity measurement is most convenient and an excellent indicator which measures the progressive increase in specific activity (for enzymes, activity in units/mg protein) (Scopes, 2013). All enzymes are proteins, and hence methods for general protein purification apply to enzymes too. Recovery steps should be such as to minimize the losses in enzyme activity. The degree of purification process will dictate the choice of separation technology to be employed (Bhatia, 2005).

3.5.1 TRADITIONAL METHODS

Traditional methods enable rapid purification of enzymes on a wide range varying from micrograms to kilograms. These techniques of purification can be subdivided into categories, namely, precipitation, chromatographic separation, and "in solution" methods which are mainly based on solubility differences (e.g., precipitation with neutral salts or organic solvents); size differences (e.g., size exclusion chromatography or dialysis); charge differences (e.g., ion exchange chromatography, or electrophoresis); and binding to specific ligands (e.g., affinity chromatography, or hydrophobic interaction chromatography) (Simpson, 2000).

3.5.2   PRECIPITATION METHODS

Precipitation method, even though is one of the oldest procedures, still finds use in many situations. Different proteins have different solubilities, and can be precipitated using various additives. Simple precipitation can be carried out sequentially or in combination using ammonium sulfate, sodium sulfate, etc., or miscible organic solvents like isopropanol, ethanol, acetone, polyethylene glycol, etc. (Scopes, 2002; 2013). In this step, initially, the bulk protein is recovered from a crude extract followed by initial resolution into manageable fractions. Mainly salts with divalent anions like ammonium sulfate are used for precipitation where the addition of salt to a solution containing the desired enzyme and other proteins results in proteins progressively precipitating. It also removes non-protein impurities present in the crude extract. The proteins precipitating through one short range of salt concentration are expected to include the majority of the desired enzyme. By collecting this precipitate by centrifugation, purification is achieved. Ammonium sulfate and organic solvents can selectively precipitate the enzyme of choice with 80–90% activity recovery (Bhatia, 2005). Isoelectric precipitation may also be adopted as protein is least soluble when there is no net charge.

3.5.3   CHROMATOGRAPHIC TECHNIQUES

Chromatography is a separation technique based on the partitioning of proteins between moving phase and stationary phase. For enzyme purification, commonly used chromatography techniques include ion exchange chromatography, adsorption chromatography, gel filtration chromatography, and affinity chromatography. Ion exchange chromatography is the most popular chromatographic method for separation of proteins and makes use of strong and weak anion and cation exchange and phosphate interaction. It is a versatile and generic tool and is suited for discovery of proteins, high-resolution purification, and industrial production of proteins (Jung-bauer & Hahn, 2009). Weakly bound ions are more suitable for enzyme adsorption as strongly bound ones may denature the enzymes due to the pH extremes (Bhatia, 2005). The basic principle involved in ion exchange chromatography is binding of charged proteins on the ion exchanger by electrostatic attraction between charged groups on the proteins and opposite charges on the exchanger. Unbound proteins are removed from the column by washing with the same medium used for pre-equilibrium. Bound proteins are eluted by passing buffer of higher ionic strength (using salts like sodium or potassium chloride) or by using a buffer of different pH. Two types of ion exchangers are in common use for separation of enzymes: Anion exchangers and cation exchangers. The most commonly used anion exchanger is diethyl-aminoethyl cellulose (DEAE cellulose) and the most commonly used cation exchanger is carboxymethyl cellulose (CM-cellulose). The basic principle underlying the adsorption chromatography is binding of the enzyme proteins by physical adsorption on the surface of the insoluble matrix. A mixture of proteins containing the desired enzyme is applied to the column, and the proteins are eluted from the column matrix by using a suitable elution buffer either having a change in ionic concentration or pH. Gel filtration chromato-graphic technique is based on the size and shape of the proteins. The protein molecules to be separated are loaded in the porous gel matrix and too large molecules are not entered in the porous matrix and are eluted out from the column. On running a gel filtration column, the largest proteins emerge first, and the smallest last. The commonly used gel filtration gels are of dextran, agarose, and polyacrylamide. Affinity chromatography is unique in purifica-tion technology since it is the only technique that enables the purification of a biomolecule on the basis of its biological function or individual chemical structure. Affinity adsorbents are designed to interact specifically with the desired enzyme, either through its active site, or with some other surface feature of the enzyme. The interacting "ligand" is chemically attached to neutral beads which constitute the adsorbent in the column. Subsequently, elution is done by treatment resulting in dissociation of the desired enzyme-ligand complex.

3.5.4   "IN SOLUTION" METHODS

In this protein separation method, while the separation takes place, the proteins remain in solution. These are in turn either electrophoretic separa-tion based on net charge, or membrane separation wherein separation is on the basis of molecular size by diffusion through pores in a membrane or in a bead particle (Scopes, 2002). Most commonly used electrophoretic technique is Polyacrylamide Gel Electrophoresis (PAGE) wherein proteins are separated on the basis of charge as well as size. When protein sample is fed into a cell fitted with a membrane, it retains high molecular weight proteins while solvent and low molecular weight molecules are filtered through the membrane. There are different kinds of membrane separation categorized by their thickness, construction, charge, or according to their origin. The different membrane separation processes include microfiltration, ultrafiltration, nanofiltration, and reverse osmosis with the most frequently used methods being nanofiltration and ultrafiltration (Szélpál et al., 2013).

3.6   SEAFOOD APPLICATIONS OF AQUATIC ENZYMES

Enzymes and their application in the food industry have gained immense focus and insights of exploration on account of their specificity and considerable activity at moderate temperatures, significant energy saving in food processing operations, and provide processes for tailor-made products under environment-friendly conditions (Haard, 1998). Initially the commercial use of enzymes in the food industry was limited confining to some fermentation and curing alone mainly based on endogenous proteases in the fish but in recent times enzymes play an important and essential role in modern food industries to produce a novel and diversified range of products for food applications (Ashie & Lanier, 2000; Shahidi & Kamil, 2001). Some of the enzyme applications include the production of commercially economic products, namely, polyunsaturated fatty acids-enriched fish oils, selective removal of skin and fish scales, in cured roe production, etc.

3.6.1   APPLICATIONS IN SURIMI

Surimi is water leached fish mince, which is further strained to remove any remaining connective tissues, lipids, and other undesirable components affecting protein denaturation (Park, 2005). It is used as a valuable, functional proteinaceous ingredient in a variety of food products. The quality of surimi products is highly affected by species-specific protein properties as well as the activity of the endogenous heat-stable proteinases, which have a gel-softening effect. Transglutaminases crosslink the gelatin involving the formation of covalent network junctions, the transglutaminase-catalyzed gelatin gels are often referred to as "chemical gels" (Babin & Dickinson, 2001; Siu et al., 2002). This cross-linkage has unique effects on protein properties, water-holding capacity, gelation capability, and thermal stability, etc., and hence, transglutaminase-catalyzed reactions can be used to modify the functional properties of surimi proteins thus, finding application in food industry. Transglutaminase was isolated from marine sources like Alaska pollock (Seki et al., 1990), red sea bream (*Pagrus major*) (Yasueda et al., 1994), etc., but it still needs commercialization.

3.6.2 SELECTIVE TISSUE DEGRADATION

In food processing, mechanical processes have obvious limitations regarding selective and gentle treatment of the raw materials resulting in quality reduction and low yield. In such situations, enzymes can be used as specific gentle tools assisting in selective removal or modification of certain tissue structures.

Descaling is a high energy and time-consuming processing operation accompanied with several concerns like incomplete scale removal, skin rupture and textural damage, poor fillet yield, loss of color, and glossiness, etc. Rather the demand for scaleless fish fillets in the market has resulted in a demand for the exploration of an alternative approach for skin removal (Benjakul et al., 2010; Venugopal, 2015). In Iceland, this mechanical process has been analyzed for replacement by a biotechnological application, using a cold-adapted fish enzyme solution for a gentler descaling. The scales from fishes like redfish or haddock were gently removed without affecting the skin or flesh by incubating the fish in an enzyme solution at 0°C followed by spraying with water (Venugopal et al., 2000). Furthermore, the enzyme treatment removes surface bacteria thereby extending the shelf life of the treated fish.

3.6.3 REMOVAL OF SQUID SKIN

The muscle and tentacles of many squid species are too tough to be accepted and manual removing of rubbery skins from both inside and outside squid belly and from the tentacles is a difficult task in squid processing industry (Shahidi & Kamil, 2001). Also, the conventional mechanical deskinning method removes only the pigmented outer skin, leaving the rubbery inner membrane intact. Using enzymes it is possible to selectively attack and remove the skin without affecting the muscle (Venugopal et al., 2000). Proteases are usually mixed with carbohydrases to facilitate skin removal. Proteolytic enzymes from squid intestines may also be used for removing the double-layered skin of the squid. Furthermore, the product made by the enzymatic method differs from mechanically and manually deskined squid in having no elastic tunics.

3.6.4   ENZYMATIC REMOVAL OF MEMBRANES AND ORGANS FROM SEAFOOD PRODUCTS

Membrane removal in the production of canned cod liver is an important task in order to prevent the contamination of seal worm, usually infested with the collagenous membrane which surrounds the liver (Stefansson, 2013). The black membrane surrounded by cod swim bladder restricts the market demand in salted cod swim bladder production; which requires ideally a white color appearance for consumer attraction. When muscle tissue of scallop is cleaned mechanically the black kidney and stomach cannot be removed completely without damaging the muscle. Also, this may not allow gonads to remain intact after cleaning which has a great demand in the market (Raa, 1990). Because of these drawbacks of the mechanical method, an enzymatic process has been developed where scallops are rinsed in a solution of enzymes obtained from fish viscera. In Norway and Iceland, methods have been devised to rinse shellfish in a solution containing enzymes which enable the easy separation of the muscle tissue without undesirable softening of the muscle (Stefansson & Steingrimsdottir, 1990).

3.6.5   EXTRACTION OF PEARL ESSENCE FROM FISH SCALE

Pearl essence is made from fish scales that contain whitish silver substance called guanine (2-amino-6-oxypurin). This compound is found in the epidermal layer on the scales of different fish species that swim near the surface of the water and is responsible for the gleaming effect of fish (Venugopal, 2011). Guanine in fish scale combines with collagen and calcium phosphate displaying a silvery white color as a result of refractive and reflex phenomena (Tanikawa); the suspension of crystalline guanine in solvent is called "pearl essence" (Venugopal & Smita, 2015). Pearl essence may be extracted from fish scales using proteolytic enzymes such as trypsin.

3.6.6 PUFA-ENRICHED FISH OILS

Nutritional importance of marine oils has received extensive attention, on account of their protective effect against coronary heart diseases, for normal brain and nervous tissue development, etc. Enzymes are widely exploited for isolation of oils and fats from seafood by-products as well as for the preparation and concentration of omega-3 polyunsaturated fatty acid (PUFA) enriched marine oils (Moore & McNeill, 1996; Wanasundara, 2011). Recovery of oil from seafood by enzymatic methods can be categorized into two, namely, protease catalyzed hydrolysis, and lipase catalyzed hydrolysis (Mtabia, 2011). Haraldsson (1990) surveyed the application of lipases to modify marine oils for preparing PUFA-enriched fish oil and predicted that this discovery might prove important for the marine oil industry.

3.6.7 PRODUCTION AND PURIFICATION OF CAVIAR AND OTHER FISH ROE

"Caviar" refers only to the riddled and cured roe of the sturgeon though recently it has been used to describe similarly treated roe of other less expensive fish species like cod, catfishes, herring, capelin, some freshwater species, etc. The roe is usually separated from the membrane sac of the ovary by either manual or mechanical means and then cured with 12–16% salt of caviar production (Bledsoe et al., 2003). The conventional riddling process is a labor intensive task carried out either manually or mechanically causing difficulty in releasing the roe particles from the supportive connective tissue of the roe sac especially in fatty fishes with several disadvantages such as labor-intensive, poor yield, short shelf life due to severe mechanical damage to the eggs, and need for experienced and skilled labor (Vilhelmsson, 1997). Enzymatic treatment using proteinases eases the riddling process in caviar production by gently separating roe from the connective tissue (Kandasamy et al., 2012). Advantages of enzymatically produced roe over those produced mechanically include better recovery, less damage to the eggs, cleanliness of product with no residues, less drip loss during thawing if the resulting caviar is frozen, less labor requirement, and good hygienic conditions of products. Whole or split-roe sacs are immersed in a water bath maintained at low temperature containing the egg-releasing enzyme and hence, enzymes having high activity at low temperatures give them unique application possibilities. After the treatment, the roe may be separated from the connective

tissue by sedimentation and the connective tissues are removed by flota-
tion. Haard and Simpson (2012) observed that fish pepsins were effective
in apparently splitting the linkages between the egg cells and the roe sack
without damaging the eggs.

3.6.8   CAROTENOID PIGMENT EXTRACTION

Carotenoproteins, by-products from shellfish processing, have the potential
for use as a colorant, flavor compound or as a feed supplement in aquaculture (Shahidi et al., 1998). Several authors have done the extraction of
carotenoid pigments with organic solvents or oil (Sachindra & Mahendrakar,
2005; Sachindra et al., 2005; Sindhu & Sherief, 2011). Extraction by these
methods reduces both the ash and chitin levels and achieves a good recovery
of pigments, but the product so obtained is devoid of protein and hence, has
decreased stability due to oxidation (Benjakul et al., 2010). Since about one-
third of the dry matter in crustacean shellfish processing discards is protein,
an enzymatic process has been developed to recover them along with the
carotenoids. By this method, the carotenoid pigment is recovered in the
form of a protein–carotenoid complex, which is more resistant to oxidation
and gives better results than free astaxanthin. The proteolytic enzymes used
to aid the extraction of carotenoprotein have been trypsin type proteases
and the extraction process recovered a carotenoprotein fraction containing
about 80% of the protein and carotenoid pigments present in shrimp offal.
Chakrabarti (2002) adopted an enzymatic method to extract carotenoprotein fraction from tropical brown shrimp shell waste. Enzyme treated crustacean materials to extract carotenoid pigments finds application in cosmetics,
consumable products, feed and food (including beverages) and additives
produced therefrom (Nielsen, 2007).

3.6.9   EXTRACTION OF SEAFOOD FLAVORS

Flavor enhancers or flavor potentiators are chemical substances that have
little or no flavor of their own but, have the ability to enhance the flavor
of the food in which it is incorporated. Enhancers used extensively worldwide include monosodium glutamate (MSG), inosine-5′-monophosphate
(5′-IMP), and guanosine-5′-monophosphate (5′-GMP) finding applications as flavor potentiators in soups, sauces, gravies, and many other
savory products. Seafood flavor is complex due to the broad range of

taxa including both vertebrate and invertebrate organisms. Endogenous enzymes that are native to the organisms are directly responsible for the pleasant, plant-like fresh raw fish flavor (Haard, 2000). Seafood flavors are in high demand for use as additives in products such as kamaboko, artificial crab, fish sausage, and cereal-based extrusion products such as shrimp chips (Kawai, 1996). Seafood flavoring from various sources of raw material can be produced by enzymatic hydrolysis using exogenous and endogenous proteolytic enzymes. Seaweed protein hydrolysate using bromelain was used as the precursor of a thermally processed seafood flavoring agent (Laohakunjit et al., 2014). Production of Seafood Flavor from Red Hake (*Urophycis chuss*) by Enzymatic Hydrolysis was tried by Imm and Lee (1999). The purpose of the enzymatic hydrolysis is to liquefy and allow the separation of bones and shells, also facilitating concentration or drying. Moreover, it permits the nucleotide and protein transformation to flavor enhancers without producing additional flavor or taste. The final products consist of pastes or powders and are natural extracts with flavoring properties.

3.6.10 VISCOSITY REDUCTION OF STICK WATER

In the fish meal industry, dissolved protein that is concentrated by evaporation can contribute to excessive viscosity and prevent efficient evaporator operation. The addition of neutral or alkaline bacterial proteinases, when the solid contents are above 20% helps in fuel savings and improved product. To achieve rapid viscosity reduction of a protein solution, proteases can be applied which reduces the viscosity of stick water (Venugopal, 2015). An endo-protease which catalyzes the hydrolysis of the polypeptide chain at random sites is preferable to an exoprotease. Work has been carried out on the application of commercial proteolytic enzyme, namely, alcalase for preparing stick water hydrolysate (Simonsen, 2003).

3.6.11 PRODUCTION OF FISHERY PRODUCTS

A number of traditional fishery products are prepared by fermentation. Endogenous proteases, rather than the action of microorganism, play the major role in transforming raw material into finished product. These products include fish sauce, ripening of fish-cured fish products, fish silage, fish protein hydrolysate, etc.

3.6.11.1   FISH SAUCE

Fish sauce has long been a traditional fermented fish product and is an important source of protein in Southeast Asia (Yu et al., 2014). This seafood is a liquid product made by storing heavily salt-preserved fish material at tropical temperatures until it is solubilized by endogenous enzymes (Aehle, 2007). Traditional fish sauce is produced by mixing 1–2 parts of salt with 3 parts of fish, followed by fermentation for 6–12 months (Peralta et al., 1996). The role of various proteolytic enzymes in fish sauce production has been investigated by many researchers who concluded that the main enzyme involved in tissue degradation appeared to be a trypsin-like enzyme and other proteases such as cathepsin B.

3.6.11.2   RIPENING OF FISH

Fermentation plays an important role in many parts of the world for the production of traditional fish products. In Southeast Asia, fermented fish products have a long history and they are of great nutritional importance (Skara et al., 2015). It is traditionally carried out by storing gutted or ungutted fish together with salt and possibly sugar in barrels for several months, in order to obtain characteristic changes in texture and flavor (Olsen & Skara, 1997). One of the main factors believed to cause the ripening is the action of endogenous enzymes (in the muscle or the intestinal tract) as well as bacteria, with the former being most significant with respect to changing texture creating a soft textured product as well as producing some of the flavors (taste-active peptides and free amino acids), and the latter aiding in the development of aroma and flavor (Nielsen & Borresen, 1997; Skara et al., 2015). Fish muscle cathepsins have also been reported to be responsible for the proteolysis (Shenderyuk & Bykowski, 1990) and this was apparently first observed by Schmidt-Nielsen in 1901. Cathepsins A and C are the major cathepsins that play an important role in proteolysis during fish fermentation.

3.6.11.3   FISH SILAGE

Fish Silage is a liquid product made from whole fish or parts of fish that are liquefied by the action of enzymes in the fish in the presence of added acid (Ghaly et al., 2013). The enzymes present in the acidic medium breakdown fish proteins into smaller soluble units while the acid helps to speed up their activity and prevent bacterial spoilage. There are two methods for the production of fish silage. The first is acid-preserved silage which is produced by the addition of inorganic or organic acid, which lowers the pH below 3–4. However, silages made with inorganic acids require a still lower pH. This pH is optimum for enzymes naturally present in the raw material used. Endogenous enzymes hydrolyze the tissue structures producing an amber liquid, which is an aqueous phase rich in small peptides and free amino acids. Pepsin plays a major role among all enzymes in preparation of silage and it can quickly degrade and liquefy minced raw material into highly nutritious protein hydrolysate (Gildberg, 2004). In fermented silage, bacterial fermentation is initiated by mixing minced or chopped raw material with a fermentable sugar, which favors the growth of lactic acid bacteria. These bacteria are usually added as a starter culture (Raghunath & Gopakumar, 2002). The lactic acid bacteria produce acids and antibiotics, which together destroy competing spoilage bacteria, and the low pH achieves the tissue hydrolysis. Hydrolysis is terminated by pasteurization at 85°C for 15 min to inactivate proteases and lipases. After heating, the silage is deoiled by decantation and centrifugation. The liquefied protein phase is acidified to pH 4 to prevent spoilage.

3.6.11.4   FISH PROTEIN HYDROLYSATES

Fish protein hydrolysates (FPH) are a mixture of proteinaceous fragments and are prepared extensively by digestion of whole fish or other aquatic animals or parts thereof using proteolytic enzymes (endogenous and exogenous) at the optimal temperature and pH required by enzymes. The hydrolytic process results in increased protein solubility and enhanced functionality as well as bioactivity (Thiansilakul et al., 2007; Klompong et al., 2007; Vignesh et al., 2012; Jeevitha et al., 2014) and hence, FPH finds various applications in the food industry including milk replacers, protein supplements, stabilizers in beverages, flavor enhancers in confectionery products (Kristinsson & Rasco, 2000), in sports nutrition, weight control diets, and nutritional supplements (Mahmoud & Cordle, 2000). However, a problem with FPH is the formation of bitter tasting hydrophobic peptides and excessive degradation of protein to peptides and amino acids (FitzGerald & O'cuinn, 2006; Wu et al., 2003). Protein hydrolysis with a combination of proteases (endo- and exo-peptidase) can minimize the bitterness in the hydrolyzed product (Imm & Lee, 1999; Liaset et al., 2000). Incorporation of additives like glutamic acid or glutamyl-rich peptides, polyphosphates, gelatin, or glycine, etc., can mask the bitterness of FPHs. Bitterness can also be eliminated by selective chromatographic separation (Lin et al., 1997) and by enzymatic treatment consisting of exopeptidases to carry out the plastein reaction (Synowiecki et al., 1996).

3.7 FUTURE AND PROSPECTS

Enzymes have emerged as vital tool in the areas of food processing, biotechnology, and other industries. Enzymatic processes are advantageous on account of their substrate and reaction specificity rendering almost complete transformation from a complex mixture of analogous molecules without byproducts under mild conditions of temperature and pH. Recent discoveries have also revealed their applicability under extreme conditions. Extreme environmental conditions in diverse habitats of the oceans have resulted in an immense biodiversity of marine organisms as well as genetic diversity. Although enzymes from plant and livestock have attracted more interest, a very few of them are actually used for industrial application. Therefore, enzymes from fish, aquatic invertebrates, and marine microbes, are gaining more and more importance with a huge potential for development and applications with industrial benefits. Even though aquatic enzymes have been considered as promising tools in conventional and few novel food applications, their potential use as a food-processing tool has to be studied with broader emphasis. Moreover, the industrial-scale recovery of marine enzymes is still under the experimental stage. It may be expected that expanding capabilities of this new area will continue to profoundly affect the seafood industries in future. Although value-added utilization of these enzymes is desirable in many food applications, the economic viability of the process and products must be realistically assessed. This is mainly because cost incurred in producing these enzymes by extraction from their natural sources being a limitation for their widespread use. More extensive research to identify the most specific and promising enzymes and to determine optimal conditions for their use is of utmost importance. Hence, the greatest challenge in food processing is to discover new enzymes, understand their kinetic and applicability to improve the functional and nutritional properties of foodstuff. Such an approach would provide the incentive for commercial developments leading to large-scale production of enzymes at a much lower cost.

KEYWORDS

- **enzymes**
- **proteins**
- **seafood**
- **application**
- **food industry**

REFERENCES

Aehle, W. *Enzymes in Industry: Production and Applications*, 3rd ed.; Wiley-VCH Verlag GmbH & Co.KGaA: Weinheim, 2007.

Anil Kumar. *Handbook of Enzymes*; Agrotech Publishing Academy: Udaipur, India, 2010.

Araujo, R.; Casal, M.; Cavaco-Paulo, A. Application of Enzymes for Textile Fibres Processing. *Biocatal. Biotransfor.* **2008,** *26* (5), 332–349.

Aryee, A. N. A.; Simpson, B. K.; Villalonga, R. Lipase Fraction from the Viscera of Grey Mullet (*Mugil cephalus*): Isolation, Partial Purification and Some Biochemical Characteristics. *Enzyme Microb. Technol.* **2007,** *40* (3), 394–402.

Ashie, I. N. A.; Lanier, T. C. Transglutaminases in Seafood Processing. In *Seafood Enzymes*; Haard, N. F., Simpson, B. K., Eds.; Marcel Dekker: New York, 2000; pp 147–190.

Babin, H.; Dickinson, E. Influence of Transglutaminase Treatment on the Thermoreversible Gelation of Gelatin. *Food Hydrocoll.* **2001,** *15,* 271–276.

Babu, J.; Pramod, W. R.; George, T. Cold Active Microbial Lipases: Some Hot Issues and Recent Developments. *Biotechnol. Adv.* **2008,** *26,* 457–470.

Bajpal, P. Application of Enzymes in the Pulp and Paper Industry. *Biotechnol. Prog.* **1999,** *15* (2), 147–157.

Benjakul, S.; Klomklao, S.; Simpson, B. K. Enzymes in Fish Processing. In *Enzymes in Food Technology*, 2nd ed.; Whitehurst, R. J., Van Oort, M., Eds.; Blackwell Publishing: Chichester, 2010; pp 211–235.

Bhatia, S. C. *Textbook of Biotechnology;* Atlantic Publishers and Distributors: Delhi, India, 2005.

Bledsoe, G. E.; Bledsoe, C. D.; Rasco, B. Caviars and Fish Roe Products. *Crit. Rev. Food Sci. Nutr.* **2003,** *43* (3), 317–356.

Bull, A. T.; Ward, A. C.; Goodfellow, M. Search and Discovery Strategies for Biotechnology: The Paradigm Shift. *Microbiol. Mol. Biol. Rev.* **2000,** *64,* 573–606.

Chakrabarti, R. Carotenoprotein from Tropical Brown Shrimp Shell Waste by Enzymatic Process. *Food Biotechnol.* **2002,** *16* (1), 81–90.

Chung, L.; Dinakarpandian, D.; Yoshida, N.; Lauer-Fields J. L.; Fields, G. B. Collagenase Unwinds Triple-helical Collagen Prior to Peptide Bond Hydrolysis. *Eur. Mol. Biol. Organ. J.* **2004,** *23,* 3020–3030.

David, W.; Michel, J. Extremozymes. *Curr. Opin. Chem. Biol.* **1999,** *3,* 39–46.

FitzGerald, R.; O'cuinn, G. Enzymatic Debittering of Food Protein Hydrolysates. *Biotechnol. Adv.* **2006,** *24* (2), 234–237.

Fu, X. T.; Kim, S. M. Agarase: Review of Major Sources, Categories, Purification Method, Enzyme Characteristics and Applications. *Mar. Drugs.* **2010,** *8* (1), 200–218.

Ghaly, A. E.; Ramakrishnan, V. V.; Brooks, M. S.; Budge, S. M.; Dave, D. Fish Processing Wastes as a Potential Source of Proteins, Amino Acids and Oils: A Critical Review. *J. Microb. Biochem. Technol.* **2013,** *5* (4), 107–129.

Ghosh, D.; Saha, M.; Sana, B.; Mukherjee, J. Marine Enzymes. *Adv. Biochem. Eng. Biotechnol.* **2005,** *96,* 189–218.

Gildberg, A. Aspartic Proteinases in Fishes and Aquatic Invertebrates. *Comp. Biochem. Physiol.* **1988,** *91,* 425–435.

Gildberg, A. Enzymes and Bioactive Peptides from Fish Waste Related to Fish Silage, Fish Feed and Fish Sauce Production. *J. Aquat. Food Prod. Technol.* **2004,** *13,* 3–11.

Gorgun, S.; Akpinar, M. A. Purification and Characterization of Lipase from the Liver of Carp, *Cyprinus carpio L.* (1758), Living in Hake Todurge (Sivas, Turkiye). *Turk. J. Fish. Aquat. Sci.* **2012,** *12,* 207–215.

Guerard, F.; Decourcelle, N.; Sabourin, C.; Floch-Laizet, C.; Le Grel, L. Recent Developments of Marine Ingredients for Food and Nutraceutical Applications: A Review. *J. Des. Sci. Halieut. Aquat.* **2010,** *2,* 21–27.

Gurung, N.; Ray, S.; Bose, S.; Rai, V. A Broader View: Microbial Enzymes and Their Relevance in Industries, Medicine, and Beyond. *BioMed. Res. Int.* **2013,** *2013.* http://dx.doi.org/10.1155/2013/329121

Haard, N. The Role of Enzymes in Determining Seafood Color, Flavor and Texture. In *Safety and Quality Issues in Fish Processing*; Bremner, H. A., Ed.; Woodhead Publishing Limited: Cambridge, England, 2000; pp 220–253.

Haard, N. F. Specialty Enzymes from Marine Organisms. *Food Technol.* **1998,** *52* (7), 64–67.

Haard, N. F.; Simpson, B. K. Proteases from Aquatic Organisms and Their Uses in the Seafood Industry. In *Fisheries Processing: Biotechnological Applications*; Martin, A. M., Ed.; Chapman and Hall: London, 2012; pp 132–154.

Haraldsson, G. G. The Applications of Lipases for Modification of Facts and Oils, Including Marine Oils. In *Advances in Fisheries Technology and Biotechnology for Increased Profitability;* Voigt, M. N., Botta, J. R., Eds.; Technomic Publishing: Lancaster, 1990; pp 337–357.

Holum, J. *Elements of General and Biological Chemistry,* 2nd ed.; Wiley: NewYork, 1968.

Imm, J. Y.; Lee, C. M. Production of Seafood Flavor from Red Hake (*Urophycis chuss*) by Enzymatic Hydrolysis. *J. Agric. Food Chem.* **1999,** *47* (6), 2360–2366.

Janson, J. C. *Protein Purification: Principles, High Resolution Methods and Applications,* 3rd ed.; Wiley Publishers: Hoboken, NJ, 2011; Vol. 54.

Jeevitha, K.; Priya, M. K.; Khora, S. S. Antioxidant Activity of Fish Protein Hydrolysates from *Sardinella longiceps*. *Int. J. Drug Dev. Res.* **2014,** *6* (4), 137–145.

Jungbauer, A.; Hahn, R. Ion-Exchange Chromatography. In *Methods in Enzymology, Guide to Protein Purification,* 2nd ed.; Burgess, R. R., Deutscher, M. P., Eds.; Elsevier: Amsterdam, Netherlands, 2009; Vol. 463, pp 349–371.

Kandasamy, N.; Velmurugan, P.; Sundarvel, A.; Jonnalagadda, R.; Bangaru, C.; Palanisamy, T. Eco-benign Enzymatic Dehairing of Goat Skins Utilizing a Protease from a *Pseudomonas fluorescens* Species Isolated from Fish Visceral Waste. *J. Clean Prod.* **2012,** *25,* 27–33.

Kanth, S. V.; Venba, R.; Madhan, B.; Chandrababu, N. K.; Sadulla, S. Studies on the Influence of Bacterial Collagenase in Leather Dyeing. *Dyes Pigm.* **2008,** *76,* 338–347.

Karan, R.; Capes, M. D.; Sarma, S. D. Function and Biotechnology of Extremophilic Enzymes in Low Water Activity. *Aquat. Biosyst.* **2012,** *8,* 4.

Kawai, T. Fish Flavor. *Crit. Rev. Food Sci. Nutr.* **1996,** *36,* 257–298.

Keisuke, I.; Michal, K. W.; Takayuki, T.; Siro, S.; Hiroyuki, O. Novel Heparin Sulfatemimetic Compounds as Antitumor Agents. *Chem. Biol.* **2004,** *11,* 367–377.

Kim, H. S.; Lee, C. G.; Lee, E. Y. Alginate Lyase: Structure, Property, and Application. *Biotechnol. Bioprocess Eng.* **2011,** *16,* 843.

Kin, S. L. Discovery of Novel Metabolites from Marine Actinomycetes. *Curr. Opin. Microbiol.* **2006,** *9,* 245–251.

Klompong, V.; Benjakul, S.; Kantachote, D.; Shahidi, F. Antioxidative Activity and Functional Properties of Protein Hydrolysate of Yellow Stripe Trevally (*Selaroides leptolepis*) as Influenced by the Degree of Hydrolysis and Enzyme Type. *Food Chem.* **2007,** *102* (4), 1317–1327.

Koga, D.; Mitsutomi, M.; Kono, M.; Matsumiya, M. Biochemistry of Chitinases. In *Chitin and Chitinases*; Jolles, P., Muzzarelli, R. A. A., Eds.; Birkhauser Verlag: Basel, Switzerland, 1999; pp 111–135.

Kojima, Y.; Shimizu, S. Purification and Characterization of the Lipase from *Pseudomonas fluorescens* HU380. *J. Biosci. Bioeng.* **2003,** *96,* 219–226.

Kristinsson, H. G.; Rasco, B. A. Fish Protein Hydrolysates: Production, Biochemical and Functional Properties. *Crit. Rev. Food Sci. Nut.* **2000,** *40* (1), 43–81.

Kumar, C.; Joo, H. S.; Koo, Y. M.; Paik, S.; Chang, C. S. Thermostable Alkaline Protease from a Novel Marine Haloalkalophilic *Bacillus clausii* Isolate. *World J. Microbiol. Biotechnol.* **2004,** *20,* 351–357.

Kumar, D.; Savitri; Thakur, N.; Verma, R.; Bhalla, T. C. Microbial Proteases and Application as Laundry Detergent Additive. *Res. J. Microbiol.* **2008,** *3,* 661–672.

Laohakunjit, N.; Selamassakul, O.; Kerdchoechuen, O. Seafood-like Flavour Obtained from the Enzymatic Hydrolysis of the Protein by-Products of Seaweed (*Gracilaria sp.*). *Food Chem.* **2014,** *158,* 162–170.

Liang, T. W.; Chen, Y. Y.; Pan, P. S.; Wang, S. L. Purification of Chitinase/chitosanase from *Bacillus cereus* and Discovery of an Enzyme Inhibitor. *Int. J. Biol. Macromol.* **2014,** *63,* 8–14.

Liaset, B.; Lied, E.; Espe, M. Enzymatic Hydrolysis of Byproducts from the Fish-filleting Industry; Chemical Characterization and Nutritional Evaluation. *J. Sci. Food Agric.* **2000,** *80,* 581–589.

Lin, S. B.; Nelles, L. P.; Cordle, C. T.; Thomas, R. L. Physical Factors Related to C18 Adsorption Columns for Debittering Protein Hydrolysates. *J. Food Sci.* **1997,** *62* (5), 946–948.

Mahmoud, M. I.; Cordle, C. T. Protein Hydrolysates as Special Nutritional Ingredients. In *Novel Macromolecules in Food Systems;* Doxastakis, G., Kiosseoglou, V., Eds.; Elsevier Science Press: Amsterdam, Netherlands, 2000; pp 181–215.

Moore, S. R.; McNeill, G. P. Production of Triglycerides Enriched in Long-chain n-3 Polyunsaturated Fatty Acids from Fish Oil. *J. Am. Oil Chem. Soc.* **1996,** *73,* 1409–1414.

Mtabia, B. Valorisation of Fish Waste Biomass through Recovery of Nutritional Lipids and Biogas. Doctoral Thesis, Lund University, Sweden, 2011.

Nayak, J.; Viswanathan Nair P. G.; Ammu, K.; Mathew, S. Lipase Activity in Different Tissues of four Species of Fish: Rohu (*Labeo rohita* Hamilton), Oil Sardine (*Sardinella longiceps* Linnaeus), Mullet (*Liza subviridis* Valenciennes) and Indian Mackerel (*Rastralliger kanagurta* Cuvier). *J. Sci. Food Agric.* **2003,** *83* (11), 1139–1142.

Niehaus, F.; Bertoldo, C.; Kähler, M.; Antranikian, G. Extremophiles as a Source of Novel Enzymes for Industrial Application. *Appl. Microbiol. Biotechnol.* **1999,** *51,* 711–729.

Nielsen, H. H.; Borresen, T. The Influence of Intestinal Proteinases on Ripening of Salted Herring. In *Seafood from Producer to Consumer, Integrated Approach to Quality*; Luten, J. B., Borresen, T., Oehlenschlager, J., Eds.; Elsevier Science: Amsterdam, Netherlands, 1997; pp 293–303.

Nielsen, P. M. Methods for Processing Crustacean Material. Patent US 7241463 B2, 2007.

Olsen, S. O.; Skara, T. Chemical Changes during Ripening of North Sea Herring. In *Seafood from Producer to Consumer, Integrated Approach to Quality;* Luten, J. B., Borresen, T., Oehlenschlager, J., Eds.; Elsevier Science: Amsterdam, Netherlands, 1997; pp 305–317.

Osama, R.; Koga, T. An Investigation of Aquatic Bacteria Capable of Utilizing Chitin as the Sole Source of Nutrients. *Lett. Appl. Microbiol.* **1995,** *21,* 288–291.

Park, J. W. *Surimi and Surimi Seafood*; Taylor and Francis Publishers: Boca Raton, FL, 2005.

Peralta, R. R.; Shimoda, M.; Osajima, Y. Further Identification of Volatile Compounds in Fish Sauce. *J. Agr. Food Chem.* **1996,** *44,* 3606–3610.

Raa, J. Biotechnology in Aquaculture and the Fish Processing Industry: A Success Story in Norway. In *Advances in Fisheries Technology for Increased Profitability;* Voigt, M. N., Botta, J. R., Eds.; Technomic Publication Company: Lancaster, 1990; pp 509–524.

Raghunath, M. R.; Gopakumar, K. Trends in Production and Utilisation of Fish Silage. *J. Food Sci. Technol.* **2002,** *39* (2), 103–110.

Sachindra, N. M.; Mahendrakar, N. S. Process Optimization for Extraction of Carotenoids from Shrimp Waste with Vegetable Oils. *Bioresource Technol.* **2005,** *96,* 1195–1200.

Sachindra, N. M.; Bhaskar, N.; Mahendrakar, N. S. Carotenoids in Crabs from Marine and Fresh Waters of India. *Lebenson. Wiss. Technol.* **2005,** *38,* 221–225.

Sarney, D. B.; Vulfson, E. N. Application of Enzymes to the Synthesis of Surfactants. *Trends Biotechnol.* **1995,** *13* (5), 164–172.

Scopes, R. K. *Enzyme Activity and Assays Encyclopedia of Life Sciences*; Macmillan Publishers Ltd,: Thiruvananthapuram, India, 2002.

Scopes, R. K. *Protein Purification: Principles and Practice*; Springer Science and Business Media: Berlin, Germany, 2013.

Seki, N.; Uno, H.; Lee, N. H.; Kimura, I.; Toyoda, K.; Fujita, T.; Arai, K. Transglutaminase Activity in Alaska Pollack Muscle and Surimi, and its Reaction with Myosin B. *Nippon Suisan Gakk.* **1990,** *56,* 125–132.

Senthilvelan, T.; Kanagaraj, J.; Mandal, A. B. Application of Enzymes for Dehairing of Skins: Cleaner Leather Processing. *Clean Technol. Environ.* **2012,** *14,* 889–897.

Shahidi, F.; Metusalach, B. J. A. Carotenoid Pigments in Seafoods and Aquaculture. *CRC Crit. Rev. Food Sci.* **1998,** *38,* 1–67.

Shahidi, F.; Kamil, J. Y. Y. A. Enzymes from Fish and Aquatic Invertebrates and Their Application in the Food Industry. *Trends Food Sci. Technol.* **2001,** *12,* 435–464.

Shenderyuk, V. I.; Bykowski, P. J. Salting and Marinating of Fish. In *Seafood: Resources, Nutritional Composition, and Preservation;* Sikorski, Z. E., Ed.; CRC Press: Boca Raton, FL,1990; pp 147–162.

Simonsen, P. S. Stick Water Hydrolyzate and a Process for the Preparation Thereof. Patent EP 0990393 A1, 2003.

Simpson, B. K. Digestive Proteinases from Marine Animals. In *Seafood Enzymes: Utilization and Influence on Post Harvest Seafood Quality;* Haard, N. F., Simpson, B. K., Eds.; Marcel Dekker Inc.: New York, 2000; pp 191–214.

Sindhu, S.; Sherief, P. M. Extraction, Characterization, Antioxidant and Anti-Inflammatory Properties of Carotenoids from the Shell Waste of Arabian Red Shrimp *Aristeus alcocki*, Ramadan 1938. *Open Conf. Proc. J.* **2011**, *2*, 95–103.

Siu, N.; Ma, C.; Mock, W.; Mine, Y. Functional Properties of Oat Globulin Modified by a Calcium-independent Microbial Transglutaminase. *J. Agric. Food Chem.* **2002**, *50*, 2666–2672.

Skara, T.; Axelsson, L.; Stefansson, G.; Ekstrand, B.; Hagen, H. Fermented and Ripened Fish Products in the Northern European Countries. *J. Ethnic Foods* **2015**, *2*, 18–24.

Smichi, N.; Fendri, A.; Chaabouni, R.; Rebah, F. B.; Gargouri, Y.; Miled, N. Purification and Biochemical Characterization of an Acid-stable Lipase from the Pyloric Caeca of Sardine (*Sardinella aurita*). *Appl. Biochem. Biotechnol.* **2010**, *162* (5), 1483–1496.

Spiegel, S.; Foster, D.; Kolesnick, R. Signal Transduction through Lipid Second Messengers. *Curr. Opin. Cell Biol.* **1996**, *8* (2), 159–167.

Stach, J. E. M.; Maldonado, L. A.; Ward, A. C.; Goodfellow, M.; Bull, A. T. New Primers for the Class Actinobacteria: Application to Marine and Terrestrial Environments. *Environ. Microbiol.* **2003**, *5*, 828–841.

Stefansson, G.; Steingrimsdottir, U. Application of Enzymes for Fish Processing in Iceland-present and Future Aspects. In *Advances in Fisheries Technology for Increased Profitability;* Voigt, M. N., Botta, J. R., Eds.; Technomic Publication: Lancaster, PA, 1990; pp 237–250.

Stefansson, G. Fish Processing. In *Enzymes in Food Processing,* 3rd ed.; Nagodawithana, T., Reed, G., Eds.; Academic Press Inc.: London, 2013; pp 459–470.

Svendsen, I. Chemical Modifications of the Subtilisins with Special Reference to the Binding of Large Substrates. A Review. *Carlsberg Res. Commun.* **1976**, *41*, 237–291.

Synowiecki, J.; Jagielka, R.; Shahidi, F. Preparation of Hydrolysates from Bovine Red Blood Cells and Their Debittering Following Plastein Reaction. *Food Chem.* **1996**, *57* (3), 435–443.

Szélpál, S.; Poser, O.; Abel, M. Enzyme Recovery by Membrane Separation Method from Waste Products of the Food Industry. *Acta Technica Corviniensis Bulletin Eng.* **2013**, *6* (2), 149–154.

Thiansilakul, Y.; Benjakul, S.; Shahidi, F. Compositions, Functional Properties and Antioxidative Activity of Protein Hydrolysates Prepared from Round Scad (*Decapterus maruadsi*). *Food Chem.* **2007**, *103* (4), 1385–1394.

Tjoelker, L. W.; Eberhardt, C.; Unger, J.; Trong, H. L.; Zimmerman, G. A.; McIntyre, T. M.; Stafforini, D. M.; Prescott, S. M.; Gray, P. W. Plasma Platelet-activating Factor Acetylhydrolase is a Secreted Phospholipase A2 with a Catalytic Triad. *J. Biol. Chem.* **1995**, *270* (43), 25481–25487.

Venugopal, V.; Lakshmanan, R.; Doke, S. N.; Bongirwar, D. R. Enzymes in Fish Processing, Biosensors and Quality Control. *Food Biotechnol.* **2000**, *14*, 21–27.

Venugopal, V. Marine Ingredients in Cosmetics. In *Marine Cosmeceuticals: Trends and Prospects;* Kim, S. K., Ed.; Taylor Francis: Boca Raton, FL, 2011; pp 211–232.

Venugopal, V. Fish Industry Byproducts as Source of Enzymes and Application of Enzymes in Seafood Processing. In *Fish Processing Byproducts-Quality Assessment and Applications;* Mahendrakar, N. S., Sachindra, N. M., Eds.; Studium Press LLC: Houston, TX, 2015; pp 136–171.

Venugopal, V.; Smita, S. L. Nutraceuticals and Bioactive Compounds from Seafood Processing Waste. In *Handbook of Marine Biotechnology;* Kim, S. K., Ed.; Springer: New York, 2015; pp 1405–1426.

Vignesh, R.; Srinivasan, M.; Jayaprabha, N.; Badhul Haq, M. A. The Functional Role of Fish Protein Hydrolysate Derived Bioactive Compounds in Cardioprotection and Antioxidative Functions. *Int. J. Pharma. Bio Sci.* **2012,** *3* (1), 560–566.

Vilhelmsson, O. The State of Enzyme Biotechnology in the Fish Processing Industry. *Trends Food Sci. Technol.* **1997,** *8,* 266–270.

Visse, R.; Nagase, H. Matrix Metalloproteinases and Tissue Inhibitors of Metalloproteinases: Structure, Function and Biochemistry. *Circ. Res.* **2003,** *92,* 827–839.

Wanasundara, U. Preparative and Industrial-scale Isolation and Purification of Omega-3 Poly Unsaturated Fatty Acids from Marine Sources. In *Handbook of Seafood Quality, Safety and Health Application;* Alasalvar, C., Shahidi, F., Miyashita, K., Wanasundara, U., Eds.; Wiley-Blackwell: Chichester, UK, 2011; pp 464–475.

Wang, J.; Jiang, X.; Mou, H.; Guan, H. Anti-oxidation of Agar Oligosaccharides Produced by Agarase from a Marine Bacterium. *J. Appl. Phycol.* **2004,** *16,* 333–340.

Webb, E. C. *Enzyme Nomenclature 1992: Recommedations of the Nomenclature Committee of the International Union of Biochemistry and Molecular Biology on the Nomenclature and Classification of Enzymes;* Academic Press: Cambridge, MA, 1992.

Wong, T. Y.; Preston, L. A.; Shiller, N. L. Alginate Lyase: Review of Major Sources and Enzyme Characteristics, Structure-function Analysis, Biological Roles, and Applications. *Annu. Rev. Microbiol.* **2000,** *54,* 289–340.

Wu, H. C.; Chen, H. M.; Shiau, C. Y. Free Amino Acids and Peptides as Related to Antioxidant Properties in Protein Hydrolysates of Mackerel (*Scomber austriasicus*). *Food Res. Int.* **2003,** *36* (9), 949–957.

Yasueda, H.; Kumazawa, Y.; Motoki, M. Purification and Characterization of a Tissuetype Transglutaminase from Red Sea Bream (*Pagrus major*). *Biosci. Biotechnol. Biochem.* **1994,** *58,* 2041–2045.

Yu, X.; Mao, X.; He, S.; Liu, P.; Wang, Y.; Xue, C. Biochemical Properties of Fish Sauce Prepared Using Low Salt, Solid State Fermentation with Anchovy by-products. *Food Sci. Biotechnol.* **2014,** *23* (5), 1497–1506.

Yuan, H. M.; Zhang, W. W.; Li, X. G. Preparation and In Vitro Antioxidant Activity of κ-carrageenan Oligosaccharides and Their Over Sulfated, Acetylated and Phosphorylated Derivatives. *Carbohydr. Res.* **2005,** *340,* 685–692.

Yuan, H. M.; Song, L. M.; Li, X. G. Immunomoduladtion and Antitumor Activity of κ-carrageenan Oligosaccharides. *Cancer Lett.* **2006,** *243,* 228–234.

Zhang, L. X.; An, R.; Wang, J. P.; Sun, N.; Zhang, S.; Hu, J. C.; Kuai, J. Exploring Novel Bioactive Compounds from Marine Microbes. *Curr. Opin. Microbiol.* **2005,** *8,* 276–281.

CHAPTER 4

UTILIZATION AND VALUE ADDITION OF CULINARY BANANA: THE POTENTIAL FOOD FOR HEALTH

PRERNA KHAWAS and SANKAR CHANDRA DEKA*

Department of Food Engineering and Technology, Tezpur University, Napaam, Tezpur 784028, Sonitpur, Assam, India

*Corresponding author. E-mail: sankar@tezu.ernet.in

CONTENTS

Abstract 70

4.1   Introduction 70

4.2   The Background and the Biology of *Musa* 72

4.3   Plantain Starch 79

4.4   Chemical Properties and Bioactive Compounds Present
      in Banana Peel 81

4.5   Encapsulation of Bioactive Compounds From Banana and
      Plantains and Its Application in Functional Food Formulation 87

4.6   Dehydration of Banana and Plantains 91

4.7   Mathematical Modeling for Thin Layer Drying of Plantains
      and Banana 93

4.8   Conclusion 96

Keywords 96

References 96

ABSTRACT

Based on our baseline research and survey, it was noticed that culinary banana (*Musa* ABB) of Assam and North East India (locally known as *kachkal*) falls under the category of fruit-vegetable and is one of the commonly consumed vegetables next to potato in the daily diet of local people. This nutritionally rich crop of local importance has not gained much attention and is under-utilized and underestimated even though its nutritional value is of worth. It is excellent source of starch which can be modified to resistant starch (RS) for its better utilization and value addition. It is considered as one of the potential storehouses of carbohydrates, starch, polyphenols, micronutrients, and functionally important bioactive compounds. Knowing the fact that peel of *kachkal* is an abundant source of cellulose, they can be considered as a potential candidate for development of reinforcing composites. Many researchers have reported their work on cellulose nanofibers from different sources, however, there is hardly any work reported on cellulose nanofibers from banana peel. This crop has immense potential for utilization and value addition in order to promote health.

4.1 INTRODUCTION

Banana plants are the world biggest herbs, grown in many countries. The annual production of banana in the world is about 139 million metric tons only second to a citrus fruit. Considering the nutritional aspect, plantains and bananas are the world's 4th leading agricultural crop (Ganapathi et al., 1999). There are two varieties of banana, that is, dessert banana (fruit) and culinary banana (vegetable). Culinary bananas, often called as plantains, look almost similar to unripe dessert bananas, but they are larger in size, more fleshy, and starchy (Emaga et al., 2008). Plantains have mostly evolved from two species *Musa acuminata* (genome A) and *Musa balbisiana* (genome B) (Stover & Simmonds, 1987). Plantains are mostly of the genomic groups of AAB, ABB, or BBB is a major staple food in many developing countries (Seenappa et al., 1986). They are considered to be one of the most important sources of energy and starchy staple food for the people of tropical humid regions (Onwuka & Onwuka, 2005). According to Doymaz (2010), bananas and plantains are rich in nutrients, starch, sugar, vitamins, potassium, calcium, sodium, and magnesium and they are nutritionally low protein food material (Offem & Njoku, 1993). Plantain and banana cultivation is attractive to farmers due to lower labor requirements compared with cassava, maize, rice, and yam (Marriott et al., 1981). As compared to the dessert banana (also known as commercial banana, which has a global distribution), culinary banana is restricted in only a few localities of the world (Mendez et al., 2003; Englberger et al., 2003). Botanical studies have revealed that this cultivar of culinary variety possesses many medicinal properties along with its pre-known health benefits (Mao et al., 2009; Kumar et al., 2012).

The maturity stages have a philosophical influence on the various physical, biochemical, functional, and morphological attributes of the fruit, which impinge on the quality of the fruits. When young, fruits develop from the female flowers appearing as slender green fingers, and on attaining maturity, the bracts slowly shed off the fruit which finally turns to be a cluster called "hand" of banana (Morton, 1987). On further maturity, its color changes from deep green to light green and finally yellow and the pulp turns light yellow from off-white. The fruit when young and immature, looks firm, taste astringent, and feels gummy with latex and slowly turns tender, soft, and sweeter when ripe. Therefore, in order to obtain good quality fruit, harvesting has to be done at the proper stage of fruit maturity. Harvesting of immature fruits may lead to poor quality fruit with inconsistent ripening while delaying harvesting affects the fruit quality and increases deterioration susceptibility (El-Ramady et al., 2015). The nutritional properties of banana and plantains alter significantly with advancement in fruit growth and maturity (Khawas et al., 2014a). The major changes are breakdown of carbohydrates and osmotic transfer from peel to pulp, conversion of starch to sugars, reduction of polyphenols, and synthesis of aromatic compounds, etc. (John & Marchal, 1995; Chen & Ramaswamy, 2002). Most of the available reports are about changes in chemical composition of dessert banana cultivars during ripening than culinary bananas (Emaga et al., 2008; Marriott et al., 1981; Cheirsilp & Umsakul, 2008; Yang & Hoffman, 1984).

Culinary variety of banana found in Assam and Northeastern region of India is called *kachkal* in local language falls under the category of fruit-vegetable and is one of the commonly consumed vegetable next to potato in the daily diet of local people (Khawas et al., 2014a). It is also used for the preparation of various traditional dishes as it is an excellent source of carbohydrates, starch, and many other functionally important bioactive compounds. Traditionally, it is used as special diets for babies, elderly, and patients with stomach problems, gout, and arthritis. Unlike any other fruits or vegetables, the biochemical compositions of culinary banana (*kachkal*) also vary with growth stage and maturity (Khawas et al., 2014a).

Coming across the difference between banana and plantains, many people tend to get confused these two varieties. Daniell (2003) stated that the word banana refers to all members of the genus *Musa* whereas plantain refers to a subset of banana. Usually, the term plantains are starchy, low in sugar variety that is cooked before serving as it is unsuitable to consume raw. It is more or less used in many savory dishes like a potato especially frying and baking (http://grabemsnacks.com/what-is-a-plantain.html). Ever since India is recognized as one of the leading centers of origin and diversity of *Musa* species at global level alongside South East Asian countries, these species are well adopted in the regions varying from tropics to humid sub-tropics and semi-arid subtropics (Singh & Uma, 2000). As reported by Wainwright and Burdon (1991), both plantain and banana are the staple foods for rural and urban consumers in India and an important source of income. Traditionally, bananas are usually eaten raw as dessert while plantain and cooking bananas, on the other hand, are habitually grown for cooking as a part of the staple diet or for processing of more durable products, such as flour that can be stored for later use. Alternatively, plantains embrace immense cultural significance in India. According to the book, "The Popular Religion and Folklore of Northern India," plantains are considered sacred. They are used in marriage ceremonies and many other religious rituals, as a tradition they hold the plantain branch and sits near the sacred fire around which the bride and groom walk (http://theindianvegan.blogspot.in/2013/03/all-about-plantain-in-india.html).

4.2 THE BACKGROUND AND THE BIOLOGY OF *Musa*

Banana belongs to the family *Musaceae* and genus *Musa* is a general term embracing a number of species or hybrids in this genus. The name *Musa* is from the Sanskrit, Moca, via its Arabic counterpart, *mauz*. Bananas descended from two wild ancestors: *M. acuminata* and *M. balbisiana* (Lehmann et al., 2002). It is originated in Southeast Asia and is the native of India and the Western Pacific (Carreel et al., 2002). It was further introduced in Africa in ancient times and taken by European explorers to the Americas and other parts of the world (De Langhe et al., 2009; Valmayor, 2001). Currently, throughout the world, it is widely cultivated in tropical and subtropical regions as one of the important staple food and commodities (Boonruangrod et al., 2009). The majority of edible banana cultivars are derived from inter and/or intra specific hybridization of diploid ($2n = 2x = 22$; AA; BB; or AB), triploid ($2n = 3x = 33$; AAA; AAB; or ABB) and tetraploid ($2n = 4x = 44$; AAAA; AAAB; AABB; or ABBB) from subspecies of *M. acuminata* Colla (A genome) and *M. balbisiana* Colla (B genome) (de Jesus et al., 2013; Pillay et al., 2006).

As per the report of Daniells (2003), India is pondering to be the major center for *M. acuminata* hybridization with native *M. balbisiana* and the region is well known for the extensive variety of AAB and ABB cultivars. *M. balbisiana* is considered to be more drought and disease resistant than *M. acuminata* and such characteristics are often found in B genome cultivars. The formation of heterogenomic triploid hybrids with the "AAA" genotype occurred within *M. acuminata* leading to the development of cultivars that encompassed the sweet bananas. Crosses of the diploid and triploid types of *M. acuminata* with *M. balbisiana* led to the formation of heterogenomic triploid hybrids that are mainly plantains (AAB) and other cooking bananas (ABB).

4.2.1 THE PRODUCTION

Banana and plantains are the fourth most important agricultural crops widely sought for and grown all over the world in tropical countries. In India, it is one of the export commodity fruits (Ganapathi et al., 1999). World banana production scenario 2013–14, indicated a global production of 106.84 million tons of banana and plantain, grown in an area of 4.034 million ha with 21.2 MTlha productivity (Indian Horticulture Database, 2014). India is the largest producer of plantain and bananas with annual production of 29.78 MT from an area of 0.83 million ha with 37 MT/ha productivity accounting 27.8% of the world's production followed by China.

4.2.2 BANANA: ITS POTENTIAL HEALTH BENEFITS

A healthy diet consists of eating a variety of foods from five food groups but in the correct proportions. These include foods containing starch, fruit, and vegetables, milk and dairy food, protein, fats, and sugars. Bananas fall into the fruit and vegetable group as well as the food group which mostly contain starch. Sweet dessert bananas are generally eaten raw (fruit) while cooking bananas and plantains are boiled, steamed, fried, or roasted (food). Any food containing carbohydrates should be the main part of our daily meals. In unripe bananas, the carbohydrates are mostly starches. In the process of ripening the starches are converted to sugars; a fully ripe banana has only

1–2% starch (Forsyth, 1980). According to UNCST (2007), eating a variety of good foods that provide nutrients helps to maintain health, feels good, and has energy. These nutrients include protein, carbohydrates, fat, water, vitamins, and minerals. Plantains and banana belong to the food group which is a reliable source of starch, energy, and dietary fiber (http://www.nairaland. com/1275850/10-health-benefits-plantains-need).

Plantain provides more calories than fruit banana. 100 g plantain consists of 122 calories, while banana has 89 calories and banana resistant starch consumption promotes lipid oxidation (Higgins, 2004). Indeed, they are very reliable sources of starch and energy ensuring food security for millions of households worldwide. Bananas have been classified as one of the antioxidant foods by Kanazawa and Sakakibara (2000). They are known as a weak primary antioxidant source, but a powerful secondary antioxidant source (Haripyaree et al., 2010; Lim et al., 2007; Yan et al., 2006). Recently, much attention has been focused on the activity of natural antioxidants present in fruits, because these components may potentially reduce the level of oxidative stress (Hassimotto, 2005), that is, preventing free radicals from damaging proteins, DNA and lipids (Isabelle et al., 2010). Besides, they are also scientifically proven for their synergistic effects and protective properties against various degenerative disorders including cancer, stroke, cardiovascular, Alzheimer's disease, and Parkinson's disease (Abdel-Hameed, 2009; Kawasaki et al., 2008). These tropical fruits have strong ability to protect themselves from the oxidative stress caused by the intense sunshine and high temperature by increasing their antioxidant levels. The antioxidant compounds identified in bananas include ascorbic acid, tocopherol, beta-carotene, phenolic groups, dopamine, and gallocatechin (Qusti et al., 2010).

Astringent taste of unripe banana is due to phenolic compounds. Bananas are also rich in dopamine antioxidant. According to Ramcharan and George (1999), the role of banana and plantain is becoming more important with the increasing emphasis today on diets that are low in sodium but high in potassium and vitamins. Both banana and plantains are a good source of potassium, vitamin A, vitamin C, vitamin B6, and low in sodium. Bananas contribute about 2.7% of the total potassium and fiber consumed by the average adult (USDAARS, 2004). According to the report of Kumar et al. (2012), bananas are an excellent source of potassium. Potassium can be found in a variety of fruits, vegetables, and even meats, however, a single banana provides with 23% of the potassium that is needed on a daily basis.

The carbohydrate type in banana is resistant starch and non-starch polysaccharides, which have low glycemic index or low digestibility (Lehmann & Robin, 2007). This property makes banana excellent ingredient for different functional and convenience foods like cookies and chips (Aparicio-Saguilan et al., 2007). The intake of sugars, fiber, vitamins, and minerals from the consumption of bananas is high, with a very low contribution to the intake of fat. Bananas, consumed cooked or raw, either as the green, half-ripe, or ripe fruit, are one of the most significant sources of calories for the human diet worldwide and orange-fleshed bananas are rich in provitamin A and other carotenoids. Provitamin A carotenoids (including beta-carotene) are important for protecting against vitamin A deficiency and anemia (as vitamin A is involved in iron metabolism). Carotenoid-rich foods may also protect against diabetes, heart disease, and certain cancers (www.traditionaltree. org). Worldwide, vitamin A deficiency is the most common form of malnutrition after protein deficiency. In parts of Asia, Africa, and Latin America, vitamin A deficiency occurs in millions of children (Sommer, 1989) and, therefore, plantain and banana could be an important source of provitamin A for the people of those regions. Foods containing high levels of carotenoids have been shown to protect against chronic disease, including certain cancers, cardiovascular disease, and diabetes, because the coloration of the edible flesh of the banana appears to be a good indicator of likely carotenoid content. Banana cultivars rich in provitamin A carotenoids may offer a potential food source for alleviating vitamin A deficiency, particularly in developing countries. Many factors are associated with the presently known food sources of vitamin A that limit their effectiveness in improving vitamin A status.

Englberger et al. (2003) stated that bananas are an ideal food for young children on account of their sweetness, texture, portion size, familiarity, availability, convenience, versatility, and cost.

4.2.3 EFFECT OF RIPENING STAGES ON THE BIOCHEMICAL AND NUTRITIONAL PROPERTIES OF BANANA

Maturation is the stage of development leading to the attainment of physiological or horticultural maturity. It is a developmental stage of the fruit on the tree, which will result in a satisfactory product after harvest. Physiological maturity refers to the stage in the development of the fruits and vegetables when maximum growth and maturation has occurred. It is usually associated with full ripening in the fruits. The physiological mature stage is followed by senescence. Commercial maturity is the state of plant organ required by a market. It commonly bears little relation to physiological maturity and may occur at any stage during the development stage. Harvest maturity may be defined in terms of physiological maturity and horticultural maturity and is the stage, at which will allow fruits and vegetables reach its peak condition by developing acceptable flavor or appearance with adequate shelf life.

The effects of different stages of ripening on the chemical composition and functional properties of plantains and banana have been studied by various researchers. According to the report of Tapre and Jain (2012), dramatic changes in banana peel color and pulp texture occur due to the rise in respiration during storage of climacteric fruits. The changes that occur during ripening are with respect to the physical, mechanical, and chemical properties of banana fruits. Skin color changes from green to yellow, firmness decreases, fruit gets softened, and starch is converted into sugar (Prabha & Bhagyalakhmi, 1998; Kajuna et al., 1997; Marriott et al., 1981). The main color changes in banana during ripening are based on the peel rather than the pulp and, hence, the color of banana peel has been used in the assessment of the stages of ripeness of banana. Knowledge of the physical and mechanical properties of banana fruit and changes in these parameters during different ripening stages is the most important attributes to design handling, sorting, peeling, processing, and packaging system. Knowing these properties of agricultural products would help design engineers to apply forces and dimensions of machine's units properly to protect fruits from bruises, injuries, decay lesions and numerous other defects that emanate as results of post-harvest processing treatments.

Tapre and Jain (2012) in their studies also reported a significant increase in moisture content of the pulp from the early development stage to mature stage. The increase in pulp moisture content during ripening may be due to carbohydrate breakdown and osmotic transfer from the peel to pulp (John & Marchal, 1995). A significant decrease in ash and protein content with increase in maturity has been reported by various authors (Loeseck, 1950; Lustre et al., 1976). Adeyemi and Oladiji (2009) reported that ash content of ripening plantain is affected by developmental stage and unripe plantain contains higher ash content compared to ripe ones. Another reason for variation in ash might be due to differential absorption capacity of minerals at different stages of development. Goswami and Borthakur (1996) observed a decline in protein content in culinary banana with maturity and attributed to protein breakdown and the resulting amino acids being utilized in gluconeogenesis. They also reported lower amount of fat content in culinary banana which was higher during early developmental stages and gradually decreased with increasing maturity. The crude fiber content in unripe and ripe plantain and increased significantly with progress of maturity. The increase in fiber content at matured stage over tender stage might be due to increase in soluble and insoluble dietary fractions (Egbebi & Bademosi, 2012). In plantains and bananas, the various reports on total carbohydrates content increased from early developmental stage to matured stage. The variation in carbohydrate contents during growth might be due to degradation of starch for synthesis of sugars (Sakyi-Dawson et al., 2008).

The most prominent chemical change that occurs during postharvest ripening of banana, cooking banana, and plantain is the hydrolysis of starch and the accumulation of sugars, that is, sucrose, glucose, and fructose (Palmer, 1971) which are responsible for sweetening the fruit as it ripens (Palmer, 1971). According to Sakyi-Dawson et al. (2008), during ripening of plantains and bananas, the starch content decreased considerably with fruit growth and development and the decrease in starch content of the cooking banana was faster than that of the plantains. In dessert bananas (e.g., Cavendish) the breakdown of starch and the synthesis of sugars are usually complete at full ripeness, while in plantains the breakdown is slower and less complete and continues in overripe and senescent fruits (Marriott et al., 1981). The decrease in starch content is explained by the degradation of starch and the formation of free sugars under the action of enzymes. Yang and Hoffman (1984) stated that during the ripening process, starch is converted into sugar, through enzymatic breakdown process.

Cordenunsi and Lajolo (1995) also reported that the disappearance of the starch reserve during banana ripening appears to be relatively rapid because of the activities of several enzymes working together. Amylase, glycosidase, phosphorylase, sucrose synthase, and invertase can act in the degradation of starch and the formation and accumulation of soluble sugars. Starch levels can vary with maturity stage of fruit variety and cultivation and ripening conditions (Loeseck, 1950). Since, the sugar content of fruits increases during ripening, it can effectively be used as an indicator of ripening or to predict the stage of fruit ripening (Pacheco-Delahaye et al., 2008). Yang and Hoffman (1984) stated that during the ripening process, starch is converted into sugar, through enzymatic breakdown process. In *Musa* AAB group, starch contains declines from 20–30% to 1–2%, but starch amount could be as high as 11% depending on plantain variety. According to Emaga et al. (2008), ripe banana pulp contains (0.7–1.2%) pectin. During ripening, insoluble proto-pectin is converted into soluble pectin that causes loosening of cell wall and texture degradation of fruit. Hence, stages of ripening of banana play important role in chemical, physical and functional properties of the fruit. Even within a cultivar, there is large plant-to-plant variation and within-plant also variation might occur in nutrient composition for fruit harvested from the same field (Shewfelt, 1990). The study reported by

Khawas et al. (2014a) on effect of growth stages on nutritional compositions of culinary banana (*Musa* ABB), revealed that this crop has potential
applications of developing numbers of value-added products, because the
antioxidant activity makes it an excellent ingredient for developing products like cookies, biscuits, bread, etc. The increased accumulation of starch
renders mature tissue a potential source for commercial starch extraction and
also the presence of considerable amount of amylose allows for developing
products which can be subjected to high temperature.

4.2.4   UTILIZATION OF BANANA AND PLANTAIN AS VALUE ADDED PRODUCTS

Even though, India is the first major producer of banana, 20–30% of the
production is wasted during every harvest season during transport due to
lack of compliance with the normal commercialized standards for fresh
fruits. Wide ranges of processing operations are employed before plantain is consumed and they include boiling, roasting, or baking, frying and
drying (Dadzie & Wainwright, 1995). Cooking bananas are used in a wide
range of food dishes of varying regional importance. There appears to be a
potential market for a wider range of snack products produced from these
commodities in the target countries, including the production of several
popular alcoholic and non-alcoholic beverages. The conversion of cooking
bananas into flour, wine, beer, and weaning food products (extruded high
protein and low cost) is the way of adding value to the crop as well as
extending the shelf life of derived foods (Khawas et al., 2016b). Flour is
an important raw material in the baking and confectionery industry. The
demand for flour in bakery products is increasing globally and banana
flour is currently being exploited in baking and complementary weaning
foods (Adeniji & Empere, 2001). New economic strategy to increase utilization of banana includes the production of banana flour when the fruit is
unripe, and to incorporate the flour into various innovative products, such
as slowly digestible cookies (Aparicio-Saguilan et al., 2007) and high-
fiber bread (Juarez-Garcia et al., 2008).

Shiau and Yeh (2001) stated that it would be possible to utilize the green
pulp as a functional ingredient in starch-rich products, such as the yellow
noodles. Saifullah et al. (2009) prepared banana pulp noodles by partial
substitution of wheat flour with green Cavendish banana pulp flour and
studied pH, color, tensile strength, and elasticity, and in vitro hydrolysis
index (HI) and estimated glycemic index (GI). In their study, they found

that banana flour noodles had higher tensile strength and elasticity modulus than control noodles. They also reported that during in vitro starch hydrolysis study, it was found that GI of banana flour noodles was lower than control noodles. Value addition of banana and plantain can be done in many different ways, such as frozen puree, juice, figs, jams, and canned banana slices (Thompson, 1995). Seasonal gluts and perishability of ripe bananas and plantains cause great economic losses and hence is of tremendous interest in the development of processing and preservation techniques for these fruits.

4.3 PLANTAIN STARCH

Starch is the storage polysaccharides of green plants and a major dietary component in all human populations (French, 1984). It is composed of a number of monosaccharides or sugar (glucose) linked together with α-D-(1-4) and/or α-D-(1-6) linkages (Sajilata et al., 2006). Starch is deposited in the fruit in the form of granules, partially crystalline, whose morphology, chemical composition, and supermolecular structure are characteristic of each particular plant species. Starch owes much of its functionality to two major high-molecular-weight carbohydrate components, amylose, and amylopectin. Amylose is essentially linear polymer in which glucose residues are α-D-(1-4) linked typically constituting 15–20% of starch. Amylopectin the major component of starch is a larger branched molecule with α-D-(1-4) and α-D-(1-6) linkages. This biopolymer constitutes an excellent raw material to modify food texture and consistency (Suma & Urooj, 2015). The amount of starch is not only important for the texture of a given food product, but starch type is equally critical (Biliaderis, 1991).

Starch, on the basis of its digestibility, has been classified into three groups, such as readily digestible starch (RDS), slowly digestible starch (SDS), and resistant starch (RS) (Englyst & Cummings, 1987). RDS is the starch fraction that causes an increase in blood glucose level after ingestion immediately, whereas, SDS is the starch fraction that is digested completely in the small intestine at a lower rate as compared to RDS. RS is the portion of starch and/or starch hydrolysis products that escape digestion in the small intestine, and enter the colon for fermentation (Sajilata et al., 2006). Extensive studies have shown RS to have physiological functions similar to those of dietary fiber (Eerlingen & Delcour, 1995). RS appears to be highly resistant to mammalian enzyme and may be classified as a component of fiber on the basis of the definitions of dietary fiber given by AACC (2000).

The diversity of the modern food industry and the enormous variety of food products being produced require starches that can tolerate a wide range of processing techniques and preparation conditions (Visser et al., 1997). These demands are met by modifying native starches with chemical, physical, and enzymatic methods which may lead to the formation of indigestible residues. The availability of such starches, therefore, deserves consideration. Four forms of RS are distinguished: RS type I is defined as physically inaccessible starch for instance in grains; type II is granular starch in raw potato and bananas; type III is retrograded starch, arising after hydrothermal treatment of starch; and type IV is considered to be a chemically modified starch (Englyst et al., 1992). Among these four types, RS type III seems to be particularly interesting because it preserves its nutritional characteristics when it is added as ingredient for cooked food. RS type III is produced by gelatinization, which is a disruption of granular structure by heating starch with excess water, and then retrogradation occurs.

The generation of RS after hydrothermal treatment is mainly due to increase interactions between starch polymers. The degree of formation of RS in foods depends not only on the type of incorporated starch and the processing conditions, but is also, influenced by the duration and conditions of storage (Goni et al., 1996). As stated by Fuentes-Zaragoza et al. (2010) unripe banana is considered as RS-richest non-processed food. The content of RS in unripe banana ranges between 47% and 57%. Because of this fact, several studies have suggested that consumption of unripe banana results in beneficial effects to human health. Rodriguez-Ambriz et al. (2010) studied on unripe banana flour where they found total starch content of 73.4%, RS content 17.5% and concluded that unripe banana is an alternative source of indigestible carbohydrates, mainly RS and dietary fiber. Therefore, when the unripe fruit is cooked, its native RS is rendered digestible.

The global trends in rising levels of obesity, diabetes, and cardiovascular disease has renewed research interest in the dietary intake of fat, protein, and carbohydrate to maintain good health. The World Health Organization (WHO) and Food and Agricultural Organization (FAO) of the United Nations stated that globally, overweight populations are a bigger problem than undernourishment and recommended people in industrialized countries base diet on low GI foods to prevent most common disease of affluence (WHO, 1998). One of the most important objectives in the dietary treatment of diabetes patients is to maintain their blood glucose level, avoid obesity, and achieve optimal lipids level. Foods containing RS generally give a low glycemic response because RS is not digested in the small intestine and instead RS passes into the large intestine where it is fermented (Sajilata, 2006).

Many tropical countries have plant species which can be used as a good source of starch; unfortunately, some of them have not been exploited. One such plant species is *kachkal* (*Musa* ABB), the only culinary banana found in the entire Assam and North-East India (Goswami & Borthakur, 1996). Starch, being the principal component of culinary bananas, can be considered as a resource for production of modern forms of consumption like processed snacks and precooked products. Therefore, efforts have been made to isolate starch from culinary banana and modifying the isolated starch into RS in order to find an alternative route to increase value addition of culinary banana by providing RS of high-quality characteristics necessary for a well-balanced diet. The utilization of RS from culinary banana should not only increase the market for culinary banana but also provide a solution to those who want to consume food with low glycemic index (GI) value. RS has received much attention for both its potential health benefits and functional properties. The study reported by Khawas and Deka (2016b) revealed that culinary banana can be a potential source for production of RS type III and justifies that culinary banana is an excellent source of RS and may be utilized as an alternative source of nutraceutical ingredient for preparing low glycemic functional foods.

4.4 CHEMICAL PROPERTIES AND BIOACTIVE COMPOUNDS PRESENT IN BANANA PEEL

The peels of a variety of fruits and vegetables have gained much attention as a natural source of antioxidants and phytochemical contents which are rich in compounds with free radical scavenging activity. Banana and plantain peels are the major agricultural wastes which have been used as medicine, animal feeds, blacking of leathers, soap making, and fillers in rubber. Banana peel being a key source of many functionally important bioactive compounds are still underutilized and very little scientific effort has been put to identify its functionality in terms of application to food and nutraceuticals. Banana peel can potentially offer new products with standardized composition for various industrial and domestic uses (Essien et al., 2005; Annadurai et al., 2002). As reported by Emaga et al. (2007; 2008), banana peel is rich source of dietary fiber (50% on a dry matter basis), protein (8–11%), crude fat (3.8–11%), lipid (2.2–10.9%) pectin, essentials amino acids (leucine, valine, phenylalanine, and threonine), polyunsaturated fatty acids mainly (linoleic acid and α-linolenic acid), and micronutrients like (potassium, phosphorous, calcium, magnesium, etc.). They also reported that all essential amino acids content is higher than FAO standard except for lysine. Pectin extracted from banana peel also contains glucose, galactose, arabinose, rhamnose, and xylose. Maturation of fruits involves, increase in soluble sugar, decrease in starch and hemicelluloses, and slight increase in protein and lipid content in fruit peel. Degradation of starch and hemicelluloses by endogenous enzymes may explain increase in soluble sugar content. Archibald (1949) in his study has reported that banana peel can also be utilized for extraction of banana oil (amyl acetate) which can be potentially used for food flavoring. Banana peels are also a good source of lignin (6–12%), pectin (10–21%), cellulose (7.6–9.6%), hemicelluloses (6.4–9.4%), and galacturonic acid.

Banana peel has been considered as a potential source of phytochemicals and antioxidants compared to its fruit (Someya et al., 2002; Kondo et al., 2005; Sulaiman et al., 2011). Someya et al. (2002) investigated the total phenolic contents to be more abundant in peel than in fruit which was consistent with the antioxidant activity. The peel extract showed stronger antioxidant activity than the fruit extract when the incubation times were compared. Gallocatechin content found in fruit peel has been reported to be in higher amount than in fruit and, therefore, the higher gallocatechin content of the banana peel may account for the better antioxidant effects. The result of Singhal and Ratra (2013) on *M. acuminata* peel extract indicated that banana peel is potential source of bioactive compounds like flavonoids and polyphenols with wide range of medicinal properties, particularly with high free radical scavenging activity. The study also reported that banana peel extracts help to increase the total leukocyte and the percentage of lymphocyte which showed good biological activities and can be effective in various diseases. Fatemeh et al. (2012) found that the total polyphenols and flavonoids contents of peel to be higher when compared to the fruit pulp during all stages of fruit ripening. Similarly, the ability of banana peel extracts to scavenge DPPH radicals was also reported to be higher which is associated with the stronger antioxidant activity.

Study on inhibition of lipid peroxidation by ethanolic extract of few varieties of banana peel by Baskar et al. (2011) reported the Poovan variety of peel extract exhibited high inhibition toward radical scavenging activity. Lipid peroxidase which is unstable and decomposes to form reactive carboxyl compounds responsible for cancer, age-related diseases, and DNA damage which can be significantly controlled by using peel of banana.

The micronutrient contents (iron and zinc) found in peels of banana is comparatively higher than that of fruit (Davey et al., 2009). The study on banana peel by Anhwange (2008) found that peel contains reasonable amount of minerals including potassium, calcium, sodium, iron, and manganese, among which potassium content was highest. The consumption of banana peel may help in the regulation of body fluids and maintain normal blood pressure. It may also help in controlling kidney failure, heart oddities, and respiratory flaw. Feming (1998) has stated that the percentage of iron content in banana peel is an ideal source for carrying oxygen to the cells and production of energy, synthesis of collagen, and for proper functioning of the immune system, cell growth, and heart.

4.4.1   BANANA PEEL PROSPECTS FOR UTILIZATION AS BY-PRODUCTS

The peel of banana represents 40% of the total weight of fresh banana (Tcho-banoglous et al., 1993) and has been underutilized and discarded as waste. Like its pulp flour counterpart, banana peel flour can potentially be used in new products with standardized composition for various industrial and domestic uses (Emaga et al., 2007). Peels are the major by-products of all fruits and vegetables obtained during processing; some studies have shown that these are good sources of polyphenols, carotenoids, and other bioactive compounds which possess various beneficial effects on human health (Zhang et al., 2005). But these wastes are either uneconomically utilized or disposed of as they are, thereby causing serious pollution problems. Of particular interest is the finding that banana peel extract contains higher antioxidant compounds than that of the fruit thus promising a more intense utilization of the peels in food and nutraceuticals. Potential applications of banana peel, however, depend on its chemical composition as well as physicochemical and functional properties (Emaga et al., 2007). Banana peel waste is a byproduct of banana processing during the production of food, such as banana chips and baby foods. The edible part of banana constitutes only 12% weight of the plant; the remaining parts become agricultural waste and cause environ-mental problems (Elanthikkal et al., 2010). However, the problem can be recovered by utilizing its high-value compounds, including the dietary fiber fraction that has a great potential in the preparation of functional foods.

Culinary banana peel at different stages of development was evaluated by Khawas and Deka (2016c) to understand the effect of fruit maturation on biochemical and functional compounds in peel at various stages and the results revealed culinary banana peel is a potential source of many function-ally important nutritional and bioactive compounds and it can also serve as a potential biomaterial in industrial applications and can add high value to this crop.

4.4.2 UTILIZATION OF PEEL AS VALUE ADDED PRODUCTS

Like its pulp flour, peel can also be used for developing a number of high-value-added products. Ramli et al. (2009) developed yellow noodles by partial substitution of wheat flour with green banana peel flour and the study reported that partial substitution of banana peel into noodles may be useful for controlling starch hydrolysis of yellow noodles. Banana peel noodles had a lower estimated glycemic index value as compared to noodles prepared with wheat flour. The modified noodle product described in their study may broaden the range of low glycemic index food products and increase utilization of waste products from banana agro-industries. Wachirasiri et al. (2009) in their research developed banana peel dietary fiber concentrate and reported that banana peel is a good source of dietary fiber exhibiting 50/100 g dry matter. The result indicated that dietary fiber concentrate obtained from banana peel provides an opportunity to enhance the functionality. The use of banana peel dietary fiber concentrate as a low-caloric functional ingredient for fiber enrichment and incorporation of them within the food system may give high value-added food products.

Amylases are well known for applications ranging from starch and food processes industry to medical applications. Krishna et al. (2012) in their study reported the potential of banana peel was evaluated for α-amylase production using the fungal culture of *Aspergillus niger* NCIM 616 in solid submerged (SmF) and solid state (SSF) fermentation. The effect of different parameters, such as substrate concentration, water content, layer thickness, and external salt addition was studied in terms of the amylase activity. The study suggested that banana peel could be used as a potential raw material for α-amylase production. Paul and Sumathy (2013) suggested that banana peel could employ as a promising substrate for the production of amylase by *Bacillus subtilis*.

Rehman et al. (2013) have confirmed that banana peel proved to be a good source for producing xylose. Xylitol is the first rare sugar that has global market for having beneficial health properties and being an alternative to current conventional sweeteners. Banana peel was used as a substrate for xylitol production by acid hydrolysis. Detoxification of peel hydrolysate by neutralization, charcoal treatment, and vacuum evaporation increased the xylitol yield. Xylitol extracted from banana peel can be used to replace sugars in different products, such as bakery and confectionary products without affecting their physicochemical characteristics and shelf stability. Jadhav et al. (2013) utilized banana peel for producing lipase an important enzyme which is extensively used in food and dairy industry for the hydrolysis of milk fat, cheese ripening, flavor enhancement and lipolysis of butterfat and cream. The authors concluded that peel of banana can potentially be used for the production of bacterial enzymes like lipase and amylase which hold an important place in food industry.

The attempt made by Byarugaba-Bazirake et al. (2014) for producing wine vinegar from banana peel took 28 days and the final product obtained had physicochemical characteristics of 6.0% (v/v) acetic acid, 5° Brix, and pH of 2.9 which complied with the standard ranges of brewed vinegar after complete fermentation. The aroma of the vinegar produced was appreciated by the consumers who were acquainted with vinegar. The study, therefore, showed that banana peel can be used as an ideal substrate for producing good-quality vinegar. This will not only increase the economic and food value of banana peel but also provides a way of utilizing banana waste. Simmonds (1966) reported that vinegar has been prepared by fermenting a mash of banana pulp and peel. Vinegar production from banana is eco-friendly and helps to minimize the cost of production.

4.4.3   EXTRACTION OF CELLULOSE FROM PEEL AND ITS UTILIZATION

Cellulose the homopolysaccharide representing about 1.5×10^{12} tons of total annual biomass production composed of glucose-glucose linkages (β-1,4-linked-glucopyranose unit) arranged in linear chains where C-1 of every glucose unit is bonded to C-4 of the next glucose molecule and nanostructures (Moon et al., 2011; Kim & Yun, 2006; Kadla & Gilbert, 2000). It is one of the most important biopolymers in existence and is derived from readily available biomass (Abraham et al., 2011). Due to its availability, biocompatibility, biodegradability, and sustainability, cellulose is widely used (Chen et al., 2011). Cellulose is composed of assemblies of nanofiber diameter ranging 2–20 nm and a length of more than a few micrometers. The nanometer-sized single fiber of cellulose is commonly referred to as nanocrystals, whiskers, nanowhiskers, microfibrillated cellulose, microfibril aggregates, or nanofibers (Chen et al., 2011; Eichhorn, 2010; Siro & Plackett, 2010). Plants and woods are the primary sources of cellulose nanofibers and their compounds encompass current area of research and compared to the commercially available fibers, nanofibers obtained from plant sources poses low density, nonabrasive, combustible, nontoxic, low cost, biodegradable, and has good thermal and mechanical properties (Eichhorn, 2010; Kalia et al., 2011; Siqueira et al., 2009).

Cellulose nanofibers unlock the door in the direction of promising research on cellulose-based nanomaterials with increasing area of potential applications including packaging material (Rodionova et al., 2011; Spence et al., 2011), transparent material (Fukuzumi et al., 2009; Iwamoto et al., 2008), paper production (Yoo & Hsieh, 2010; Henriksson et al., 2008), and biomedical applications (Cherian et al., 2011; Czaja et al., 2007). Cellulose nanofibers could also be used as a rheological modifier in foods, paints, cosmetics, and pharmaceutical products (Turbak et al., 1984). Many studies have been done on isolation and characterization of cellulose nanofibers from various sources like wood fibers (Abe et al., 2007), cotton (de Morais et al., 2010), potato tuber cells (Dufresne et al., 2000), prickly pear fruits (Habibi et al., 2008), lemon and maize (Rondeau-Mouro et al., 2003), soybean (Wang & Sain, 2007), wheat straw and soy hulls (Alemdar & Sain, 2008), coconut husk fibers (Rosa et al., 2010), branch-barks of mulberry (Li et al., 2009), pineapple leaf fibers (Cherian et al., 2010), banana rachis (Zuluaga et al., 2009), pea hull fiber (Chen et al., 2009), and sugar beet (Dinand et al., 1999). Cellulose nanofibers can be extracted from the cell walls by chemical and/ or mechanical treatment, such as cryocrushing (Chakraborty et al., 2006), grinding (Abe et al., 2007; Abe & Yano, 2010; Abe et al., 2009), high pressure homogenization (Pelissari et al., 2014), acid hydrolysis (Liu et al., 2010; Elazzouzi-Hafraoui et al., 2008), biological treatment, like enzyme assisted hydrolysis (Tibolla et al., 2014; Paakko et al., 2007) chemical treatment combined with ultrasonication (Khawas & Deka, 2016a).

Use of ultrasonic technique for isolation of cellulose nanofibers is an emerging method and has been extensively used by various researchers (Chen et al., 2011; Cheng et al., 2010; Cheng et al., 2009; Wang & Cheng, 2009). During the process of ultrasonic treatment ultrasound energy is transferred to cellulose chains through a process called cavitation, which refers to the formation, growth, and violent collapse of cavities in water (Chen et al., 2011). The energy provided by cavitation in this so-called sonochemistry is approximately 10–100 kJ/mol, which is within the hydrogen bond energy scale (Tischer et al., 2010). Thus, the ultrasonic impact can gradually disintegrate the micron-sized cellulose fibers into nanofibers.

In the process of development of value-added food products from culinary banana, the peel is a waste material of various fruit and vegetables processing units located in Northeast India. Therefore, it is possible to obtain banana peel sufficiently and application depends on its chemical compositions (Emaga et al., 2007). Literature revealed hitherto unexploited of this biomaterial in terms of its value addition and has enormous potential for its industrial use. Hence, the use of this biomaterial will not only help in increasing value addition but also help the environment from pollution free. Knowing the fact that, peel of culinary banana is an abundant source of cellulose, it can be considered as a potential candidate for development of reinforcing composites.

Cellulose nanopaper (CNP) obtained from CNF is similar to conventional paper but constitutes a network of nanofibers having pore size in the range of nanometer. It possesses excellent mechanical properties with high optical transparency, low thermal expansion and has good oxygen barrier characteristics (Hu et al., 2013; Sehaqui et al., 2012). It has potential to be used as a strong sheet-like material as well as light-weight reinforcement material in biocomposites. Irimia-Vladu (2014) used CNP to replace traditional glass and plastic in energy devices and termed as "Green" electronics to produce eco-friendly electronics by developing new green and efficient routes. The structure of CNP shows tightly packed nanofibers and interacts strongly with each other, contributing outstanding mechanical properties and thus CNP from CNF are considered as green and potential alternative for multiple applications depending on the requirements of final use applications including food packaging (Urruzola et al., 2014). Peel of banana is an excellent source of cellulosic fiber which can be used as a biomaterial, as peel represents 40% of total fruit weight of banana. The utilization of this cellulose-rich biomass would not only increase value of this agro-waste but also help to overcome environmental pollution issues. Khawas et al. (2016b) developed CNP from culinary banana peel using chemical treatment combined with high-intensity ultrasonication and the developed CNP exhibited high crystallinity, thermal, mechanical, and electrical stability, and considered as one of the potent renewable reinforcement agents for use in the field of food packaging industries.

4.5 ENCAPSULATION OF BIOACTIVE COMPOUNDS FROM BANANA AND PLANTAINS AND ITS APPLICATION IN FUNCTIONAL FOOD FORMULATION

In the last decades, consumer demands in the field of food production have changed considerably. Consumers more and more believe that foods contribute directly to their health (Mollet & Rowland, 2002). Today foods are not intended to only satisfy hunger and to provide necessary nutrients for humans but also to prevent nutrition-related diseases and improve physical and mental well-being (Menrad, 2003). Functional foods that contain bioactive components may provide desirable health benefits beyond basic nutrition and play important roles in the prevention of lifestyle-related diseases (Wang & Bohn, 2015).

Numerous food products are marketed with enhanced quantities of bioactive food compounds and these products are collectively referred to as functional foods (Crowe & Murray, 2013). They are one of the fastest growing sectors of the food industry due to increasing demand from consumers for foods that promote health and well-being (Mollet & Lacroix, 2007).

Functional foods must generally be made available to consumers in forms that are consumed within the usual daily dietary pattern of the target population group. Consumers expect functional foods to have good organoleptic qualities (e.g., good aroma, taste, texture, and visual aspects) and to be of similar qualities to the traditional foods in the market (Kwak & Jukes, 2001; Klahorst, 2006). Bioactive food components are components in foods or dietary supplements, other than those necessary to meet the basic nutritional needs, which are responsible for changes in health status (Studdert et al., 2011). In this case, it is important to understand that the bioactive compounds are not nutrients (Kris-Etherton et al., 2004) even if they are contained in foods or their constituents. Bioactive compounds are non-nutritional constituents that typically occur in small quantities in foods. Generally, these compounds are found in millions of species of plants, animals, marine organisms, and microorganisms, and can be obtained by extraction and biotechnological methods. Extracted bioactive compounds can be incorporated to produce new functional foods, enhancing its shelf life, nutritional quality, and increasing consumer acceptance of these commodities. Among the most used bioactive compounds are antioxidants, antimicrobials, probiotics, and flavors, in addition to nutraceutical substances (Ayala-Zavala et al., 2007; Muranyi, 2013).

The incorporation of bioactive compounds into food products provides many advantages in food preservation and contributes to the development of functional foods. However, bioactive compounds may have certain disadvantages, such as off flavors and an early loss of functionality (Ayala-Zavala et al., 2007; Silva-Weiss et al., 2013). Phenolic compounds, which are highly researched for their antioxidant and anti-inflammatory activities, have a basic structure containing at least one aromatic ring with several hydroxyl groups. In general, these compounds are relatively unstable in comparison to other secondary plant metabolites. Given the potential health benefits of bioactive food compounds and the emphasis of the nutrition community on food first behaviors, functional foods formulated with bioactive compounds are being developed with matrices to improve compound stability, bioactivity, and bioavailability such that they are easily degraded during food processing and digestion (Crowe & Murray, 2013; Manach et al., 2004; Stinco, 2013). However, in some foods, polyphenols exist as conjugates with proteins resulting in soluble and insoluble protein-polyphenol complexes which have a stabilizing effect on polyphenols (Parada & Aguilera, 2007). Polyphenolic compounds when included in a food product, they may impart an astringent or bitter taste, or introduce a degree of brown coloring. For example, grape seed extracts are difficult to incorporate into functional foods due to their dark brown color and low water solubility (Wang & Bohn, 2015). In addition, the challenges for applications of polyphenols in food system are the initial protection of the bioactivity of the polyphenols, as they may lose their antioxidative properties or bioactive functionalities during processing of food, due to their sensitivity to oxygen, temperature, light (Ottaway, 2008), and to the gastrointestinal tract environment (pH and enzymes) (Bell, 2011). Furthermore, the development of appropriate formulations to increase solubility of polyphenols according to a specific food matrix is necessary.

In today's world, much attention has been focused on the activity of natural antioxidants present in fruits and vegetables, because potentially these components may reduce the level of oxidative stress (Hassimotto et al., 2005), that is, preventing free radicals from damaging proteins, DNA and lipids (Isabelle et al., 2010). Besides, these compounds are also scientifically proven for their synergistic effects and protective properties against various degenerative disorders including cancer, stroke, cardiovascular, Alzheimer's and Parkinson's diseases (Abdel-Hameed, 2009; Giasson et al., 2002; Kawasaki et al., 2008; Ndhlala et al., 2006). Natural antioxidants may be added to a wide range of processed food, such as baked goods, biscuits, chewing gum, dry snacks, fruit drinks, mayonnaise, meat products, etc. (Lopez-Cordoba et al., 2014). The aqueous extracts obtained from plants, exhibit low stability and some of it has an unpleasant flavor, therefore, only a few extracts are currently employed in the food industry (Kosaraju et al., 2008; Makris & Rossiter, 2000). To protect and/or preserve the important bioactive ingredients from adverse environmental conditions like light, moisture, oxygen and to overcome undesirable interactions with the carrier food matrix encapsulation by cocrystallization technique has become an actual and cost-effective choice (Lopez-Cordoba et al., 2014; Onwulata, 2011). Encapsulation by cocrystallization in sucrose matrix is a relatively new and simple method which offers an economical and flexible alternative for handling and preserving various active components used in the food industry (Jackson & Lee, 1991). Cocrystallization with sucrose could improve the solubility, dispersibility, wettability, anticaking, anti-dusting, anti-separation, homogeneity, flowability, and stability of food materials (Lopez-Cordoba et al., 2014).

Various studies have been reported on this technique dealing with flavors, natural extracts, essential oils, honey, glucose, and fructose (Beristain et al., 1996; Bhandari et al., 1998; Bhandari & Hartel, 2002; Maulny et al., 2005; Sardar & Singhal, 2013).

Looking at the present scenario of bioavailability of polyphenols and antioxidants, encapsulation technology becomes an actual choice and is a promising technique that can solve the disadvantages of the use of bioactive compounds as food additives (Khawas & Deka, 2016d). Encapsulation is a rapidly expanding technology with a lot of potential applications in areas including pharmaceutical and food industries. Encapsulation may be defined as a process to entrap one substance (active agent) within another substance (wall material). The encapsulated substance, except active agent, can be called the core, fill, active, internal, or payload phase. The substance that is encapsulating is often called the coating, membrane, shell, capsule, carrier material, external phase, or matrix (Wandrey et al., 2009; Fang & Bhandari, 2010). Encapsulation is a useful tool to improve delivery of bioactive molecules (e.g., antioxidants, minerals, vitamins, phytosterols, lutein, fatty acids, and lycopene) and living cells (e.g., probiotics) into foods (Wandrey et al., 2009; Vos et al., 2010).

The main objective of encapsulation is to protect the core material from adverse environmental conditions, such as undesirable effects of light, moisture, and oxygen, thereby contributing to an increase in the shelf life of the product, and promoting a controlled liberation of the encapsulate (Shahidi & Han, 1993). In the food industry, the encapsulation process can be applied for a variety of reasons as summarized by Desai and Park (2005) as follows:

1. protection of the core material from degradation by reducing its reactivity to its outside environment;
2. reduction of the evaporation or transfer rate of the core material to the outside environment;
3. modification of the physical characteristics of the original material to allow easier handling;
4. tailoring the release of the core material slowly over time, or at a particular time;
4. to mask an unwanted flavor or taste of the core material;
6. dilution of the core material when only small amounts are required while achieving uniform dispersion in the host material;

7. to help separate the components of the mixture that would otherwise react with one another. Food ingredients of acidulants, flavoring agents, sweeteners, colorants, lipids, vitamins, and minerals, enzymes and microorganisms, are encapsulated using different technologies.

Natural antioxidants may be added to a wide range of processed foods. Co-crystallization is an encapsulation process in which the crystalline structure of sucrose is modified from a perfect to an irregular agglomerated crystal, to provide a porous matrix in which a second active ingredient can be incorporated (Chen et al., 1988). Spontaneous crystallization of supersaturated sucrose syrup is achieved at high temperature (above 120°C) and low moisture (95–97°Brix). If a second ingredient is added at the same time, the spontaneous crystallization results in the incorporation of the second ingredient into the void spaces inside the agglomerates of the microsized crystals with a size less than 30 mm (Bhandari et al., 1998). The main advantages of cocrystallization are improved solubility, wettability, homogeneity, dispersibility, hydration, anticaking, stability, and flowability of the encapsulated materials (Beristain et al., 1996). Other advantages are that the core materials in a liquid form can be converted to a dry powdered form without additional drying, and the products offer direct tableting characteristics because of their agglomerated structure and thus offer significant advantages to the candy and pharmaceutical industries (Desai & Park, 2005). In this technique, the granular product obtained possesses a very low hygroscopicity, a good fluidity, and a better stability. Furthermore, the co-crystallization offers a good economic alternative and remains a flexible technique because of its simplicity.

4.6   DEHYDRATION OF BANANA AND PLANTAINS

Drying is an alternative method to preserve the food quality and to reduce post-harvest losses. Generally, drying of foods is characterized by two separate phases: the constant rate and the falling rate periods. During drying, moisture present in the food diffuses from the internal to the food surface and evaporates into the air stream and at the same time, the heat is transferred from the air to the food. When moisture is removed, the volume of food decreases. Moisture gradient occurring inside the food during drying generates stresses in the cellular structure of the food resulting in the structure collapse which respond to the physical changes of shape and dimension or the volume change of material (Pan et al., 2008).

Hot air drying is one of the most widely used methods for food preserva-tion. The advantage of dehydrated foods is that decreased moisture content reduces thermodynamic water activity, thus preventing the growth of micro-organisms that cause spoilage reactions (Babalis & Belessiotis, 2004). Due to the complexity of food, drying can occur simultaneously by a different mechanism. Although drying processing effectively extends the shelf life of agricultural products, loss of sensory and nutritive qualities is considered inevitable during traditional drying process due to the undesirable textural and biochemical changes, higher drying rate, lower drying temperature and oxygen deficient processing environment, etc., these characteristics may help to improve the quality and nutritive value other dried products (Watson & Harper, 1988).

Presently, vacuum drying has been applied to dry various food materials, the vacuum drying kinetics of many fruits and vegetables has been inves-tigated, and the effect of vacuum drying conditions on the drying process and the qualities of dried products have been evaluated (Arevalo-Pinedo & Murr, 2007). The most common method used in the drying of heat-sensitive products is the vacuum drying method. Vacuum drying is widely used to dry various heat-sensitive products in which the color, structure, and vitamins are impaired with increasing temperatures (Methakhup et al., 2005). Vacuum drying results in better product quality with respect to characteristics, such as flavor, fragrance, and rehydration (Drouzas & Schubert, 1996). Vacuum drying also has advantages, such as a reduction in processing temperature, improvement in the drying rate, and a reduction in shriveling (Montgomery et al., 1998). The vacuum drying process has been successfully used for the drying of fruit, vegetables, and heat-sensitive products. A high-quality product is obtained due to the retention of flavor and nutritive value in the structure of materials. Vacuum reduces the boiling point of water keeping the product temperature low, as well as creating a pressure gradient that enhances the drying rate. Load size, power level, and vacuum pressure influ-ence drying rate.

Rajkumar et al. (2007) found in his study that dehydration of food materials, especially fruits and vegetables, with antioxidant properties, is a difficult food process operation, mainly because of undesirable changes in quality of the dehydrated products. Further direct exposure to solar radiation results in undesirable color changes, lowering quality of the dried products significantly. Therefore, use of hot air in controlled cabinet drying through convective air is far more rapid and provides uniformity and hygiene for industrial food drying processes become inevitable. According to Drouzas and Schubert (1996), vacuum drying is a potential dehydration technique mainly used for heat-sensitive food products due to drying at lower pressure and temperature under low oxygen environment. Better product quality, such as taste, flavor, or rehydration ratio can be achieved by high degree vacuum treatment. The lower pressure allows the drying temperature to be reduced and results in higher quality than the classical convective air-drying process at atmospheric pressure (Jaya & Das, 2003). Modeling the drying process and predicting the drying behavior under different conditions is necessary to have a better drying understanding of the mechanism. Fick's law of diffusion has been used to describe the drying kinetics of fruits during falling rate period (Garcia et al., 1988). Few studies have been done on banana drying Microwave/air assisted drying was done (Maskan, 1999). The thin layer drying characteristics of banana were studied by applying different pre-treatments, namely, blanching chilling and freezing and blanching and freezing combined and studied using different two-term exponential models.

4.7 MATHEMATICAL MODELING FOR THIN LAYER DRYING OF PLANTAINS AND BANANA

The dynamics of food drying process involves simultaneous heat and mass transfer, where water is transferred by diffusion from inside of the food material toward the air-food interface, and from the interface to the air stream by convection. Heat is transferred by convection from the air to the air-food interface and by conduction to the interior of food (Maroulis et al., 1995). This phenomenon has been modeled with different levels of complexity. Most of the mathematical models have presented physical and thermal properties of the drying material as a function of time for any air temperature and air velocity; therefore, several models have been necessary. In particular, the solution of these equations must allow for the prediction of the process parameters as a function of input parameters, air temperature, and air velocity at any time in the dryer equipment (Gunhan et al., 2005). Due to the dynamic behavior and complexity of the drying process of the agricultural products, mathematical and regression approaches are considered faulty. Thus, a model should be stable and precise enough in case uncertainty exists in the inputs. Artificial intelligence methods, such as neural networks, have gained momentum and are suitable for identifying plant and fruit responses (Banakar et al., 2007). Modeling of drying of banana has been reported by Khawas et al. (2014b; 2015) using Fick's law of diffusion. The optimization of dehydration processes in the agro-food industry has led to choosing the technological variables involved in the process itself.

Nowadays, many researchers have investigated artificial neural networks (ANNs) as the artificial learning tool in a wide range of biotechnology applications including optimization of bioprocesses and enzyme production from microorganisms (Ricca et al., 2012). ANN is biologically inspired and mimics human brain. They are consisting of a large number of simple processing elements named neurons. These neurons are connected with connection link. Each link has a weight that multiplied with transmitted signal in network. Each neuron has an activation function to determine the output. There are many kinds of activation function. Usually, nonlinear activation functions, such as sigmoid and step functions are used. ANNs are trained by experience when applied a new input to the network it can generalize from past experiences and produce a new result (Hanbay et al., 2008).

The simple structure of ANN normally consists of an input layer, a hidden layer, and an output layer. By applying algorithms that mimic the processes of real neurons, the network can learn to solve many types of problems. From the perspectives of process modeling, ANN has been applied to solve complex engineering problems where it is difficult to develop models from the fundamental principles, particularly when dealing with non-linear systems which also exist in bioconversion process (Ricca et al., 2012). It provides a mathematical alternative to the quadratic polynomial for representing data derived from statistically designed experiments. ANN is also able to handle a large amount of data to approximate functions to any desired degree of accuracy, thus making it attractive as empirical model (Panda et al., 2007).

The choosing of an appropriate neural network topology (i.e., number of hidden neurons, learning rate, and momentum) that strongly affects predictability of network is critical and usually carried out by trial and error method. Genetic algorithm (GA) as an optimization technique can be used for overcoming this limitation of neural network. GA is inspired by the natural selection principles and Darwin's species evolution theory. GA offers several advantages over the conventional optimization method, such as less susceptibility to be stuck at local minima, requiring little knowledge of the process is optimized and capability to find the optimum conditions when the search space is very large (Versace et al., 2004).

GA is one of the search methods and optimization techniques which aim at an optimal value of a complex objective function by simulating biological evolutionary processes based, as in genetics, on crossover and mutation. The principles of GA are based on natural competitions of beings for appropriating limited natural source (Erenturk & Erenturk, 2007). Morimoto et al.

(1997) combined an artificial neural network and GA method for optimal control of the fruit-storage process. They also used GA for optimizing heat treatment for fruit during the storage. They determined the drying characteristic of carrot under different drying air conditions and to estimate its dynamical drying behavior using the neural network while a mathematical model of drying procedure was optimized using the GA.

Chen-Hua et al. (2011) stated that at present, ANN is widely used in science. They not only could learn by inspection of data rather than having to be told what to do but also could construct a suitable relationship between input data and the target responses without any need for a theoretical model with which to work. Therefore, ANN could be applied to almost every aspect in food processing, from raw material assessing, thermal processing, freezing, fermentation, enzymatic hydrolysis, ultra-filtrating and drying to composition detecting, quality assessing, and safety evaluating. Accurate modeling and control of food processing operations could be beneficial in increasing the process efficiency and maintaining the uniform quality of the final product. Hernandez (2009) optimized the operating conditions for heat and mass transfer in mango and cassava drying by means of ANN inverse, taking into account air temperature, air velocity, and shrinkage as a function of moisture content, time and air humidity as input parameters. Fathi et al. (2011) reported that ANNs have turned out as a powerful method for numerous practical applications in food science. Neural network models are constructed by interconnecting many non-linear computational adaptive processing elements, known as neurons or nodes, operating in parallel, and arranged in patterns similar to biological networks.

According to Morimoto et al. (1997), GA as an optimization technique can be used for overcoming this limitation of neural network. GA is inspired by the natural selection principles and Darwin's species evolution theory. GA offers several advantages over the conventional optimization method, such as less susceptibility to be stuck at local minima, requiring little knowledge of the process being optimized and capability to find the optimum conditions when the search space is very large. Khawas et al. (2016a) reported that ANN predicted the quality attributes of vacuum-dried culinary banana undergoing various drying temperatures. A comparison of the performance of ANN and RSM with their modeling, prediction, and optimization using the experimental data for vacuum dehydration process of culinary banana revealed ANN models are found to be capable of better predictions for responses (rehydration ratio, scavenging activity, non-enzymatic browning, and hardness) compared to RSM.

4.8 CONCLUSION

Culinary banana both pulp and peel can be a potential source of many functionally important nutrients and bioactive compounds, and other important compounds of interest that could be isolated and can be incorporated into high-value foods.

KEYWORDS

- **banana plants**
- *Musa*
- **cocrystallization**
- **amylase**
- **banana peel**

REFERENCES

Abdel-Hameed, E. S. S. Total Phenolic Contents and Free Radical Scavenging Activity of Certain *Egyptian ficus* Species Leaf Samples. *Food Chem.* **2009,** *114* (4), 1271–1277.

Abe, K.; Yano, H. Comparison of the Characteristics of Cellulose Microfibril Aggregates Isolated from Fiber and Parenchyma Cells of Moso Bamboo (*Phyllostachys pubescens*). *Cellulose* **2010,** *17* (2), 271–277.

Abe, K., et al. High-strength Nanocomposite Based on Fibrillated Chemi-thermomechanical Pulp. *Compos. Sci. Technol.* **2009,** *69* (14), 2434–2437.

Abe, K., et al. Obtaining Cellulose Nanofibers with a Uniform Width of 15 nm from Wood. *Biomacromolecules* **2007,** *8* (10), 3276–3278.

Abraham, E., et al. Extraction of Nanocellulose Fibrils from Lignocellulosic Fibres: A Novel Approach. *Carbohydr. Polym.* **2011,** *86* (4), 1468–1475.

Adeniji, T. A.; Empere, C. E. The Development, Production and Quality Evaluation of Cake Made from Cooking Banana Flour. *Global J. Pure Appl. Sci.* **2001,** *7* (4), 633–635.

Adeyemi, O. S.; Oladiji, A. T. Compositional Changes in Banana (*Musa* sp) Fruits during Ripening. *Afr. J. Biotechnol.* **2009,** *8* (5), 858–859.

Alemdar, A.; Sain, M. Isolation and Characterization of Nanofibers from Agricultural Residues-Wheat Straw and Soy Hulls. *Bioresour. Technol.* **2008,** *99* (6), 1664–1671.

Anhwange, B. A. Chemical Composition of *Musa sapientum* (Banana) Peels. *J. Food Technol.* **2008,** *6* (6), 263–266.

Annadurai, G., et al. Use of Cellulose-based Wastes for Adsorption of Dyes from Aqueous Solution. *J. Hazard. Mater.* **2002,** *92* (3), 263–274.

Aparicio-Saguilan, A., et al. Slowly Digestible Cookies Prepared from Resistant Starch-ich Lintnerized Banana Starch. *J. Food Compos. Anal.* **2007**, *20* (3–4), 175–181.

Archibald, J. G. Nutrient Composition of Banana Skins. *J. Dairy Sci.* **1949**, *32* (11), 969–971.

Arevalo-Pinedo, A.; Murr, F. E. X. Influence of Pre-treatments on the Drying Kinetics during Vacuum Drying of Carrot and Pumpkin. *J. Food Eng.* **2007**, *80* (1), 152–156.

Ayala-Zavala, J. F. N., et al. Agro-industrial Potential of Exotic Fruit Byproducts as a Source of Food Additives. *Food Res. Int.* **2007**, *44* (7), 1866–1874.

Babalis, S. J.; Belessiotis, V. G. Influence of the Drying Conditions on the Drying Constants Moisture Diffusivity during the Thin-layer Drying of Figs. *J. Food Eng.* **2004**, *65* (3), 449–458.

Banakar, A., et al. Comparative Study of Wavelet Based Neural Network and Neuro-fuzzy Systems. *Int. J. Wavelets Multiresolut. Inf. Process* **2007**, *5* (6), 879–906.

Baskar, R., et al. Antioxidant Potential of Peel Extracts of Banana Varieties (*Musa sapientum*). *Food Nutr. Sci.* **2011**, *2* (10), 1128–1133.

Bell, L. N. Stability Testing of Nutraceuticals and Functional Foods. In *Handbook of Nutraceuticals and Functional Foods*; Wildman, R. E. C., Eds.; CRC Press: New York, 2011; pp 501–516.

Beristain, C. I., et al. Encapsulation of Orange Peel Oil by Cocrystallization. *LWT Food Sci. Technol.* **1996**, *29* (7), 645–647.

Bhandari, B. R.; Hartel, R. W. Cocrystallization of Sucrose at High Concentration in the Presence of Glucose and Fructose. *J. Food Sci.* **2002**, *67* (5), 1797–1802.

Bhandari, B. R., et al. Co-crystallization of Honey with Sucrose. *LWT Food Sci. Technol.* **1998**, *31* (2), 138–142.

Biliaderis, C. G. The Structure and Interactions of Starch with Food Constituents. *Can. J. Physiol. Pharmacol.* **1991**, *69* (1), 60–78.

Boonruangrod, R., et al. Elucidation of Origin of the Present Day Hybrid Banana Cultivars Using the 5'ETS rDNA Sequence Information. *Mol. Breed.* **2009**, *24* (1), 24–77.

Byarugaba-Bazirake, G. W., et al. The Technology of Producing Banana Wine Vinegar from Starch of Banana Peels. *Afr. J. Food Sci. Technol.* **2014**, *5* (1), 1–5.

Carreel, F., et al. Ascertaining Maternal and Paternal Lineage within Musa by Chloroplast and Mitochondrial DNA RFLP Analyses. *Genome* **2002**, *45* (4), 679–692.

Chakraborty, A., et al. Reinforcing Potential of Wood Pulp-derived Microfibres in a PVA Matrix. *Holzforschung* **2006**, *60* (1), 53–58.

Cheirsilp, B.; Umsakul, K. Processing of Banana Based Wine Products Using Pectinase and Alpha-amylase. *J. Food Process. Eng.* **2008**, *31* (1), 78–90.

Chen, A. C., et al. Cocrystallization: An Encapsulation Process. *Food Technol.* **1988**, *42* (11), 87–90.

Chen, C. R.; Ramaswamy, H. S. Color and Texture Change Kinetics in Ripening Bananas, *LWT Food Sci. Technol.* **2002**, *35* (5), 415–419.

Chen, W., et al. Individualization of Cellulose Nanofibers from Wood Using High-intensity Ultrasonication Combined with Chemical Pretreatments. *Carbohydr. Polym.* **2011**, *83* (4), 1804–1811.

Chen, Y., et al. Bionanocomposites Based on Pea Starch and Cellulose Nanowhiskers Hydrolyzed from Pea Hull Fibre: Effect of Hydrolysis Time. *Carbohydr. Polym.* **2009**, *76* (4), 607–615.

Cheng, Q., et al. Novel Process for Isolating Fibrils from Cellulose Fibers by High-intensity Ultrasonication. II. Fibril Characterization. *J. Appl. Polym. Sci.* **2010**, *115* (5), 2756–2762.

Cheng, Q., et al. Poly (vinyl alcohol) Nanocomposites Reinforced with Cellulose Fibrils Isolated by High Intensity Ultrasonication. *Compos Part A: Appl. Sci. Manufac.* **2009,** *40* (2), 218–224.

Chen-Hua, et al. In *Artificial Neural Network in Food Processing,* Proceedings of the 30th Chinese Control Conference, Yantai, China, 2011; pp 2687–2692.

Cherian, B. M., et al. Cellulose Nanocomposites with Nanofibres Isolated from Pineapple Leaf Fibres for Medical Applications. *Carbohydr. Polym.* **2011,** *86* (4), 1790–1798.

Cherian, B. M., et al. Isolation of Nanocellulose from Pineapple Leaf Fibres by Steam Explosion. *Carbohydr. Polym.* **2010,** *81* (3), 720–725.

Cordenunsi, B. R.; Lajolo, F. M. Starch Breakdown during Banana Ripening: Sucrose Synthase and Sucrose Phosphate Synthase. *J. Agric. Food Chem.* **1995,** *43* (2), 347–351.

Crowe, K. M.; Murray, E. Deconstructing a Fruit Serving: Comparing the Antioxidant Density of Select Whole Fruit and 100% Fruit Juices. *J. Acad. Nutr. Diet.* **2013,** *113* (10), 1354–1358.

Czaja, W. K., et al. The Future Prospects of Microbial Cellulose in Biomedical Applications. *Biomacromolecules* **2007,** *8* (1), 1–12.

Dadzie, B. K. K.; Wainwright, H. Plantain Utilization in Ghana. *Tropical Sci.* **1995,** *35* (4), 405–410.

Daniells, J. W. Bananas and Plantains–the crops and their importance. In *Encyclopedia of Food Science and Nutrition*; London: Elsevier Science, 2003; pp 372–378.

Davey, M. W., et al. Genetic Variability in *Musa* fruit Provitamin A Carotenoids, Lutein and Mineral Micronutrient Contents. *Food Chem.* **2009,** *115* (3), 806–813.

de Jesus, N. O., et al. Genetic Diversity and Population Structure of *Musa* Accessions in *Ex Situ* Conservation. *BMC Plant Biol.* **2013,** *13* (41), 1–22.

De Langhe, E., et al. Why Bananas Matter: An Introduction to the History of Banana Domestication. *Ethnobot. Res. Appl.* **2009,** *7* (1), 165–177.

de Morais, T. E., et al. Cellulose Nanofibers from White and Naturally Colored Cotton Fibers. *Cellulose* **2010,** *17* (3), 595606.

Desai, K. G. H.; Park, H. J. Recent Developments in Microencapsulation of Food Ingredients. *Drying Technol.* **2005,** *23,*1361–94.

Dinand, E., et al. Suspensions of Cellulose Microfibrils from Sugar Beet Pulp. *Food Hydrocoll.* **1999,** *13* (3), 275–283.

Doymaz, I. Evaluation of Mathematical Models for Prediction of Thin-layer Drying of Banana Slices. *Int. J. Food Prop.* **2010,** *13* (3), 486–497.

Drouzas, A. E.; Schubert, H. Microwave Application in Vacuum Drying of Fruits. *J. Food Eng.* **1996,** *28* (2), 203–209.

Dufresne, A., et al. Cellulose Microfibrils from Potato Tuber Cells: Processing and Characterization of Starch-cellulose Microfibril Composites. *J. Appl. Polym. Sci.* **2000,** *76* (14), 2080–2092.

Eerlingen, R. C.; Delcour, J. A. Formation, Analysis, Structure and Properties of Type III Enzyme Resistant Starch. *J. Cereal Sci.* **1995,** *22* (2), 129–138.

Egbebi, A. O.; Bademosi, T. A. Chemical Compositions of Ripe and Unripe Banana and Plantain. *Int. J. Trop. Med. Public Health* **2012,** *1* (1), 1–5.

Eichhorn, S. J. Review: Current International Research into Cellulose Nanofibres and Nanocomposites. *J. Mater. Sci.* **2010,** *45* (1), 1–33.

Elanthikkal, S., et al. Cellulose Microfibres Produced from Banana Plant Wastes: Isolation and Characterization. *Carbohydr. Polym.* **2010,** *80* (3), 852–859.

Elazzouzi-Hafraoui, S., et al. The Shape and Size Distribution of Crystalline Nanoparticles Prepared by Acid Hydrolysis of Native Cellulose. *Biomacromolecules* **2008,** *9* (1), 57–65.

El-Ramady, H. R., et al. Postharvest Management of Fruits and Vegetables Storage. *Sustainable Agric. Rev.* **2015,** *15,* 65–152.

Emaga, T. H., et al. Dietary Fibre Components and Pectin Chemical Features of Peels during Ripening in Banana and Plantain Varieties. *Bioresour. Technol.* **2008,** *99* (10), 4346–4354.

Emaga, T. H., et al. Effects of the Stage of Maturation and Varieties on the Chemical Composition of Banana and Plantain Peels. *Food Chem.* **2007,** *103* (2), 590–600.

Englberger, L., et al. Carotenoid-rich Bananas: A Potential Food Source for Alleviating Vitamin A Deficiency. *Food Nutr. Bull.* **2003,** *24* (4), 303–318.

Englyst, H. N.; Cummings, J. H. Digestion of the Polysaccharides of Potato in the Small Intestine of Man. *Am. J. Clin. Nutr*. **1987,** *45* (2), 423–431.

Englyst, H. N., et al. Classification and Measurement of Nutritionally Important Starch Fractions. *Eur. J. Clin. Nutr.* **1992,** *46* (2), 33–50.

Erenturk, S.; Erenturk, K. Comparison of Genetic Algorithm and Neural Network Approaches for the Drying Process of Carrot. *J. Food Eng.* **2007,** *78* (3), 905–912.

Essien, J. P., et al. Studies on Mould Growth and Biomass Production Using Waste Banana Peel. *Bioresour. Technol.* **2005,** *96* (13), 1451–1455.

Fatemeh, S. R., et al. Total Phenolics, Flavonoids and Antioxidant Activity of Banana Pulp and Peel Flours: Influence of Variety and Stage of Ripeness. *Int. Food Res. J.* **2012,** *19* (3), 1041–1046.

Fathi, M., et al. Application of Image Analysis and Artificial Neural Network to Predict Mass Transfer Kinetics and Color Changes of Osmotically Dehydrated Kiwifruit. *Food Bioprocess Technol.* **2011,** *4* (8), 1357–1366.

Feming, D. J. Dietary Determination of Iron Stones in a Free-living Elderly Population. The Framingham Heart Study. *Am. J. Clin. Nutr.* **1998,** *67,* 722–733.

Forsyth, W. G. C. Banana and Plantain. In *Tropical and subtropical fruits;* Nagy, S., Shaw, P. E., Eds.; AVI Publishing: Westport, CT, 1980; pp 258–278.

French, D. Organization of Starch Granules. In *Starch: Chemistry and Technology,* 2nd ed.; Whistler, R. L., BeMiller, J. N., Paschall, E. F., Eds.; Academic Press: New York, London, 1984; pp 183–247.

Fuentes-Zaragoza, E., et al. Resistant Starch as Functional Ingredient: A Review. *Food Res. Int.* **2010,** *43* (4), 931–942.

Fukuzumi, H., et al. Transparent and High Gas Barrier Films of Cellulose Nanofibers Prepared by TEMPO-mediated Oxidation. *Biomacromolecules* **2009,** *10* (1), 162–165.

Ganapathi, T. R., et al., Somatic Embryogenesis and Plant Regeneration from Male Flower Buds of Banana. *Curr. Sci.* **1999,** *76,* 1128–1231.

Garcia, R., et al. Drying of Bananas Using Microwave and Air Ovens. *Int. J. Food Sci. Technol.* **1988,** *23* (1), 73–80.

Giasson, B. I., et al. The Relationship between Oxidative/nutritive Stress and Pathological Inclusions in Alzhemier's and Parkinson's Diseases. *Free Radic. Biol. Med.* **2002,** *32* (12), 1264–1275.

Goni, I., et al. Analysis of Resistant Starch: A Method for Foods and Food Products. *Food Chem.* **1996,** *56* (4), 445–449.

Goswami, B.; Borthakur, A. Chemical and Biochemical Aspects of Developing Culinary Banana (*Musa* ABB) 'Kachkal'. *Food Chem.* **1996,** *55* (2), 169–172.

Gunhan, T., et al. Mathematical Modeling of Drying of Bay Leaves. *Energy Convers. Manage.* **2005,** *46* (11–12), 1667–1679.

Habibi, Y., et al. Morphological and Structural Study of Seed Pericarp of *Opuntia ficus-indica* Prickly Pear Fruits. *Carbohydr. Polym.* **2008,** *72* (1), 102–112.

Hanbay, D., et al. An Expert System Based on Wavelet Decomposition and Neural Network for Modeling Chua's Circuit. *Expert Syst. Appl.* **2008,** *34* (4), 2278–2283.

Haripyaree, A., et al. Evaluation of Antioxidant Properties of Some Wild Edible Fruits Extracts by Cell Free Assays. *Electron J. Environ. Agric. Food Chem.* **2010,** *9* (2), 345–450.

Hassimotto, N. M., et al. Antioxidant Activity of Dietary Fruits, Vegetables, and Commercial Frozen Fruit Pulps. *J. Agric. Food Chem.* **2005,** *53* (8), 2928–2935.

10 Health Benefits of Plantains You Need To Know–Health–Nairaland. http://www.nairaland.com/1275850/10-health-benefits-plantains-need. (accessed Feb 6, 2014).

Henriksson, M., et al. Cellulose Nanopaper Structures of High Toughness. *Biomacromolecules* **2008,** *9* (6), 1579–1585.

Hernandez, J. A. Optimum Operating Conditions for Heat and Mass Transfer in Foodstuffs Drying by Means of Neural Network Inverse. *Food Control* **2009,** *20* (4), 435–438.

Higgins, J. A., et al. Resistant Starch Consumption Promotes Lipid Oxidation. *Nutr. Metab.* **2004,** *1* (1), 1–8.

Hu, L., et al. Transparent and Conductive Paper from Nanocellulose Fibers. *Energy Environ. Sci.* **2013,** *6* (2), 513–518.

Indian Horticulture Database, National Horticulture Board, Ministry of Agriculture, Government of India. Aristo Printing Press, India, 4–5, 2014. http://nhb.gov.in/area-pro/NHB_Database_2015.pdf. (accessed Oct 29, 2015).

Irimia-Vladu, M. "Green" Electronics: Biodegradable and Biocompatible Materials and Devices for Sustainable Future. *Chem. Soc. Rev.* **2014,** *43* (2), 588–610.

Isabelle, M., et al. Antioxidant Activity and Profiles of Common Fruits in Singapore. *Food Chem.* **2010,** *123* (1), 77–84.

Iwamoto, S., et al. The Effect of Hemicelluloses on Wood Pulp Nanofibrillation and Nanofiber Network Characteristics. *Biomacromolecules* **2008,** *9* (3), 1022–1026.

Jackson, L. S.; Lee, K. Microencapsulation and the Food Industry. *LWT Food Sci. Technol.* **1991,** *24* (4), 289–297.

Jadhav, S., et al. Lipase Production from Banana Peel Extract and Potato Peel Extract. *Int. J. Res. Pure Appl. Microbiol.* **2013,** *3* (1), 11–13.

Jaya, S.; Das, H. A Vacuum Drying Model for Mango Pulp. *Drying Technol.* **2003,** *21* (7), 1215–1234.

John, P.; Marchal, J. Ripening and Biochemistry of the Fruit. In *Banana and Plantains;* Gowen, S., Ed.; Chapman and Hall: London, 1995; pp 434–467.

Juarez-Garcia, E., et al. Composition, Digestibility and Application in Bread Making of Banana Flours. *Plant Foods Hum. Nutr.* **2006,** *61* (3), 131–137.

Kadla, J. F.; Gilbert, R. D. Cellulose Structure: A Review. *Cellul. Chem. Technol.* **2000,** *34* (3–4), 197–216.

Kajuna, S. T. A. R., et al. Textural Changes of Banana and Plantain Pulp during Ripening. *J. Sci. Food Agric.* **1997,** *75* (2), 244–250.

Kalia, S., et al. Cellulose-based Bio- and Nanocomposites: A Review. *Int. J. Polym. Sci.* **2011,** *2011,* 1–35.

Kanazawa, K.; Sakakibara, H. High Content of a Dopamine, a Strong Antioxidant, in Cavendish Banana. *J. Agric. Food Chem.* **2000,** *48* (3), 844–848.

Kawasaki, B. T., et al. Targeting Cancer Stem Cells with Phytochemicals. *Mol. Interv.* **2008,** *8* (4), 174—184.

Khawas, P.; Das, A. J.; Deka, S. C. Production of Renewable Cellulose Nanopaper from Culinary Banana (*Musa* ABB) Peel and Its Characterization. *Ind. Crops Prod.* **2016b,** *86,* 102–112. http://dx.doi.org/10.1016/j.indcrop.2016.03.028.

Khawas, P.; Dash, K. K.; Das, A. J.; Deka, S. C. Drying Characteristics and Assessment of Physicochemical and Microstructural Properties of Dried Culinary Banana Slices. *Int. J. Food Eng.* **2015,** *11* (5), 667–678.

Khawas, P.; Deka, S. C. Isolation and Characterization of Cellulose Nanofibers from Culinary Banana Peel Using High-intensity Ultrasonication Combined with Chemical Treatment. *Carbohydr. Polym.* **2016a,** *137,* 608–616.

Khawas, P.; Deka, S. C. Effect of Modified Resistant Starch of Culinary Banana on Physicochemical, Functional, Morphological, Diffraction and Thermal Properties. *Int. J. Food Prop.* **2016b,** *20* (1), 133–150. DOI: 10.1080/10942912.2016.1147459.

Khawas, P.; Deka, S. C. Comparative Nutritional, Functional, Morphological and Diffractogram Study on Culinary Banana (*Musa* ABB) Peel at Various Stages of Development. *Int. J. Food Prop.* **2016c,** *19* (12), 2832–2853. DOI: 10.1080/10942912.2016.1141296.

Khawas, P.; Deka, S. C. Encapsulation of Natural Antioxidant Compounds from Culinary Banana by Cocrystallization. *J. Food Process Preserv.* **2016d,** *41* (1), e13033. DOI:10.1111/jfpp.13033.

Khawas, P.; Das, A. J.; Sit, N.; Badwaik, L. S.; Deka, S. C. Nutritional Composition of Culinary *Musa* ABB at Different Stages of Development. *Am. J. Food Sci. Technol.* **2014a,** *2* (3), 80–87 (accessed Feb 5, 2014)."

Khawas, P.; Das, A. J.; Dash, K. K.; Deka, S. C. Thin Layer Drying Characteristics of *kachkal* Banana Peel (*Musa* ABB) of Assam, India. *Int. Food Res. J.* **2014b,** *21* (3) 1011–1018.

Banana Peel (Musa ABB) of Assam, India. Int. Food Res. J. 2014b, 21 (3) 1011–1018.

Khawas, P.; Dash, K. K.; Das, A. J.; Deka, S. C. Modeling and Optimization of the Process Parameters in Vacuum Drying of Culinary Banana (*Musa* ABB) Slices by Application of Artificial Neural Network and Genetic Algorithm. *Drying Technol.* **2016a,** *34* (4), 491–503.

Kim, J.; Yun, S. Discovery of Cellulose as a Smart Material. *Macromolecules.* **2006,** *39* (12), 4202–4206.

Klahorst, S. J. Flavour and Innovation Meet. *World of Food Ingredients*; 2006; pp 26–30.

Kondo, S., et al. Preharvest Antioxidant Activities of Tropical Fruit and the Effect of Low Temperature Storage on Antioxidants and Jasmonates. *Postharvest Biol. Technol.* **2005,** *36* (3), 309–318.

Kosaraju, S. L., et al. Delivering Polyphenols for Healthy Ageing. *Nutr. Diet.* **2008,** *65* (3), S48–S52.

Kris-Etherton, P. M., et al. Bioactive Compounds in Nutrition and Health-research Methodologies for Establishing Biological Function: The Antioxidant and Anti-inflammatory Effects of Flavonoids on Atherosclerosis. *Annu. Rev. Nutr.* **2004,** *24,* 511–538.

Krishna, R. P., et al. Banana Peel as Substrate for α-amylase Production Using *Aspergillus niger* NCIM 616 and Process Optimization. *Indian J. Biotechnol.* **2012,** *11* (3), 314–319.

Kumar, S. K. P., et al. Traditional and Medicinal Uses of Banana. *J. Pharmacogn. Phytochem.* **2012,** *1* (3), 51–63.

Kwak, N. S.; Jukes, D. J. Functional Foods. Part 2: The Impact on Current Regulatory Terminology. *Food Control.* **2001,** *12* (2), 109–117.

Lehmann, U.; Robin, F. Slowly Digestible Starch-Its Structure and Health Implications: A Review. *Trends Food Sci. Technol.* **2007,** *18* (7), 346–355.

Lehmann, U., et al. Characterization of Resistant Starch Type III from Banana (*Musa acuminata*). *J. Agric. Food Chem.* **2002**, *50* (18), 5236–5240.

Li, R., et al. Cellulose Whiskers Extracted from Mulberry: A Novel Biomass Production, Carbohydrate Polymers. *Carbohydr. Polym.* **2009**, *76* (1), 94–99.

Lim, Y. Y., et al. Antioxidant Properties of Several Tropical Fruits: A Comparative Study. *Food Chem.* **2007**, *103* (3), 1003–1008.

Liu, H., et al. Fabrication and Properties of Transparent Polymethylmethacrylate/cellulose Nanocrystals Composites. *Bioresour. Technol.* **2010**, *101* (14), 5685–5692.

Loeseck, W. H. Chemical Changes during Ripening of Bananas. In *Chemistry, Physiology and Technology*; Interscience: New York, NY, 1950; Vol. 4, pp 67–118.

Lopez-Cordoba, A., et al. Yerba Mate Antioxidant Powders Obtained by Cocrystallization: Stability during Storage. *J. Food Eng.* **2014**, *124*, 158–165.

Lustre, A. O., et al. Physico-chemicalchanges in "SABA" Bananas during Normal and Acetylene-induced Ripening. *Food Chem.* **1976**, *1* (2), 125–132.

Makris, D. P.; Rossiter, J. T. Heat-induced, Metal-catalyzed Oxidative Degradation of Quercetin and Rutin (quercetin 3-O-rhamnosylglucoside) in Aqueous Model Systems. *J. Agric. Food Chem.* **2000**, *48* (9), 3830–3838.

Manach, C., et al. Polyphenols: Food Sources and Bioavailability. *Am. J. Clin. Nutr.* **2004**, *79* (5), 727–747.

Mao, A. A., et al. Plant Wealth of Northeast India with Reference to Ethnobotany. *Indian J. Tradit. Knowl.* **2009**, *8* (1), 96–103.

Maroulis, Z. B., et al. Heat and Mass Transfer Modeling in Air Drying of Foods. *J. Food Eng.* **1995**, *26* (1), 113–130.

Marriott, J., et al. Starch and Sugar Transformation During the Ripening of Plantains and Bananas. *J. Sci. Food Agric.* **1981**, *32* (10), 1021–1026.

Maskan, M. Microwave/Air and Microwave Finish Drying of Banana. *J. Food Eng.* **2000**, *44* (2), 71–78.

Maulny, A. P. E., et al. Physical Properties of Cocrystalline Sugar and Honey. *J. Food Sci.* **2005**, *70* (9), E567–E572.

Mendez, C. D. M. V., et al. Content of Free Phenolic Compounds in Bananas from Tenerife (Canary Islands) and Ecuador. *Eur. Food Res. Technol.* **2003**, *217* (4), 287–290.

Menrad, K. Market and Marketing of Functional Food in Europe. *J. Food Eng.* **2003**, *56* (2–3), 181–188.

Methakhup, S., et al. Effects of Drying Methods and Conditions on Drying Kinetics and Quality of Indian Gooseberry Flake. *LWT Food Sci. Technol.* **2005**, *38* (6), 579–587.

Mollet, B.; Lacroix, C. Where Biology and Technology Meet for Better Nutrition and Health. *Curr. Opin. Biotechnol.* **2007**, *18*, 154–155.

Mollet, B.; Rowland, I. Functional Foods: At the Frontier between Food and Pharma. *Curr. Opin. Biotechnol.* **2002**, *13* (5), 483–485.

Montgomery, S. W., et al. Vacuum Assisted Drying of Hydrophilic Plates: Static Drying Experiments. *Int. J. Heat Mass Transfer.* **1998**, *41* (4–5), 735–744.

Moon, R. J., et al. Cellulose Nanomaterials Review: Structure, Properties and Nanocomposites. *Chem. Soc. Rev.* **2011**, *40*, 3941–3994.

Morimoto, T., et al. An Intelligent Approach for Optimal Control of Fruit-storage Process Using Neural Networks and Genetic Algorithms. *Comput. Electron. Agric.* **1997**, *18* (2–3), 205–224.

Morton, J., Ed. Banana. In *Fruits of Warm Climates*; Creative Resource Systems, Inc.: Miami, FL, 1987; pp 29–46.

Muranyi, P. Functional Edible Coatings for Fresh Food Products. *J. Food Process. Technol.* **2013,** *4* (1), e114.

Ndhlala, A. R., et al. Antioxidant Properties and Degrees of Polymerization of Six Wild Fruits. *Sci. Res. Essays* **2006,** *1* (3), 87–92.

Offem, J. O.; Njoku, P. C. Mineral Distribution in the Fruits of the Plantain Plant (*Musa paradisiaca*) in Relation to Mode and Degree of Maturation. *Food Chem.* **1993,** *48* (1), 63–68.

Onwuka, G. I.; Onwuka, N. D. The Effects of Ripening on the Functional Properties of Plantain and Plantain Based Cake. *Int. J. Food Prop.* **2005,** *8* (2), 347–353.

Onwulata, C. I. Encapsulation of New Active Ingredients. *Annu. Rev. Food Sci. Technol.* **2011,** *3* (1), 183–202.

Ottaway, P. B. *Food Fortification and Supplementation-Technological, Safety and Regulatory Aspects,* 1st ed.; Woodhead Publishing: Sawston UK, 2008.

Paakko, M., et al. Enzymatic Hydrolysis Combined with Mechanical Shearing and High-pressure Homogenization for Nanoscale Cellulose Fibrils and Strong Gels. *Biomacromolecules.* **2011,** *8* (6), 1934–1941.

Pacheco-Delahaye, E., et al. Production and Characterization of Unripe Plantain (*Musa paradisica* L) Flours. *Interciencia.* **2008,** *33* (4), 290–296.

Palmer, J. K. The Banana. In *The Biochemistry of Fruits and Their Products*; Hulme, A. C., Ed.; Academic Press: London, 1971, Vol. 2, pp 65–105.

Pan, Z., et al. Study of Banana Dehydration Using Sequential Radiation Heating and Freeze-drying. *LWT Food Sci. Technol.* **2008,** *41* (10), 1944–1951.

Panda, B. P., et al. Fermentation Process Optimization. *Res. J. Microbiol.* **2007,** *2* (3), 201–208.

Parada, J.; Aguilera, J. M. Food Microstructure Affects the Bioavailability of Several Nutrients. *J. Food Sci.* **2007,** *72* (2), 21–32.

Paul, S. M.; Sumathy, H. J. V. Production of α-amylase from Banana Peels with *Bacillus subtilis* Using Solid State Fermentation. *Int. J. Curr. Microbiol. Appl. Sci.* **2013,** *2* (10), 195–206.

Pelissari, M. F., et al. Isolation and Characterization of Cellulose Nanofibers from Banana Peels. *Cellulose.* **2013,** *21* (1), 417–432.

Pillay, M., et al. Ploidy and Genome Composition of *Musa* Germplasm at the International Institute of Tropical Agriculture (IITA). *Afr. J. Biotechnol.* **2006,** *5* (13), 1224–1232.

Prabha, T. N.; Bhagyalakhmi, N. Carbohydrate Metabolism in Ripening Banana Fruit. *Phytochem.* **1998,** *48* (6), 915–919.

Qusti, S. Y., et al. Free Radical Scavenger Enzymes of Fruit Plant Species Cited in Holy Quran. *World Appl. Sci. J.* **2010,** *9* (3), 338–344.

Rajkumar, P., et al. Drying Kinetics of Tomato Slices in Vacuum Assisted Solar and Open Sun Drying Methods. *Drying Technol.* **2007,** *25* (7), 1349–1357.

Ramcharan, C.; George, C. Growing Banana and Plantain in Virgin Islands. *Farmers Bulletins No.* **11,** FAO, 1999; pp 13–14.

Ramli, S., et al. Utilization of Banana Peels as a Functional Ingredient in Yellow Noodle. *As. J. Food Ag-Ind.* **2009,** *2* (3), 321–329.

Rehman, S., et al. Biotechnological Production of Xylitol from Banana Peel and Its Impact on Physicochemical Properties of Rusks. *J. Agric. Sci. Technol.* **2013,** *15* (4), 747–756.

Ricca, R. N., et al. The Potential of Artificial Neural Network (ANN) in Optimizing Media Constituents of Citric Acid Production by Solid State Bioconversion. *Int. Food Res. J.* **2012,** *19* (2), 491–497.

Rodionova, G., et al. Mechanical and Oxygen Barrier Properties of Films Prepared from Fibrillated Dispersions of TEMPO-oxidized Norway Spruce and Eucalyptus Pulps. *Cellulose.* **2011,** *19* (3), 705–711.

Rodriguez-Ambriz, S. L., et al. Characterization of a Fibre-rich Powder Prepared by Liquefaction of Unripe Banana Flour. *Food Chem.* **2008,** *107* (4), 1515–1521.

Rondeau-Mouro, C., et al. Structural Features and Potential texturising Properties of Lemon and Maize Cellulose Microfibrils. *Carbohydr. Polym.* **2003,** *53* (3), 241–252.

Rosa, M. F., et al. Cellulose Nanowhiskers from Coconut Husk Fibers: Effect of Preparation Conditions on Their Thermal and Morphological Behavior. *Carbohydr. Polym.* **2010,** *81* (1), 83–92.

Saifullah, R., et al. Utilization of Green Banana Flour as a Functional Ingredient in Yellow Noodle. *Int. Food Res. J.* **2009,** *16* (3), 373–379.

Sajilata, M. G., et al. Resistant Starch-A Review. *Compr. Rev. Food Sci. Food Saf.* **2006,** *5* (1), 1–17.

Sakyi-Dawson, E., et al. Biochemical Changes in New Plantain and Cooking Banana Hybrids at Various Stages of Ripening. *J. Sci. Food Agric.* **2008,** *1* (15), 2724–2729.

Sardar, B. R.; Singhal, R. S. Characterization of Cocrystallized Sucrose Entrapped with Cardamom Oleoresin. *J. Food Eng.* **2013,** *117* (4), 521–529.

Seenappa, M., et al. Availability of L-ascorbic Acid in Tanzanian Banana. *J. Food Sci. Technol.* **1986,** *23* (5), 293–295.

Sehaqui, H., et al. Cellulose Nanofiber Orientation in Nanopaper and Nanocomposites by Cold Drawing. *ACS Appl. Mater. Interfaces.* **2012,** *4* (2), 1043–1049.

Shahidi, F.; Han, X. Q. Encapsulation of Food Ingredients. *Crit. Rev. Food Sci. Nutr.* **1993,** *33,* 501–547. https://doi.org/10.1080/10408399309527645.

Shewfelt, R. L. Sources of Variation in the Nutrient Content of Agricultural Commodities from the Farm to the Consumer. *J. Food Qual.* **1990,** *13* (1), 37–54.

Shiau, S. Y.; Yeh, A. I. Effects of Alkali and Acid on Dough Rheological Properties and Characteristics of Extruded Noodles. *J. Cereal Sci.* **2001,** *33* (1), 27–37.

Silva-Weiss, A., et al. Natural Additives in Bioactive Edible Films and Coatings: Functionality and Applications in Foods. *Food Eng. Rev.* **2013,** *5* (4), 200–216.

Simmonds, N. W. *Bananas.* 2nd ed.; Longmans: London, 1966.

Singh, H. P.; Uma, S. In *Genetic diversity of banana in India*, Proceedings of the Conference on Challenges for Banana Production and Utilization in 21st Century. Singh, H. P., Chadha, K. L., Eds.; India, 2000; pp 136–156.

Singhal, M.; Ratra, P. Antioxidant Activity, Total Flavonoid and Total Phenolic Content of *Musa acuminata* Peel Exctracts. *Global J. Pharmacol.* **2013,** *7* (2), 118–122.

Siqueira, G., et al. Cellulose Whiskers Versus Microfibrils: Influence of the Nature of the Nanoparticle and Its Surface Functionalization on the Thermal and Mechanical Properties of Nanocomposites. *Biomacromolecules* **2009,** *10* (2), 425–432.

Siro, I.; Plackett, D. Microfibrillated Cellulose and New Nanocomposite Materials: A Review. *Cellulose* **2010,** *17* (3), 459–494.

Someya, S., et al. Antioxidant Compounds from Bananas (*Musa* Cavendish). *Food Chem.* **2002,** *79* (3), 351–354.

Sommer, A. New Imperatives for an Old Vitamin (A). *J. Nutr.* **1989,** *119,* 96–100.

Spence, K. L., et al. Water Vapor Barrier Properties of Coated and Filled Microfibrillated Cellulose Composite Films. *BioResour.* **2011,** *6* (4), 4370–4388.

Stinco, C. M. Industrial Orange Juice Debittering: Impact on Bioactive Compounds and Nutritional Value. *J. Food Eng.* **2013,** *116* (1), 155–161.

Stover, R. H.; Simmonds, N. W. *Bananas (Tropical Agriculture Series),* 3rd ed.; Longman: London, UK, 1987; p 468.

Studdert, V. P., et al. *Saunders Comprehensive Veterinary Dictionary,* 4th ed.; Elsevier Health Sciences: London, UK, 2011; p 79.

Sulaiman, F. S., et al. Correlation between Total Phenolic and Mineral Contents with Antioxidant Activity of Eight Malaysian Bananas (*Musa* sp.). *J. Food Compos. Anal.* **2011,** *24* (1), 1–10.

Suma, P. F.; Urooj, A. Isolation and Characterization of Starch from Pearl Millet (*Pennisetum typhoidium*) Flours. *Int. J Food Prop.* **2015,** *18* (12), 2675–2687.

Tapre A. R.; Jain R. K. Study of Advanced Maturity Stages of Banana. *Int. J. Adv. Eng. Res. Stud.* **2012,** *1* (3), 272–274.

Tchobanoglous, G.; Theisen, H.; Vigil, S. *Integrated Solid Waste Management: Engineering Principals and Management Issues*; McGraw-Hill Inc.: Boston, MA, 1993.

The Indian Vegan: All about plantain in India. http://theindianvegan.blogspot.in/2013/03/all-about-plantain-in-india.html. (accessed Feb 6, 2014).

Thompson, A. K. *Banana Processing. In Bananas and Plantains*; Gowen, S., Ed.; Chapman and Hall: London, UK, 1995; pp 481–492.

Tibolla, H., et al. Cellulose Nanofibers Produced from Banana Peel by Chemical and Enzymatic Treatment. *LWT Food Sci. Technol.* **2014,** *59* (2), 1311–1318.

Tischer, P. C. S. F., et al. Nanostructural Reorganization of Bacterial Cellulose by Ultrasonic Treatment. *Biomacromolecules* **2010,** *11* (5), 1217–224.

Turbak, A. F.; Snyder, F. W.; Sandberg, K. R. *Microfibrillated Cellulose, a New Cellulose Product: Properties, Uses, and Commercial Potential*. Proceedings of the Ninth Cellulose Conference, Applied Polymer Symposium, Sarko, A., Ed.; Wiley: New York, 1983; pp 815–827.

UNCST. The Biology of Bananas and Plantains Uganda National Council for Science and Technology (UNCST) in Collaboration with Program for Biosafety Systems (PBS). US Agency for International Development (USAID) funded Project. 2007; pp 1–3.

Urruzola, I., et al. Nanopaper from Almond (*Prunus dulcis*) Shell. *Cellulose.* **2014,** *21* (3), 1619–1629.

USDAARS. US Department of Agriculture, Agricultural Research Service 2004. USDA National Nutrient Database for Standard Reference, Release 17. http://www.nal.usda.gov/fnic/foodcomps. (accessed Oct 20, 2015).

Valmayor, R. V. Classification and Characterization of *Musa exotica, M. alinsanaya* and *M. acuminata* ssp. *errans. Infomusa.* **2001,** *10* (2), 35—39.

Versace, M., et al. Predicting the Exchange Traded Fund DIA with a Combination of Genetic Algorithms and Neural Networks. *Expert Syst. Appl.* **2004,** *27* (3), 417–425.

Visser, R. G. F., et al. Comparison between Amylose-free and Amylose Containing Potato Starches. *Starch Starke.* **1997,** *49* (11), 438–443.

Vos, P.; Faas, M. M.; Spasojevic, M.; Sikkema, J. Review: Encapsulation for Preservation of Functionality and Targeted Delivery of Bioactive Food Components. *Int. Dairy J.* **2010,** *20,* **292–302.**

Wachirasiri, P., et al. The Effects of Banana Peel Preparations on the Properties of Banana Peel Dietary Fibre Concentrate. *Songklanakarin J. Sci. Technol.* **2009,** *31* (6), 605–611.

Wainwright, H.; Burdon, J. N. Problems and Prospects for Improving the Postharvest Technology of Cooking Bananas, *Postharvest News Inf.* **1991,** *2* (4), 249–253.

Wandrey, C.; Bartkowiak, A.; Harding, S. E. Materials for Encapsulation. In *Encapsulation Technologies for Food Active Ingredients and Food Processing*; Zuidam, N. J., Nedovic, V. A., Eds.; Springer: Dordrecht, The Netherlands; 2009; pp 31–100.

Wang, B.; Sain, M. Isolation of Nanofibers from Soybean Source and Their Reinforcing Capability on Synthetic Polymers. *Compos. Sci. Technol.* **2007,** *67* (11–12), 2521–2527.

Wang, L.; Bohn, T. Health-promoting Food Ingredients and Functional Food Processing, Nutrition, Well-being and Health. 2012. http://www.intechopen.com/books/nutrition-well-being-and-health/health-promoting-food-ingredients-development-and-processing (accessed Sep 23, 2015).

Wang, S.; Cheng, Q. A Novel Process to Isolate Fibrils from Cellulose Fibers by High-intensity Ultrasonication, Part 1: Process Optimization. *J. Appl. Polym. Sci.* **2009,** *113* (2), 1270–1275.

Watson, E. L.; Harper, J. C. *Elements of Food Engineering.* 2nd ed.; AVI: New York, 1988.

What is a Plantain? Plantains vs. Bananas? Gaurment Plantain Chips and Snacks. http://grabemsnacks.com/what-is-a-plantain.html. (accessed Feb 5, 2014).

WHO. Food and Agriculture Organization/World Health Organization. *Carbohydrates in Human Nutrition*; Report of a Joint FAO/WHO Expert Consultation. FAO: Rome, 1998.

www.traditionaltree.org. *Traditional Trees of Pacific Islands. Their Culture, Environment and Use.* (accessed March 15, 2015).

Yan, L. Y., et al. Antioxidant Properties of Guava Fruit: Comparison with Some Local Fruits. *Sunway Acad. J.* **2006,** *3,* 9–20.

Yang, S. F.; Hoffman, N. E. Ethylene Biosynthesis and Its Regulation in Higher Plants. *Annu. Rev. Plant Physiol.* **1984,** *35,* 155–189.

Yoo, S.; Hsieh, J. S. Enzyme-assisted Preparation of Fibrillated Cellulose Fibers and Its Effect on Physical and Mechanical Properties of Paper Sheet Composites. *Ind. Eng. Chem. Res.* **2010,** *49* (5), 2161–2168.

Zhang, P., et al. Banana Starch: Production, Physicochemical Properties, and Digestibility-A Review. *Carbohydr. Polym.* **2005,** *59* (4), 443–458.

Zuluaga, R., et al. Cellulose Microfibrils from Banana Rachis: Effect of Alkaline Treatments on Structural and Morphological Features. *Carbohydr. Polym.* **2009,** *76* (1), 51–59.

CHAPTER 5

MODIFICATION OF CULINARY BANANA RESISTANT STARCH AND ITS APPLICATION

PRERNA KHAWAS and SANKAR CHANDRA DEKA*

Department of Food Engineering and Technology, Tezpur University, Napaam, Tezpur 784028, Sonitpur, Assam, India

Corresponding author. E-mail: sankar@tezu.ernet.in

CONTENTS

Abstract 108

Effect of Modified Resistant Starch of Culinary Banana on Physicochemical, Functional, Morphological, Diffraction, and Thermal Properties

5.1   Introduction 108

5.2   Materials and Methods 110

5.3   Results and Discussion 114

5.4   Conclusion 131

Effect of Partial Replacement of Wheat Flour with Type III Resistant Starch and Flour of Culinary Banana on the Chemical Composition, Textural Properties, and Sensory Quality of Brown Bread

5.5   Introduction 133

5.6   Materials and Methods 134

5.7   Results and Discussion 139

5.8   Conclusion 151

Keywords 151

References 152

ABSTRACT

This chapter deals with the isolation, characterization, and modification of culinary banana starch. The isolated starch was further modified to resistant starch (RS) using hydrothermal as well as debranching enzyme methods. The physicochemical characterization revealed that culinary banana starch is a mixture of A and B type polymorphs and functional groups present are typical bands of C-type starch with a mixture of spherical and elliptical granules. In partial replacement of wheat flour with RS and culinary banana flour (KF) for developing brown bread revealed that replacement of wheat flour with 10% RS and 10% KF is the best combination from the standpoint of various quality parameters for making brown bread. The effect of fortification on quality attributes of brown bread revealed that incorporation of RS and KF up to 10% in substituting wheat flour resulted the bread with high-quality attributes in terms of nutrients, yield, texture, color, and sensory analysis with highest score of consumer acceptance. The results justified that culinary banana is an excellent source of RS and may be utilized as an alternative source of nutraceutical ingredient for preparing low glycemic functional foods.

The chapter has been divided into two sections for better discussions of the content.

- A) Effect of modified resistant starch of culinary banana on physicochemical, functional, morphological, diffraction, and thermal properties
- B) Effect of partial replacement of wheat flour with type III resistant starch and flour of culinary banana on the chemical composition, textural properties, and sensory quality of brown bread

Effect of Modified Resistant Starch of Culinary Banana on Physicochemical, Functional, Morphological, Diffraction, and Thermal Properties

5.1 INTRODUCTION

Starch is the storage polysaccharides of green plants and a major dietary component in all human populations (French, 1984). It is composed of a number of monosaccharides or sugar (glucose) linked together with α-D-(1-4) and/or α-D-(1-6) linkages (Sajilata et al., 2006). Starch is deposited in the fruit in the form of partially crystalline granules. The morphology, chemical composition, and supermolecular structure are characteristic to particular plant species. Starch owes much of its functionality to two major high-molecular-weight carbohydrate components, namely, amylose and amylopectin. Amylose is an essentially linear polymer in which glucose residues are α-D-(1-4) linked typically constituting 15–20% of starch. Amylopectin the major component of starch is a larger branched molecule with α-D-(1-4) and α-D-(1-6) linkages. This biopolymer constitutes an excellent raw material to modify food texture and consistency (Suma & Urooj, 2015). The amount of starch is not only important for the texture of a given food product, but starch type is equally critical (Biliaderis, 1991).

On the basis of its digestibility, starch has been classified into three groups, such as readily digestible starch (RDS), slowly digestible starch (SDS), and resistant starch (RS) (Englyst & Cummings, 1987). RDS is the starch fraction that causes an increase in blood glucose level after ingestion immediately, whereas SDS is the starch fraction that is digested completely in the small intestine at a lower rate as compared to RDS. RS is the portion of starch and/or starch hydrolysis products that escape digestion in the small intestine and enter the colon for fermentation (Sajilata et al., 2006). Extensive studies have shown RS to have physiological functions similar to those of dietary fiber (Eerlingen & Delcour, 1995). RS appears to be highly resistant to mammalian enzyme and may be classified as a component of fiber on the basis of the definitions of dietary fiber given by AACC (2000).

The diversity of the modern food industry and the enormous variety of food products being produced require starches that can tolerate a wide range of processing techniques and preparation conditions (Visser et al., 1997). These demands are met by modifying native starches with chemical, physical, and enzymatic methods which may lead to the formation of indigestible residues. The availability of such starches, therefore, deserves consideration. Four forms of RS are distinguished: RS type I is defined as physically inaccessible starch for instance in grains; type II is granular starch in raw potato and bananas; type III is retrograded starch, arising after hydrothermal treatment of starch; and type IV is considered to be a chemically modified starch (Englyst et al.,1992). Among these four types, RS type III seems to be particularly interesting because it preserved its nutritional characteristics when it is added as ingredient for cooked food. RS type III is produced by gelatinization, which is a disruption of granular structure by heating starch with excess water and then retrogradation occurs.

5.2   MATERIALS AND METHODS

5.2.1   RAW MATERIALS

Unripe culinary bananas at optimum harvesting stage were harvested and collected from Assam, India. Prior to the starch isolation, the samples were cleaned thoroughly under running tap water followed by rinsing with distilled water and pat dried using a clean cloth.

5.2.2   STARCH ISOLATION

The starch was isolated following the method described by Bello-Perez et al. (2001). In brief, the fruits were peeled and cut into 5–6 cm cubes and immediately rinsed in sodium sulfite solution (1.22 g/L). The homogenate was consecutively sieved through screens numbers 50 and 100 mesh until the washing water was clean. The starch milk was centrifuged at 10,000 rpm for 30 min and the white-starch sediments were dried in a convection oven at 40°C for 24 h, ground with a mortar-pestle and passed through 100 mesh sieve and stored at ambient temperature in sealed containers.

5.2.3   CHEMICAL ANALYSIS

Proximate composition analysis was carried out where moisture content, ash, crude fiber fat, and protein content were determined according to AOAC (2010) methods. The pH of starch dispersion (8% w/v) was measured by using a pH meter. Total amylose content was determined as per the method described by McGrance et al. (1998).

The RS content in the culinary banana starch was determined by an enzymatic method (AOAC, 2010). The samples were incubated with pancreatic α-amylase and amyloglucosidase (AMG) for 16 h at 37°C. The reaction was terminated by the addition of an equal volume of ethanol and the RS was recovered as a pellet on centrifugation. This was then washed twice with ethanol (50% v/v) and centrifuged. The RS in the pellet was dissolved in 2 M KOH by vigorously stirring on an ice-water bath. This solution was neutralized with acetate buffer and the starch was quantitatively hydrolyzed to glucose with AMG. The glucose was quantified with glucose oxidase/peroxidase reagent (GOPOD), which gave a measure of the RS content of the sample.

5.2.4   FUNCTIONAL PROPERTIES

5.2.4.1   WATER HOLDING CAPACITY, STARCH SWELLING POWER, AND SOLUBILITY

Water holding capacity was determined as described by Hallgren (1985). Starch pastes 5% (w/v) were heated to 60, 70, 80, and 90°C for 15 min with shaking every 5 min period. Tubes were centrifuged at 3000 g for 15 min, the supernatant was decanted, and the tubes were then weighed, and the gain in weight was used to calculate the water holding capacity. Starch suspension (40 mL of 1% w/v) was taken in previously weighed 50 mL flask and was heated from 50 to 90°C for 30 min. The flask was removed and left for cooling to room temperature and centrifuged for 15 min at 3000 rpm. The supernatant decanted and the swollen granules weighed. A 10 mL sample was taken from the supernatant, placed in a crucible, and dried in a convection oven at 120°C for 4 h to constant weight. Percentage solubility and swelling power were calculated as:

$$\% \text{ Solubility} = \frac{\text{Weight of dried starch}}{\text{Sample weight}} \times 100$$

$$\text{Swelling power}(\%) = \frac{\text{weight of swollen granules}}{\text{sample weight} - \text{weight of dissolved starch}} \times 100$$

5.2.4.2   FREEZE-THAW STABILITY AND PASTE CLARITY

About 5% starch was dissolved at 95°C for 30 min with continuous stirring and the freeze-thaw stability of the starch paste was studied by four alternate freezing and thawing of 5 mL of 5% starch pastes (freezing for 18 h at −22°C and 6 h thawing at room temperature, respectively), following the method of Jeong and Lim (2003) followed by centrifugation at 50 for 10 min. The percentage of water separated after the freeze-thaw cycle was measured as weight of exudates to the weight of paste.

Following the method described by Bello-Perez et al. (2001) starch sample (0.2 g) was suspended in 5 mL of water in screw cap tubes and placed in a boiling water bath for 30 min. The tubes were thoroughly shaken every 5 min. After cooling the tubes to room temperature, the percentage transmittance at 650 nm was determined against a water blank in a spectrophotometer (Spectrascan UV-2600, Thermo Fisher Scientific, Nasik, India).

Stability and clarity of starch pastes were determined at both room temperature (30°C ± 2°C) and at 4°C at 24, 48, and 72 h.

5.2.4.3   PASTING PROPERTIES

Pasting properties of starches were evaluated in Rapid Visco-Analyzer (RVA-4, Newport Scientific, Sydney, Australia). An 8% slurry was given a programed heating and cooling cycle set for 23 min, where the sample was held at 30°C for 1 min, heated to 95°C in7.5 min, further held at 95°C for 5 min before cooling to 50°C within 7.5 min, and holding at 50°C for 2 min. The speed was 960 rpm for the first 10 s, then 160 rpm for the remainder of the experiment. Peak viscosity (PV), hold viscosity (HV), final viscosity (FV), Breakdown viscosity (BV), setback viscosity (SV), and pasting temperature (PT) of starches were measured and measurements were replicated four times.

5.2.5   STRUCTURAL ANALYSIS OF STARCH

5.2.5.1   X-RAY DIFFRACTION

The X-ray diffraction was obtained from a D/max 2500 X-ray diffractometer (Rigaku Miniflex, Japan) at room temperature using CuKα radiation (λ = 0.15418 nm). A conventional X-ray tube was set to 30 kv acceleration potential and 15 mA current. Data were collected from 2θ of 5–50° (θ being the angle of diffraction) with a scan speed of 8° 2θ/min. The starch sample was dried at 50°C to constant moisture (10%) in a vacuum oven and 50 mg samples were added to the slide for packing prior to X-ray scanning.

5.2.5.2   FOURIER TRANSFORMS INFRARED (FTIR) SPECTRA

An infrared spectrum of starch was measured using KBr disk method. The dry sample was blended with KBr in a ratio starch/KBr 1:4. The blend was pressed to obtain a pellet and introduced in the spectrometer (Nicolet Instruments 410 FTIR equipped with KBr optics and a DTGS detector). Each spectrum was analyzed in the range of resolution from 400 to 4000 cm^{-1} with a resolution of 4 cm^{-1} and total of 64 scans were collected.

5.2.5.3   MORPHOLOGICAL ANALYSIS BY SEM

Starch granules were observed under a scanning electron microscope (JEOL JSM 6390 LV, USA) operating at an accelerating voltage of 15 kv. A small portion of starch sample was assembled on metallic stubs with double-sided tape and coated with a thin layer of gold. Magnification was taken at 500 X and shape and size of the starch granules were observed.

5.2.5.4   THERMAL CHARACTERISTICS BY THERMOGRAVIMETRIC ANALYSIS

Thermal degradation behavior of starch was evaluated using thermogravi-metric analysis (TGA) (Shimadzu, TGA-50, North America). The thermal stability of each sample was conducted at 25–600°C with constant heating rate of 10°C/min under nitrogen atmosphere.

5.2.6   DEVELOPMENT OF TYPE III RESISTANT STARCH (RS) FROM CULINARY BANANA STARCH

5.2.6.1   AUTOCLAVING AND COOLING METHOD

The RS samples were prepared following the method of Berry (1986) with slight modification by suspending 10 and 20% (w/v) of culinary banana starch in 250 mL of water and autoclaved using 15 psi pressure at 120°C for 30 min followed by cooling at 4°C for 24 h. After three repetitions of the autoclaving and cooling cycles, the samples were freeze-dried (Model NO. LDF-5512, Daihan Lab Tech Co., Ltd, South Korea), ground into fine particles by using mechanical grinder (Fritsch, Germany), passed through 100 mesh screen sieve, and stored until further analysis.

5.2.6.2   ENZYME DEBRANCHING METHOD

Following the method of RS production by enzyme debranching (Morales-Medina et al., 2014) RS was developed by suspending culinary banana starch (20%) in sodium acetate buffer (pH 5.0) and the slurry was subjected to repeated thermal and enzymatic treatment. The samples were given thermal treatment at 100°C for 15 min, diluted further using 10% sodium acetate buffer, and mixed thoroughly. For the enzyme debranching process, pullulanase (EC 3.2.1.41) from *Klebsiella pneumoniae* (5%) was added to the homogenized suspension of starch gel and the hydrolysis reaction was maintained for 24 h at 60°C. The thermal treatment at 100°C for 15 min was repeated in between enzyme debranching process. The developed RS was freeze-dried ground and stored for further analysis.

5.2.6.3   CHEMICAL ANALYSIS OF RS

Moisture content, ash, crude fiber fat, and protein content were determined according to AOAC methods (AOAC, 2010). The pH of modified starch dispersion (8% w/v) was measured by using a pH meter. The amylose content in developed RS was measured following the method described by McGrance et al. (1998) and the RS content in the starch modifications was determined by an enzymatic method (AOAC, 2010).

5.2.6.4   STATISTICAL ANALYSIS

Experiments were carried out in three replicates. The data analysis tool "Microsoft Excel" was used for statistical analysis. Data were subjected to ANOVA and Fisher's least significant difference (LSD) was used to separate means.

5.3   RESULTS AND DISCUSSION

5.3.1   CHEMICAL COMPOSITION OF CULINARY BANANA STARCH

The yield of starch isolated from culinary banana was 16.00% with a high purity of 96.00% (Table 5.1). Waliszewski et al. (2003) isolated starch from *Valery* variety of bananas and the yield of starch reported by the author were much higher (33.8%). Bello-Perez et al. (1999) reported two yields of starch from macho and criollo banana 43.8 and 11.8%, respectively. The high difference in the starch yield between two varieties may be due to the texture of banana fruit and its maturity stages of the fruit used (Perez-Sira, 1997). The moisture content of culinary banana starch was found to be 10.90%. According to the reports of Mweta et al. (2008), the moisture content of the starches ranged from 8.96 to 11.93% and falls within the acceptable range for storage and marketing without deterioration in quality of starches. The data of this study (Table 5.1) revealed that the culinary banana starch contained 0.35% ash, 0.31% protein, 0.27% fiber, and 0.50% fat. The content of ash reported in this study was higher than that of plantain starch (0.02%) reported by Perez-Sira (1997). The high ash content in culinary banana starch may be indicative of presence of more minerals like potassium and magnesium. A result of protein content is comparable with results of Mweta et al. (2008) in case of cassava starch. The higher fat content (0.50%) reported in this study is perhaps the reason for its resistance to amylolysis due to the formation of amylose-lipid-complex (Bertolini et al., 2005). The pH value obtained for the culinary banana starch was recorded to be 6.70 which are within the pH range of 3–9 obtained for most starches used in the pharmaceutical, cosmetics, and food industries (Omojola et al., 2010). As most normal starches contain 20–30% of amylose, 34.10% amylose content of culinary banana starch revealed a non-waxy starch type. Amylose content in *Valery* banana starch (40.7%) (Waliszewski et al., 2003) and was different for Cavendish banana starch (19.5%) (Ling et al., 1982). The high amylose starch is much more resistant to digestive enzymes than the low amylose starch. The RS content in the starch sample studied was 18.88% which was on the higher side as compared to the content of RS (17.5%) in banana flour reported by Ovando-Martinez et al. (2009). The photograph of starch obtained from culinary banana is presented in Figure 5.1.

TABLE 5.1 Chemical Composition (%) of Culinary Banana Starch.

Starch yield	Starch purity	Ash	Crude fiber	Fat	Protein	pH	Amylose	Resistant starch content
16.00 ± 0.12	96.00 ± 0.08	0.35 ± 0.01	0.27 ± 0.09	0.50 ± 0.02	0.31 ± 0.03	6.70 ± 0.03	34.10 ± 0.02	18.88 ± 0.05

Results are mean of three replicates ± SD.

 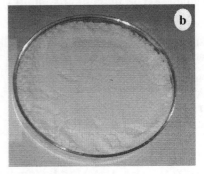

FIGURE 5.1 **(See color insert.)** (a) Culinary banana at matured edible stage and (b) starch from culinary banana.

5.3.2   FUNCTIONAL PROPERTIES OF CULINARY BANANA STARCH

5.3.2.1   WATER HOLDING CAPACITY, STARCH SWELLING POWER, AND SOLUBILITY

This study revealed that water-holding capacity of culinary banana starch (Table 5.2) increased with rise in temperature. The maximum water holding capacity (42.24%) was observed at 90°C. Water holding capacity of culinary banana starch was in accordance with the findings of Bello-Perez et al. (2000) for "criollo" and "macho" starches. The swelling behavior of starch is an indication of the water absorption characteristics of the granules during heating (Hoover & Ratnayake, 2002). Generally, the solubility and swelling profiles show a general trend of increase with increase in temperature and expansion of starch mainly depends on degree of gelatinization (Kannadhason & Muthukumarappan, 2010). The swelling and solubility profile of culinary banana starch are presented in Table 5.2. The starch sample swelled slowly up to 70°C and above it, the starch granules swelled rapidly due to the breakage of intermolecular hydrogen bonds in amorphous region (Bello-Perez et al., 1999). However, the starch exhibited a restricted swelling pattern, which may be attributed to the fat content, as fats are known to inhibit swelling by forming insoluble complexes with the linear fraction of starch (Betancur et al., 2001). Similar range of swelling power was also reported for kernel starch (Cai et al., 2015) and white and yellow plantain starches (Nwokocha & Williams, 2009). Lower swelling power of culinary banana starch is also reflection of more stable granular structure within the starch molecule. A similar pattern was also observed for solubility (Table 5.2). This might be result of the swollen starch granules allowing amylose exudation. The maximum solubility (9.00%) was observed at 90°C which

TABLE 5.2   Water Holding Capacity, Swelling, and Solubility Profile of Culinary Banana Starch.

Parameters	Temperature (°C)			
	60	**70**	**80**	**90**
Water holding capacity (%)	12.64 ± 0.09	20.42 ± 0.07	36.26 ± 0.01	42.24 ± 0.01
Swelling (gH$_2$O/g dry samples)	2.30 ± 0.12	4.50 ± 0.04	8.90 ± 0.00	12.80 ± 0.06
Solubility (%)	1.55 ± 0.01	3.51 ± 0.52	7.11 ± 0.08	9.00 ± 0.17

Results are mean of three replicates ± SD.

is lower than cornstarch (15.80%) at the same temperature (Thomas & Atwell, 1999). The lower values of starch solubility at low temperatures might be due to the semicrystalline structure of the starch granules and the hydrogen bonds formed between hydroxyl groups within the starch molecules. As the temperature increased, the solubility increased due to breaking of starch granules and exposure of hydrophilic groups to water (Eliasson & Gudmundsson, 1996).

5.3.3 FREEZE-THAW STABILITY AND PASTE CLARITY

Exudation of water from frozen gels (syneresis) indicates the retrogradation behavior of the cooked starch pastes. Freeze-thaw stability measures the amount of water released from the gels during storage by degree of syneresis and is an important factor to be considered when formulating refrigerated and frozen foods (Baker & Rayas-Duarte, 1998). Culinary banana starch gel was unstable during different freezing and thawing cycles releasing 24.13–42.58% of the water (Fig. 5.2a). The amount of water separated from the gels during freezing increased with storage time. This result suggests that banana starch is not desirable for frozen products. The low freeze-thaw stability of culinary banana starch might have been affected by the amylose and amylopectin content. Baker and Rayas-Duarte (1998) have reported this behavior for corn starches mentioning low freezing-thawing gel stability for corn and amaranth starches.

Starch gel clarity is a much desirable functionality of starches for its utilization in food industries since it directly influences brightness and opacity in foods. The starch sample experienced low paste clarity (3.2–1.2% light transmittance) during storage time (Fig. 5.2b) and the results are in line with previous reports of Bello-Pérez et al. (2000) who reported that banana starch forms an opaque gel and with increase in storage period the transmittance value decreases. This reduction in transmittance is due to retrogradation tendency of starch pastes which means that under refrigerated condition banana starch have tendency to retrograde. Opaqueness of starch paste has been attributed to various factors, such as granule swelling, granule remnants, leached amylose and amylopectin, amylose-amylopectin chain-length, intra-or intermolecular bonding, and presence of lipids (Jacobson et al., 1997). Since the paste clarity of culinary starch is very low, it could be used in food products that do not require transparency.

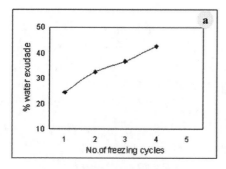 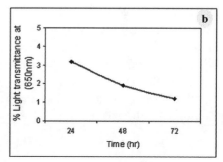

FIGURE 5.2 **(See color insert.)** (a) Freeze-thaw stability of culinary banana starch and (b) Paste-clarity of culinary banana starch.

5.3.4 PASTING PROPERTIES

Pasting properties of culinary banana starch are presented in Figure 5.3. The starch sample exhibited a high (81.80°C) pasting temperature which indicates that the starch is highly resistant toward swelling and it favorably supports our findings for swelling power. The pasting temperature is an indication of the gelatinization temperature of the starches; which indicates culinary banana starches have high gelatinization temperature. The culinary banana starch showed a peak viscosity of 4507 cP and reflects the ability of starch granules to swell freely before their physical breakdown (Kayisu et al., 1981). Peak viscosity occurs at the equilibrium point between granule swelling and polymer leaching, which causes an increase in viscosity and granule rupture and polymer alignment because of mechanical shear. The starch exhibited high setback viscosity during cooling indicating that it retrograded highly which might be due to the effect of amylose and amylopectin contents. The starch with high amylose could undergo the retrogradation process faster than the starch with low amylose content (Goheen & Wool, 1991). During the holding temperature at 95°C accompanied with shear, a decrease in the viscosity of the starch pastes was observed, resulting in the breakdown of some swollen starch granules. The culinary banana starch also experienced a low breakdown (447 cP) viscosity which is also indicative of lower degree of swelling and subsequent disintegration. Kayitsu et al. (1981) reported that the high pasting temperature, as high peak viscosity of the starches as shown are highly resistant to swelling and rupturing and exhibiting restricted swelling.

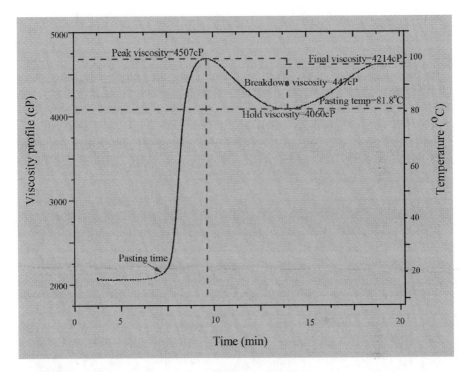

FIGURE 5.3   Pasting properties of culinary banana starch.

5.3.5   STRUCTURAL ANALYSIS

5.3.5.1   X-RAY DIFFRACTION

X-ray diffraction is one of the most effective methods for evaluating the structure of starch and determining the crystalline form of starch (Blazek & Gilberta, 2011). X-ray diffraction provides an elucidation of the long-range molecular order, typically termed as crystalline, which is due to ordered arrays of double helices formed by the amylopectin side chains (Perez & Bertoft, 2010). The culinary banana starch exhibited strong diffraction peaks at 15.0 and 17.09° (2θ) and one very broad peak at 23° (2θ). The wide-angle X-ray diffractogram revealed that culinary banana starch (Fig. 5.4) is a mixture between the A- and B-type polymorphs. According to the report of Yu et al. (2013), C-type starch pattern has been considered a mixture of both A- and B-types because its X-ray diffraction pattern can be resolved as a combination of the previous two. The results of XRD reported in this study are in agreement with Chang et al. (1991) and Waliszewski et al. (2003) as they have also assigned a C type diffraction pattern for banana starches. The percentage crystallinity of starch sample was recorded to be 27.45% which is comparatively higher than the reported values of crystallinity index of different verities of banana starches studied by Soares et al. (2011). The higher crystallinity index of culinary banana starch may be attributed to the ability of enzyme penetration inside the granule and this penetration is dependent on the presence of pores as well as the lamellar organization of starch.

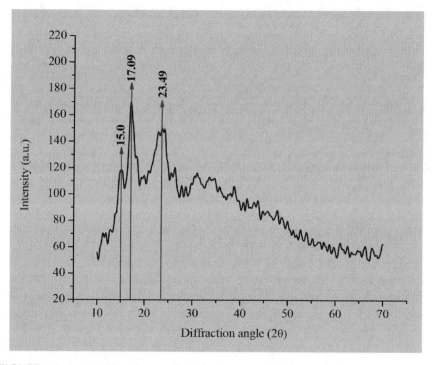

FIGURE 5.4   X-ray diffractogram of culinary banana starch.

5.3.5.2   FOURIER TRANSFORMS INFRARED (FTIR) SPECTRA

The FTIR spectrum of culinary banana starch is illustrated in Figure 5.5. The characteristics absorption bands appeared at 575, 785, 855, and 929 cm^{-1} may be attributed to anhydroglucose ring stretching vibrations. The carbohydrate nature of the starch sample was confirmed by the spectra observed near the wave numbers 1156, 1369, 1417, 1646, 2925, and 3429 cm^{-1} (Sanchez-Rivera et al., 2010). An extremely broadband appeared at 3429 cm^{-1} is attributed to hydrogen bonded hydroxyl groups. The sharp band observed 2925 cm^{-1} is characteristics of O—H and H—C—H bond stretching associated with the methine ring hydrogen atom (Gallant et al., 1997). The band at 1646 cm^{-1} is related to COO$^-$stretching vibration in a carbohydrate group (Zeng et al., 2011; Fan et al., 2012). Peaks observed at 1417 cm^{-1} and 1369 cm^{-1} were attributable to the bending modes of H—C—H and C—H symmetric bending of CH$_3$ (Zeng et al., 2011; van Soest, et al., 1994). The signal at 1156 cm^{-1} could be attributed to C—O bond stretching (Sanchez-Rivera et al., 2010). The additional characteristics absorption bands at 929, 855, 785, 707, and 575 cm^{-1} are due to the entire anhydroglucose ring stretching vibrations (Sanchez-Rivera et al., 2010). The peak observed at around 1002 cm^{-1} is associated with the crystalline and amorphous structure of starch (Zeng et al., 2011; Fan et al., 2012). The FTIR spectra observed in this study resembles the spectra obtained for plantain and banana starches studied by various authors (Perez-Sira, 1997; Sanchez-Rivera et al., 2010; Van Soest et al., 1995).

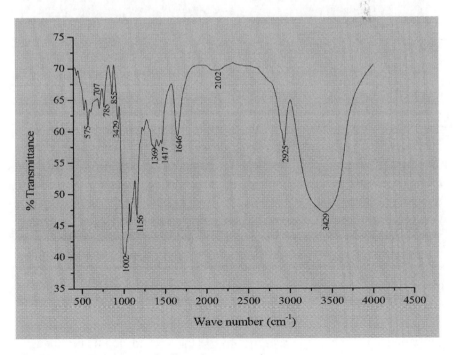

FIGURE 5.5   FTIR spectra of culinary banana starch.

5.3.5.3   MORPHOLOGICAL ANALYSIS BY SEM

The scanning electron micrographs (Fig. 5.6) of the starch revealed that the culinary banana starch granules appeared as a mixture of spherical and elliptical shaped with granule size ranged from 7.55 to 68.00 μm. Eggleston et al. (Eggleston et al., 1992) reported that the plantain starch had a broad range of granule size (7.8–61.3 μm) for plantain and is lower than our findings for culinary banana starch granules. The surface of the starch sample appeared to be smooth and as reported by Kayiasu et al. (1981). It could be indicated that the isolation process was efficient and it did not cause damage to starch granules.

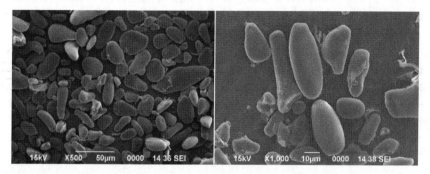

FIGURE 5.6   Scanning electron micrograph of culinary banana starch.

5.3.5.4   THERMAL STABILITY BY TGA

The thermal stability of culinary banana starch with increasing temperature at 10°C/min was evaluated by thermogravimetric (TG) and the corresponding derivative (DTG) curves (Fig. 5.7). From the TGA curve, it is revealed that thermal degradation of starch showed three distinct weight losses, the first weight loss initiated at 25.15–247.35°C corresponds to the initial vaporization of residual water. The second weight loss at around 265.39–331.34°C and is due to the pyrolysis of starch and the weight loss observed at the temperature range of 360–480°C may be due to decomposition of cellulose, hemicelluloses and lignin present in the starch sample (Vega et al., 1996). The region at where second degradation or pyrolysis of starch occurred a sharp and well-defined peak of DTG curve (Fig. 5.7) was observed which on further increase in temperature showed asymmetric appearance (shoulder formation). The plausible reason may be the second degradation stage is quite complex and probably that may be divided into sub-stages which were specifically defined by the formation of shoulder at the end of second decomposition process. The temperature and degradation rate increased with increase in heating rate which is a characteristic behavior for the thermally stimulated processes in the solid state (Jankovi, 2013).

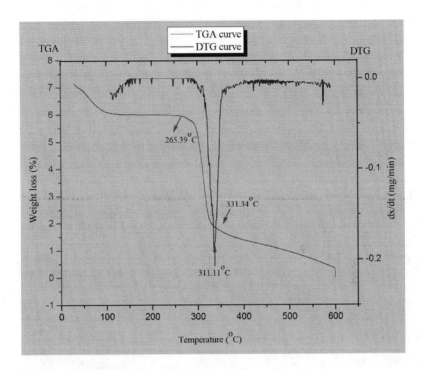

FIGURE 5.7 TGA and DTG curves of culinary banana starch.

5.3.6 DEVELOPMENT OF TYPE III RESISTANT STARCH (RS) FROM CULINARY BANANA STARCH

5.3.6.1 RS PRODUCTION BY HYDROTHERMAL METHOD

The production of RS type III by autoclaving and cooling is mainly due to the retrogradation of the amylose fraction (Fig. 5.8a). The four sets of RS obtained after autoclaving cooling cycle were coded as RS_{10-A} (10% starch, 4°C storage), RS_{10-B} (10% starch, −20°C storage), RS_{20-A} (20% starch 4°C storage), and RS_{20-B} (10% starch, −20°C storage). The generation of RS type III by hydrothermal treatment depends mainly on the number of heating and cooling cycles, starch type, storage time and temperature, etc. The amylose content and the amount of water are directly correlated to the yield of RS

(Haralampu, 2000). The effect of starch concentration and the number of autoclaving and cooling cycles on the yield of RS type III are presented in Table 5.3. This study showed that the yield of RS increased with the increase starch concentration. Similarly, the yield of RS content also increased with the number of cycles. However, the effect of starch concentration RS yield was considerably lower than the number of autoclaving and cooling cycles. After the third autoclaving and cooling cycle the RS content increased to 23.31%. This gradual rise in the yield of RS after repeated heating and cooling cycles might be associated with a decrease in the hydrolysis limit of pancreatic α-amylase and the increased RS type III formation. Haralampu (2000) determined RS values between 2.5 and 21.3% for diverse starch sources and with the Berry method (1986) the RS values ranged between 2.8 and 31% (Pomeranz & Sievert, 1990) which is in favorably supports our findings.

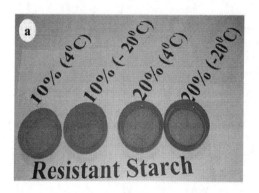 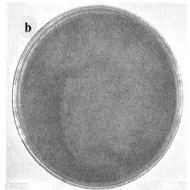

FIGURE 5.8 (See color insert.) Resistant starch (RS) obtained by (a) Hydrothermal process and (b) Enzyme debranching.

TABLE 5.3 Effect of Starch Concentration and Number of Autoclaving and Cooling Cycles on RS Content.

Starch concentration (% [w/v])	No. of cycles	RS content (%)
10	1	9.45 ± 0.75[a]
10	2	10.23 ± 0.06[b]
10	3	12.30 ± 0.09[d]
20	1	11.98 ± 0.14[c]
20	2	14.00 ± 0.01[e]
20	3	23.31 ± 0.67[f]

Results are mean of three replicates ± SD; mean followed by same superscript small letters within a column are not significantly different ($p > 0.05$).

5.3.6.2   EFFECT OF ENZYME CONCENTRATION ON STARCH DEBRANCHING

An increased degree of debranching enables the chains to align, aggregate, and hence form perfectly crystalline structures, thereby leading to the formation of more RS (Guraya et al., 2001). Two sets of RS obtained after enzyme debranching were coded as RS_{Ez-A} (4°C storage) and RS_{Ez-B} (−20°C storage). Berry (1986) reported that debranching of potato amylopectin with pullulanase before subjecting it to heating and cooling cycles substantially increased the RS type III content, and was attributed to an increase in the content of linear starch chains resulting from debranching. This pattern demonstrated that the substrate level needed higher enzyme concentration for starch debranching and maximum RS (31.17%) yield was obtained when debranched with 5% enzyme concentration.

5.3.6.3   EFFECT OF STORAGE TEMPERATURE ON RS CONTENT

The RS obtained from hydrothermal treatment after third cycle of cooling and autoclaving as well as RS obtained after enzyme debranching were subjected to two different storage temperatures (4°C and −20°C) and their effect on the yield of RS type III are presented in Table 5.4. The results revealed that low-temperature storage enhanced the formation of RS. The percentage of RS content gradually increased at lower level of storage condition. The RS obtained from 10% starch increased from 12.30% (stored at 4°C) to 15.43% (stored at −20°C). Similarly, the RS obtained from 20% starch increased to 26.42% when stored at −20°C. The RS obtained after enzyme debranching showed very little increase in the RS content from 30.21 to 31.17% with respect to storage condition. This supports the general behavior that RS yield increases during storage, especially during low-temperature storage.

TABLE 5.4   Effect of Storage Temperature on RS Content.

Resistant starch (RS)	Sample code	Storage (°C)	RS (%)
Autoclaving and cooling (10%)	RS_{10-A}	4	12.30 ± 0.05^a
Autoclaving and cooling (10%)	RS_{10-B}	−20	15.43 ± 0.07^b
Autoclaving and cooling (20%)	RS_{20-A}	4	23.79 ± 0.0^c
Autoclaving and cooling (20%)	RS_{20-B}	−20	26.42 ± 0.12^d
Enzyme debranched (24 h)	RS_{Ez-A}	4	30.20 ± 0.0^e
Enzyme debranched (24 h)	RS_{Ez-B}	−20	31.17 ± 0.01^f

Results are mean of three replicates ± SD; mean followed by same superscript small letters within a column are not significantly different ($p > 0.05$).

5.3.7 CHEMICAL ANALYSIS

The chemical analysis of RS obtained after undergoing various treatments was evaluated and presented in Table 5.5. There was a significant difference in the moisture content of RS obtained at various isolating conditions. The moisture content was in the range of 4.65–5.97% in all the samples which are in the considerable range for long-term storage of product without any microbial decomposition. The variation in the moisture content might be attributed to the linear chains produced during the different treatments which sometimes increased water-binding properties (Shin et al., 2003). The ash content varied from 0.14 to 0.22% in all the samples obtained and not much significant difference among the treatments given. The reason behind may be possibly the temperature used during autoclaving process is not high enough to digest the minerals present in the banana starch (Aparicio-Saguilan et al., 2005). The content of crude fiber in all the RS obtained showed significant difference among the samples. The highest crude fiber recorded was 6.15% in case of enzymatically debranched RS stored at −20ºC and the minimum was 4.23% in case of RS obtained by autoclaving and cooling method followed by storage at 4°C.

TABLE 5.5   Chemical Compositions (%) of Culinary Banana RS.

Sample code	Moisture content (wb)	Ash	Crude fiber	Fat	Protein
RS_{10-A}	5.63 ± 0.07^c	0.16 ± 0.95^c	4.23 ± 0.07^a	0.13 ± 0.21^a	0.26 ± 0.75^d
RS_{10-B}	4.65 ± 0.09^a	$0.18 \pm 0.09^{a,\,b}$	5.26 ± 0.56^b	0.11 ± 0.09^a	0.35 ± 0.08^c
RS_{20-A}	5.97 ± 1.03^f	0.15 ± 0.32^a	5.79 ± 0.23^c	0.15 ± 0.05^a	0.24 ± 0.09^d
RS_{20-B}	4.92 ± 1.21^b	0.14 ± 0.67^d	5.81 ± 0.04^d	1.21 ± 0.78^d	0.21 ± 0.04^d
RS_{Ez-A}	5.33 ± 0.99^d	0.22 ± 0.09^a	6.01 ± 0.01^e	0.78 ± 1.07^c	0.41 ± 0.12^b
RS_{Ez-B}	5.01 ± 1.06^c	0.21 ± 0.05^a	6.15 ± 0.75^f	0.64 ± 0.76^b	0.46 ± 0.08^a

Results are mean of three replicates $\pm$ SD; mean followed by same superscript small letters within a column are not significantly different ($p > 0.05$).

The fat content varied from 0.11 to 0.78% and did not have significant difference among the samples obtained by autoclaving and cooling cycle stored at 10°C, but samples stored at −20°C showed the significant difference. On the other hand, enzymatically debranched RS showed statistical difference among them. Protein varied from 0.21 to 0.46% with significant difference among samples. The protein content obtained after the autoclaving and cooling cycle was comparatively lesser than that of RS obtained from enzyme debranching. Since the heat treatment given during autoclave cycle denatures proteins and saponified lipids which may become solubilized (Aparicio-Saguilan et al., 2005). The results of chemical analysis obtained in this study corroborate with the findings of Aparicio-Saguilan et al. (**2005**) in case of RS obtained from linterized banana.

5.3.7.1   SCANNING ELECTRON MICROSCOPY

The modified RS (RS_{10-B}, $RS_{20-B,}$ and RS_{Ez-B}) were studied under electron microscope in order to evaluate the microstructure and examined how the treatment affected the microstructure (Figs. 5.9a–c). The RS obtained by hydrothermal treatment (autoclaving and cooling) of 10% starch sample and stored for 24 h (Fig. 5.9a) is not prominent in its structure as it did not undergo significant morphological changes compared to RS obtained from 20% starch (Fig. 5.9b). The structure of starch is hazy and there was no proper conversion of RS from starch samples. On the other hand, with

FIGURE 5.9   Scanning electron micrograph of culinary banana RS: (a) (RS_{10-B} = 10%, −20ºC); (b) (RS_{20-B} = 20%, −20ºC), and c) (RS_{Ez-B} = 24 h, −20°C).

increase in the concentration of starch sample from 10 to 20%, the structural difference was evident and the modification of starch to RS with
higher percentage of 20% was more proper and clearly visible. This could
be due to partial gelatinization of starch which occurs appropriately during
retrogradation process for RS with higher amount of starch (Ovando-
Martinez et al., 2013). The structural morphology of RS obtained after
enzyme debranching is illustrated in Figure 5.9c and the modification of
starch to RS was more distinguished and composed of round to irregular
shaped granules. The resistance of enzymatically modified starch changes
in the composition and structure of starch particle upon modification
(Leszczynski, 2004).

5.3.7.2   FTIR SPECTRA OF RS

The changes in molecular structure during the starch modification process
were studied using FTIR (Fig. 5.10). When FTIR spectra of RS was
compared to the spectra of culinary banana starch as illustrated in Figure

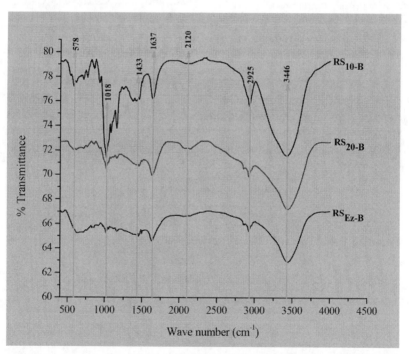

FIGURE 5.10   FTIR spectra of RS obtained after hydrothermal (RS$_{10-B}$ = 10%, −20°C and
RS$_{20-B}$ = 20%, −20°C) and enzyme debranched (RS$_{Ez-B}$ = 24 h, −20°C).

5.5 revealed that bands in case of RS_{10-B} were almost similar to the spectra of starch sample before retrogradation which suggests that the molecular structure of RS_{10-B} was not altered during retrogradation process. There were seven major spectra were observed in the region of 500–3500 cm^{-1} in all the RS studied located at 3446, 2925, 2120, 1637, 1433, 1018, and 578 cm^{-1}. But the bands were sharper in case of RS_{10-B} decreased in RS_{20-B} and RS_{Ez-B}. The sharp band observed at 1018 cm^{-1} may be related to the crystalline starch and water content. Stretching vibrations of the hydrogen-bonded hydroxyl groups at 3446 cm^{-1}, O—H and H—C—H bond at 2925 cm^{-1}, COO— at 1637 cm^{-1}, bending modes of H—C—H at 1433 cm^{-1} and anhydroglucose ring at 578 cm^{-1} bring the visible difference in spectral bands during retrogradation and modification of starch structurally (van Soest et al., 1994).

5.3.7.3   THERMOGRAVIMETRIC ANALYSIS

The TGA results are illustrated by a three-step weight loss curves of TG and derivative DTG (Fig. 5.11). The results were compared with TGA-DTG curves of culinary banana starch where RS showed endothermic peak at around 337.91°C (RS_{10-B}), 334.34°C (RS_{20-B}), and 457.08°C (RS_{Ez-B}) which were comparatively higher to that of starch sample which may be because of the treatment is given in order to modify the starch to RS. The different treatments (hydrothermal and enzyme debranching) given to the starch sample for modification to RS may lead to the formation of crystallites and amounts of double helices with different stabilities (Zhou et al., 2014). The endothermic transition of starch during modification is generally influenced by the interactions between amylose-amylose, amylose-amylopectin, and amylose-lipid content (Shin et al., 2003). From the DTG curve (Fig. 5.11a–c), it is evident that the sharp interval of weight loss indicates the presence of large amounts of compounds, such as homopolysaccharides (Di-Medeiros et al., 2014). The RS_{EZ-B} showed faster rate of decomposition at higher temperature with higher initial degradation temperature 457.08°C with lower weight loss compared to RS_{10-B} and RS_{20-B} which may be because of crystal formation during starch modification in RS_{EZ-B} was different (Zhou et al, 2014). Our findings are in line with the reports of Zhou et al. (2014) in case of RS obtained from rice starch.

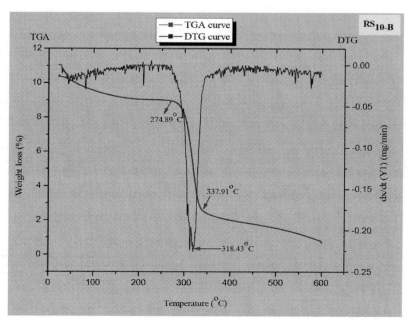

FIGURE 5.11a   TGA and DTG curves of RS obtained after hydrothermal (RS_{10-B} = 10%, −20ºC).

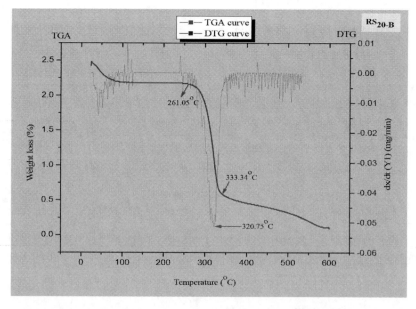

FIGURE 5.11b   TGA and DTG curves of RS obtained after hydrothermal (RS_{20-B} = 20%, −20ºC).

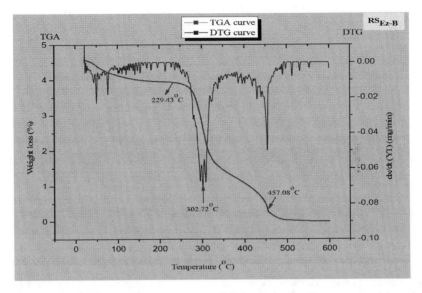

FIGURE 5.11c TGA and DTG curves of RS obtained after enzyme debranched $(RS_{Ez-B} = 24 \text{ h}, -20°C)$.

5.4   CONCLUSION

The starch isolated from culinary banana was further modified to RS type III by hydrothermal and enzyme debranching treatment and revealed marked changes in physicochemical, functional morphological, and thermal properties when compared both the starches. The yield of isolated starch was 16% with a purity of 96% and the amount of RS content was 18.88%. The isolated starch experienced a restricted swelling and solubility profile which was unstable during freezing and thawing cycles. The starch exhibited a high pasting temperature (81.80°C) indicating high resistant toward swelling with a peak viscosity of 4507 cP reflecting the ability of starch granules to swell freely. The XRD study clearly revealed culinary banana starch is a mixture of A and B type polymorphs and FTIR spectra evinced the functional groups present are typical bands of C-type starch with a mixture of spherical and elliptical granules. TGA behavior showed pyrolysis of starch occurred between the temperature range of 265.39 and 331.34°C. Further the starch was modified to RS due to retrogradation of the amylose fraction. Temperature cycling and incubation at certain temperature and storage time enhanced the formation of RS. The modified RS were analyzed using scanning electron microscopy (SEM), FTIR, DSC, and TGA to determine the structural changes. The significant morphological changes were observed with increase in starch concentration and in enzyme debranched RS elicited more distinguished modification. The FTIR structural changes were less in hydrothermal treatment compared to use of debranching enzyme. In addition, debranching enzyme treatment revealed higher rate of initial decomposition with lower weight loss due to crystal formation and modification indicated structural changes and is dependent on type of treatment given. Therefore, it is prudent to justify that culinary banana is an excellent source of RS and may be utilized as an alternative source of nutraceutical ingredient for preparing low glycemic functional foods. The generation of RS after hydrothermal treatment is mainly due to increase interactions between starch polymers. The degree of formation of RS in foods depends not only on the type of incorporated starch and the processing conditions but also influenced by the duration and conditions of storage (Goni, et al., 1996). As stated by Fuentes-Zaragoza et al. (2010), unripe banana is considered as RS-richest non-processed food. The content of RS in unripe banana ranges between 47 and 57%. Because of this fact, several studies have suggested that consumption of unripe banana results in beneficial effects to human health. Rodriguez-Ambriz et al. (2008) studied on unripe banana flour where they found total starch content of 73.4%, RS content 17.5% and concluded that unripe banana is an alternative source of indigestible carbohydrates, mainly RS and dietary fiber. Therefore, when the unripe fruit is cooked, its native RS is rendered digestible.

The global trends in rising levels of obesity, diabetes, and cardiovascular disease has renewed research interest in the dietary intake of fat, protein, and carbohydrate to maintain good health. The World Health Organization (WHO) and Food and Agricultural Organization (FAO) of the United Nations stated that globally, overweight populations are a bigger problem than undernourishment and recommended people in industrialized countries base diet on low GI foods to prevent most common disease of affluence (WHO, 1998). One of the most important objectives in the dietary treatment of diabetes patients is to maintain their blood glucose level, avoid obesity, and achieve optimal lipids level. Foods containing RS generally give a low glycemic response because RS is not digested in the small intestine and instead RS passes into the large intestine where it is fermented (Sajilata et al., 2006).

Many tropical countries have plant species which can be used as a good source of starch; unfortunately, some of them have not been exploited. One such plant species is *kachkal* (*Musa* ABB), the only culinary banana found in the entire Assam and North-East India (Goswami & Borthakur,

1996). Starch is the principal component of culinary bananas can be considered as a resource for production of modern forms of consumption like processed snacks and precooked products. Therefore, in this study, an effort has been made to isolate starch from culinary banana and modifying the isolated starch into RS in order to find an alternative route to increase value addition of culinary banana by providing RS of high-quality characteristics necessary for a well-balanced diet. The utilization of RS from culinary banana should not only expand the market for it but also provide a solution to those who want to consume food with low glycemic index (GI) value.

Effect of Partial Replacement of Wheat Flour with Type III Resistant Starch and Flour of Culinary Banana on the Chemical Composition, Textural Properties, and Sensory Quality of Brown Bread

5.5 INTRODUCTION

In today's world, food is not just consumed to satisfy hunger and provide necessary nutrients but also to prevent nutrition-related diseases and improve physical and mental well-being. The concept of food has experienced radical transformation toward nutritional, sensory, health maintenance, psychophysical well-being, and prevention of diseases (Menrad, 2003; Roberfroid, 2002). Increasing evidence shows that many of the chronic health conditions could be prevented or restrained by dietary changes. The global trends in rising levels of obesity, diabetes, and cardiovascular disease has refueled consumer and research interest in the dietary intake of fat, protein, and carbohydrate to maintain good health. Taking into consideration the obese and diabetics, the choice of food in a diet should be taken into consideration for not only its chemical composition but also for its ability to influence postprandial glycemia (Riccardi et al., 2008). Postprandial glycemia is a normal physiological response which refers to the elevation of blood glucose concentrations after consumption of food and varies in magnitude and duration.

The treatment of people with diabetes includes the glycemic index (GI) as a helpful additional indicator regarding the appropriate carbohydrate-containing foods for inclusion in the diet (Jenkins et al., 2012). In addition, choosing food according to its GI could help in metabolic activities. Foods rich in fiber are generally considered as low GI foods and its benefit includes lower postprandial glucose and insulin response, reduce insulin resistance and improve lipid profile (Riccardi et al., 2008). Foods containing resistant starch (RS) gives low GI response because RS is a type of a starch that is not digested in the small intestine instead to the large intestine where it is fermented, thus it resists digestion (Englyst et al., 2007). This is the reason, why there is no spike in the blood glucose or insulin level when we consume food enriched with RS. The RS intake of 6–12 g in a meal is being observed to beneficial effects on postprandial glucose and insulin level and in addition, it has been reported that consumption of 20 g of RS in a day would promote benefits in digestive health including fecal bulking (Murphy et al., 2008).

The variety of culinary is among one of the few sources of RS. The excellent amount of starch present in it may be modified to RS for its better utilization and value addition. Bread covers large part of daily human nutrition and also plays an important role in the diet of an ill person aiding their treatment (Dewettinck et al., 2008). Ready to eat (RTE) foods also known as convenience food has made people's life easier in today's fast-moving world and certainly baked products including bread fall under this category. Keeping in view the immense importance of this crop this study was attempted to develop a brown bread incorporated with RS modified from culinary banana. Hence, the developed brown bread favors low GI health beneficial brown bread enriched with RS. This study will not only help in value addition but also help in identifying the hidden potential of underutilized culinary banana in order to extend the market for this fruit.

5.6   MATERIALS AND METHODS

5.6.1   RAW MATERIALS

Unripe culinary bananas were harvested from Assam, India.

5.6.2   STARCH ISOLATION AND MODIFICATION INTO RS

The starch was isolated following the method described by Bello-Perez et al. (Bello-Perez et al., 2000). Therefore, for development of RS enriched brown bread, we have modified RS using hydrothermal treatment to economize the production of RS.

5.6.3   PREPARATION OF BROWN BREAD INCORPORATED WITH RS

The materials required for preparing RS brown bread (RSB) were, commercially available whole wheat flour (Brand ITC Aashirvaad; having composition of moisture content 9.23%, protein 4%, sugar 2%, and dietary fiber 5%), culinary banana flour (KF) (having moisture content 7.34%, starch 22.66%; protein 3.99%; and fiber 1.66%), modified type III RS from culinary banana starch (moisture content 4.92%, fiber 5.81%, and protein 1.21%), 6% moist yeast, 3% sugar, 2% salt, 3% refined oil, and 4% egg white powder and Lukewarm water. The combinations of ingredients (wheat flour, RS, and KF) are presented in Table 5.6. The selection of ingredients and quantity was determined after conducting preliminary experiments, and optimized bread making procedure as described by Tsatsaragkou et al. (2012; 2014). The amount of water used for mixing of dough was kept constant (150 mL/100 g ingredient) for all combinations and altogether 16 sets of brown bread were prepared with one control (control brown bread (BB-C)).

TABLE 5.6   Brown Bread Ingredients and Combination (g/100 g).

Sl. No	Sample code	Wheat flour	Resistant starch (RS)	Culinary banana flour (KF)
1	BB-C	100	0	0
2	RSB-5	95	5	0
3	RSB-10	90	10	0
4	RSB-15	85	15	0
5	KFB-10	90	0	10
6	KFB-20	80	0	20
7	KFB-30	70	0	30
8	RSKFB-5-10	85	5	10
9	RSKFB-10-10	80	10	10
10	RSKFB-15-10	75	15	10
11	RSKFB-5-20	75	5	20
12	RSKFB-10-20	70	10	20
13	RSKFB-15-20	65	15	20
14	RSKFB-5-30	65	5	30
15	RSKFB-10-30	60	10	30
16	RSKFB-15-30	55	15	30

5.6.4   PREPARATION OF BROWN BREAD

The procedure followed for brown bread preparation includes mixing of all dry ingredients properly followed by addition of previously active yeast. The dough was prepared using 150 mL of Lukewarm water and mixed properly for 5 min to obtain homogenous mixer using hand beater (Philips, HR 1459 300-Watt, India). The dough was kept for proofing at ambient temperature (28 ± 2°C) for 50 min and the bread was baked in a preheated oven at 180°C for 25 min (CS Aerotherm rotary rack oven-B1300, India). The bread loaves were cooled to 25°C and sealed in polyethylene zip lock bags for 12 h before determination of chemical, physical, and sensory properties. Photographs of the steps involved in the development of brown bread are illustrated in Figure 5.12.

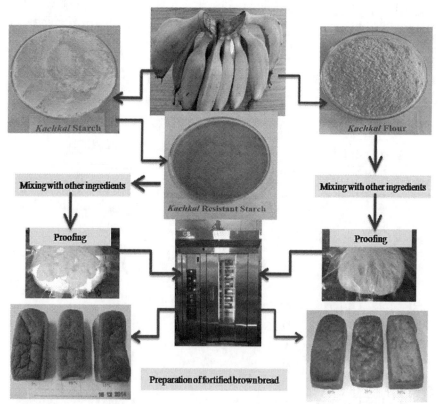

FIGURE 5.12   Preparation of brown bread fortified with resistant starch (RS) and flour (KF) from culinary banana.

5.6.5   CHEMICAL COMPOSITION

The proximate compositions of brown bread were determined according to AOAC (2010) methods. The amylose content and RS content were determined by McGrance et al. (McGrance et al., 1998) and an enzymatic method described in AOAC (2010), respectively.

5.6.6   PHYSICAL PROPERTIES

5.6.6.1   PROOFING OF DOUGH

The dough proofing was measured for each combination following the method of Cordoba (2010). About 10 mL of dough was taken in 100 mL measuring cylinder and kept for 60 min at 35°C and increase in the height of dough was checked every 10 min.

5.6.6.2   WATER RETENTION, WEIGHT, AND VOLUME OF BAKED BREAD LOAVES

After baking, bread samples were cooled to 25°C and yield of baked bread was calculated as the ratio of weight of sample before and after baking using the following formula:

$$\% \text{ water retention} = \frac{W_a}{W_b} \times 100 ,$$

where W_a and W_b are the weights of bread loaf after and before baking, respectively. The bread samples cooled to 25°C was weighed using electrical weighing balance having 0.1 g accuracy (Sumo Digi Tech, India). The specific volume (cm^3/g) of bread samples was evaluated following rapeseed displacement method of AACC (2000).

$$\text{Specific volume} = \frac{\text{loaf volume}}{\text{loaf weight}} .$$

5.6.6.3   TEXTURE PROFILE ANALYSIS OF BREAD CRUMB

Bread samples were analyzed for their extensibility and firmness using texture analyzer (TA-HD plus, Stable Micro System, UK) equipped with

50 N load cell. The sample size of thickness 15 mm was taken from the center of the loaf and compressed at 40% of its initial height with a 25 mm diameter cylindrical probe at a pre-test speed of 1 mm/s, test speed of 1.7 mm/s and post-test speed of 10 mm/s. The data acquisition rate was collected at 200 pps (points per second) with relaxation time of 5 s, force 10 g expressed the resistance of crumb to the penetrating probe and represented crumb firmness. The relative elasticity of crumb was evaluated by taking a crumb cube of $2 \times 2 \times 2$ cm^3 (length $\times$ width $\times$ height) from the center of the loaf. A uniaxial compression test with subsequent relaxation phase that lasted for 4 min was applied at 25% compression within the viscoelastic region. The relative elasticity (R_{EL}) of crumb (i.e., the force with which the crumb resists the defined mechanical stress during compression) was derived from the recorded force-time diagram and calculated using the following equation (Tsatsaragkou et al., 2015):

$$R_{EL}(\%) = \left(\frac{F_{res}}{F_{max}} \right) \times 100 ,$$

where F_{max} is the maximum force at 25% compression of the crumb and F_{res} is the residual force after 240 s relaxation phase (N). The textural properties evaluated were hardness, adhesiveness, cohesiveness, springiness, and chewiness.

5.6.6.4   CRUMB COLOR

The crumb color of bread samples was evaluated using tristimulus color parameters L^* (Lightness), a^* (redness to greenness), and b^* (yellowness to blueness) in digital colorimeter (Ultrascan VIS, Hunterlab, USA).

5.6.6.5   SENSORY EVALUATION

The sensory evaluation of brown bread was done using 9-point hedonic scale. A judging panel of 25 judges comprising of both trained and untrained members were selected for the study. Judges have explained the definition of quality attributes selected for sensory evaluation, score sheet, and methods of scoring. They were advised to rinse their mouth with water between tasting the consecutive samples (Jaya & Das, 2003). The major sensory parameters namely, appearance and color were judged visually, texture was judged through sense of touch, taste, and mouthfeel was judged by the perceptible character and eating quality.

5.6.6.6   STATISTICAL ANALYSIS

Experiments were carried out in three replicates. The data analysis tool "Microsoft Excel" was used for statistical analysis. Data were subjected to ANOVA and Fisher's least significant difference (LSD) was used to analyze means.

5.7   RESULTS AND DISCUSSION

5.7.1   EFFECT OF RS AND KF ON CHEMICAL COMPOSITION OF BROWN BREAD

In our preliminary experiments, we studied the effect of hydrothermal and enzyme debranching treatments on modification of RS, where we found that the RS content in native culinary banana starch was 18.88%, which on modification its yield increased to 26.42% in hydrothermal treatment and 31.13% in enzymatic treatment.

The isolated starch from culinary banana had yield of 16% with 96% purity which was further modified to type III RS by hydrothermal treatment with three repeated autoclaving and cooling cycles. The yield of modified RS was 26.42% with 91.32% purity and its amylose content was 57.98%. The chemical compositions of 16 sets of bread prepared are presented in Table 5.7. Water plays very important role in bread making procedure namely, for dough formation, fluidity of dough, and also acts as a medium for food transport to the yeast through cell membrane. In addition, water helps in starch and sugar hydrolysis resulting in gelatinization of starch during baking and also activates enzyme which forms macromolecules in dough responsible for rheological properties of dough. Hence, the amount of water added is solely responsible for final quality of bread in terms of textural and physicochemical properties (Gil et al., 1997). The moisture content in all 16 sets of brown bread was ranged between 31.31 and 36.72% and there was not much difference among the samples. The moisture content recorded is in the general range of moisture present in bread. The moisture in the range of 35–45% in case of gluten-free bread from carob flour has been reported by Tsatsaragkou et al. (2012). The protein content was in the range of 8.01–11.96% and varied among the sample while fat (3.11–4.97%) did not vary significantly for all the brown bread samples. The use of 3% oil and 4% egg white powder as shortening and protein supplements not only improved the crumb appearance and flavor of brown bread but also affected the final quality and softness (Stauffer, 1999). According to the

TABLE 5.7 Chemical Composition of Fortified Brown Bread (g/100 g).

Sample code	Moisture content	Protein	Fat	Crude fiber	Carbohydrates	Amylose	Resistant starch
BB-C	35.68±2.35[be]	8.32 ± 0.9[ac]	3.51 ± 0.01[a]	6.32 ± 0.45[af]	15.73 ± 0.97[cdej]	3.82 ± 0.09[a]	0.43 ± 0.01[a]
RSB-5	33.46 ± 2.44[a]	8.01 ± 0.18[ab]	3.31 ± 0.01[a]	7.32 ± 0.48[def]	12.65 ± 0.76[c]	4.01 ± 0.06[ac]	4.8 ± 0.03[eg]
RSB-10	33.59 ± 3.56[a]	8.58 ± 0.78[ab]	3.22 ± 0.12[abc]	8.86 ± 0.61[k]	9.55 ± 0.88[b]	6.67 ± 0.08[bd]	9.65 ± 0.08[h]
RSB-15	34.52 ± 3.87[b]	8.72 ± 0.97[abc]	3.47 ± 0.18[a]	9.97 ± 0.36[l]	7.65 ± 0.73[a]	7.35 ± 0.14[bcef]	12.98 ± 0.07[lm]
KFB-10	35.67 ± 2.49[bd]	9.28 ± 0.86[d]	3.73 ± 0.06[a]	5.54 ± 0.44[abc]	17.87 ± 0.61[fghij]	8.32 ± 0.47[bcd]	1.23 ± 0.13[b]
KFB-20	36.72 ± 3.28[bf]	9.79 ± 0.07[e]	3.88 ± 0.05[a]	5.90 ± 0.67[a]	18.43 ± 0.98[ijkm]	9.28 ± 0.49[defh]	2.17 ± 0.01[c]
KFB-30	33.57 ± 3.46[a]	10.32 ± 0.85[efghi]	4.01 ± 0.08[abc]	6.57 ± 0.89[abchi]	20.55 ± 1.03[lm]	10.33 ± 0.31[ghij]	2.96 ± 0.04[df]
RSKFB-5-10	32.34 ± 2.01[ad]	9.33 ± 0.76[df]	3.38 ± 0.06[a]	6.43 ± 0.18[abg]	14.32 ± 0.87[cdghi]	8.67 ± 0.76[deg]	5.02 ± 0.07[e]
RSKFB-10-0	33.74 ± 2.48[a]	9.86 ± 0.62[egh]	3.42 ± 0.02[ad]	7.06 ± 0.39[abcdj]	13.11 ± 0.65[cef]	9.70 ± 0.08[fgi]	9.88 ± 0.09[hij]
RSKFB-15-0	31.37 ± 2.11[a]	10.27 ± 0.99[efg]	3.57 ± 0.05[a]	7.85 ± 0.77[fghi]	11.03 ± 0.77[cd]	10.91 ± 0.65[hik]	13.54 ± 0.11[mno]
RSKFB-5-20	32.89 ± 2.38[ae]	10.03 ± 0.95[ef]	3.61 ± 0.09[a]	6.39 ± 0.28[a]	15.97 ± 0.86[def]	11.57 ± 0.06[ij]	5.25 ± 0.02[ef]
RSKFB-10-0	33.78 ± 1.46[af]	10.86 ± 0.85[ik]	3.54 ± 0.16[a]	7.11 ± 0.37[de]	16.51 ± 0.83[defgk]	12.27 ± 0.75[jkm]	9.71 ± 0.06[hik]
RSKFB-15-0	34.87 ± 2.49[bc]	11.00 ± 0.98[ij]	3.76 ± 0.27[a]	7.97 ± 0.29[ghij]	12.92 ± 0.65[c]	12.82 ± 0.54[kl]	12.97 ± 0.07[l]
RSKFB-5-30	31.39 ± 2.16[abc]	11.32 ± 0.99[ijkm]	3.97 ± 0.08[ab]	6.12 ± 0.11[ade]	19.91 ± 0.54[ikl]	13.38 ± 0.51[lmo]	5.64 ± 0.09[fgij]
RSKFB-10-0	33.05 ± 2.13[a]	11.73 ± 0.89[jl]	4.19 ± 0.64[abcd]	7.43 ± 0.83[defg]	17.66 ± 0.97[fghij]	14.66 ± 0.73[n]	10.46 ± 0.18[jkmno]
RSKFB-15-0	32.48 ± 3.19[a]	11.96 ± 0.84[lm]	4.97 ± 0.19[e]	7.75 ± 0.76[fgh]	16.93 ± 0.39[efgh]	15.36 ± 0.15[no]	13.22 ± 0.11[lmn]

Values reported as mean ± SD of three replications; mean followed by same small letter superscripts within a column are not significantly different ($p > 0.05$).

report of Demiralp et al. (2000), during dough mixing lipid entrenched into the protein matrix interacts and resulting in the viscoelastic properties of gluten network which is important for dough expansion and retention of gas during proofing. The contents of crude fiber, carbohydrates, and amylose were recorded in the range of 5.54–9.79%, 7.65–20.55%, and 3.82–4.01%, respectively, and all these three parameters varied significantly among the samples. Crude fiber was observed higher in samples incorporated with RS which proves that RS is also considered as fiber in food (Birt et al., 2013). The carbohydrate content was less in RS incorporated brown bread and is because of the partial replacement of wheat flour with RS and decreased in amount of carbohydrate present in the bread. The RS content varied from 0.43% to 13.54% among the samples and the lowest amount was recorded in BB-C and differed significantly. The photographs of developed brown bread samples are presented in Figure 5.13.

5.7.2 PHYSICAL PROPERTIES OF BROWN BREAD

5.7.2.1 PROOFING OF DOUGH

The increase in volume of bread dough was measured at every 10 min interval and is illustrated in Figure 5.14. The increase in loaf volume with respect to time was observed in all samples. As the amount of water use for dough preparation was fixed (150 mL), the similar trend in rising in dough volume in all samples was seen. The lowest increase in bread loaf volume was measured in brown bread prepared by incorporating 30% KF while highest dough proofing was recorded in brown bread prepared with the combination of 10% RS and 10% KF. Proofing of bread also termed as dough maturing or ripening is mainly attributed to the action of yeast. At this, stage starch breaks down and fermentation occurs as yeast produces carbon dioxide gas that causes the gluten network to expand which leaves an open cellular structure with the gases trapped inside loaf. Additionally, yeast also produces alcohol which influences the colloidal nature of the flour protein and alters the interfacial tension within the dough (Giannou et al., 2003). Correctly proofed dough results in optimum rheological properties of final product with absolute volume and crumb characteristics. The carbon dioxide produced during fermentation helps in dough expansion and in addition, it dissolves in the aqueous phase of dough and carbonic acid is formed which lowers the pH of the dough (Beuchat, 1987). The results of this study reveal that all the dough samples absorbed water considerably and increased in fermentation capability.

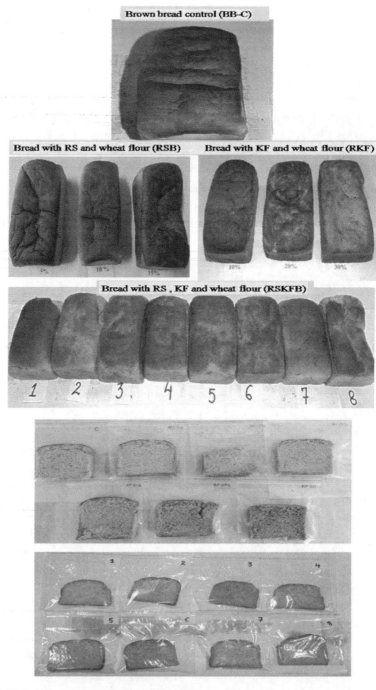

FIGURE 5.13 Fortified brown bread samples.

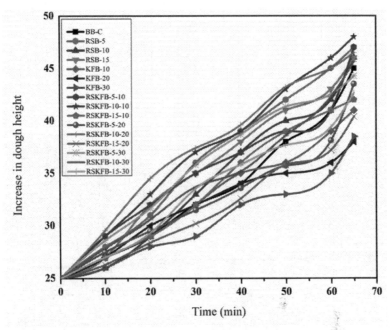

FIGURE 5.14 **(See color insert.)** Proofing of dough prior to baking.

5.7.2.2 WATER RETENTION, LOAF WEIGHT, AND VOLUME OF BAKED PRODUCT

The measure of water absorbing or holding capacity of bread is generally considered as water retention of baked products. In this study, the recorded water retention was in the range of 78–87% in all the 16 sets of bread samples (Fig. 5.15). Similarly, the total initial weight (W_b) of brown bread ingredients was ~500 g in all samples which after baking, weighed in the range of 393–436 g (W_a) and is illustrated in Table 5.8. It was observed that BB-C weighed heaviest with 436 g and RSKFB-15-30 (brown bread incorporated with 15% RS and 30% KF) had minimum weight of 393 g. The amount of water used for dough preparation was constant (150 mL) for all samples. The possible reason behind this may be there was a non-uniform absorption of water in RSKFB-15-30 bread and in addition, it may also be attributed to amount of water used that helped insufficient stretching of gluten present in wheat flour in presence of oil and yeast, thus yielding loaf with greater volume in BB-C. In case of RSB-5 RSB-10, RSKFB-5-10, and RSKFB-10-10 showed no significant difference with BB-C and had higher yield and loaf weight because of uniform absorption of water in dough. As reported by

Mandala et al. (2009), the water retention in baked product usually decreases with ingredient substitution as well as when higher amount of moisture is present in dough. In this study, the decrease in yield may be mainly because of the gluten dilution effect by substitution (Krishnan et al., 1987).

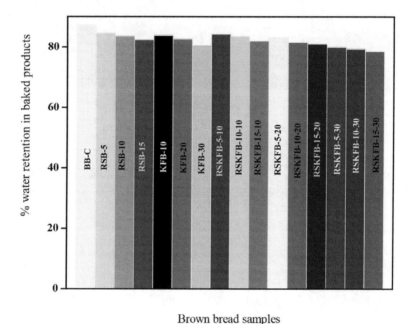

Brown bread samples

FIGURE 5.15   (See color insert.) Percentage water retention in baked products.

Loaf volume is one of the key quality parameters in baked products which provides the quantitative measurement of baking performance. The specific volume of bread samples is a characteristic quality parameter as it indicates dough inflating ability (Giannou & Tzia, 2007). The rheological and textural properties of bread dough are significantly affected by loaf volume and specific volume. The greater the bread volume, softer will be the bread with minimum hardness. The loaf volume was recorded in the range of 642.88–322.08 cm³ and specific volume was 1.47–0.81 cm³ g⁻¹ in all 16 sets of brown bread samples (Table 5.8). Loaf volume and specific volume are directly related to the weight of the bread and likewise, the bread samples with highest percentage of substitute ingredients (15% RS with 30% KF) showed lowest values of loaf volume and specific volume. This is also associated with gluten dilution effect as well as low protein network in the dough (Rossell et al., 2001). In this study, as gluten (wheat flour) is substituted with starch (RS and high starch KF) which might have created weak interaction between gluten and starch (Oates, 2001). On the other hand, when the gluten is substituted with high dietary fiber RS up to 10% with 10% KF, the final product was of fine quality with higher loaf weight, volume, and specific volume. This may be attributed to the balance interaction between gluten and non-gluten network took place which causes sufficient production of carbon dioxide gas during proofing, and thus the brown bread was of better quality.

TABLE 5.8   Volume of Baked Fortified Brown Bread.

Sample code	Loaf weight (g)	Loaf volume (cm^3)	Specific volume (cm^3 g^{-1})
BB-C	436 ± 2.46^p	642.88 ± 4.56^n	1.47 ± 0.95^{fg}
RSB-5	423 ± 2.95^{no}	634.23 ± 4.37^m	1.49 ± 0.91^g
RSB-10	418 ± 2.67^{kl}	612.50 ± 4.42^l	1.46 ± 0.94^f
RSB-15	412 ± 2.01^{gh}	603.26 ± 4.23^k	1.46 ± 0.96^f
KFB-10	419 ± 2.55^{lmo}	428.64 ± 3.22^h	1.02 ± 0.86^d
KFB-20	413 ± 2.23^{hikl}	392.76 ± 3.38^f	0.95 ± 0.05^{bc}
KFB-30	403 ± 1.75^d	374.91 ± 3.65^d	0.93 ± 0.07^b
RSKFB-5-10	417 ± 2.17^{jk}	602.33 ± 4.29^k	1.41 ± 0.87^{ef}
RSKFB-10-10	421 ± 2.33^n	597.42 ± 4.76^j	1.44 ± 0.96^{egf}
RSKFB-15-10	410 ± 2.01^{fgi}	588.57 ± 4.28^i	1.43 ± 0.63^e
RSKFB-5-20	416 ± 2.22^{jm}	417.39 ± 3.87^g	1.00 ± 0.08^d
RSKFB-10-20	408 ± 1.75^{fh}	381.75 ± 3.85^e	0.93 ± 0.05^b
RSKFB-15-20	405 ± 1.65^{deg}	374.28 ± 3.33^d	0.92 ± 0.07^{bc}
RSKFB-5-30	400 ± 1.55^{ce}	338.76 ± 3.12^c	0.84 ± 0.64^a
RSKFB-10-30	397 ± 1.34^b	331.57 ± 3.56^b	0.83 ± 0.54^a
RSKFB-15-30	393 ± 1.10^a	322.08 ± 3.71^a	0.81 ± 0.12^a

Values reported as mean $\pm$ SD of three replications; mean followed by same small letter superscripts within a column are not significantly different ($p > 0.05$).

5.7.2.3   TEXTURE PROFILE ANALYSIS

The bread structure is usually solid colloid, where gluten forms interconnected network with numerous pockets of carbon dioxide distributed uniformly throughout the loaf. The baking resulted in characteristic honeycomb texture of final product which can be analyzed using texture analysis. Texture analysis is one of the most important quality parameters primarily concern with measurement of mechanical, rheological, and sensory properties of developed brown bread (Krupa et al., 2010). In this study, the textural properties studied were hardness, adhesiveness, springiness, cohesiveness, and chewiness. The results presented in Table 5.9 revealed that the textural properties were affected by percentage of substituting of ingredients used. The textural quality was poor when wheat flour was substituted with 30% KF. Brown read incorporated with 10% RS and 10% KF had minimum hardness of 3.41 N while maximum hardness was observed in RSKFB-15–30 (16.42 N). The adhesiveness ranged from −0.03 to −0.43 g.s was not observed in all samples. Hence, the minimum negative values confirm that developed brown bread is typically not adhesive (http://texturetechnologies.com/texture-profile-analysis/texture-profile-analysis.php).

The springiness or elasticity property of developed brown bread ranged 0.81–1.08 and was minimum and maximum in sample incorporated with 30% KF and RSKFB-10-10, respectively. In order to have good quality bread, springiness should be higher, and substitution of wheat flour with RS and KF up to 10% favored good quality of bread. Lowest springiness value in KFB-30 confirms low elasticity and expansion of dough (Sankhon et al., 2013). Cohesiveness property of bread is negatively related to the hardness, which means lesser the hardness, higher will be the cohesiveness value and confirmed by our results. The minimum cohesiveness (0.36) was found in RSKFB-15–30 which recorded highest hardness value (16.42 N) and on contrary RSKFB-10-10 had maximum cohesiveness value of 0.59 with minimum hardness value of 3.14 N. Low cohesiveness value is indicating to loss of intermolecular attraction between bread ingredients with loss of moisture (Valcarcel-Yamani & Da Silva, 2013). The chewiness property was directly proportional to the hardness, RSKFB-10-10 exhibited lowest chewiness (192.17 N) while maximum chewiness value was observed in RSKFB-15-30 (731.84 N). Chewiness is associated with moisture absorption and uniform distribution and in developed brown bread substituted with 10% RS and KF confirmed uniform interaction of starch and gluten with uniform moisture distribution in dough.

Relative elasticity is the most important sensory and textural properties of bread which are mainly dependent on the quality and the proportion of different ingredients used. The softer breadcrumb with higher elasticity is the consumers' requirement. Sample RSBKF-10-10 showed highest relative elasticity and did not differ significantly with control (Fig. 5.16). Samples containing higher amount of KF (30%) significantly reduced the relative elasticity of brown bread and may attribute to non-uniform distribution of water in dough because of weak interaction between gluten and starch. Our results are in line with the findings of Tsatsaragkou et al. (2012) and Mandala et al. (2009).

TABLE 5.9 Texture Property of Brown Bread Fortified with RS and KF.

Sample code	Hardness (N)	Adhesiveness (g.s)	Springiness	Cohesiveness	Chewiness (N)
BB-C	6.11 ± 0.01[ce]	ND	1.00 ± 0.04[ef]	0.47 ± 0.07[cd]	302.45 ± 5.87[e]
RSB-5	3.96 ± 0.37[de]	ND	0.99 ± 0.01[e]	0.57 ± 0.04[gh]	310.69 ± 3.40[h]
RSB-10	6.96 ± 0.75[de]	−0.43 ± 0.02[h]	1.01 ± 0.06[fh]	0.57 ± 0.08[gh]	206.87 ± 2.58[b]
RSB-15	8.59 ± 0.77[g]	−0.12 ± 0.03[d]	0.89 ± 0.07[bcd]	0.50 ± 0.03[efij]	393.04 ± 3.56[g]
KFB-10	4.71 ± 0.02[b]	−0.17 ± 0.01[f]	1.01 ± 0.06[fh]	0.39 ± 0.01[ac]	311.95 ± 4.61[f]
KFB-20	5.70 ± 0.05[cd]	−0.03 ± 0.01[a]	0.95 ± 0.05[cdf]	0.44 ± 0.01[bce]	412.69 ± 4.88[h]
KFB-30	8.90 ± 0.71[ghj]	−0.05 ± 0.02[b]	0.81 ± 0.04[a]	0.39 ± 0.02[ac]	290.39 ± 3.45[d]
RSKFB-5-10	3.83 ± 0.01[a]	−0.08 ± 0.04[ce]	1.08 ± 0.08[gh]	0.59 ± 0.06[ij]	232.63 ± 2.54[c]
RSKFB-10-10	3.41 ± 0.03[a]	ND	1.00 ± 0.07[ef]	0.59 ± 0.15[ij]	192.17 ± 2.11[a]
RSKFB-15-10	9.72 ± 0.86[ik]	ND	0.99 ± 0.05[e]	0.48 ± 0.02[d]	480.58 ± 3.97[i]
RSKFB-5-20	6.74 ± 0.54[cd]	−0.12 ± 0.01[de]	0.93 ± 0.07[bc]	0.46 ± 0.01[cf]	297.27 ± 2.34[ef]
RSKFB-10-20	7.86 ± 0.91[fh]	−.40 ± 0.01[g]	1.07 ± 0.06[g]	0.58 ± 0.04[hi]	509.65 ± 3.19[j]
RSKFB-15-20	10.33 ± 0.97[ij]	ND	1.01 ± 0.09[fh]	0.43 ± 0.07[bd]	389.56 ± 2.16[g]
RSKFB-5-30	10.93 ± 0.94[jk]	ND	1.07 ± 0.11[g]	0.49 ± 0.03[eh]	592.69 ± 5.29[k]
RSKFB-10-30	13.75 ± 1.13[l]	ND	0.99 ± 0.07[e]	0.49 ± 0.01[de]	693.16 ± 5.46[l]
RSKFB-15-30	16.42 ± 1.46[m]	ND	0.92 ± 0.13[b]	0.36 ± 0.02[a]	731.84 ± 6.15[m]

Values reported as mean ± SD of five replications; mean followed by same small letter superscripts within a column are not significantly different ($p > 0.05$).

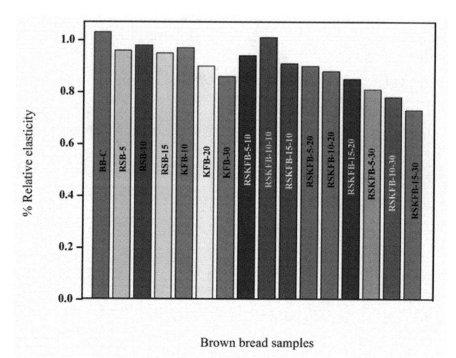

Brown bread samples

FIGURE 5.16   **(See color insert.)** Relative elasticity of brown bread samples.

5.7.2.4   COLOR ATTRIBUTES

One of the most important quality attributes which defines the acceptability of baked bread by consumer is the color. During baking of bread, development of color in crust occurs at the later stages of baking which is generally used to check the completion of baking process (Mundt & Wedzicha, 2007). Therefore, follow up of crust color during baking is necessary to check the defined color of bread as required by consumers (Ureta et al., 2014). Color development in bread is a function of moisture content, baking time, and baking temperature. Color of brown bread samples widely varied with respect to substituted ingredients (Table 5.5) and all the samples were compared to control, which showed L^*, a^*, and b^* values of 56.59, 2.45, and 12.27, respectively (Table 5.10). The L^* and b^* values of bread incorporated with RS (RSB-5, 10, and 15) decreased while a^* value increased with increasing percentage of RS from 5 to 15%. On the contrary, with increase in the percentage of KF L^* and b^* values increased significantly and a^* value decreased considerably. The color of the RS (brown) and the KF (white) also substantially affected the crumb color of the developed brown bread (Fig. 5.14). However, brown bread prepared by incorporating the combination of RS and KF in the balanced proportion did not affect the color very much. The overall color of all bread samples was found appealing from the consumer point of view which has been further confirmed by sensory analysis (Table 5.6).

TABLE 5.10   Color Attributes of Fortified Brown Bread.

Sample code	L^*	a^*	b^*
BB-C	56.59 ± 2.12^{kln}	2.45 ± 0.76^{jk}	12.27 ± 1.25^{k}
RSB-5	58.39 ± 2.44^{mn}	2.07 ± 0.79^{df}	11.72 ± 1.01^{j}
RSB-10	52.16 ± 2.31^{e}	2.14 ± 0.38^{e}	9.73 ± 0.97^{fg}
RSB-15	50.12 ± 2.15^{c}	2.16 ± 0.92^{eg}	8.99 ± 0.99^{de}
KFB-10	42.44 ± 1.76^{a}	1.69 ± 0.97^{c}	5.11 ± 0.87^{a}
KFB-20	47.93 ± 2.34^{b}	1.40 ± 0.99^{b}	7.37 ± 0.91^{c}
KFB-30	48.94 ± 1.76^{bd}	1.31 ± 0.95^{a}	8.15 ± 0.97^{d}
RSKFB-5-10	52.71 ± 2.48^{ef}	2.23 ± 0.34^{fg}	9.47 ± 0.94^{f}
RSKFB-10-10	53.15 ± 3.29^{fg}	2.19 ± 0.37^{ef}	9.22 ± 1.31^{efg}
RSKFB-15-10	50.87 ± 2.88^{cdfg}	1.70 ± 0.99^{c}	6.49 ± 0.47^{b}
RSKFB-5-20	57.29 ± 3.56^{lm}	2.39 ± 0.87^{ik}	10.66 ± 1.28^{i}
RSKFB-10-20	57.36 ± 2.51^{m}	2.33 ± 0.75^{hj}	10.17 ± 1.97^{h}
RSKFB-15-20	54.29 ± 1.87^{h}	2.44 ± 0.96^{j}	8.61 ± 0.94^{df}
RSKFB-5-30	57.96 ± 3.76^{m}	2.65 ± 0.97^{m}	10.86 ± 0.99^{i}
RSKFB-10-30	55.08 ± 2.15^{ilm}	2.47 ± 0.29^{kl}	9.31 ± 0.95^{f}
RSKFB-15-30	56.37 ± 2.79^{k}	2.41 ± 0.64^{ijl}	8.28 ± 0.65^{de}

Values reported as mean $\pm$ SD of five replications; mean followed by same small letter superscripts within a column are not significantly different ($p > 0.05$).

5.7.2.5   SENSORY ANALYSIS

In preparation of bread, the sensory properties are altered by role of baking as it directly relates to palatability, taste, aroma, and texture. The results of sensory analysis of the developed brown bread samples are presented in Table 5.11 and the sensory attributes differed significantly among samples. The sensory attributes of RSB-5, RSB-10, RSKFB-5-10, and RSKFB-10-10 were highest in all sensory properties with good overall acceptability. The RSKFB-10-10 scored the highest overall acceptability of 8 points which

TABLE 5.11 Sensory Attributes of Fortified Brown Bread.

Sample code	Appearance	Color	Texture	Taste	Mouthfeel	Overall acceptability
BB-C	8.75 ± 0.87[jk]	8.88 ± 0.05[m]	8.75 ± 0.14[m]	8.0 ± 0.07[gh]	8.55 ± 0.85[j]	8.0 ± 0.57[h]
RSB-5	8.35 ± 0.75[g]	8.35 ± 0.07[k]	7.55 ± 0.07[i]	8.17 ± 0.05[gh]	8.0 ± 0.71[h]	7.85 ± 0.21[gh]
RSB-10	8.55 ± 0.86[hi]	8.55 ± 0.08[l]	7.75 ± 0.31[k]	8.55 ± 0.16[i]	8.25 ± 0.44[i]	7.65 ± 0.45[g]
RSB-15	8.0 ± 0.76[f]	7.85 ± 0.07[h]	6.75 ± 0.03[e]	8.12 ± 0.07[g]	7.89 ± 0.58[hi]	7.0 ± 0.33[f]
KFB-10	8.0 ± 0.75[f]	7.15 ± 0.05[e]	7.44 ± 0.07[h]	7.55 ± 0.04[f]	8.15 ± 0.18[hi]	7.50 ± 0.36[gh]
KFB-20	6.65 ± 0.71[d]	6.57 ± 0.03[d]	7.0 ± 0.16[f]	5.32 ± 0.01[d]	4.25 ± 0.36[d]	5.0 ± 0.05[d]
KFB-30	5.34 ± 0.90[b]	4.50 ± 0.06[a]	5.15 ± 0.11[c]	4.58 ± 0.03[c]	3.55 ± 0.06[b]	3.55 ± 0.07[c]
RSKFB-5-10	8.56 ± 0.99[hi]	8.0 ± 0.76[i]	7.65 ± 0.85[j]	8.48 ± 0.19[i]	8.25 ± 0.59[i]	7.85 ± 0.27[gh]
RSKFB-10-10	8.85 ± 0.79[jk]	8.15 ± 0.17[j]	7.85 ± 0.56[l]	8.55 ± 0.42[i]	8.31 ± 0.17[i]	8.0 ± 0.95[h]
RSKFB-15-10	8.0 ± 0.54[f]	7.50 ± 0.11[g]	7.0 ± 0.09[f]	8.18 ± 0.65[h]	7.45 ± 0.08[g]	6.55 ± 0.38[e]
RSKFB-5-20	7.67 ± 0.78[e]	7.45 ± 0.07[g]	7.15 ± 0.18[g]	6.45 ± 0.19[e]	6.57 ± 0.16[ef]	5.0 ± 0.02[d]
RSKFB-10-20	7.66 ± 0.43[e]	7.50 ± 0.52[g]	7.26 ± 0.05[h]	6.56 ± 0.18[e]	6.85 ± 0.04[ef]	5.21 ± 0.07[d]
RSKFB-15-20	7.75 ± 0.56[e]	7.25 ± 0.13[f]	6.35 ± 0.01[d]	6.55 ± 0.25[e]	6.70 ± 0.08[e]	5.0 ± 0.05[d]
RSKFB-5-30	5.61 ± 0.58[c]	5.90 ± 0.02[c]	4.15 ± 0.07[a]	3.35 ± 0.02[a]	3.00 ± 0.01[a]	2.55 ± 0.01[b]
RSKFB-10-30	5.35 ± 0.55[b]	5.58 ± 0.04[b]	4.56 ± 0.05[b]	4.15 ± 0.06[b]	4.25 ± 0.02[d]	2.00 ± 0.01[a]
RSKFB-15-30	5.1 ± 0.75[a]	5.55 ± 0.01[b]	4.17 ± 0.02[a]	3.25 ± 0.01[a]	3.85 ± 0.05[c]	2.25 ± 0.04[a]

Values reported as Mean ± SD of 25 replications (judges); mean followed by same small letter superscripts within a column are not significantly different ($p > 0.05$).

was same as that of control sample. The increased percentage of KF for substituting wheat flour substantially decreased the scores of sensory attributes and overall acceptability was low. Results revealed that sensory attributes were not affected by incorporation of RS and KF up to 10%. One of the reasons of non-accepting of higher percentage incorporation of KF may be the presence of specific smell of the flour which is attributed by high amount of mineral salts present in it. Hence, from the sensory point of view, substitution of wheat flour up to 10% RS and 10% KF can be used successfully for making brown bread without affecting the consumers' acceptability.

5.8 CONCLUSION

Results of this study revealed that replacement of wheat flour with 10% RS and 10% KF is the best combination from the standpoint of various quality parameters for making brown bread. Addition of RS and KF up to 10% significantly improved the quality characteristics of brown bread with balanced nutrient composition, soft texture, better chewiness, high water retention, and loaf volume and higher consumer acceptance in terms of appearance, color, texture, and taste. Though samples namely, RSB-5, RSB-10, KFB-10, RSKFB-5-10, and RSKF-10-10 showed uniform distribution and absorption of water in bread loaf with balance interaction between gluten and non-gluten network, however, RSKFB-10-10 was the best from the consumers' acceptability. Hence, the developed brown bread favors low GI health beneficial brown bread enriched with RS. This study will not only help in value addition but also help in identifying the hidden potential of underutilized culinary banana in order to extend the market for this fruit.

KEYWORDS

- **resistant starch**
- **culinary banana**
- **FTIR**
- **X-ray diffraction**
- **brown bread**

REFERENCES

AACC. *Approved Methods of the AACC*, 10th ed.; AACC: St. Paul, MN, 2000.

AOAC. *Official Methods of Analysis of the Association of Official Analytical Chemists*, 18th ed.; AOAC International: Washington, DC, 2010.

Saguilan, A. A.; Huicochea, E. F.; Tovar, J.; Suárez, F. G.; Meraz, F. G.; Pérez, L. A. B. Resistant Starch-rich Powders Prepared by Autoclaving of Native and Lintnerized Banana Starch: Partial Characterization. *Starch Starke* **2005,** *57* (9), 405–412.

Baker, L. A.; Rayas-Duarte, P. Freeze-thaw Stability of Amaranth Starch and the Effects of Salt and Sugars. *Cereal Chem.* **1998,** *75* (3), 301–307.

Biliaderis, C. G. The Structure and Interactions of Starch with Food Constituents. *Can. J. Physiol. Pharmacol.* **1991,** *69* (1), 60–78.

Bello-Perez, L. A., et al. Isolation and Partial Characterization of Banana Starches. *J. Agric. Food Chem.* **1999,** *47* (3), 854–857.

Bello-Perez, L. A., et al. Some Structural, Physicochemical and Functional Studies of Banana Starches Isolated from Two Varieties Growing in Guerrero, Mexico. *Starch Starke* **2000,** *52* (2–3), 68–73.

Bello-Perez, L. A., et al. Functional Properties of Corn, Banana and Potato Starch Blends. *Acta Cient. Venez.* **2001,** *52* (1), 62–67.

Berry, C. S. Resistant Starch: Formation and Measurement of Starch that Survives Exhaustive Digestion with Amylolytic Enzymes during the Determination of Dietary Fiber. *J. Cereal Sci.* **1986,** *4* (4), 301–314.

Bertolini, C. A., et al. Some Rheological Properties of Sodium Caseinate-starch Gels. *J. Agric. Food Chem.* **2005,** *53* (6), 2248–2254.

Betancur, A. D. A., et al. Physicochemical and Functional Characterization of Baby Lima Bean (*Phaseolus lunatus*) Starch. *Starch Starke* **2001,** *53* (5), 219–226.

Beuchat, L. R. *Food and Beverage Mycology*; Van Nostrand Reinhold: New York, 1987; pp 255–256.

Birt, D. F., et al. Resistant Starch: Promise for Improving Human Health. *Adv. Nutr.* **2013,** *4,* 587–601.

Blazek, J.; Gilberta, E. P. Application of Small-angle X-ray and Neutron Scattering Techniques to the Characterization of Starch Structure: A Review. *Carbohydr. Polym.* **2011,** *85* (2), 281–293.

Cai, J., et al. Physicochemical Properties of Ginkgo Kernal Starch. *Int. J. Food Prop.* **2015,** *18* (2), 380–391.

Chang, S. M., et al. X-ray Diffraction Patterns of Some Taiwan Native Starches. *Bull. Inst. Chem. Acad. Sin.* **1991,** *38,* 91–98.

Cordoba, A. Quantitative Fit of a Model for Proving of Bread Dough and Determination of Dough Properties. *J. Food Eng.* **2010,** *96* (3), 440–448.

Demiralp, H., et al. Effects of Oxidizing Agents and Defatting on the Electrophoretic Patterns of Flour Proteins during Dough Mixing. *Eur. Food Res. Technol.* **2000,** *211* (5), 322–325.

Dewettinck, K., et al. Nutritional Value of Bread: Influence of Processing, Food Interaction and Consumer Perception. *J. Cereal Sci.* **2008,** *48* (2) 243–257.

Di-Medeiros, M. C., et al. Rheological and Biochemical Properties of *Solanum lycocarpum* Starch. *Carbohydr. Polym.* **2014,** *104,* 66–72.

Eerlingen, R. C.; Delcour, J. A. Formation, Analysis, Structure and Properties of Type III Enzyme Resistant Starch. *J. Cereal Sci.* **1995,** *22* (2), 129–138.

Eggleston, G., et al. Physicochemical Studies on Starches Isolated from Plantain Cultivars, Plantain Hybrids and Cooking Bananas. *Starch Starke* **1992,** *44* (4), 121–128.

Eliasson, A. C.; Gudmundsson, M. Starch: Physicochemical and Functional Aspects, In *Carbohydrates in Food;* Eliasson, A. C., Eds.; Marcel Dekker: New York, 1996; pp 431–503.

Englyst, H. N.; Cummings, J. H. Digestion of the Polysaccharides of Potato in the Small Intestine of Man. *Am. J. Clin. Nutr*. **1987,** *45* (2), 423–431.

Englyst, H. N., et al. Classification and Measurement of Nutritionally Important Starch Fractions. *Eur. J. Clin. Nutr.* **1992,** *46* (2), 33–50.

Englyst, K. N., et al. Nutritional Characterization and Measurement of Dietary Carbohydrates. *Eur. J. Clin. Nutr.* **2007,** *61* (1), S19–S39.

French, D. Organization of Starch Granules. In *Starch: Chemistry and Technology*, 2nd ed.; Whistler, R. L., BeMiller, J. N., Paschall, E. F., Eds.; Academic Press: New York, London, 1984; pp 183–247.

Fuentes-Zaragoza, E., et al. Resistant Starch as Functional Ingredient: A Review. *Food Res. Int.* **2010,** *43* (4), 931–942.

Fan, D., et al. Determination of Structural Changes in Microwaved Rice Starch Using Fourier Transform Infrared and Raman Spectroscopy. *Starch Starke* **2012,** *64* (8), 598–606.

Gallant, D. J., et al. Microscopy of Starch: Evidence of a New Level of Granule Organization. *Carbohydr. Polym.* **1997,** *32* (3–4), 177–191.

Giannou, V., et al. Quality and Safety Characteristics of Bread Made from Frozen Dough. *Trends Food Sci. Technol.* **2003,** *14* (3), 99–108.

Giannou, V.; Tzia, C. Frozen Dough Bread: Quality and Textural Behavior during Prolonged Storage-Prediction of Final Product Characteristics. *J. Food Eng.* **2007,** *79* (3), 929–934.

Gil, M. J., et al. Effect of Water Content and Storage Time on White Pan Bread Quality: Instrument Evaluation. *LWT Food Sci. Technol.* **1997,** *205,* 268–273.

Goheen, S. M.; Wool, R. P. Degradation of Polyethylene-starch Blends in Soil. *J. Appl. Polym. Sci.* **1991,** *42* (10), 2691–2701.

Goni, I., et al. Analysis of Resistant Starch: A Method for Foods and Food Products. *Food Chem.* **1996,** *56* (4), 445–449.

Goswami, B.; Borthakur, A. Chemical and Biochemical Aspects of Developing Culinary Banana (*Musa* ABB) "*Kachkal.*" *Food Chem.* **1996,** *55* (2), 169–172.

Guraya, H. S., et al. Effect of Enzyme Concentration and Storage Temperature on the Formation of Slowly Digestible Starch from Cooked Debranched Rice Starch. *Starch Starke* **2001,** *53* (3–4), 131–139.

Hallgren, L. Physical and Structural Properties of Cereals, Sorghum in Particular, In *Relation to Milling Methods and Product Use;* Carlsberg Research Laboratory, Technical University of Denmark: Copenhagen, Denmark, 1985; Vol. 1.

Haralampu, S. G. Resistant Starch: A Review of the Physical Properties and Biological Impact of RS$_3$. *Carbohydr. Polym.* **2000,** *41* (3), 285–292.

Hoover, R.; Ratnayake, W. S. Starch Characteristics of Black Bean, Chick Pea, Lentil, Navy Bean and Pinto Bean Cultivars Grown in Canada. *Food Chem.* **2002,** *78,* 489–498.

http://texturetechnologies.com/texture-profile-analysis/texture-profile-analysis.php. An Overview of Texture Profile Analysis (TPA). (accessed Oct 1, 2015).

Jacobson, M. R., et al. Retrogradation of Starches from Different Botanical Sources. *Cereal Chem.* **1997,** *74* (5), 511–518.

Jankovi, B. Thermal Characterization and Detailed Kinetic Analysis of Cassava Starch Thermo-oxidative Degradation. *Carbohydr. Polym.* **2013,** *95* (2), 621–629.

Jaya, S.; Das, H. Sensory Evaluation of Mango Drinks Using Fuzzy Logic. *J. Sens. Stud.* **2003**, *18* (2), 163–176.

Jenkins, D. J., et al. Effect of Legumes as Part of a Low Glycemic Index Diet on Glycemic Control and Cardiovascular Risk Factors in Type 2 Diabetes Mellitus: A Randomized Controlled Trial. *Arch. Intern. Med.* **2012**, *172* (2), 16531660.

Jeong, H. Y.; Lim, S. T. Crystallinity and Pasting Properties of Freeze-thawed High Amylose Maize Starch. *Starch Starke.* **2003**, *55* (11), 511–517.

Kannadhason, S.; Muthukumarappan, K. Effect of Starch Sources on Properties of Extrudates Containing DDGS. *Int. J. Food Prop.* **2010**, *13* (5), 1012–1034.

Kayisu, K., et al. Characterization of Starch and Fiber of Banana Fruit. *J. Food Sci.* **1981**, *46* (6), 1885–1890.

Krishnan, P. G., et al. Effect of Commercial Oat Bran on the Characteristics and Composition of Bread. *Cereal Chem.* **1987**, *64* (1), 55–58.

Krupa, U., et al. Bean Starch as Ingredient for Gluten-free Bread. *J. Food Process. Preserv.* **2010**, *34* (S2), 501–518.

Leszczynski, W. Resistant Starch-classification, Structure, Production. *Pol. J. Food Nutr. Sci.* **2004**, *13–54* (SI 1), 37–50.

Ling, L. H., et al. Physical Properties of Starch from Cavendish Banana Fruit. *Starch Starke.* **1982**, *34* (6), 184–188.

Mandala, I., et al. Influence of Frozen Storage on Bread Enriched with Different Ingredients. *J. Food Eng.* **2009**, *92* (2), 137–145.

McGrance, S. J., et al. A Simple and Rapid Colorimetric Method for the Determination of Amylose in Starch Products. *Starch Starke* **1998**, *50* (4), 158–163.

Menrad, K. Market and Marketing of Functional Food in Europe. *J. Food Eng.* **2003**, *56* (2–3), 181–188.

Morales-Medina, R., et al. Produciton of Resistant Starch by Enzymatic Debranching in Legume Flours. *Carbohydr. Polym.* **2014**, *101* (30), 1176–1183.

Mundt, S.; Wedzicha, B. L. A Kinetic Model for Browning in the Baking of Biscuits: Effects of Water Activity and Temperature. *LWT Food Sci. Technol.* **2007**, *40* (6), 1078–1082.

Murphy, M. M., et al. Resistant Starch Intakes in the United States. *J. Am. Diet. Assoc.* **2008**, *108* (1), 67–78.

Mweta, D. E., et al. Some Properties of Starches from Cocoyam (*Colocasia esculenta*) and Cassava (*ManihoteEsculenta* crantz.) Grown in Malawi. *Afr. J. Food Sci.* **2008**, *2*, 102–111.

Nwokocha, L. M.; Williams, P. A. Some Properties of White and Yellow Plantain (*Musa paradisiaca*, Normalis) Starches. *Carbohydr. Polym.* **2009**, *76* (1), 133–138.

Oates, C. G. Bread Microstructure, In *Bread Staling;* Chinachoti, P., Vodovot, Z., Eds.; FI CRC Press: Boca Raton, FL, 2001; pp 149–162.

Omojola, M. O., et al. Isolation and Physic-chemical Characterization of Cola Starch. *Afr. J. Food Agric. Nutr. Dev.* **2010**, *10* (7), 2884–2990.

Ovando-Martinez, M., et al. Unripe Banana Flour as an Ingredient to Increase the Indigestible Carbohydrates of Pasta. *Food Chem.* **2009**, *113* (1), 121–126.

Ovando-Martinez, M., et al. Effect of Hydrothermal Treatment on Physicochemical and Digestibility Properties of Oat Starch. *Food Res. Int.* **2013**, *52* (1), 17–25.

Perez-Sira, E. Characterization of Starch Isolated from Plantain (*Musa Paradisiaca normalis*). *Starch Starke* **1997**, *49* (2), 45–49.

Perez, S.; Bertoft, E. The Molecular Structures of Starch Components and Their Contribution to the Architecture of Starch Granules: A Comprehensive Review. *Starch Starke* **2010**, *62* (8), 389–420.

Pomeranz, Y.; Sievert, D. Purified Resistant Starch Products and Their Preparation. WO 9,015,147, University of Washington, 1990.

Roberfroid, M. B. Global View on Functional Foods: European Perspectives. *Br. J. Nutr.* **2002,** *88* (2), S133–S138.

Rodriguez-Ambriz, S. L., et al. Characterization of a Fibre-rich Powder Prepared by Lique-faction of Unripe Banana Flour. *Food Chem.* **2008,** *107* (4), 1515–1521.

Riccardi, G., et al. Role of Glycemic Index and Glycemic Load in the Healthy State, in Prediabetes, and in Diabetes. *Am. J. Clin. Nutr.* **2008,** *87* (1), 269S–274S.

Rossell, C. M., et al. Influence of Hydrocolloids on Dough Rheology and Bread Quality. *Food Hydrocoll.* **2001,** *15* (1), 75–81.

Sajilata, M. G., et al. Resistant Starch-a Review. *Compr. Rev. Food Sci. Food Saf.* **2006,** *5* (1), 1–17.

Sanchez-Rivera, M. M., et al. Acetylation of Banana (*Musa paradisiaca L.*) and Maize (*Zea mays L.*) Starches Using a Microwave Heating Procedure and Iodine as Catalyst: Partial Characterization. *Starch Starke* **2010,** *62* (3–4), 155–164.

Sankhon, A., et al. Application of Resistant Starch in Bread: Processing, Proximate Compo-sition and Sensory Quality of Functional Bread Products from Wheat Flour and African Locust Bean (*Parkia biglobosa*) Flour. *Agric. Sci.* **2013,** *4* (5B), 122–129.

Shin, M., et al. Hot-water Solubilities and Water Sorptions of Resistant Starches at 25°C. *Cereal Chem.* **2003,** *80* (5), 564–566.

Soares, C. A., et al. Plantain and Banana Starches: Granule Structural Characteristics Explain the Differences in Their Starch Degradation Patterns. *J. Agric. Food Chem.* **2011,** *59* (12), 6672–6681.

Stauffer, C. E. *Fats and Oils*; Eagan Press Handbook Series: St. Paul, MN, 1999; pp 61–79.

Suma, P. F.; Urooj, A. Isolation and Characterization of Starch from Pearl Millet (*Pennisetum typhoidium*) Flours. *Int. J Food Prop.* **2015,** *18* (12), 2675–2687.

Thomas, D. J.; Atwell, W. A. Gelatinization, Pasting and Retrogradation, in *Starches: Prac-tical Guides for the Food Industry*; Thomas, D. J., Atwell, W. A., Eds.; Eagan Press: St Paul, MN, 1999; pp 25–29.

Tsatsaragkou, K., et al. Mathematical Approach of Structural and Textural Properties of Gluten Free Bread Enriched with Carob Flour. *J. Cereal Sci.* **2012,** *56* (3), 603–609.

Tsatsaragkou, K., et al. Development of Gluten Free Bread Containing Carob Flour and Resistant Starch. *LWT Food Sci. Technol.* **2014,** *58* (1), 124–129.

Tsatsaragkou, K., et al. Rheological, Physical, and Sensory Attributes of Gluten-free Rice Cakes Containing Resistant Starch. *J. Food Sci.* **2015,** *80* (2), 341–348.

Ureta, M. M., et al. Baking of Muffins: Kinetics of Crust Color Development and Optimal Baking Time. *Food Bioprocess Technol.* **2014,** *7* (11), 3208–3216.

Visser, R. G. F., et al. Comparison between Amylose-free and Amylose Containing Potato Starches. *Starch Starke* **1997,** *49* (11), 438–43.

Valcarcel-Yamani, B.; Da Silva, L. S. C. Quality Parameters of Some Brazilian Panettones. *Braz. J. Pharm. Sci.* **2013,** *49* (3), 511–519.

van Soest, J. J. G., et al. Short-range Structure in (partially) Crystalline Potato Starch Deter-mined with Attenuated Total Reflectance Fourier-transform IR Apectroscopy. *Carbohydr. Res.* **1995,** *279,* 201–214.

van Soest, J. J. G., et al. Retrogradation of Potato Starch as Studied by Fourier Transform Infrared Spectroscopy. *Starch Starke* **1994,** *46* (12), 453–457.

Vega, D., et al. Thermogravimetric Analysis of Starch-based Biodegradable Blends. *Polym. Bull.* **1996,** *37* (2), 229–235.

WHO. Food and Agriculture Organization/World Health Organization. *Carbohydrates in Human Nutrition*; Report of a Joint FAO/WHO Expert Consultation: FAO, Rome, 1998.

Waliszewski, K. N., et al. Changes of Banana Starch by Chemical and Physical Modification. *Carbohydr. Polym.* **2003,** *52* (3), 237–242.

Yu, H., et al. Structure and Physicochemical Properties of Starches in Lotus (*Nelumbo nucifera Gaertn.*) Rhizome. *Food Sci. Nutr.* **2013,** *1* (4), 273–283.

Zeng, J., et al. Comparison of A and B Starch Granules from Three Wheat Varieties. *Molecules* **2011,** *16* (12), 10570–10591.

Zhou, Y., et al. Structure Characterization and Hypoglycemic Effects of Dual Modified Resistant Starch from Indica Rice Starch. *Carbohydr. Polym.* **2014,** *103,* 81–86.

PART II

Potential Source of Proteins, Amino Acids, and Oils, and Safety of Food and Food Products

CHAPTER 6

GELATIN FROM COLD WATER FISH SPECIES AND THEIR FUNCTIONAL PROPERTIES

NOR FAZLIYANA MOHTAR[1] and CONRAD O. PERERA[2*]

[1]School of Fisheries and Aquaculture Sciences, Universiti Malaysia Terengganu, Kuala Terengganu 21030, Malaysia

[2]School of Chemical Sciences, Food Science Programme, The University of Auckland, Private Bag 92019, Auckland 1142, New Zealand

[*]Corresponding author. E-mail: conradperera@gmail.com

CONTENTS

Abstract 160
6.1   Introduction 160
6.2   Fish Gelatin 173
6.3   Cross-Linking of Fish Gelatin 182
6.4   Functional Properties of Cross-Linked Fish Gelatin 183
6.5   Conclusion 189
Keywords 190
References 190

ABSTRACT

Fish gelatin is generally known to have low gel strength and low melting point. Cold water fish gelatins are known to possess the lowest melting points and least gel strengths. One exception is hoki fish (*Macruronus novaeze-landiae*) gelatin, which has gel strength of 179 g force to penetrate 4 mm of the gel under standard conditions. Most cold-water fish gelatins melt at about 10–12°C. However, hoki fish gelatin melts at 16.6°C. This value is still low compared to bovine gelatin (melting point 26.9 ± 0.65°C, gel strength 273 ± 16.1 g force) or porcine gelatin (melting point 29.1 ± 0.55°C, gel strength 307 ± 8.37 g force) used in the food industry. This is because fish collagen is lacking in proline and hydroxyproline-rich regions of the collagen molecules. Therefore, fish gelatins from cold water fish species are difficult to use in the food industry as a functional ingredient. However, chemical and enzymatic modifications of the gelatin molecules were shown to have improved gel strength and melting point. Enzymatic modification using transglutaminase was found to be the best modifying agent for hoki gelatin giving a melting point and gel strength comparable to those of bovine gelatin that will enable wide use in the food industry for Halal and Kosher food applications.

6.1   INTRODUCTION

Gelatin is a heat-denatured form of collagen that can be extracted from any collagenous material (Bailey & Light, 1989). Most commercial gelatins are commonly derived from mammalian sources, especially porcine skins, bovine bones, and hides. The biggest end-use market for gelatin is known to be in the food and beverage industry, estimated at 54% of the total market share in 2013 (Grand View Research, 2014). The food and beverage industry in Asia-Pacific are predicted to increase at a compound annual growth rate (CAGR) of 4.3% for the period 2014 through 2020 (Grand View Research, 2014). The global demand for gelatin has been increasing over the years. The annual world production of gelatin is approximately 326,000 tons, with the highest amount of 46% derived from porcine skins, followed by bovine hides (29.4%), bones (23.1%), and other sources (1.5%) (Karim & Bhat, 2009). Mammalian gelatin is generally preferred to gelatin derived from marine sources, due to its superior functional properties, such as gel strength and melting point (Cho et al., 2005). However, there are alternative sources that could be used in gelatin manufacture, including by-products from fishery and aquaculture industries. In fact, the outbreak of bovine spongiform encephalopathy (BSE) disease in cows has resulted in researchers tending to shift their attention away from gelatin derived from land animals, given that gelatin derived from fish is not associated with the risk of such disease outbreaks. Fish gelatin also meets the Halal and Kosher regulations.

According to the latest FAO (2014), fishery statistic of the world, the total wild capture in the past 6 years has remained the same, but inland aquaculture has increased by more than 40% and the total fish harvest for human consumption has reached 136 million metric tons per annum. Depending on the fish species, the filleting waste can be as much as 60% or more of the weight of the raw whole fish, with almost 40% of such waste is in the form of skins and bones (Crapo et al., 2004). These waste materials are usually processed into low-value products, such as fish meal and fertilizers but more commonly are being discarded into the sea in large factory ships. One of the ways to utilize some of this waste material is to produce gelatin as an alternative to replace commercial mammalian gelatin. Apart from the scientific papers published by the authors (Mohtar et al., 2010; Mohtar et al., 2011; Mohtar et al., 2014), to the author's knowledge there is no other published information available on the extraction of gelatin from hoki (*Macruronus novaezelandiae*) fish skin. Hoki is New Zealand's largest fishery resource and is found in deep cold waters of the Southern Ocean. There are a number of studies on the extraction of fish gelatin from other cold-water fish species, such as Alaskan pollock (Chiou et al., 2006), brownstripe red and bigeye snapper (Jongjareonrak et al., 2006), cod (Gudmundsson & Hafsteinsson, 1997; Fernández-Díaz et al., 2001; Gómez-Guillén et al., 2002), hake (Gómez-Guillén et al., 2002), mackerel (Badii & Howell, 2006), megrim (Montero & Gómez-Guillén, 2000), salmon (Arnesen & Gildberg, 2007), and a number of other fish species. This chapter will review the properties of hoki fish gelatin from a functional point of view.

6.1.1   FISH SKIN

The skin of a fish shapes the external form covering the body through which most of the contacts with the environment are made. The skin performs many important functions in fish but it mainly functions as a protective organ. The skin is the first line of defense against disease and protection against adverse environmental factors (Lagler et al., 1967). There are two basic layers of a fish skin, the epidermis and the dermis. The outer layer of the epidermis is the cuticle which consists of a mucopolysaccharide layer (with a thickness of approximately 1 μm), normally produced by surface epithelial cells. The thickness of the cuticle layer differs significantly between species. This layer is composed of specific immunoglobulins, lysozymes, and fatty acids which also have anti-pathogenic activity (Roberts, 2012). The epidermis is composed of some layers of moist and flattened cells (Lagler et al., 1967). The deepest layer of the epidermis surrounds a zone of active cell multiplication and the thickness of this layer depends on the age of the species, the site, as well as the reproductive cycle.

The dermis contains blood vessels, nerves, cutaneous sense organs, and connective tissue with relatively few cells (Lagler et al., 1967). It is much thicker and more complex than the epidermis and is composed of two layers, namely the stratum spongiosum and the stratum compactum. The stratum spongiosum is located below the epidermis and contains a loose network of collagen fibers and pigment cells, while the stratum compactum consists of collagenous dense matrix that provides strength to the skin (Roberts, 2012). The tissue of the hypodermis layer is located below the dermis and is more flaccid than the dermis, and more vascular than the stratum compactum. It is a common site for the growth of infectious processes (Roberts, 2012). The dermis of the fish plays a vital function in the formation of scales. The scales mostly overlap with the free edge to the tail. This helps in minimizing the resistance of the fish body to the water. Bony ridge scales first occur in the dermis layer of fish as small aggregations of cells, predominantly forming on the caudal peduncle where they start to develop. Such an aggregation forms a scale platelet that exhibits different sizes in different species during their first appearance. Then, the ridges are deposited on the surface of the growing scale. The deepest part of the scale, called the plate, is composed of successive layers of parallel fibers that contribute to the strength of the scales (Lagler et al., 1967).

6.1.2   WHAT IS GELATIN?

Gelatin is a high molecular weight polypeptide obtained by thermal denaturation of collagen, which is the parent molecule that is obtained from the skins, white connective tissues, and the bones of animals (Balian & Bowes, 1977; de Man, 1999; Belitz et al., 2004). Gelatin is one of the most widely applied hydrocolloids commonly used in foods. It is a unique, natural, and multifunctional ingredient which functions as a gelling and thickening agent, structure enhancer, and foam stabilizer (Gudmundsson, 2002; Zhou & Regenstein, 2005; Yang et al., 2007; Karim & Bhat, 2009). Gelatin extracted from different raw materials has been used for clarifying wines and juices (Taylor, 1997; Hickman et al., 2000) and also in casings for meat products (Hood, 1987). The properties of the collagen from which gelatin is obtained significantly influence the properties of the gelatin. As a thermoreversible gel, gelatin has the ability to soften and turned into a liquid when heated. Subsequently, it returns to the gel form when the solution is cooled (Stainsby, 1987). In addition, gelatin is also able to melt completely in the mouth, which results in a pleasant mouthfeel and release of flavor. Texture and mouth-feel are considered to be important factors that influence the sensory perception of a food product.

6.1.3   COLLAGEN

Collagen is the most abundant protein of animal source (de Man, 1999), composed of three polypeptide chains known as α-chains. Each of the α-chains contains long sequences of repeating tri-peptides based on the general structure Gly-X-Y, where Gly is glycine, X is commonly proline (Pro), and Y can correspond to any amino acid but is often the hydroxyproline (Hyp) (Bailey & Light, 1989). The sequence of amino acids in the α-chains is shown in Figure 6.1.

-Gly-Pro-Lys-Gly-Asp-His-Gly-Pro-Phe-Gly-Pro-Hyp-Gly-Val-Hyp-Gly-Pro-Ala-Gly-Pro-
Hyp-Gly-Pro-Hyp-Gly-Pro-Hyp-Gly-Pro-Ser-Gly-Pro-Hyp-Gly-Ser-Hyp-

FIGURE 6.1   Amino acid sequences in the polypeptide chains. Gly: glycine; Pro: proline; Lys: lysine; Asp: aspartic acid; His: histidine; Phe: phenylalanine; Val: valine; Hyp: hydroxyproline; Ala: alanine; Ser: serine; His: histidine.

Glycine is the smallest amino acid which can pack tightly in the center of the triple helix structure. The distance between each third glycine residue is known to be 0.87 nm (Yonath & Traub, 1969; Piez, 1984; Nimni & Harkness, 1988). The two imino acids, namely, proline (Pro) and hydroxyproline (Hyp), contribute to the stability of the triple helix of collagen, as they control the rotation of the polypeptide backbone. In fact, a high content of hydroxyproline in the triple helical structure indicates that this residue plays an important function in stabilizing the triple helical conformation of collagen molecules (Sakakibara et al., 1973). The sequence of glycine-proline-hydroxyproline (Gly-Pro-Hyp) in the polypeptide chains makes up about 10% of the collagen molecule. Each α-chain in the triple helix structure contains about 1000 amino acids, having about three amino acids per turn. The triple helices of collagen are right handed and formed by three parallel left-hand helical α-chains (Beck & Brodsky, 1998). The triple helix structure of collagen and the arrangement of amino acid sequences are shown in Figure 6.2.

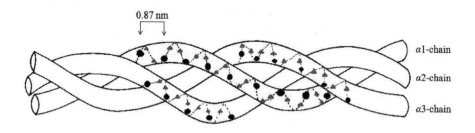

● Glycine

▲ Predominantly imino acids

FIGURE 6.2 **(See color insert.)** Illustrative drawing of the collagen triple helix structure and the arrangement of amino acids (Gly-X-Y) in the collagen molecule (Modified from Nimni and Harkness, 1988).

The α-chains are stabilized by the formation of hydrogen bonds between the N—H group of glycine and the carbonyl oxygen atoms of other peptide groups (Fig. 6.3). Such hydrogen bonds can also be formed between different α-chains. These hydrogen bonds (one per Gly-X-Y triplet) are aligned perpendicular to the helical axis. They attach the α-chains together and provide the main stabilizing strength of the triple helix structure. Collagen molecules link closely from one end to the other by the presence of hydrogen bonds to form collagen fibers. These collagen fibers are sometimes arranged in parallel to give strength in tendons or they may be highly branched and disordered as found in the skins of animals (Foegeding et al., 1996).

The basic structural unit of the collagen triple helix is called tropocollagen, having a molecular weight of approximately 330 kDa, a length of 300 nm, and a diameter of 1.5 nm (Belitz et al., 2004), as illustrated in Figure 6.4. It consists of the triple helical oligomer with short non-helical domains, known as telopeptides, which are mainly the large globular regions at both N- and C-terminals (Bailey & Light, 1989). The N-terminal carries a free amine group (—NH$_2$) while the C-terminal end carries a free carboxyl group (—COOH). These terminal regions are rich in amino acids which assist in

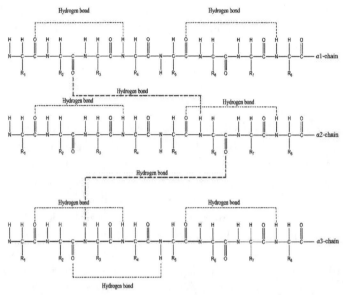

FIGURE 6.3 Schematic diagram illustrating the formation of hydrogen bonds in polypeptide α-chains of the collagen molecule. Dotted lines in black (---) represent the hydrogen bond formed between the same chains. Dotted lines in gray (---) represent the hydrogen bond formed between different α-chains.

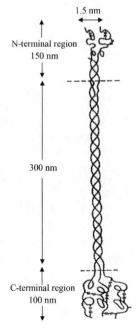

FIGURE 6.4 General structure of tropocollagen (Modified from Bailey and Light, 1989).

the formation of fibrils. However, they do not form triple helical structures like the main body of the collagen molecule due to the absence of glycine in every third position of the terminal chains.

6.1.3.1 TYPES OF COLLAGEN

The α-chains of collagen with different primary structure are numbered with Arabic numerals, such as α1, α2, and α3 in which they can be similar or different within the same molecule (Fig. 6.2). These three α-chains can be either identical (homotrimers) or a mixture of two or three genetically distinct chains (heterotrimers). The individual α-chains are denoted by Roman numerals, such as Type I, Type II, Type III, and Type IV. For instance, a collagen molecule comprised two identical α1 (I) chains and one α2 (I) chain is given as [α1 (I)]$_2$α2 (I). The chains with the same Arabic number of different types of collagen are not identical. For example, the chain of α1 (I) in Type I collagen consists of a different primary structure to that of α1 (II) in Type II collagen. The polypeptide α-chains that make up the collagen molecule vary in size and composition from different types of collagen. There are at least 11 known types of collagens that are different in size and composition as given in Table 6.1; however, there is a structural similarity between all the types of collagens and they can be broadly classified into three groups depending on their functions in the muscle, as explained below (Miller, 1985).

6.1.3.2 FIBRIL-FORMING COLLAGENS

The terminology of fibril and fiber is used to discuss their formation in connective tissues and extracellular matrix (ECM) of the cells. Three polypeptide α-chains combine and coil tightly around each other to form a stable triple helix of collagen (Fig. 6.5). The triple helices of collagen molecules further assemble to form collagen fibrils that are staggered in arrangement and are held together by intramolecular and intermolecular covalent bonds. These covalent bonds stabilize the side-by-side packing of the collagen molecules and generate a strong fibril. The units of collagen fibrils are arranged into fibers as shown in Figure 6.5 (Schmitt et al., 1942).

Type I, Type II, Type III, Type V, and Type XI collagens, known as the common fibrous collagens, are scattered universally around the connective tissues in the body. These collagens are long (300 nm) with rod-like

TABLE 6.1 Collagen Types and Their Distributions in Animal Tissues (Modified from Bailey and Light, 1989).

Type	Class	Molecular length (nm)	Chain composition	Molecular composition	Tissue distribution
I	Fibrillar	300	$\alpha1(I)$, $\alpha2(I)$	$[\alpha1(I)]_2\alpha2(I)$	Intramuscular, skin, tendon, bone, dentine
II	Fibrillar	300	$\alpha1(II)$	$\alpha1(II)_3$	Cartilage, disc, vitreous humor
III	Fibrillar	300	$\alpha1(III)$	$\alpha1(III)_3$	Intramuscular, skin, vascular, intestine
IV	Network	420	$\alpha1(IV)$, $\alpha2(IV)$	$[\alpha1(IV)]_2\alpha2(IV)$	Basement membranes
V	Fibrillar	300	$\alpha1(V)$, $\alpha2(V)_2$, $\alpha3(V)$	Various combinations of $\alpha1(V)[\alpha2(V)]_2$ and $\alpha3(V)$	Intramuscular, skin, embryonic tissues
VI	Network	105	$\alpha1(VI)$, $\alpha2(VI)_2$, $\alpha3(VI)$	$\alpha1(VI)\alpha2(VI)\alpha3(VI)$	Vascular system
VII	Anchoring fibrils	450	$\alpha1(VII)$	$[\alpha1(VII)]_3$	Skin, amniotic membrane
VIII	Network	Not known	$\alpha1(VIII)$, $\alpha2(VIII)$	$[\alpha1(VIII)]_2\alpha2(VIII)$ $[\alpha1(VIII)]_3$, $[\alpha2(VIII)]_3$	Aortic endothelium
IX	FACIT[a]	200	$\alpha1(IX)$, $\alpha2(IX)$, $\alpha3(IX)$	$\alpha1(IX)\alpha2(IX)\alpha3(IX)$	Cartilage
X	Network	150	$\alpha1(X)$	$\alpha1(X)_3$	Cartilage
XI	Fibrillar	300	$\alpha1(XI)$, $\alpha2(XI)$, $\alpha3(XI)$	$\alpha1(XI)\alpha2(XI)\alpha3(XI)$	Cartilage

[a]Fibril associated collagens with interrupted triple helices.

molecules which assemble in a parallel, quarter staggered, over-lapped arrangement to form fibrils possessing a characteristic band pattern with a periodicity of 67 nm (Bailey et al., 1998). They are characterized by a large continuous triple helical domain of approximately 300 nm. These collagens are synthesized in the muscle as procollagens composed of three α-chains, which contain mostly terminal globular N- and C-terminal propeptides. The terminal peptides are cleaved prior to their incorporation into fibrils. After the cleavage, mature Type I, Type II, and Type III collagens essentially form a triple helix with only short non-helical terminal peptides (telopeptides).

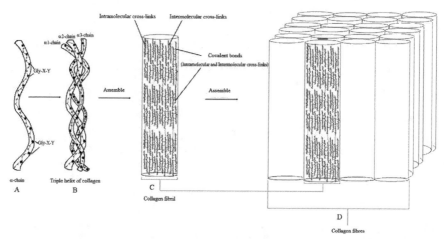

FIGURE 6.5   Formation of collagen fibers. (A) Polypeptide α-chain of collagen with the characteristic repetitive amino acid sequence Glycine-Proline-Hydroxyproline (Gly-X-Y). (B) Three polypeptide α-chains coil together and form a triple helix structure of collagen. (C) These collagen molecules then align along the helix axis and group as a bundle to form the collagen fibril. Covalent bonds occur within (intramolecular cross-link) and between (intermolecular cross-link) a triple helical of collagen molecules in fibrils. (D) These collagen fibrils can align laterally to form bundles of collagen fibers.

In contrast, Type V and Type XI collagens retain large N-terminal domains that are not cleaved prior to their incorporation into fibrils and that may play an important role in regulating the diameter of fibrils (Morris & Bachinger, 1987; Linsenmayer et al., 1993). Type I collagen is the most important structural protein in vertebrates, being responsible for the mechanical strength of skin, tendon, bone, dentine, and other tissues (Bailey & Light, 1989; Bailey et al., 1998). It consists of one-third glycine, but contains no tryptophan and cysteine, and is very low in tyrosine and histidine. Most fish skin collagens are known to be of Type I collagen (Nagai et al., 2001; Nagai et al., 2002).

Type II collagen is the major structural protein of cartilage. It is a homo-polymer of three α1-chains and contains three times the amount of hydroxy-lysine found in Type I collagen. It has a more specific distribution mainly in cartilaginous tissues (Bailey & Light, 1989). Type III collagen assembles into fibrils with Type I collagen in many tissues, particularly blood vessel walls and skins. It is mostly found in embryonic skins, scar tissue, lungs, blood vessels, and also with intra-organ connective tissues (Hulmes, 2008), which are composed of identical α1-chains with intramolecular and intermolecular covalent bonds. Type V and Type XI are identified as minor collagens and are commonly found in small quantities within Type I and Type II collagen fibrils (Boot-Handford & Tuckwell, 2003). Type V is a fibrous-type collagen that is able to form fibers in vivo and contains α1, α2, and α3-chains.

6.1.3.3   BASEMENT MEMBRANE COLLAGENS

Type IV and Type VII collagens belong to the category of network collagens. These form networks and usually found in basement membranes, mostly within thin layers of specialized ECM supporting epithelial or endothelial cells, muscle fibers, and peripheral nerves. They have multiple interruptions in the Gly-X-Y sequence which serve as flexible hinges between more rigid triple helical domains. The more abundant Type IV collagen forms polyg-onal two-dimensional (2D) networks, which constitute an insoluble scaffold of basement membranes. It contains two different α-chains and the molecule is longer than Type I, Type II, or Type III collagens, being about 400 nm in length (Bailey & Light, 1989). Type VII collagen is one of the common mammalian collagens, which anchors the basement membrane to collagen fibers in the supporting connective tissue.

6.1.3.4   OTHER COLLAGENS

These collagens are found in multiple tissues and have a wide range of different functions. Type VI, Type VIII, and Type X are known as the short-chain collagens with continuous triple helical domains. These colla-gens show significant homology but different tissue distribution. Type VI collagen is recognized to be important in muscle tissues, while Type VIII and Type X are important in cornea and cartilage, respectively. Type XIII collagen is known as a trans-membrane collagen and contains the transmem-brane domains. Type IX is referred to as fibril-associated collagens with interrupted triple helices (FACIT) collagens, based on their primary structure and function (Wu et al., 1987; Fitzgerald & Bateman, 2001; Tuckwell, 2002). Type VI collagen is commonly referred to as a filamentous structure, generally scattered in the vascular system, cartilage, and cornea (Bailey & Light, 1989). The other collagen types occur in minor amounts and are usually found associated with the major variants.

6.1.4   CONVERSION OF COLLAGEN TO GELATIN

A pre-treatment process is required to breakdown the collagen into a suitable form for the extraction of gelatin. This allows the loss of the ordered structure of the native insoluble collagen, resulting in a swollen but still insoluble collagen (Stainsby, 1987). Successive heat treatments cleave the hydrogen and covalent bonds to destabilize the triple helix, resulting in helix-to-coil transition and conversion into soluble gelatin (Djabourov et al., 1993), as illustrated in Figure 6.6. Upon cooling, the α-chains reform short collagen-like helices between neighboring monomers, thereby developing an extensive and interconnected network, and further resulting in the formation of a gel (Fig. 6.7). The denaturation of the collagen triple helix is a two-stage process with polypeptide separation as the first step and loss of triple helix as the second step. The degree of conversion of collagen to gelatin is associated with the conditions of the pre-treatment and the extraction process, depending on the pH, ionic strength, extraction temperature, and extraction time (Johnston-Banks, 1990; Karim & Bhat, 2009).

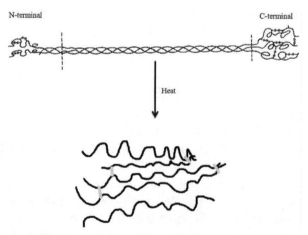

FIGURE 6.6   Denaturation process of collagen to gelatin.

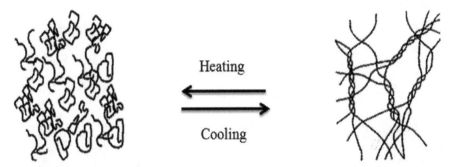

FIGURE 6.7   Gelatin solution forms a gel upon cooling due to the formation of triple helices stabilized by intermolecular hydrogen bonds, whereas heating the gelatin gels resulting to the reverse process.

6.1.5   FUNCTIONAL PROPERTIES OF GELATIN COMPARED TO OTHER HYDROCOLLOIDS

Gelatin is a widely used hydrocolloid in food applications, providing a large number of possibilities to product developers (Schrieber & Gareis, 2007). It is able to form high viscosity solutions in warm water, which sets to a gel on cooling. Also, it has the ability to form and stabilize emulsions, avoid recrystallization, bind by adhesion, stabilize suspensions, clarify wines, and juices, as well as help in the formation of foams and films. It can also be used to reduce the calorific value of foods by increasing the water content or substituting part of the sugar that is normally used (Oesterle, 2004). Gelatin has some of the best functional properties compared to other hydrocolloids for food applications. The comparison of properties between gelatin and other hydrocolloids is given in Table 6.2.

Gelatin is commonly used in the food industry not only as a gelling agent, but also as a stabilizing and an emulsifying agent. For instance, during some cheese manufacturing processes, casein usually loses its emulsifying properties after fermentation; the use of gelatin compensates for this loss and prevents the creaming of milk fat. Apart from that, the application of gelatin in the dairy industry consists of enhancing and stabilizing the texture of casein gels that are usually destroyed during the stirring process, especially in yoghurt-based products (Schrieber & Gareis, 2007). In ice creams and whipped desserts, gelatin is added as a foam stabilizer which helps in the distribution of ice crystals and decreases the melting rate of water in case of temperature fluctuation during storage. The new trends in consuming healthy foods have resulted in the increase in the production of low-fat

TABLE 6.2 Comparison of Frequently Used Hydrocolloids (Modified from Schrieber and Gareis, 2007).

Hydrocolloid	Gel formation	Thickening effect	Transparency of the gel	Cold water solubility	pH stability	Thermoreversibility of the gel
Gelatin	+++	++	+++	0	++	+++ Thermoreversible, difference in melting/gelling temperature low
Agar-agar	+++	+++	+	0	++	++ Thermoreversible, difference in melting/gelling temperature high
Alginates	+++	+++	+++	+++	+	With calcium
Carrageenan	+++	++	++	++	++	With cations
Carboxymethyl cellulose (CMC)	0	+++	–	+++	++	0
Gum arabic	0	+	–	+++	++ (pH 4–9)	0
Hydroxypropylmethyl cellulose (HPMC)	+++	+++	+	+++	+++ (pH 1–10)	Gel formation on heating
Locust bean starches	0	++	–	+	++	0
Modified starches	+++	+++	+	0 With the exception of physically modified starches	++ (pH 3–11)	+++
Native starches	+++	+++	+	0	+	+++
Pectin	+++	++	+++	++	+	+++

Qualitative assessment: 0: none; +: low; ++: medium; +++: high.

products. Gelatin is usually added to low-fat margarine to prevent syneresis and phase-separation processes (Schrieber & Gareis, 2007). Its ability to act as a fat emulsifier improves the structure of margarine. Other uses of gelatin in food applications include thickener for syrups and in the production of granules and capsules for health food products (Schrieber & Gareis, 2007).

6.2 FISH GELATIN

Gelatin can be obtained from different marine sources. There are three categories of species that have been used in the production of gelatin, namely marine invertebrates, sea mammals, and fish (Boran & Regenstein, 2010). Fish gelatin exhibits inferior functional properties compared to those extracted from land animals. Usually, the gelatin extracted from cold water fish, form gels below 10°C, and are liquid at room temperature (Norland, 1990). They can be used in applications that do not require high-melting point, such as frozen or refrigerated products that are normally consumed quickly upon removal from the fridge. Gelatin with low melting points could also be used in confectionery products to enhance the rate of melting of water in gel desserts (Karim & Bhat, 2009). As a protein, gelatin is low in calories compared to fat and melts in the mouth to provide excellent sensory properties, thus making it ideal for use in low-fat products. Gelatin extracted from warm water fishes such as tilapia and grass carp can have a range of gel strengths from 180 to 330 g, and melting points from 25°C to 27°C (Kasankala et al., 2007; Songchotikunpan et al., 2008), thus making them a good substitute for mammalian gelatin. However, those extracted from cold water fish species are known to have low gel strengths and low melting points (Mohtar et al., 2010; Mohtar et al., 2011; Mohtar et al., 2014; Chiou et al., 2006; Gudmundsson & Hafsteinsson, 1997; Fernández-Díaz et al., 2001; Gómez-Guillén et al., 2002).

6.2.1 EXTRACTION OF GELATIN

Extraction of gelatin involves either acid or alkaline pre-treatment followed by extraction in warm water, which converts the collagen to gelatin. Further clarification procedures, such as filtration, concentration, drying, and milling are conducted prior to packaging (Schrieber & Gareis, 2007). The heat treatment prior to extraction hydrolyzes the hydrogen bonds and the intramolecular and intermolecular covalent bonds of the collagen molecules. Such changes result in the conversion of the triple helix structure of collagen to a more amorphous form called gelatin (Dai et al., 2006). The properties of gelatin are affected by the extraction conditions and the molecular weight of the subunits in the collagen molecule (Veis, 1964). There are two types of gelatin, namely, type A and type B. Type A gelatin is produced from acid-treated collagen, while type B gelatin is produced from alkali-treated collagen. Type A gelatin has an isoelectric point of 7–9 based on the severity and duration of the acid treatment of the collagen, which causes limited hydrolysis of the asparagine and glutamine amino acid side chains. On the other hand, the alkali extraction for type B gelatin involves a lengthy treatment of the raw material with an alkali solution. This leads to hydrolysis of the asparagine and glutamine side chains to glutamic and aspartic acid (Veis, 1964) that produced gelatin with an isoelectric point of 4.8–5.2 (Schrieber & Gareis, 2007). Acidic treatment is most suitable for the less covalently cross-linked collagens found in fish or pig skins, while alkaline treatment is suitable for the more complex collagens found in bovine bones, hides, and skins.

The extraction of fish gelatin involves acid or alkaline pre-treatment of the skins prior to the extraction in warm water. There are three main steps involved in the gelatin manufacturing process, namely the pre-treatment of the raw materials, extraction of the gelatin, followed by the purification and drying process. The general flow diagram for the extraction of gelatin from cold water fish skins is shown in Figure 6.8.

The pre-treatment of raw materials involves the removal of non-collagenous materials by alkaline solution to increase purity of the extracted gelatin. This process is important as it helps to inactivate the protease involved in degradation of collagen and to breakdown the interchain cross-links, thus contributing to the solubilization of collagen (Johns & Courts, 1977; Zhou & Regenstein, 2005; Regenstein & Zhou, 2007). There are several key parameters that may affect the properties of gelatin. Such parameters are concentration, time, and temperature used during the pre-treatment time (Yang et al., 2007). Therefore, optimization of each parameter could contribute to their superior properties, such as high in yield and gel strength values. The purification of extracted fish gelatin could be carried out using the dialysis, electrodialysis, ultrafiltration, and ultra-centrifugation methods (Chakravorty & Singh, 1990; Nachod, 2012). After the purification process, fish gelatin is filtered to remove any insoluble matters, such as fat, unextracted collagen fibers, and other unwanted residues. The drying of fish gelatin could be carried out using three different methods, such as freeze-drying, hot-air drying, and spray drying (Kwak et al., 2009). There are several processing parameters that greatly affect the properties of fish gelatin, namely the pre-treatment processes, extraction temperature, extraction time, and the storage

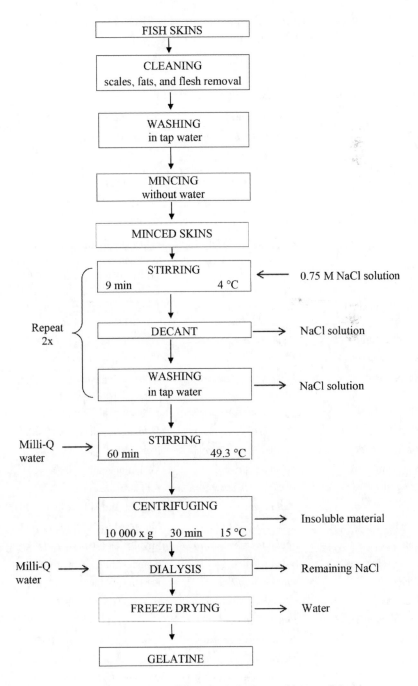

FIGURE 6.8 Flow diagram of extraction of gelatin from cold water fish skins.

of the raw materials (Ledward, 1986; Karim & Bhat, 2009; Boran & Regenstein, 2010). Fish gelatin has been extracted using a number of different methods. These are given in Table 6.3.

TABLE 6.3　Available Literatures on the Extraction of Fish Gelatin.

Species	Scientific name	References
African catfish	*Clarias gariepinus*	Sanaei et al. (2013); Alexandre et al. (2014)
Alaska Pollock	*Theragra chalcogramma*	Zhou and Regenstein (2005); Zhou et al. (2006)
Amur sturgeon	*Acipenser schrenckii*	Mehdi et al. (2014)
Arabic shaari	*Lethrinus microdon*	Ghalib et al. (2011)
Asian redtail catfish	*Hemibagrus nemurus*	Ratnasari et al. (2013)
Atlantic cod	*Gadus morhua*	Gudmundsson and Hafsteinsson (1997); Fernández-Díaz et al. (2001); Gómez-Guillén et al. (2002)
Atlantic mackerel	*Scomber scombrus*	Khiari et al. (2011); Barry-Ryan et al. (2013)
Atlantic salmon	*Salmo salar*	Arnesen and Gildberg (2007)
Bigeye snapper	*Lutjanus vitta*	Nalinanon et al. (2008)
Bigeye snapper	*Priacanthus fayenus*	Kittiphattanabawon et al. (2005); Sukkwai et al. (2011)
Black kingfish	*Ranchycentron canadus*	Killekar et al. (2012)
Black tilapia	*Oreochromis mossambicus*	Jamilah and Harvinder (2002)
Blacktip shark	*Carcharhinus limbatus*	Phanat et al. (2010)
Blue shark	*Prionace glauca*	Yoshimura et al. (2000)
Brownstripe red snapper	*Priacanthus macracanthus*	Jongjareonrak et al. (2006)
Catla	*Catla catla*	Chandra and Shamasundar (2015)
Channel catfish	*Ictalurus punctatus*	Yang et al. (2007); Liu et al. (2008)
Dog shark	*Scoliodon sorrakowah*	Shyni et al. (2014)
Dover sole	*Solea vulgaris*	Gómez-Guillén et al. (2002)
Flounder	*Platichthys flesus*	Fernández-Díaz et al. (2001)
Grass carp	*Ctenopharyngodon idella*	Kasankala et al. (2007)
Grey triggerfish	*Balistes capriscus*	Kemel et al. (2011)
Hake	*Merluccius merluccius*	Gómez-Guillén et al. (2002)
Hake	*Merluccius productus*	Wang et al. (2015)
Halibut	*Hippoglossus stendepsis*	Wang et al. (2015)
Herrings	*Clupea herengus*	Kołodziejska et al. (2008)

TABLE 6.3   *(Continued)*

Species	Scientific name	References
Hoki	*Macruronus novaezelandiae*	Mohtar et al. (2010); Mohtar et al. (2011)
Horse mackerel	*Trachurus trachurus*	Badii and Howell (2006)
Indian Mackerel	*Rastrelliger kanagurta*	Irwandi et al. (2009)
Megrim	*Lepidorhombrus boscii*	Montero and Gómez-Guillén (2000)
Milkfish	*Chanos chanos*	Wang et al. (2015)
Nile perch	*Lates niloticus*	Muyonga et al. (2004)
Nile tilapia	*Oreochromis niloticus*	Songchotikumpan et al. (2008); Wang et al. (2015)
Pangas catfish	*Pangasius pangasius*	Ratnasari et al. (2013)
Rainbow trout	*Onchorhynchus mykiss*	Tabarestani et al. (2010)
Red tilapia	*Oreochromis niloticus*	Jamilah and Harvinder (2002)
Redbelly tilapia	*Tilapia zillii*	Wuyin et al. (2014)
Rohu	*Labeo rohita*	Shyni et al. (2014)
Seabass	*Lates calcarifer*	Sittichoke et al. (2014); Sittichoke et al. (2015); Thanasak and Soottawat (2015)
Shark	*Isurus oxyrinchus*	Cho et al. (2004); Kittiphattanabawon et al. (2010)
Shortfin scad	*Decapterus macrosoma*	Cheow et al. (2007)
Silver catfish	*Pangasius sutchi*	Normah et al. (2015)
Sin croaker	*Johnius dussumieri*	Cheow et al. (2007)
Skate	*Raja kenojei*	Cho et al. (2006)
Skipjack tuna	*Katsuwonus pelamis*	Shyni et al. (2014)
Splendid squid	*Loligo formosana*	Muralidharan et al. (2012)
Sturgeon	*Acipenser schrenckii*	Nikoo et al. (2011)
Unicorn leatherjacket	*Aluterus monoceros*	Ahmad and Benjakul (2010)
Yellowfin tuna	*Thunnus albacares*	Cho et al. (2005); Chiou et al. (2006); Rahman et al. (2008)
Whitecheek shark	*Carcharhinus dussumieri*	Kharyeki et al. (2011)

6.2.2   PROPERTIES OF FISH GELATIN

The physico-chemical properties of gelatin are affected by the source of gelatin and by the age of the animals (Ledward, 1986; Boran & Regenstein,

2010). However, in the case of fish, age will not affect the gel properties as much as for land animals because the fish collagens do not form cross-links as the animal ages as in the case of land animals (Haard, 1992). There are three important properties of gelatin, namely, gel strength, melting point, and amino acids composition (Karim & Bhat, 2009).

6.2.2.1   GEL STRENGTH

Gel strength is one of the most important properties of a gelatin gel. It is defined as the weight in grams (g) that is required for a specified plunger to depress the surface of a thermostated gel to a defined depth under standard conditions (Schrieber & Gareis, 2007). The standard measurement of gel strength is determined using the Bloom test, which consists of performing a well-defined protocol at a given gelatin concentration (6.67%), temperature of 10°C, and maturation time of 18 h. The gel strength measured under these conditions is expressed in the normalized "Bloom value." Mammalian gelatins usually have a high range of gel strengths from 200 to 400 g, while fish gelatins exhibit much lower gel strengths (Gómez-Guillén et al., 2002; Chiou et al., 2006; Zhou et al., 2006; Arnesen & Gildberg, 2007; Mohtar et al., 2010; Mohtar et al., 2011; Mohtar et al., 2014). Gelatins with a range of gel strengths from 250 to 300 g are the most desirable in food applications (Holzer, 1996). Apart from factors relating to the source of the raw materials, the extraction condition of gelatin also affects its gel strength. For instance, Gudmundsson and Hafsteinsson (1997) have reported a low gel strength of cod gelatin extracted using sulfuric acid and citric acid at concentrations higher than 0.02 and 0.05 M, respectively. The observation of low gel strength indicates that the gel-forming ability of gelatin is sensitive to acid and alkali hydrolysis of the collagen.

6.2.2.2   MELTING POINT

As a thermoreversible gel, gelatin gels will start to melt when the temperature increases above a certain point, defined as the gel melting point (Karim & Bhat, 2009). It is also determined as a critical temperature at which gelatin begins to soften. This unique behavior allows gelatin products to encapsulate and release flavors to give optimum organoleptic properties. The main drawback of fish gelatins is that their gels tend to be less stable and have poor rheological properties than gelatins from land mammals, and this may limit their field of application. This is generally true in the case of cold-water fish species, such as cod, salmon, and Alaska pollock. Nevertheless, studies have shown that tropical and sub-tropical warm-water fish species (tilapia, Nile perch, and catfish) might have similar rheological properties and thermostability to those of mammal gelatins, depending on the species, type of raw material, and processing conditions (Gilsenan & Ross-Murphy, 2000; Karim & Bhat, 2009; Gómez-Guillén et al., 2009; Rawdkuen et al., 2010). The melting points of mammalian gelatins are higher (26–37°C) than those of gelatins extracted from fish (11°C to 27°C) (Karim & Bhat, 2009). There are two methods commonly used in the determination of melting point, namely the slip-point and the rheological methods (Boran & Regeinstein, 2010; Mohtar et al., 2010; Mohtar et al., 2014).

6.2.2.3   AMINO ACID COMPOSITION

The amino acid composition of gelatin depends largely on the species of fish. There are three important amino acids in gelatin, namely glycine (Gly), proline (Pro), and hydroxyproline (Hyp). Gelatin consists of large amounts of glycine although the amino acid content varies with the species (Gilsenan & Ross-Murphy, 2000). The chemical structures of the three amino acids are shown in Figure 6.9.

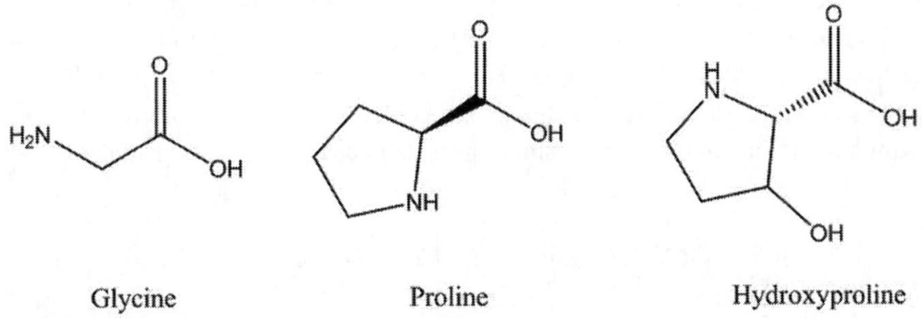

| Glycine | Proline | Hydroxyproline |

FIGURE 6.9   Chemical structures of important amino acids in gelatin.

The imino acids (proline & hydroxyproline) composition is known as a key determinant of the gel strength and melting point properties (Johnston-Banks, 1990), where a low content of imino acids will result in low gel strength and melting point of gelatin. The imino acid content of hoki skin gelatin (18.6/100 g) is higher than that reported for salmon (16.6/100 g), cod (15.4/100 g) (Arnesen & Gildberg, 2007), and bigeye snapper (14.43/100 g)

(Binsi et al., 2009). The presence of the imino acids contributes to the stabilization of the triple helix structure of collagens (Ikoma et al., 2003). Mammalian gelatin has higher imino acids content than that of fish with approximately 200 and 150 residues per 1000 total amino acid residues, respectively (Piez & Gross, 1960). Gelatin with high imino acids content demonstrates better gelling properties than those with low content of proline and hydroxyproline. Therefore, collagen from different sources may have different functional properties due to the differences in the imino acid content. The amino acid composition of several cold water fish species compared with warm water fish species and mammalian gelatins is shown in Table 6.4.

6.2.3 ISSUES WITH FISH GELATIN

The market share of fish gelatin is still developing compared to mammalian gelatin. Two basic factors are found to affect the large-scale utilization of fish gelatin and they are discussed below.

6.2.3.1 INFERIOR PROPERTIES OF FISH GELATIN

As mentioned before, fish gelatin has lower gel strength and melting point due to the lack of proline and hydroxyproline-rich regions of the collagen molecules (Ledward, 1986; Norland, 1990). The total amount of Gly-Pro-Hyp sequence mostly affects the stability of the triple helix structure of collagens (Burjandze, 2000). Gelatin from fish exhibits a low amount of the Gly-Pro-Hyp sequence; therefore, it behaves as a viscous liquid at room temperature which limits its use in many applications, especially in the food industry.

6.2.3.2 EFFECT OF DIFFERENT EXTRACTION CONDITIONS ON GEL PROPERTIES

The method of extraction varies with species. Variation in processing conditions, such as extraction temperature and time may affect the properties of fish gelatin (Cho et al., 2005). For instance, extraction at high temperatures can lead to damage of the structure of gelatin, resulting in its lower gel-forming properties. The occurrence of degraded protein subunits may hinder the correct organization of the α-chains, thereby reducing the formation of nucleation sites (Normand et al., 2000). Thus, optimization of conditions

TABLE 6.4 Amino Acid Composition (Amino Acid Residues Per 1000 Residues) of Several Cold Water Fish Species Compared with Warm Water Fish and Commercial Gelatins.

Amino acid	Alaska pollock	Cod	Hake	Hoki	Mackerel	Megrim	Sole	Tilapia	Bovine	Porcine
Ala	108	96	119	57	124	123	122	123	112	112
Arg	51	56	54	–	55	54	55	47	46	49
Asx	51	52	49	34	49	48	48	48	46	46
Cys	0	0	–	–	–	–	–	0	–	0
Glx	74	78	74	93	72	72	72	69	71	72
Gly	358	344	331	463	334	350	352	347	333	330
His	8	8	10	13	6	8	8	6	5	4
Hyl	6	6	5	–	6	5	5	8	6	6
Hyp	55	50	59	38	66	60	61	79	98	91
Ile	11	11	9	12	9	8	8	8	12	10
Leu	20	22	23	13	23	21	21	23	23	24
Lys	26	29	28	113	26	27	27	25	28	27
Met	16	17	15	10	14	13	10	9	6	4
Phe	12	16	15	12	14	14	14	13	12	14
Pro	95	106	114	55	108	115	113	119	129	132
Ser	63	64	49	60	44	41	44	35	37	35
Thr	25	25	22	–	27	20	20	24	17	18
Trp	0	0	–	–	–	–	–	0	–	0
Tyr	3	3	4	17	3	3	3	2	2	3
Val	18	18	19	10	20	18	17	15	20	26
Imino acid	150	156	173	93	174	175	174	198	227	223
References	Zhou et al. (2006)	Gómez-Guillén et al. (2002)	Gómez-Guillén et al. (2002)	Mohtar (2013)	Kimura (1983)	Gómez-Guillén et al. (2002)	Gómez-Guillén et al. (2002)	Sarabia et al. (2000)	Eastoe and Leach (1977)	Eastoe and Leach (1977)

Alaskan pollock, cod, hake, and hoki are cold water fish species. Mackerel, megrim, sole, tilapia are the examples of warm water fish species. Bovine and porcine are the commercial gelatins.

for extraction of fish gelatin would identify the significant variables from the insignificant ones. Knowledge of the optimum extraction conditions will help in maximizing and improving the properties of gelatin obtained, which would further lead to the development of better end-products (Boran & Regeinstein, 2010; Mohtar et al., 2010).

6.3   CROSS-LINKING OF FISH GELATIN

Cross-linking of proteins involves the formation of covalent bonds between polypeptide chains within a protein or between different protein molecules (Feeney & Whitaker, 1988). Gelation of unmodified gelatin occurs by intra-molecular and intermolecular association at different parts in the molecule leading to the formation of "junction zones" and a three-dimensional (3D) branched network (Gilsenan & Ross-Murphy, 2000). To minimize some of the problems associated with the inferior properties of fish gelatin, modification through chemical and enzymatic cross-linking was carried out (Mohtar et al., 2011; Mohtar et al., 2014). The rationale was that the incorporation of such cross-linking agents to gelatin will increase the gel strength and elasticity, as well as the melting point of gelatin gels, and further improve the formation of well-developed food products (Fernández-Diaz et al., 2001; Kołodziejska et al., 2004).

6.3.1   CHEMICAL CROSS-LINKING

The use of chemical cross-linking agents, such as glutaraldehyde and genipin in the modification of food proteins has been exploited in several studies (Strauss & Gibson, 2004; Chiou et al., 2006). Glutaraldehyde has been extensively used for chemical modification of proteins, such as in the production of biopolymers (Hernández-Muñoz et al., 2004; Marquie et al., 1995; Ustunol & Mert, 2004) and the formation of multimeric compounds after cross-links with ribonuclease, bovine serum albumin (BSA), and casein (Meade et al., 2003; Silva et al., 2004). The chemical modifications are generally expensive and have not often been approved for use in food products (Singh, 1991), thus making this approach less popular than the enzymatic approach. Mohtar (2013) used glutaraldehyde and natural compounds genipin, and caffeic acid to modify hoki skin gelatin and found that the gel strength, expressed as the force in grams to penetrate 4 mm of the gel under standard conditions could be increased from 179 (no modification) to 231 g (glutaraldehyde), 229 (caffeic acid), and 221 g (genipin), respectively. The melting point was also increased from 18.6°C to 21.9°C (glutaraldehyde), 21.6°C (caffeic acid), and 20.5°C (genipin), respectively. However, the improved gel strength and melting points were still significantly lower than those of bovine or porcine gelatin.

6.3.2 ENZYMATIC CROSS-LINKING

The use of enzymes in modifying the functional properties of food proteins is of interest to researchers due to their natural behavior. They also fall into the category of food processing aids and are not required to be listed as food ingredients.

The enzymatic cross-linking using transglutaminase (TGase) has been conducted in several food products, such as gelatin, dairy, meat, soy products, and pasta to improve their physical and textural properties through the formation of ε-(γ-glutamyl) lysine bonds and by intramolecular, or intermolecular covalent cross-linking of protein molecules (Nielsen, 1995; Motoki & Seguro, 1998; Larré et al., 2000; Lorenzen, 2000; Kuraishi et al., 1997; Kuraishi et al., 2001; Gómez-Guillén et al., 2001; Fernández-Diaz et al., 2001). The effect of added TGase on the properties of gelatin gels is dependent on several factors, such as the concentration of enzyme used, incubation time and temperature, as well as the source of gelatin (Sakamoto et al., 1994; Jongjareonrak et al., 2006; Noziah et al., 2009). Therefore, it is vital to determine the optimal conditions under which the cross-linking takes place, prior to the preparation of the gels. Mohtar et al. (2013) optimized the conditions used for cross-linking of hoki gelatin using purified TGase from commercial TGase obtained from Ajinomoto Company. They found that the maximum gel strength was obtained when purified TGase was added at a rate of 3.33 mg per gram of gelatin at 37°C and incubated for 30 min. The gel strength of TGase cross-linked hoki gelatin increased from 179 to 278 g and the melting point also increased from 18.6°C to 26°C. These values are similar to those of bovine gelatin. Thus, a significant improvement in gelling properties of hoki gelatin could be affected by the use of TGase enzyme.

6.4   FUNCTIONAL PROPERTIES OF CROSS-LINKED FISH GELATIN

The emulsifying properties of gelatins showed that porcine gelatin exhibited superior emulsion properties to those of bovine and hoki gelatins

(Mohtar, 2013). Porcine gelatin showed a better stabilization of the oil droplets against coalescence compared to bovine and hoki gelatins. This phenomenon could be explained by the difference in molecular weight subunits of the gelatins using SDS-PAGE. The bovine and porcine gelatins demonstrated a higher number of low molecular weight subunits than hoki gelatin. Mohtar (2013) found that hoki gelatin exhibited two intense (190 and 100 kDa) and two faint bands (50 and 39 kDa).

6.4.1   GEL STRENGTH

Mohtar (2013) found that when 0.133 M of glutaraldehyde was added to hoki gelatin, the gel strength increased from 179 to 231 ± 0.85 g. However, an optimum gel strength of 211 ± 0.52 g was obtained when 0.044 M of genipin was added and 229 ± 1.09 g was obtained when 0.111 M of caffeic acid was added to hoki gelatin. It could be seen that the highest gel strength was observed at a lower concentration of genipin, whereas the optimum gel strength with added glutaraldehyde and caffeic acid were obtained at higher concentrations. The gel strengths of gelatins with added glutaraldehyde were significantly higher than those with added genipin and caffeic acid, at the concentrations of 0.044 and 0.111 M ($p < 0.05$), while at other concentrations they were not statistically different at $p < 0.05$.

6.4.2   MELTING POINT

Similar trends to gel strengths were observed in the melting points obtained with different cross-linking agents. Hoki gelatin with added glutaraldehyde at 0.133 M exhibited the highest melting point (21.9°C ± 0.14°C), followed by caffeic acid at 0.111 M (21.6°C ± 0.10°C), and genipin at 0.044 M (20.5°C ± 0.05°C). These differences in gel strengths and melting points of gelatin cross-linked with different cross-linking agents could be due to their different reaction mechanisms.

6.4.3   FOAMING PROPERTIES

The foaming properties of gelatins were assessed by a modified procedure described by Tay et al. (2006) at room temperature (~20°C). Specifically, this analysis was expected to describe the formation of foam and its stability in different gelatin solutions at a given concentration and at various time intervals. A volume of 6 mL of 0.1% gelatin solution was taken in a measuring cylinder of 10 mm diameter, previously prepared by washing in concentrated chromic acid, rinsed well in distilled water, and dried. Compressed air at a constant flow rate of 90 cm^3 per minute was sparged into the gelatin solution for 140 s. The bubbling was done from the bottom of the column using a glass Pasteur pipette, 230 mm in length and with a diameter of 1 mm. The volume of liquid incorporated in the foam was determined by measuring the difference between the volume of the liquid remaining in the column at a given time of sparging and the initial volume (6 mL). The experiment was conducted in triplicate.

6.4.4   EMULSIFYING PROPERTIES

As in the case of gelling properties, the emulsifying capacity of gelatin from fish species is frequently lower than those from mammalian species. Apart from the distribution of charge, an important criterion in selecting a suitable gelatin type is its emulsifying capacity, because, at the same temperature and concentration, the higher the emulsifying capacity, better the protective sheath around the oil droplets (Schrieber & Gareis, 2007). For example, the emulsifying capacity of tuna fin gelatin is lower than that of pigskin gelatin at the same protein concentration (Aewsiri et al., 2008). With both types of gelatin, the emulsion capacity increased with increasing protein concentration from 2% to 5%, with high protein concentrations facilitating more protein adsorption at the interface. However, oil-in-water emulsions could be prepared using a relatively low concentration of gelatin (0.05%), extracted from the skin of bigeye snapper fish (Binsi et al., 2009). The above authors attributed the higher value of emulsifying capacity recorded by increasing gelatin concentrations to a higher degree of polypeptide unfolding during the shearing involved in the emulsifying process. Besides protein concentration, the molecular weight could be a key factor influencing the ability of gelatin to form and stabilize oil-in-water emulsions. In this respect, low molecular weight fish gelatin (~55 kDa) emulsion contained more large droplets and exhibited more oil destabilization than high molecular weight fish gelatin (~120 kDa) (Surh et al., 2006). These authors also observed the presence of a small population of large droplets in the emulsions after homogenization, which was attributed to the relatively low surface activity of fish gelatin compared with globular proteins such as beta-lactoglobulin.

6.4.5   RHEOLOGICAL PROPERTIES OF FISH GELATIN

Rheological properties play a vital role in food process design, assessment, quality control, storage, and modeling processes (Dogan & Kokini, 2006). Rheological measurements can also be used as an indicator of food product quality. The potential functionality of gelatin depends largely on its thermal and rheological properties (Giménez et al., 2005). Fish gelatin generally tends to exhibit poor rheological properties (low gel strength and low melting point) compared to bovine or porcine gelatin (Giménez et al., 2005), thus limiting its use in food applications, which require gel formation at room temperature or above. Knowledge of the rheological characteristics of fish gelatin is sparse and it is crucial to gather more information on these gelatins especially if they are to be manufactured on a larger scale than is done today. Indeed, the use of rheological methods can provide a better quantitative way of characterizing gelatins and their use in different food products.

Small deformation rheology is a test that is used to examine the linear viscoelastic behavior of a food material by investigating the relationship between stress and strain of the material (Bourne, 2002). Small deformation rheology is a non-destructive test that is able to provide rheological information for the whole gelation process; it is also able to study the solution–gel transition of fish gelatin. Three main parameters are measured by small deformation rheology, namely the elastic modulus (G'), loss modulus (G''), and loss tangent δ (tan δ). These dynamic viscoelastic rheological parameters can be defined as follows (Mirsaeedghazi et al., 2008):

$$G' = \frac{\tau_o}{\gamma_o}\cos\delta \tag{6.1}$$

$$G'' = \frac{\tau_o}{\gamma_o}\sin\delta \tag{6.2}$$

$$\tan\delta = \frac{G''}{G'}, \tag{6.3}$$

where τ_o represents the amplitude of the resulting stress and γ_o is the amplitude of the strain input. A strain is applied to a gelatin gel and the resulting stress is measured thereby G' and G''. G'' characterizes the elastic behavior of the gelatin gel and is related to the energy stored during deformation. In contrast, G'' characterizes the viscous behavior of the gel and is related to energy dissipation during deformation. For a perfectly elastic material, G'' is zero given that all energy is stored and so stress and strain is in phase

($\delta = 0°$), whereas for a perfectly viscous material, G'' is zero as all energy is dissipated, thus stress and strain are out of place by 90° (Dogan & Kokini, 2006). Tan δ is the ratio of loss modulus to elastic modulus and characterizes the relative viscous to elastic responses in a viscoelastic material. A large tan δ value implies that the material has more liquid-like characteristics, while a small tan δ value indicates the more solid-like behavior.

Large deformation rheology test measures the non-linear behavior of food materials by applying large stresses which damage the sample permanently. This can yield important quality characteristics with regard to functional properties of the gels, such as shaping, handling, and cutting. The main purpose of a large deformation rheology test is to collect rheological information on the breaking strain after the gel is formed. One of the main types of large deformation tests is the stress overshoot experiment or the constant shear rate test, where the yield stress (σ_{yield}) and yield strain (γ_{yield}) parameters are determined. As such, the results obtained from both small and large deformation tests can offer useful information on the theoretical aspects of the gel structure.

Time sweep, that is the measurement of G' and G'' as a function of time, is used for determining the gelation kinetics of fish gelatins. The time sweep measurements of John Dory, hoki, ling, and salmon gelatins were measured by Mohtar (2013). She found that qualitatively all fish gelatins exhibited liquid behavior with very low G' values at the early stage of gelation (Fig. 6.10). After a critical time is reached, the proteins in the gelatins start to aggregate, where a gel begins to form and G' increased markedly as gelation time progressed. This is an indication that aggregation and the formation of a gel lead to a sharp increase in G'.

She also did time sweep measurements and found that the curve of G' as a function of gelation time was different for each of the fish gelatin samples. Although qualitatively, John Dory, hoki, and salmon gelatins demonstrated similar elastic responses, ling gelatin showed the lowest G' values (Fig. 6.10). The G' values of hoki gelatin increased faster with an increase in gelation time, compared to those of dory, ling, and salmon gelatins (hoki > salmon > dory > ling).

For gelatin samples of each fish species and at all applied frequencies, the elastic modulus (G') values were greater than the loss modulus (G'') (Fig. 6.11). This is an indication that the gelatin samples are gelled and a network is formed. The protein molecule coils into a triple helices structure when a gelatin solution is subjected to the cooling process (Giraudier et al., 2004). According to studies by Gómez-Guillén et al. (2002), Simon et al. (2003), Haug et al.

(2004) and Haug and Draget (2009), gelatin containing a higher concentration of helical structures will exhibit a higher elastic modulus (G').

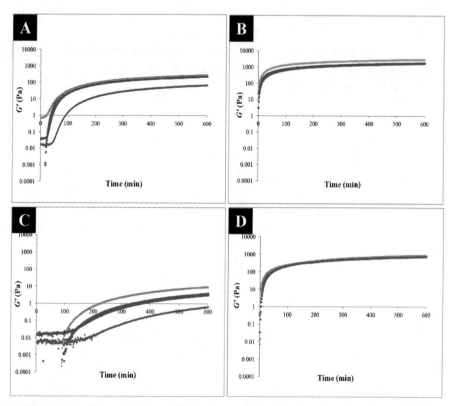

FIGURE 6.10   (See color insert.) Elastic modulus, G', as a function of gelation time for 5 % (w/w) of (A) dory, (B) hoki, (C) ling, and (D) salmon gelatins containing various concentration levels of TGase. Concentrations of TGase were: control (■), TGase 1.6 mg per g (■), TGase 3.33 mg per g (■), and TGase 5 mg per g (■).

To further characterize the nature of the gel formed, the terminology of "weak-gel" and "strong-gel" is used (Lapasin & Pricl, 1995). A strong-gel will exhibit a specific behavior of G' and G' as a function of frequency, where first, G' and G'' are nearly independent of the frequency, and second, G' is higher than G'' by more than tenfold. Conversely, in the case of a "weak-gel", G' and G'' will vary with the frequency and in addition G' will be higher than G'' by less than tenfold (Fig. 6.10). Mohtar (2013) found that based on the values of G' and G'', hoki gelatin has a stronger gel compared to salmon, dory, and ling gelatins (hoki > salmon > dory > ling) in agreement with the time sweep measurements.

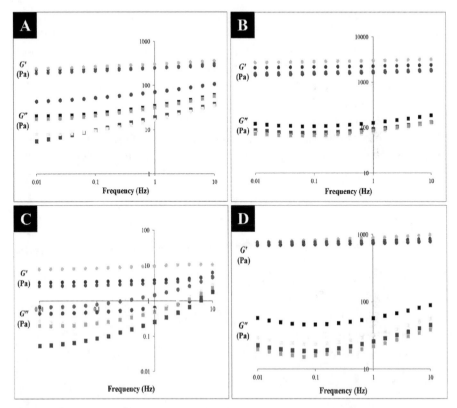

FIGURE 6.11   (See color insert.) Elastic modulus, G' (circle symbols) and loss modulus, G'' (square symbols) as a function of frequency for 5 % (w/w) of (A) dory, (B) hoki, (C) ling, and (D) salmon gelatins containing various concentration levels of TGase. Gelatin concentration was 5 % (w/w) and TGase concentrations were: Control (•,■), TGase 1.6 mg per g (•,■), TGase 3.33 mg per g (•,■), and TGase 5 mg per g •,■). Please note that the Y-axis scales are different for the different fish gelatins. This was done to allow the reader to see the differences between the different curves of the same fish gelatin.

6.5   CONCLUSION

Gelatin from cold water fish species is known to have poor gel properties compared to those from land animals, such as bovine or porcine. It is generally recognized that the poor gel properties are due to their lower imino acid content in their gelatin molecules compared to the land animals. Enzyme-modified hoki skin gelatin exhibited gel properties similar to those of bovine gelatin. As stated in the Grand View Research report (2014), the use of gelatin in applications, such as food and beverage, pharmaceuticals, and nutraceuticals is expected to grow at an increasing rate over the next five years. Thus, there is a potential application of gelatin extracted from the skin of hoki to replace commercial mammalian gelatins, based on its modified properties. The use of fish gelatin as a potential alternative to mammalian gelatin to meet the demand for Kosher and Halal markets could further increase the demand for fish gelatin. Even if fish gelatin could partially replace the commercial mammalian gelatins, because of the increased demands, it might become a compatible product offering competitive properties to other hydrocolloids, as well as alleviating the problems of waste disposal during fish processing.

KEYWORDS

- **gelatin**
- **collagen**
- **hydrocolloids**
- **intramolecular**
- **glutaraldehyde**

REFERENCES

Ahmad, M.; Benjakul, S. Extraction and Characterisation of Pepsin-solubilised Collagen from the Skin of Unicorn Leatherjacket (*Aluterus monocerous*). *Food Chem.* **2010,** *120,* 817–824.

Aewsiri, T.; Benjakul, S.; Visessanguan, W.; Tanaka, M. Chemical Compositions and Functional Properties of Gelatin from Pre-cooked Tuna Fin. *Int. J. Food Sci. Technol.* **2008,** *43* (4), 685–693.

Alexandre, A. T. A.; Fabíola, C. B.; Carline, M.; Ivane, B. T.; Nilson, E. S. African Catfish *(Clarias gariepinus)* Skin Gelatin: Extraction Optimization and Physical–Chemical Properties. *J. Food Res. Int.* **2014,** *65,* 416–422.

Arnesen, J. A.; Gildberg, A. Extraction and Characterisation of Gelatine from Atlantic Salmon (*Salmo salar*) Skin. *Bioresour. Technol.* **2007,** *98,* 53–57.

Badii, F.; Howell, N. K. Fish Gelatin: Structure, Gelling Properties and Interaction with Egg Albumen Proteins. *Food Hydrocoll.* **2006,** *20* (5), 630–640.

Bailey, A.; Light, N. Molecular and Fibre Structure of Collagen. In *Connective Tissue in Meat and Meat Products;* Bailey, A., Light, N., Eds.; Elsevier Applied Science: New York, 1989; pp 25–35.

Bailey, A. J.; Paul, R. G.; Knott, L. Mechanisms of Maturation and Aging of Collagen. *Mech. Aging Dev.* **1998,** *106,* 1–56.

Balian, G.; Bowes, J. H. The Structure and Properties of Collagen. In *The Science and Technology of Gelatin;* Ward, A. G., Courts, A., Eds.; Academic Press: London, 1977; pp 1–31.

Barry-Ryan, C.; Khiari, Z.; Rico, D.; Martin-Diana, A. Comparison between Gelatines Extracted from Mackerel and Blue Whiting Bones after Different Pre-treatments. *J. Food Chem.* **2013,** *139,* 347–354.

Beck, K.; Brodsky, B. Supercoiled Protein Motifs: The Collagen Triplehelix and the Alpha-helical Coiled Coil. *J. Struct. Biol.* **1998,** *122,* 17–29.

Belitz, H. D.; Grosch, W.; Schieberle, P. *Food Chemistry,* 3rd ed.; Springer: New York, 2004.

Binsi, P. K.; Shamasundar, B. A.; Dileep, A. O.; Badii, F.; Howell, N. K. Rheological and Functional Properties of Gelatin from the Skin of Bigeye Snapper (*Priacanthus hamrur*) Fish: Influence of Gelatin on the Gel Forming Ability of Fish Mince. *Food Hydrocoll.* **2009,** *23,* 132–145.

Boot-Handford, R. P.; Tuckwell, D. S. Fibrillar Collagen: The Key to Vertebrate Evolution? A Tale of Molecular Incest. *Bioessay* **2003,** *25,* 142–151.

Boran, G.; Regenstein, J. M. Fish Gelatine. *Adv. Food Nutr. Res.* **2010,** *60,* 119–143.

Bourne, M. C. *Food Texture and Viscosity: Concept and Measurement,* 2nd ed.; Academic Press: California, 2002.

Burjandze, T. V. New Analysis of the Phylogenic Change of Collagen Thermostability. *Biopolymers* **2000,** *53,* 523–528.

Chakravorty, B.; Singh, D. P. Concentration and Purification of Gelatin Liquor by Ultrafiltration. *Desalination* **1990,** *78,* 279–286.

Chandra, M. V.; Shamasundar, B. A. Rheological Properties of Gelatine Prepared from the Swim Bladders of Freshwater Fish *Catla catla. J. Food Hydrocoll.* **2015,** *48,* 47–54.

Cheow, C. S.; Norizah, M. S.; Kyaw, Z. Y.; Howell, N. K. Preparation and Characterization of Gelatins from the Skins of Sin Croaker (*Johnius dussumieri*) and Shortfin Scad (*Decapterus macrosoma*). *Food Chem.* **2007,** *101,* 386–391.

Chiou, B. S.; Avena-Bustillos, R. J.; Shey, J.; Yee, E.; Bechtel, P. J.; Imam, S. H.; Glenn, G. M.; Orts, W. J. Rheological and Mechanical Properties of Cross-linked Fish Gelatines. *Polymer* **2006,** *47,* 6379–6386.

Cho, S. M.; Gu, Y. S.; Kim, S. B. Extracting Optimization and Physical Properties of Yellowfin Tuna (*Thunnus albacares*) Skin Gelatine Compared to Mammalian Gelatines. *Food Hydrocoll.* **2005,** *19,* 221–229.

Crapo, C.; Paugt, B.; Babbitt, J. Recoveries and Yield from Pacific Fish and shellfish. Marine Advisory Bulletin No 37. 2004. http://www.alaskaseafood.org/industry/qc/documents/RecoveriesandYieldsbooklet.pdf (accessed on March 21, 2016).

Dai, C. A.; Chen, Yi. F.; Liu, M. W. Thermal Properties Measurements of Renatured Gelatin Using Conventional and Temperature Modulated Differential Scanning Calorimetry. *J. Appl. Polym. Sci.* **2006,** *99,* 1795–1801.

de Man, J. M. Proteins: Animal Proteins. In *Principles of Food Chemistry;* de Man, J. M., Ed.; Aspen Publishers, Inc.: Gaithersburg, 1999; pp 147–149.

Djabourov, M.; Lechaire, J.; Gaill, F. Structure and Rheology of Gelatine and Collagen Gels. *Biorheology* **1993,** *30,* 191–205.

Dogan, H.; Kokini, J. L. Rheological Properties of Foods. In *Handbook of Food Engineering;* Heldman, D. R., Lund, D. B., Eds.; CRC Press: New York, 2006; pp 1–124.

Eastoe, J. E.; Leach, A. A. Chemical Constitution of Gelatin. In *The Science and Technology of Gelatin;* Ward, A. G., Courts, A., Eds.; Academic Press: New York, 1977; pp 73–107.

FAO. The State of World Fisheries and Aquaculture. Food and Agricultural Organization of the United Nations: Rome, Italy, 2014.

Feeney, R. E.; Whitaker, J. R. Importance of Cross-linking Reactions in Proteins. *Adv. Cereal Sci. Technol.* **1988,** *IX,* 21–43.

Fernández-Díaz, M. D.; Montero, P.; Gómez-Guillén, M. C. Gel Properties of Collagens from Skins of Cod (*Gadua morhua*) and Hake (*Merluccius merluccius*) and Their Modification by the Coenhancers Magnesium Sulphate, Glycerol and Transglutaminase. *Food Chem.* **2001,** *74,* 161–167.

Fitzgerald, J.; Bateman, J. F. A New FACIT of the Collagen Family: COL21A1. *FEBS Lett.* **2001,** *505,* 275–280.

Foegeding, E. A.; Lanier, T. C.; Hultin, H. O. Characteristics of Edible Muscle Tissues. In *Food Chemistry;* Fennema, O. R., Ed.; Marcel Dekker, Inc.: New York, 1996; pp 902–906.

Ghalib, A.; Mohammad, S. R.; Ahmed, A.; Nejib, G. Thermal Characteristics of Gelatine Extracted from Shaari Fish Skin. *J. Therm. Anal. Calorim.* **2011,** *104,* 593–603.

Gilsenan, P. M.; Ross-Murphy, S. B. Rheological Characterization of Gelatines from Mammalian and Marine Sources. *Food Hydrocoll.* **2000,** *14,* 191–195.

Giménez, B.; Turnay, J.; Lizarbe, M. A.; Montero, P.; Gómez-Guillén, M. C. Use of Lactic Acid for Extraction of Fish Skin Gelatine. *Food Hydrocoll.* **2005,** *19,* 941–950.

Giraudier, S.; Hellio, D.; Djabourov, M.; Larreta-Garde, V. Influence of Weak and Covalent Bonds on Formation and Hydrolysis of Gelatin Networks. *Biomacromolecules* **2004,** *5,* 1662–1666.

Gómez-Guillén, M. C.; Sarabia, A. I.; Solas, M. T.; Montero, P. Effect of Microbial Transglutaminase of the Functional Properties of Megrim (*Lepidorhombus boscii*) Skin Gelatine. *J. Sci. Food Agric.* **2001,** *81,* 665–673.

Gómez-Guillén, M. C.; Turnay, J.; Fernández-Díaz, M. D.; Ulmo, N.; Lizarbe, M. A.; Montero, P. Structural and Physical Properties of Gelatine Extracted from Different Marine Species: A Comparative Study. *Food Hydrocoll.* **2002,** *16,* 25–34.

Gómez-Guillén M. C.; Pérez-Mateos, M.; Gómez-Estaca, J.; López-Caballero, E.; Giménez, B.; Montero, P. Fish Gelatin: A Renewable Material for Developing Active Biodegradable Films. *Trends Food Sci. Technol.* **2009,** *20,* 3–16.

Gudmundsson, M.; Hafsteinsson, H. Gelatine from Cod Skins as Affected by Chemical Treatments. *J. Food Sci.* **1997,** *62,* 37–47.

Gudmundsson, M. Rheological Properties of Fish Gelatine. *J. Food Sci.* **2002,** *67,* 2172–2176.

Grand View Research. 2014. http://www.grandviewresearch.com/industry-analysis/gelatin-market-analysis (accessed March 21, 2016).

Haard, N. F. Control of Chemical Composition and Food Quality Attributes of Cultured Fish. *Food Res. Int.* **1992,** *25,* 289–307.

Haug, I. J.; Draget, K. I.; Smidsrød, O. Physical and Rheological Properties of Fish Gelatine Compared to Mammalian Gelatine. *Food Hydrocoll.* **2004,** *18,* 203–213.

Haug, I. J., Draget, K. I. Gelatine. In *Handbook of Hydrocolloids;* Philips, G. O., Williams, P. A., Eds.; Woodhead Publishing Ltd.: Cambridge, 2009; pp 67–86.

Hernández-Muñoz, P.; Villalobos, R.; Chiralt, A. Effect of Cross-linking Using Aldehydes on Properties of Glutenin-rich Films. *Food Hydrocoll.* **2004,** *18,* 403–411.

Hickman, D.; Sims, T. J.; Miles, C. A.; Bailey, A. J.; de Mari, M.; Koopmans, M. Isinglass/collagen: Denaturation and Functionality. *J. Biotechnol.* **2000,** *79,* 245–257.

Holzer, D. Gelatine Production. US Patent 5,484,888, 1996.

Hood, L. L. Collagen in Sausage Casings. In *Advances in Meat Research, Collagen as a Food;* Pearson, A. M., Dutson, T. R., Bailey, A. J. Eds.; Van Norstrand Reinhold Company: New York, 1987; pp 109–129.

Hulmes, D. J. S. Collagen Diversity, Synthesis, and Assembly. In *Collagen, Structure and Mechanics;* Fratzl, P. Ed.; Springer: New York, 2008; pp 16–22.

Ikoma, T.; Kobayashi, H.; Tanaka, J.; Walsh, D.; Mann, S. Physical Properties of Type (I) Collagen Extracted from Fish Scales of *Pagrus ajor* and *Oreochromis niloticas*. *Int. J. Biol. Macromole.* **2003,** *32,* 199–204.

Irwandi, J.; Faridayanti, S.; Mohamed, E. S. M.; Hamzah, M. S.; Torla, H. H.; Che Man, Y. B. Extraction and Characterization of Gelatin from Different Marine Fish Species in Malaysia. *J. Int. Food Res.* **2009,** *16,* 381–389.

Jamilah, B.; Harvinder, K. G. Properties of Gelatins from Skins of Fish-black Tilapia (Oreochromis Mossambicus) and Red Tilapia (Oreochromis Nilotica). *Food Chem.* **2002,** *77,* 81–84.

Johnston-Banks, F. A. Gelatine. In *Food Gels;* Harris, P., Ed.; Elsevier Applied Sciences: New York, 1990; pp 233–289.

Johns, P.; Courts, A. Relationship Between Collagen and Gelatin. In *The Science and Technology of Gelatin*; Ward, A. G., Courts, A., Eds.; Academic Press: London, 1977; pp 138–177.

Jongjareonrak, A.; Benjakul, S.; Visessanguan, W.; Tanaka, M. Skin Gelatine from Bigeye Snapper and Brownstripe Red Snapper: Chemical Compositions and Effect of Microbial Transglutaminase on Gel Properties. *Food Hydrocoll.* **2006,** *20,* 1216–1222.

Karim, A. A.; Bhat, R. Fish Gelatines: Properties, Challenges, and Prospects as an Alternative to Mammalian Gelatines. *Food Hydrocoll.* **2009,** *23,* 563–576.

Kasankala, L. M.; Xue, Y.; Weilong, Y.; Hong, S. D.; He, Q. Optimization of Gelatine Extraction from Grass Carp (*Catenopharyngodon idella*) Fish Skin by Response Surface Methodology. *Bioresour. Technol.* **2007,** *98,* 3338–3343.

Kemel, J.; Rafik, B.; Ali, B.; Noomen, H.; Ahmed, B.; Moncef, N. Chemical Composition and Characteristics of Skin Gelatin from Grey Triggerfish (*Balistes capriscus*). *J. Food Sci. Technol.* **2011,** *44,* 1965–1970.

Kharyeki, M. E.; Rezaei, M.; Motamedzadegan, A. The Effect of Processing Conditions on Physico Chemical Properties of White Cheek Shark (Carcharhinus Dussumieri) Skin Gelatin. *Int. J. Aquat. Res.* **2011,** *3,* 63–69.

Khiari, Z.; Rico, D.; Martin-Diana, A. B.; Barry-Ryan, C. The Extraction of Gelatine from Mackerel (*Scomber scombrus*) Heads with the Use of Different Organic Acids. *J. Fish Sci.* **2011,** *5,* 52–63.

Killekar, V. C.; Koli, J. M.; Sharangdhar, S. T.; Metar, S. Y. Functional Properties of Gelatin Extracted from Skin of Black Kingfish (Ranchycentron Canadus). *Indian J. Fundamental Appl. Life Sci.* **2012,** *2* (3), 106–111.

Kimura, S. Vertebrate Skin Type I Collagen: Comparison of Bony Fishes with Lamprey and Calf. *Comp. Biochem. Physiol.* **1983,** *74B,* 525–528.

Kittiphattanabawon, P.; Benjakul, S.; Visessanguan, W.; Shahidi, F. Comparative Study on Characteristics of Gelatin from the Skins of Rownbanded Bamboo Shark and Blacktip Shark as Affected by Extraction Conditions. *Food Hydrocoll.* **2010,** *24* (2–3), 164–171.

Kittiphattanabawon, P.; Benjakul, S.; Visessanguan, W.; Nagai T.; Tanaka, M. Characterisation of Acid-soluble Collagen from Skin and Bone of Bigeye Snapper (Priacanthus Tayenus). *Food Chem.* **2005,** *89,* 363–372.

Kołodziejska, I.; Kaczorowski, K.; Piotrowska, B.; Sadowska, M. Modification of the Properties of Gelatine from Skins of Baltic Cod (*Gadus morhua*) with Transglutaminase. *Food Chem.* **2004,** *86,* 203–209.

Kołodziejska, I.; Skierka, E.; Sadowska, M.; Kołodziejski, W.; Niecikowska, C. Effect of Extracting Time and Temperature on Yield of Gelatin from Different Fish Offal. *Food Chem.* **2008,** *107* (2), 700–706.

Kuraishi, C.; Sakamoto, H.; Yamazaki, K.; Susa, Y.; Kuhara, C.; Soeda, T. Production and Restructured Meat Using Microbial Transglutaminase Without Salt and Cooking. *J. Food Sci.* **1997,** *62,* 488–490.

Kuraishi, C.; Yamazaki, K.; Susa, Y. Transglutaminase: Its Utilization in the Food Industry. *Food Rev. Int.* **2001,** *17,* 221–246.

Kwak, K. S.; Cho, S. M.; Ji, C. I.; Lee, Y. B.; Kim, S. B. Changes in Functional Properties of Shark (*Isurus oxyrinchus*) Cartilage Gelatin Produced by Different Drying Methods. *Int. J. Food Sci. Technol.* **2009,** *44,* 1480–1484.

Lagler, K. F.; Bardach, J. E.; Miller, R. R. Skin. In *Ichthyology*; Lagler, K. F., Bardach, J. E., Miller, R. R., Eds.; John Wiley and Sons, Inc.: New York, 1967; p 108.

Lapasin, R.; Pricl, S. *Rheology of Industrial Polysaccharides: Theory and Applications;* Aspen Publishers Inc.: New York, 1995.

Larré, C.; Denery-Papini, S.; Popineau, Y.; Deshayes, G.; Desserme, C.; Lefebvre, J. Biochemical Analysis and Rheological Properties of Gluten Modified by Transglutaminase. *Cereal Chem.* **2000,** *77,* 32–38.

Ledward, D. A. Gelation of Gelatine. In *Functional Properties of Food Macromolecules;* Mitchell, J. R., Ledward, D. A., Eds.; Elsevier Applied Science Publisher: London, 1986; pp 171–201.

Linsenmayer, T. F.; Gibney, E.; Igoe, F.; Gordon, M. K.; Fitch, J. M.; Fessler, L. I.; Birk, D. E. Type V Collagen: Molecular Structure and Fibrillar Organization of the Chicken Alpha 1(V) NH_2-terminal Domain, a Putative Regulator of Corneal Fibrillogenesis. *J. Cell Biol.* **1993,** *121,* 1181–1189.

Lorenzen, P. C. Techno-functional Properties of Transglutaminase-treated Milk Proteins. *Milchwissenschaft* **2000,** *55,* 667–670.

Marquie, C.; Aymard, C.; Cuq, J. L.; Guilbert, S. Biodegradable Packaging Made from Cottonseed Flour: Formation and Improvement by Chemical Treatments with Gossypol, Formaldehyde and Glutaraldehyde. *J. Agric. Food Chem.* **1995,** *43,* 2762–2767.

Meade, S. J.; Miller, A. G.; Gerrard, J. A. The Role of Dicarbonyl Compounds in Non-enzymatic Crosslinking: A Structure-activity Study. *Bioorganic Med. Chem.* **2003,** *11,* 853–862.

Mehdi, N.; Soottawat, B.; Mohanad, B.; Masood, A.; Abdoulaye, I. C.; Na, Y.; Xueming, X. Physicochemical Properties of Skin Gelatine from Farmed Amur Sturgeon (*Acipenser schrenckii*) as Influenced by Acid Pre-treatment. *J. Food Biosci.* **2014,** *5,* 19–26.

Miller, E. J. The Structure of Fibril-forming Collagens. *Ann. NY Acad. Sci.* **1985,** *460,* 1–13.

Mirsaeedghazi, H.; Emam-Djomeh, Z.; Mousavi, S. M. A. Rheometric Measurement of Dough Rheological Characteristics and Factors Affecting it. *Int. J. Agric. Biol.* **2008,** *10,* 112–119.

Mohtar, N. F.; Perera, C. O.; Quek, S. Y. Optimization of Gelatine Extraction from Hoki (*Macruronus novaezelandiae*) Skins and Measurement of Gel Strength and SDS-PAGE. *Food Chem.* **2010,** *122,* 307–313.

Mohtar, N. F.; Perera, C. O.; Quek, S. Y. Utilization of Gelati from NZ Hoki (*Macruronus novaezelandiae*) Fish Skins. *Int. Food Res. J.* **2011,** *18* (3), 1062–1066.

Mohtar, N. F.; Perera, C. O.; Quek, S. Y.; Hemar, Y. Optimization of Gelatine Gel Preparation from New Zealand Hoki (*Macruronous novaezelandiae*) Skins and the Effect of Transglutaminase Enzyme on the Gel Properties. *Food Hydrocoll.* **2013,** *31,* 204–209.

Mohtar, N. F. Extraction and Characterisation of New Zealand Hoki (*Macruronus novaez-elandiae*) Skin Gelatine. PhD Thesis Submitted to the University of Auckland, New Zealand, 2013.

Mohtar, N. F.; Perera, C. O.; Hemar, Y. Chemical Modification of New Zealand Hoki (*Macruronus novaezelandiae*) Skin Gelatin and Its Properties. *Food Chem.* **2014,** *155*, 64–73.

Montero, P.; Gómez-Guillén, M. C. Extracting Conditions for Megrim (*Lepidorhombus boscii*) Skin Collagen Affect Functional Properties of the Resulting Gelatine. *J. Food Sci.* **2000,** *65*, 434–438.

Morris, N. P.; Bachinger, H. P. Type XI Collagen is a Heterotrimer with the Composition (1 alpha, 2 alpha, 3 alpha) Retaining Non-triple-helical Domains. *J. Biol. Chem.* **1987,** *262*, 11345–11350.

Motoki, M.; Seguro, K. Transglutaminase and its Use for Food Processing. *LWT Food Sci. Technol.* **1998,** *9*, 204–210.

Muralidharan, N.; Soottawat, B.; Thummanoon, P.; Ponusa, S.; Hideki, K. Characteristics and Functional Properties of Gelatin from Splendid Squid (*Loligo formosana*) Skin as Affected by Extraction Temperatures. *J. Food Hydrocoll.* **2012,** *29*, 389–397.

Muyonga, J. H.; Colec, C. G. B.; Duodub, K. G. Extraction and Physicochemical Characterisation of Nile Perch (Lates Niloticus) Skin and Bone Gelatine. *Food Hydrocoll.* **2004,** *8*, 581–592.

Nachod, F. C. Ion Exchange: Theory and Application. Academic Press Inc.: New York, 2012; pp 364–365.

Nagai, T.; Yamashita, E.; Taniguchi, K.; Kanamori, N.; Suzuki, N. Isolation and Characterisation of Collagen from the Outer Skin Waste Material of Cuttlefish (*Sepia lycidas*). *Food Chem.* **2001,** *72*, 425–429.

Nalinanon, S.; Benjakul, S.; Visessanguan, W.; Kishimura, H. Improvement of Gelatin Extraction from Bigeye Snapper Skin Using Pepsinaided Process in Combination with Protease Inhibitor. *Food Hydrocoll.* **2008,** *22* (4), 615–622.

Nagai, T.; Arakai, Y.; Suzuki, N. Collagen of the Skin of Occelate Puffer Fish (*Takifugu rubripes*). *Food Chem.* **2002,** *78*, 173–177.

Nielsen, P. M. Reactions and Potential Industrial Application of Transglutaminase. Review of Literature and Patents. *Food Biotechnol.* **1995,** *9*, 119–156.

Nikoo, M.; Xu, X.; Benjakul, S.; Xu, G.; Ramirez-Suarez, J. C.; Ehsani, A.; Kasankala, L. M.; Duan, X.; Abbas, S. Characterization of Gelatine from the Skin of Farmed Amur Sturgeon *Acipenser schrenckii. Int. Aquat. Res.* **2011,** *3*, 135–145.

Nimni, M. E.; Harkness, R. D. Molecular Structures and Functions of Collagen. In *Collagen;* Nimni, M. E., Ed.; CRC Press: Boca Raton, 1988; pp 1–78.

Norland, R. E. Fish Gelatine. In *Advances in Fisheries Technology and Biotechnology for Increased Profitability;* Voight, M. N., Botta, J. K., Eds.; Technomic Publishing Co.: Lancaster, 1990; pp 325–333.

Normah, I.; Muhammad Fahmi, I. Physicochemical Characteristics of Gummy Added with Sutchi Catfish (*Pangasius hypophthalmus*) Gelatin. *Int. Food Res. J.* **2015,** *22* (3), 1059–1066.

Normand, V.; Muller, S.; Ravey, J. C.; Parker, A. Gelation Kinetics of Gelatine: A Master Curve and Network Modelling. *Macromolecules* **2000,** *33*, 1063–1071.

Norziah, M. H.; Al-Hassan, A.; Khairulnizam, A. B.; Mordi, M. N.; Norita, M. Characterization of Fish Gelatine from Surimi Processing Wastes: Thermal Analysis and Effect of Transglutaminase on Gel Properties. *Food Hydrocoll.* **2009,** *23*, 1610–1616.

Oesterle, A. M. Low Sugar Alternatives for Jams and Jellies. 2004. http://nchfp.uga.edu/tips/summer/Lowsugar_JamsJelliesST.html (accessed March 14, 2016).

Phanat, K.; Soottawat, B.; Wonnop, V.; Fereidoon, S. Comparative Study on Characteristics of Gelatin from the Skins of Brownbanded Bamboo Shark and Blacktip Shark as Affected by Extraction Conditions. *J. Food Hydrocoll.* **2010**, *24,* 164–171.

Piez, K. A.; Gross, J. The Amino Acid Composition of Some Fish Collagens: The Relations Between Composition and Structure. *J. Biol. Chem.* **1960**, *235,* 995–998.

Piez, K. A. Molecular and Aggregate Structure of the Collagens. In *Extracellular Matrix Biochemistry;* Piez, K. A., Reddi, A. H., Eds.; Elsevier: New York, 1984; pp 1–39.

Rahman, M. S.; Al-Saidi, G. S.; Guizani, N. Thermal Characterisation of Gelatin Extracted from Yellowfin Tuna Skin and Commercial Mammalian Gelatin. *Food Chem.* **2008**, *108,* 472–481.

Ratnasari, I.; Yuwono, S. S.; Nusyam, H.; Widjanarko, S. B. Extraction and Characterization of Gelatine from Different Fresh Water Fishes as Alternative Sources of Gelatine. *J. Int. Food Res.* **2013**, *20,* 3085–3091.

Rawdkuen, S.; Sai-Ut, S.; Benjakul, S. Properties of Gelatin Films from Giant Catfish Skin and Bovine Bone: A Comparative Study. *Eur. Food Res. Technol.* **2010**, *231* (6), 907–916.

Regenstein, J.; Zhou, P. Collagen and Gelatine from Marine By-Products. In *Maximising the Value of Marine By-Products;* Shahidi, F., Ed.; Woodhead Publishing Limited: Cambridge, 2007; pp 279–303.

Roberts, R. J. Fish Pathology, 4th ed.; Blackwell Publishing Ltd.: Hoboken, NJ, 2012. (Online) DOI: 10.1002/9781118222942.

Sakakibara, S.; Inouye, K.; Shudo, K.; Kishida, Y.; Kobayashi, Y.; Prockop, D. J. Synthesis of (Pro-Hyp-Gly) n of Defined Molecular Weights. Evidence for the Stabilization of Collagen Triple Helix by Hydroxypyroline. *Biochim. Biophys. Acta.* **1973**, *303,* 198–202.

Sakamoto, H.; Kumazama, Y.; Motoki, M. Strength of Protein Gels Prepared with Microbial Transglutaminase as Related to Reaction Conditions. *J. Food Sci.* **1994**, *59,* 866–871.

Sanaei, A. V.; Mahmoodani, F.; See, S. F.; Yusop, S. M.; Babji, A. S. Processing Optimization and Characterization of Gelatin from Catfish (*Clarias gariepinus*) Skin. *J. Sains Malays.* **2013**, *42,* 1697–1705.

Sarabia, A. I.; Gómez-Guillén, M. C.; Montero, P. The Effect of Added Salts on the Visco-elastic Properties of Fish Skin Gelatin. *Food Chem.* **2000**, *70,* 71–76.

Schmitt, F. D.; Hall, C. E.; Jakns, M. A. Electron Microscopy Investigations of the Structure of Collagens. *J. Cell. Comp. Physiol.* **1942**, *20,* 11–33.

Schrieber, R.; Gareis, H. *Gelatine Handbook;* Wiley-VCH GmbH and Co.: Weinhem, 2007.

Shyni, K.; Hema, G. S.; Ninan, G.; Mathew, S.; Joshy, C. G.; Lakshmanan, P. T. Isolation and Characterization of Gelatin from the Skins of Skipjack Tuna (*Katsuwonus pelamis*), Dog Shark *(Scoliodon sorrakowah)*, and Rohu (*Labeo rohita*). *J. Food Hydrocoll.* **2014**, *39,* 68–76.

Silva, C. J. S. M.; Sousa, F.; Gubitz, G.; Cavaco-Paulo, A. Chemical Modifications on Proteins Using Glutaraldehyde. *Food Technol. Biotechnol.* **2004**, *42,* 51–56.

Simon, A.; Grohens, Y.; Vandanjon, L.; Bourseau, P.; Balnois, E.; Levesque, G. A Comparative Study of the Rheological and Structural Properties of Gelatine Gels of Mammalian and Fish Origins. *Macromol. Symp.* **2003**, *203,* 331–338.

Singh, H. Modification of Food Proteins by Covalent Crosslinking. *Trends Food Sci. Technol.* **1991**, *2,* 196–200.

Sittichoke, S.; Soottawat, B.; Hideki, K. Characteristics and Gel Properties of Gelatin from Skin of Seabass *(Lates calcarifer)* as Influenced by Extraction Conditions. *J. Food Chem.* **2014**, *152,* 276–284.

Sittichoke, S.; Soottawat, B.; Hideki, K. Molecular Characteristics and Properties of Gelatin from Skin of Seabass with Different Sizes. *J. Int. Biol. Macromol.* **2015**, *73,* 146–153.

Songchotikunpan, P.; Tattiyakul, J.; Supaphol, P. Extraction and Electrospinning of Gelatine from Fish Skin. *Int. J. Biol. Macromol.* **2008**, *42,* 247–255.

Stainsby, G. Gelatine Gels. In *Advances in Meat Research;* Pearson, A. M., Dutson, T. R., Bailey, A. J., Eds.; Van Nostrand Reinhold: New York, 1987; pp 209–222.

Strauss, G.; Gibson, S. M. Plant Phenolics as Cross-linkers of Gelatine Gels and Gelatine-based Coacervates for Use as Food Ingredients. *Food Hydrocoll.* **2004**, *18,* 81–89.

Sukkwai, S.; Kijroongrojana, K.; Benjakul, S. Extraction of Gelatin from Bigeye Snapper (Priacanthus Tayenus) Skin for Gelatin Hydrolysate Production. *Int Food Res J.* **2011**, *18* (3), 1129–1134.

Surh, J.; Decker, E. A.; McClements, D. J. Properties and Stability of Oil-in-water Emulsions Stabilized by Fish Gelatin. *Food Hydrocoll.* **2006**, *20* (5), 596–606.

Tabarestani, S. H.; Maghsoudlou, Y.; Motamedzadegan, A.; Sadeghi Mahoonak, A. R. Optimization of Physico-chemical Properties of Gelatin Extracted from Fish Skin of Rainbow Trout (Onchorhynchus Mykiss). *Bioresour Technol.* **2010**, *101* (15), 6207–6214.

Tay S. L.; Kasapis, S.; Perera, C. O.; Barlow, P. J. Functional and Structural Properties of 2S Soy Protein in Relation to Other Molecular Protein Fractions. *J. Agric. Food Chem.* **2006**, *54* (16), 6046–6053.

Taylor, R. Collagen Finings and Preparations Thereof. United States Patent 5703211, 1997.

Thanasak, S.; Soottawat, B. Physico-chemical Properties and Fishy Odor of Gelatin from Seabass (*Lates calcarifer*) Skin Stored in Ice. *J. Food Biosci.* **2015**, *10,* 59–68.

Tuckwell, D. Identification and Analysis of Collagen Alpha 1(XXI), a Novel Member of the FACIT Collagen Family. *Matrix Biol.* **2002**, *21,* 63–66.

Ustunol, Z.; Mert, B. Water Solubility, Mechanical, Barrier, and Thermal Properties of Cross-linked Whey Protein Isolate-based Films. *J. Food Sci.* **2004**, *69,* FEP129–FEP133.

Veis, A. *The Macromolecular Chemistry of Gelatine;* Academic Press: New York, 1964.

Wang, T. Y.; Hsieh, C. H.; Hung, C. C.; Jao, C. L.; Chen, M. C.; Hsu, K. C. Fish Skin Gelatin Hydrolysates as Dipeptidyl Peptidase IV Inhibitors and Glucagon-like Peptide-1 Stimulators Improve Glycaemic Control in Diabetic Rats: A Comparison between Warm and Cold-water Fish. *J. Funct. Foods.* **2015**, *19,* 330–340.

Wu, J. J.; Eyre, D. R.; Slayter, H. S. Type VI Collagen of the Intervertebral Disc. Biochemical and Electron-microscopic Characterization of the Native Protein. *Biochem. J.* **1987**, *248,* 373–381.

Wuyin, W.; Huibin, Z.; Wenjin, S. Characterization of Edible Films Based on Tilapia (*Tilapia zillii*) Scale Gelatin with Different Extraction pH. *J. Food Hydrocoll.* **2014**, *41,* 19–26.

Yang, H.; Wang, Y.; Jiang, M.; Oh, J. H.; Herring, J.; Zhou, P. 2-step Optimization of the Extraction and Subsequent Physical Properties of Channel Catfish (*Ictalurus punctatus*) Skin Gelatine. *J. Food Sci.* **2007**, *72,* C188–C195.

Yonath, A.; Traub, W. Polymers of Tripeptides as Collagen Models. IV. Structure Analysis of Poly(L-prolyl-glycyl-L-proline). *J. Mole. Biol.* **1969**, *43,* 461–477.

Yoshimura, K.; Terashima, M.; Hozan, D.; Shirai, K. Preparation and Dynamic Viscoelasticity Characterization of Alkali-solubilized Collagen from Shark Skin. *J. Agric. Food Chem.* **2000**, *48,* 685–690.

Zhou, P.; Regenstein, J. M. Effects of Alkaline and Acid Pretreatments on Alaska Pollock Skin Gelatine Extraction. *J. Food Sci.* **2005**, *70,* C392–C396.

Zhou, P.; Mulvaney, S. J.; Regenstein, J. M. Properties of Alaska Pollock Skin Gelatin: A Comparison with Tilapia and Pork Skin Gelatins. *J. Food Sci.* **2006**, *71,* C313–C321.

CHAPTER 7

FISH PROCESSING WASTE PROTEIN HYDROLYSATE: A RICH SOURCE OF BIOACTIVE PEPTIDES

GIRIJA G. PHADKE[1*], L. NARASIMHA MURTHY[1],
ASIF U. PAGARKAR[2], FAISAL R. SOFI[3], and K. K. SABHA NISSAR[3]

[1]*Mumbai Research Centre of CIFT—Central Institute of Fisheries Technology, Vashi 400703, Navi Mumbai, Maharashtra, India*
[2]*Marine Biological Research Station, Near Murugwada, Pethkilla, Ratnagiri 415612, Maharashtra, India*
[3]*Department of Fish Processing Technology, College of Fishery Science, Sri Venkateshwara Veterinary University, Muthukur 524344, Andhra Pradesh, India*
Corresponding author. E-mail: girija_cof@yahoo.com

CONTENTS

Abstract 200
7.1   Introduction 200
7.2   Biologically Active Peptides 200
7.3   Fish Processing Waste for Production of Bioactive
       Protein Hydrolysates 201
7.4   Mechanism of Hydrolysis 202
7.5   Hydrolysis: Influencing Factors 202
7.6   Use of Fish Waste for Enzymatic Hydrolysis 203
7.7   Bioactivities of Fish Waste Protein Hydrolysates 203
7.8   Research Gaps, Opportunities, and Challenges in
       Utilization of Fish Waste as a Source for the Production
       of Fish Protein Hydrolysate 210
7.9   Conclusion 211
Keywords 211
References 211

ABSTRACT

There is an increasing demand for fish and fishery products in the world. In order to keep the pace with demand, it is important to explore the utilization of byproducts of desirable species and low economic value fish species. Isolating proteins from such sources is a conscientious effort in using the aquatic resources wisely along with added value benefits for raw material with low economic value. In this chapter, an attempt has been made to review derivation of peptides from seafood processing waste by enzymatic hydrolysis and their bioactive properties such as antihypertensive, antioxidant, anticoagulant, and antimicrobial activity.

7.1   INTRODUCTION

Fish is considered as a rich source of valuable proteins. Fish is consumed in fresh or processed forms. Fish processing generates more than 50–70% of total as waste which includes various parts such as fish frames, visceral organs, skin, head, etc. (Chalamaiah et al., 2012). Although, the proportion of the total fishery production that is processed remained relatively stable over the last decade, the steady increase in fish production resulted into increase in the total bulk of processed fish products. The proportion of waste generated is currently discarded or processed into low-value products such as fishmeal. Processing of large quantities of fish and other aquatic organisms produces a corresponding large bulk of byproducts and wastes which are high in biological oxygen demand (BOD) and chemical oxygen demand (COD) which are likely to produce adverse effects on coastal and marine environments (Islam et al., 2004). Considerable amounts of fish processing waste generated currently impose a cost burden on the seafood industry in terms of waste disposal, with little benefit generated. Sustainable use of these processing wastes and low-value fish for the recovery and hydrolysis of proteins rich in bioactive peptides is one of the viable options for better utilization (He et al., 2013).

7.2   BIOLOGICALLY ACTIVE PEPTIDES

Proteins and native peptides in food and living things enable the effective functioning of an organism. Peptides possessing biological or nutritional properties improve the functioning of an organism's body as well as enhance food quality. These peptides are referred to as "bioactive peptides" (Hartmann & Meisel, 2007) which are of short sequences of amino acids, inactive within the sequence of the parent protein and may be released by proteolytic hydrolysis using commercially available enzymes or proteolytic microorganisms and fermentation (Vercruysse et al., 2005). Bioactive peptides may be readily present in foods or protein sources as unbound compounds, or they may only be released into their active forms upon cleavage from a native protein source. The release of functional and potent bioactive peptides from intact proteins in various studies has been achieved by methods such as enzymatic hydrolysis (Je et al., 2007). Beneficial effects of bioactive peptides from different sources are depicted in Figure 7.1.

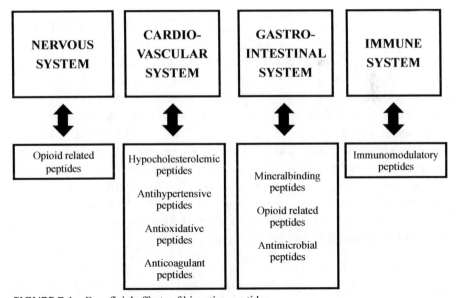

FIGURE 7.1   Beneficial effects of bioactive peptides.

7.3   FISH PROCESSING WASTE FOR PRODUCTION OF BIOACTIVE PROTEIN HYDROLYSATES

The enzymatic hydrolysis is an effective approach for the recovery of valuable components like bioactive peptides as well as the improvement of the functional and nutritional properties of protein from fish by-products without prejudicing its nutritive value (Nalinanon et al., 2011). In the production of protein hydrolysates, high molecular weight proteins hydrolyze to give low molecular weight peptides by addition of water (Shirai & Ramirez, 2011). A wide range of hydrolysates with different physical, chemical, and biological characters can be produced using different enzymes, different fish species as substrates, different proteolytic conditions such as pH, temperature, enzyme to substrate ratio and time (Kristinsson & Rasco, 2000). Fish protein hydrolysates can be used in food systems, comparable with other pertinent protein hydrolysates. Protein hydrolysis releases peptide sequences which could initiate, enhance a particular function, reduce the activity of a protein or terminate its function altogether (Nalinanon et al., 2011).

7.4   MECHANISM OF HYDROLYSIS

Hydrolysates are proteins that are chemically or enzymatically broken down into peptides of varying sizes (Adler-Nissen, 1986). Protein hydrolysis allows fish proteins for improvement of nutritional, functional, and bioactive properties. Chemical hydrolysis includes use of acid or alkali for hydrolysis of proteins, whereas in enzymatic hydrolysis, peptide bonds in proteins are hydrolyzed by use of enzymes from various sources. Enzymes such as proteases catalyze only one specific reaction and function by forming a complex with the substrate whose transformation they catalyze. Proteases are characterized further by their hydrolyzing mechanism into endoproteinases or exopeptidases. The endoproteinases cleave/hydrolyze the peptide bonds within protein molecules, usually at specific residues to produce relatively large peptides whereas exopeptidases cleave/hydrolyze the terminal peptide bonds to produce relatively smaller peptides.

7.5   HYDROLYSIS: INFLUENCING FACTORS

Enzymatic hydrolysis of fish waste proteins is one of the efficient ways to recover potent bioactive peptides (Thiansilakul et al., 2007). There are various factors which determine the potential of peptides to display bioactive properties. These include the protein source, the specificity of enzyme used in protein hydrolysis, the amino acid composition and peptide sequences (Je et al., 2007). The physicochemical conditions of the reaction media, including time, temperature, pH, and enzyme/substrate ratio must be optimized for the activity of the enzyme.

Proteases from different sources are commonly used to obtain a more selective hydrolysis as they are specific for peptide bonds adjacent to certain amino acid residues. Many proteases have been used for the generation of bioactive peptides from proteins. Proteases from animal, plant, and microbial sources have been used for enzymatic protein hydrolysis. Many bioactive peptides have been experimentally generated using various commercial proteases (Korhonen & Pihlanto, 2003). Qian et al. (2007) used chymotrypsin, papain, neutrase, and trypsin for the hydrolysis of tuna dark muscle under the optimal pH and temperature conditions for the respective enzymes. The antioxidant property of a peptide is said to be closely linked to the amino acid components and their sequence within the peptide (Ren et al., 2008). The degree of hydrolysis as well as the size or molecular weight (Ren et al., 2008) of the peptides has been said to be a contributory factor to the expression of bioactive property by peptides.

7.6   USE OF FISH WASTE FOR ENZYMATIC HYDROLYSIS

Filleting waste generate significant quantity of fish frames which are normally discarded as industrial by-products in the fish processing plants. Attempts have been made to produce protein hydrolysates by several researchers to extract the commercially valuable products from various by-products such as Alaska Pollack frame protein hydrolyzed with mackerel intestine crude enzyme (Je et al., 2005), yellowfin sole (*Limanda aspera*) frame protein hydrolyzed with alcalase, neutrase, pepsin, papain, α-chymotrypsin, trypsin, and previously extracted tuna pyloric caeca crude enzyme (Rajapakse et al., 2005), tuna frame protein hydrolyzed using alcalase, neutrase, pepsin, papain, α-chymotrypsin, and trypsin (Lee et al., 2010), pink perch frame waste protein hydrolysate using papain and bromelain (Phadke et al., 2016) and many more.

7.7   BIOACTIVITIES OF FISH WASTE PROTEIN HYDROLYSATES

Bioactive compounds are essential and non-essential compounds that occur in nature, are part of the food chain, and can be shown to have an effect on human health. The bioactivities, namely, antihypertensive activity, antioxidant activity, antimicrobial activity, and anticoagulant activity have been discussed in this chapter.

7.7.1 ANTIHYPERTENSIVE ACTIVITY

7.7.1.1 HYPERTENSION, RENNIN–ANGIOTENSIN SYSTEM AND ACE-I INHIBITION

Hypertension affects about a quarter of the world's population and is a major, yet controllable, risk factor in cardiovascular disease and related complications, the biggest cause of death. Hypertension has numerous deleterious effects on the body, significantly increasing the risk of coronary artery disease, stroke, cardiac arrhythmia, heart failure, and abnormal renal function, and producing many other complications related to structural damage to the cardiovascular system (Murray & FitzGerald, 2007). Hypertension is mainly treated by lifestyle modification and pharmacological treatment with antihypertensive drugs (Hermansen, 2000).

Within the body, the major physiological pathways involved in the regulation of blood pressure (BP) are the rennin–angiotensin system (RAS), the kinin–nitric oxide system (KNOS), the rennin–chymase system (RCS), and neutral endopeptidase system (NEPS). The RAS is perhaps the most important of various humoral vasoconstrictor and vasodilator mechanisms implicated in BP regulation. It regulates BP, electrolyte balance, renal, neuronal, and endocrine functions associated with cardiovascular control in the body.

7.7.1.2 MECHANISM OF ACTION OF RENIN–ANGIOTENSIN SYSTEM AND ACE-I INHIBITION

In the RAS, through the action of kallikrein, renin is released from the precursor compound prorenin (Ondetti & Cushman, 1982). Renin cleaves angiotensinogen to release angiotensin I (Ang I), a decapeptide (Asp-Arg-Val-Tyr-Ile-His-Pro-Phe-His-Leu). The ACE hydrolyzes the Ang I by removing the C-terminal dipeptide His-Leu to give an octapeptide, angiotensin II (Ang II) (Asp-Arg-Val-Tyr-Ile-His-Pro-Phe), a potent vasoconstrictor (FitzGerald et al., 2004). ACE converts angiotensin I (Asp-Arg-Val-Tyr-Ile-His-Pro-Phe-His-Leu) into a powerful vasoconstrictor, angiotensin II, by removing the C-terminal dipeptide His-Leu. In addition, ACE inactivates the vasodilator bradykinin (Hartmann & Meisel, 2007). Therefore, the inhibition of ACE is considered to be a useful approach in the treatment of hypertension.

Currently, antihypertensive medications such as captopril and enalapril, marketed under the trade names Accupril, Altace, Capoten, Lotensin, Monoril, Prinvil, Vasotec, and Zestril, stabilize BP without removing the root cause, which is as yet unknown (Ahhmed & Muguruma, 2010). Although, these synthetic ACE inhibitory (ACEI) drugs have demonstrated their usefulness, they are not entirely without side effects, such as cough, loss of taste, renal impairment and angioneurotic edema (Cooper et al., 2006). Therefore, search for natural safe and more effective ACEI agents, as alternative ones, is necessary for the prevention and treatment of hypertension.

7.7.1.3 ACE INHIBITORY ACTIVITY OF FISH PROTEIN HYDROLYSATES

Chemical base medications may have harmful side effects for individuals. The search of natural ACE inhibitors as alternatives for synthetic medications is of great interest to in order to overcome the side effects. Therefore, current research is now focused on ACEI peptides from natural sources such as milk, eggs, meat, fish, and plants (Vercruysse et al., 2005). ACEI peptides have been isolated and characterized from a number of fish sources, either from the muscle proteins or from waste and discards from fish processing. ACEI activity of various enzymatic hydrolysates produced from fish waste (Je et al., 2004; Jung et al., 2006; Bougatef et al., 2008; Phadke et al., 2016) has been well documented. Work carried out on antihypertensive properties of fish protein hydrolysates by few researchers is summarized in Table 7.1.

TABLE 7.1 Antihypertensive Properties of Fish Protein Hydrolysates.

Substrate	Enzymes	ACE inhibitory activity (%) or (IC_{50} value)	References
Yellowfin sole frame	Chymotrypsin	Ultra filtration fraction action —	Jung et al. (2006)
		Fraction I: 47.6% ACE inhibition	
		Fraction II: 34.5% ACE inhibition	
		Fraction III: 68.8% ACE inhibition	
Sardinella by-products	Protease-K	47.4% ACE inhibition	Bougatef et al. (2008)
	Alcalase	43.0% ACE inhibition	
	Sardine visceral	63.2% ACE inhibition	
	Enzyme	55.8% ACE inhibition	
	Chymotrypsin	13.2% ACE inhibition	
	Protease ES 1		

TABLE 7.1   *(Continued)*

Substrate	Enzymes	ACE inhibitory activity (%) or (IC$_{50}$ value)	References
Tilapia	Cryotin	62–71% ACE inhibition	Raghavan and
	Flavourzyme	66–73% ACE inhibition	Kristinsson (2009)
Jellyfish	Papain	IC$_{50}$: 6.56 µM	Kim et al. (2011)
Fresh water calm by-products	Pepsin	IC$_{50}$: 0.23 mg/mL	Sun et al. (2011)
Giant jellyfish	Alcalase	39.61% ACE inhibition	Yoon et al. (2011)
	Flavourzyme	36.36% ACE inhibition	
	Neutrase	62.29% ACE inhibition	
	Papain	76.73% ACE inhibition	
	Protamex	70.01% ACE inhibition	
	Trypsin	68.01% ACE inhibition	
Pink perch fish frame waste	Papain	69% ACE inhibition	Phadke et al. (2016)
	Bromelain		

7.7.2   ANTIOXIDANT ACTIVITY

7.7.2.1   OXIDATION MECHANISM AND ITS EFFECT ON FOOD

The term oxidation refers to the loss of electrons from a molecule, an addition of oxygen or the loss of hydrogen from a compound (Parkin & Domadoran, 2003) which is a vital process in aerobic organisms, particularly in vertebrates and humans although it leads to the formation of free radicals. Oxidation primarily occurs on unsaturated fats by a free radical-mediated process. The radicals interact with molecular oxygen to form lipid peroxy radicals. These radicals can abstract a hydrogen atom from adjacent unsaturated fatty acids and produce a hydroperoxide and a new lipid radical, which causes the continuation and acceleration of the chain reaction. The formation of reactive oxygen species (ROS) including free radicals such as superoxide anion radicals (O_2^-), hydroxyl radicals (OH˙), and non-free radical species such as hydrogen peroxide (H_2O_2) and singlet oxygen ($^1O_2$), is an unavoidable consequence in the body's normal use of oxygen during respiration (Bernardini et al., 2011).

This process is of importance especially to the food industry and health sectors due to its possible detrimental effects such as off flavors, odors, dark colors, and formation of potentially toxic products (Wong, 1992). Lipid oxidation and its prevention using bioactive peptides is area of interest among the researchers. In food systems, lipid oxidation causes deterioration in food quality due to rancidity and formation of toxic compounds. Lipid oxidation in food products can be controlled by reducing metal ions and minimizing exposure to light and oxygen using packaging, as well as by employing correct levels of natural and synthetic antioxidants.

7.7.3 ANTIOXIDANTS

The word "antioxidant" has gained increased attention lately due to mass media coverage of its health benefits. Antioxidants are used to prolong the shelf life and maintain the nutritional quality of lipid-containing foods and to modulate the consequences of oxidative damage in the human body. Halliwell and Gutteridge (1999) defined an antioxidant as "any substance that, when present at low concentration compared with those of an oxidizable substrate, significantly delays or prevents oxidation of that substrate." Antioxidants are compounds which prevent or minimize oxidation by mechanisms such as free radical scavenging, reducing oxidants, or by mopping pro-oxidants such as iron. The role of bioactive peptides cannot be over emphasized in today's world, where people are increasingly becoming health conscious. Consumers of food products are showing an interest in the components that are present in their processed food and have a tendency to go in for those which are perceived as natural. These tendencies have been attributed to the adverse effects elicited by extreme levels of synthetic food additives and preservatives.

Butylated hydroxyanisole (BHA), tertiary-butyl-hydroquinone (TBHQ) and butylated hydroxytoluene (BHT) are some examples of synthetic antioxidants while α-tocopherol, ascorbate, and carotenoids are some examples of naturally occurring antioxidants. These are however, regulated due to suspicions of carcinogenic effects and as such limits have been placed on how much of these synthetic compounds can be used as preservatives. Because of potential toxic effects of synthetic antioxidants, natural antioxidants are preferred by consumers and processors. Therefore, additional sources of natural antioxidants including protein hydrolysates, peptides, and amino acids from plant and animal sources are currently being investigated. Numerous antioxidant peptides have been isolated from various sources which include plants, fishes, and mammals. Generally, the ability of a peptide to stabilize free radicals is due to its ability to donate electrons to the free radical or absorb the free radicals electron in order to reduce its reactivity. Peptides rich in histidine, cysteine, and methionine have been proposed as effective free radical scavengers (Ren et al., 2008). Antioxidants are widely used in dietary supplements to boost health and reduce the risk of diseases such as cancer and coronary heart disease. In addition, antioxidants have many industrial uses, such as preservatives in food and cosmetics and the prevention of rubber and gasoline degradation.

Although, synthetic antioxidants are very effective in food systems, their application in the food industry has declined due to consumer safety concerns and demand for all natural products (Ito et al., 1986).

7.7.3.1   ANTIOXIDANT ACTIVITY OF FISH PROTEIN HYDROLYSATES

Proteins from various plant and animal sources and their hydrolysis products, individual peptides, and amino acids are documented to exhibit antioxidant activity against the peroxidation of lipids and fatty acid. Peptides derived from enzymatic digestion of various fish proteins are reported to possess antioxidant activity based on the nature, size, and composition of the different peptide fractions and the protease specificity. Lipid peroxidation in foods affects the nutritive value and may cause disease following the consumption of products that could potentially cause a toxic reaction (Je et al., 2007). Antioxidative properties of round scad and yellow stripe trevally were influenced by the degree of hydrolysis and the enzyme type used (Klompong et al., 2007). Antioxidant activities of fish protein hydrolysates for the reported work by few researchers have been summarized in Table 7.2.

TABLE 7.2   Antioxidant Properties of Fish Protein Hydrolysates.

Substrate	Enzymes	Bioactive properties studied and peptide sequence	References
Striped catfish frame meat	Papain Bromelain	DPPH radical scavenging activity (90%), ferric reducing antioxidant power assay	Tanuja et al. (2012)
Goby muscle proteins	Alcalase	DPPH radical scavenging activity and reducing power assay	Nasri et al. (2012)
Rastrelliger kanagurta backbone	Pepsin Papain	DPPH radical scavenging activity (36–46%)	Sheriff et al. (2014)
Salmon protein hydrolysate	Pepsin	DPPH Radical scavenging activity (55%)	Girgih et al. (2013)
Pink perch frame waste hydrolysate	Papain Bromelain	DPPH free radical scavenging activity (up to 90%), ferric reducing antioxidant power	Phadke et al. (2016)

7.7.3.2   ANTICOAGULANT ACTIVITY OF FISH PROTEIN HYDROLYSATES

A series of processes occur which also could contribute to coronary artery blockage during blood coagulation if not regulated. During platelet activation, numerous platelets aggregate, this is accompanied by conversion of prothrombin to thrombin, a serine protease known to convert soluble fibrinogen to its insoluble form, fibrin. As more thrombin is produced, the conversion of fibrinogen to fibrin is facilitated. The combined effect of vasoconstriction and blocking of coronary arteries by fibrin complex formation within the blood vessel leads to myocardial ischemia and heart attacks. Therefore, regulation of platelet aggregation under these conditions will limit vasoconstriction and subsequently reduce the risk of myocardial ischemia. Rajapakse et al. (2005) purified anticoagulant from fish protein hydrolysate from yellowfin sole. It inhibited the activated coagulation factor XII by forming an inactive complex regardless of Zn^{2+}.

7.7.4   ANTIMICROBIAL ACTIVITY

Antimicrobials in food are used for inhibiting the growth of food spoilage microbes. Organic acids have been the most widely used antimicrobial agents and their mode of action has been by altering cell membrane permeability to substrates and by creating intolerable pH conditions for microbial growth (Back et al., 2009). Examples of these acids are sorbic acid, acetic acid, and citric acid. Organic acids like sorbic acid, although a very good preservative, is known to undergo degradation and yield compounds of possible toxicity like acetaldehyde when exposed to aqueous conditions. Therefore, the use of antimicrobial peptides is a potentially safer option which eliminates such occurrences. Upon enzymatic hydrolysis of proteins, peptides with different amino acid sequence and varying size have been reported to possess the antimicrobial activity. Anti-microbial peptides have been reported to exhibit a broad range of activities against various microorganisms including Gram-positive and Gram-negative bacteria, virus, fungi, and protozoa. They possess a wide spectrum range of activity against both bacteria and fungi. Anti-microbial peptides, also known as nature's antibiotics, are a group of bioactive peptides that exert anti-microbial activities and are naturally occurring or derived from hydrolysis of proteins.

Several microorganisms, including Gram-positive and Gram-negative bacteria, as well as fungi, represent the major causes of various human infections. Most antimicrobial peptides have been reported to be cationic in nature; meaning, they each possess a net positive charge due to positively charged amino acid groups like lysine and arginine (Brogden, 2005). Mode of action of antimicrobial peptides is believed to be by membrane pore formation and followed by penetration which results in the release of microbial cellular components and subsequent destruction of the cell.

7.8   RESEARCH GAPS, OPPORTUNITIES, AND CHALLENGES IN UTILIZATION OF FISH WASTE AS A SOURCE FOR THE PRODUCTION OF FISH PROTEIN HYDROLYSATE

In the recent years, there is considerable attention about fish protein hydrolysates among the researchers due to high amount of protein with good source of amino acids and biologically active peptides. Functional and antioxidant properties of fish processing waste protein hydrolysates have been well documented (Chalamaiah et al., 2012; Phadke et al., 2016) but further research is required in line with the commercialization of such hydrolysates with due consideration of the economic aspects as well. It is necessary to optimize the extent to which fish waste proteins can be hydrolyzed in order to maintain the optimal balance between functional and bioactive properties of these hydrolysates. Very few studies have been focused on the application of protein hydrolysates in food formulations. It is necessary to carry out further research related to safety of these hydrolysates. There are permissible limits for pathogen when raw or cooked fish are considered but for the production of protein hydrolysates from fish waste, standard permissible limits have not been formulated yet. During the enzymatic hydrolysis and post-hydrolysis process, more number of harmful microbes is killed but there is paucity of information available on safety of fish waste protein hydrolysates based on scientific studies. Microbial analysis and allergen analysis are vital to complete the food safety profile of protein hydrolysate before commercialization. Further, as these hydrolysates are from seafood source, monitoring of histamine levels are also mandatory. Though plenty of in vitro studies have been conducted on the bioactivity of protein hydrolysate, the fate of these functional compounds in the gastrointestinal tract, their absorption and bioavailability are to be thoroughly exploited.

7.9 CONCLUSION

Fish processing operation generates large amounts of waste that could be utilized for production of bioactive fish waste protein hydrolysate. In order to find out the natural alternatives to the synthetic drugs, research on bioactive protein hydrolysates from fish waste has paved attention of researchers. Various factors directly or indirectly influence the hydrolysis process as well as bioactive properties of the fish waste protein hydrolysates produced which include temperature, hydrolysis time, enzyme to substrate ratio, pH, substrate used, and degree of hydrolysis. Fish protein hydrolysates are currently used as health supplements or as nutraceuticals under various brand names. However, use of fish waste for production of biologically active protein hydrolysates is possible on industrial scale after comprehensive evaluation of its safety aspects.

KEYWORDS

- **waste utilization**
- **fish waste protein hydrolysate**
- **bioactive peptides**
- **antihypertensive activity**
- **antioxidant activity**
- **antimicrobial activity**

REFERENCES

Adler-Nissen, J. *Enzymic Hydrolysis of Food Proteins;* Elsevier Applied Science Publishers: Barking, UK, 1986.

Ahhmed, A. M.; Muguruma, M. A Review of Meat Protein Hydrolysates and Hypertension. *Meat Sci.* **2010,** *86,* 110–118.

Back, S. Y.; Jin, H. H.; Lee, S. Y. Inhibitory Effect of Organic Acids against *Enterobacter sakazakii* in Laboratory Media and Liquid Foods. *Food Control.* **2009,** *20* (10), 867–872.

Bernardini, R. D.; Harnedy, P.; Declan-Bolton, D.; Kerry, J.; O'neill, O.; Mullen, E. Antioxidant and Antimicrobial Peptidic Hydrolysates from Muscle Protein Sources and By-products: Review. *Food Chem.* **2011,** *124,* 1296–1307.

Bougatef, A.; Nedjar-Arroume, N.; Ravallec-Ple, R.; Leroy, Y.; Guillochon, D.; Barkia, A.; Nasri, M. Angiotensin I-converting Enzyme (ACE) Inhibitory Activities of Sardinella

(*Sardinella aurita*) By-products Protein Hydrolysates Obtained by Treatment with Microbial and Visceral Fish Serine Proteases. *Food Chem.* **2008,** *111,* 350–356.

Brogden, K. A. Antimicrobial Peptides: Spore Formers or Metabolic Inhibitors in Bacteria? *Nat. Rev. Microbiol.* **2005,** *3,* 238–250.

Chalamaiah, M.; Dinesh Kumar, B.; Hemalatha, R.; Jyothirmayi, T. Fish Protein Hydrolysates: Proximate Composition, Amino Acid Composition, Antioxidant Activities and Applications: A Review. *Food Chem.* **2012,** *135* (4), 3020–3038.

Cooper, W. O.; Hernandez-Diaz, S.; Arbogast, P. G.; Dudley, J. A.; Dyer, S.; Gideon, P. S.; Hall, K.; Ray, W. A. Major Congenital Malformations after First-trimester Exposure to ACE Inhibitors. *N. Engl. J. Med.* **2006,** *354,* 2443–2451.

FitzGerald, R. J.; Murray, B. A.; Walsh, D. J. The Emerging Role of Dairy Proteins and Bioactive Peptides in Nutrition and Health. *J. Nutr.* **2004,** *134,* 980–988.

Girgih, A. T.; Udenigwe, C. C.; Hasan, F. M.; Gill, T. A.; Aluko, R. E. Antioxidant Properties of Salmon (*Salmo salar*) Protein Hydrolysate and Peptide Fraction Isolated by Reverse-phase HPLC. *Food Res. Int.* **2013,** *52,* 315–322.

Halliwell, B.; Gutteridge, J. M. C. *Free Radicals in Biology and Medicine,* 4th ed.; Oxford University Press: Oxford, UK, 2007.

Hartmann, R.; Meisel, H. Food-derived Peptides with Biological Activity: From Research to Food Applications. *Curr. Opin. Biotechnol.* **2007,** *18,* 163–169.

He, S.; Franco, C.; Zhang, W. Functions, Applications and Production of Protein Hydrolysates from Fish Processing Co-products (FPCP). *Food Res. Int.* **2013,** *50,* 289–297.

Hermansen, K. Diet, Blood Pressure and Hypertension. *Br. J. Nutr.* **2000,** *83,* S113–S119.

Islam, M. D.; Khan, S.; Tanaka. M. Waste Loading in Shrimp and Fish Processing Effluents: Potential Source of Hazards to the Coastal and Nearshore Environments. *Marine Poll. Bull.* **2004,** *49,* 103–110.

Ito, N.; Hirose, M.; Fukushima, S.; Tsuda, H.; Shirai, T.; Tatematsu, M. Studies on Antioxidants: Their Carcinogenic and Modifying Effects on Chemical Carcinogenic Compounds. *Food Chem. Toxicol.* **1986,** *24,* 1099–1102.

Je, J. Y.; Oian, Z. J.; Byun, H. G.; Kim, S. K. Purification and Characterization of an Antioxidant Peptide Obtained from Tuna Backbone Protein by Enzymatic Hydrolysis. *Process Biochem.* **2007,** *42* (5), 840–846.

Je, J. Y.; Park, P. J.; Kim, S. K. Antioxidant Activity of a Peptide Isolated from Alaska Pollock (*Theragra chalcogramma*) Frame Protein Hydrolysate. *Food Res. Int.* **2005,** *38,* 45–50.

Je, J. Y.; Park, P. J.; Kwon, J. Y.; Kim, S. K. Novel Angiotensin I Converting Enzyme Inhibitory Peptide from Alaska Pollack (*Theragra chalcogramma*) Frame Protein Hydrolysate. *J. Agric. Food Chem.* **2004,** *52* (26), 7842–7845.

Jung, W. K.; Mendis, E.; Je, J. Y.; Park, P. J.; Son, B. W.; Kim, H. C.; Choi, Y. K.; Kim, S. K. Angiotensin I-converting Enzyme Inhibitory Peptide from Yellowfin Sole (*Limanda aspera*) Frame Protein and its Antihypertensive Effect in Spontaneously Hypertensive Rats. *Food Chem.* **2006,** *94* (1), 26–34.

Kim, Y.; Lim, C.; Yeun, S.; Lee, M.; Moon, H.; Cho, H.; Yoon, N.; Yoon, H.; Park, H.; Lee, D. Dipeptide (Tyr-Ile) Acting as an Inhibitor of Angiotensin-I-converting Enzyme (ACE) from the Hydrolysate of Jellyfish *Nemopilema nomurai. Fish Aquatic. Sci.* **2011,** *14* (4), 283–288.

Klompong, V.; Benjakul, S.; Kantachote, D.; Shahidi, F. Antioxidative Activity and Functional Properties of Protein Hydrolysate of Yellow Stripe Trevally (*Selaroides leptolepis*) as Influenced by the Degree of Hydrolysis and Enzyme Type. *Food Chem.* **2007,** *102,* 1317–1327.

Korhonen, H.; Pihlanto, A. Food-derived Bioactive Peptides: Opportunities for Designing Future Foods. *Curr. Pharm. Des.* **2003,** *9,* 1297–1308.

Kristinsson, H. G.; Rasco, B. A. Fish Protein Hydrolysates: Production, Biochemical, and Functional Properties. *Crit. Rev. Food Sci. Nutr.* **2000,** *40* (1), 43–81.

Lee, H. S.; Qian, Z. J.; Kim, S. K. A Novel Angiotensin I Converting Enzyme Inhibitory Peptide from Tuna Frame Protein Hydrolysate and Its Antihypertensive Effect in Spontaneously Hypertensive Rats. *Food Chem.* **2010,** *118,* 96–102.

Murray, B. A.; Fitzgerald, R. J. Angiotensin Converting Enzyme Inhibitory Peptides Derived from Food Proteins: Biochemistry, Bioactivity and Production. *Curr. Pharm. Des.* **2007,** *13,* 773–791.

Nalinanon, S.; Benjakul, S.; Kishimura, H.; Shahidi, F. Functionalities and Antioxidant Properties of Protein Hydrolysates from the Muscle of Ornate Threadfin Bream Treated with Pepsin from Skipjack Tuna. *Food Chem.* **2011,** *124,* 1354–1362.

Nasri, R.; Amor, I. B.; Bougatef, A.; Nedjar-Arroume, N.; Dhulster, P.; Gargouri, J.; Chaabouni, M. K.; Nasri, M. Anticoagulant Activities of Goby Muscle Protein Hydrolysates. *Food Chem.* **2012,** *133,* 835–841.

Ondetti, M. A.; Cushman, D. W. Enzymes of the Renin-Angiotensin System and Their Inhibitors. *Ann. Rev. Biochem.* **1982,** *51,* 283–308.

Parkin, K. L.; Damodaran, S. Oxidation of Food Components. In *Encyclopedia of Food Science and Nutrition,* 2nd ed.; Caballero, B., Trugo, L. C., Finglas, P. M., Eds.; Academic Press: San Diego, CA, 2003; pp 4288–4294.

Phadke, G. G.; Elavarasan, K.; Shamasundar, B. A. Bioactive and Functional Properties of Protein Hydrolysates from Fish Frame Processing Waste Using Plant Proteases. *Environ. Sci. Pollut. Res.* **2016,** *23* (24), 24901–24911.

Qian, Z. J.; Jung, W. K.; Lee, S. H.; Byun, H. G.; Kim, S. K. Antihypertensive Effect of an Angiotensin I-converting Enzyme Inhibitory Peptide from Bullfish (*Rana catesbeiana* shaw) Muscle Protein in Spontaneously Hypertensive Rats. *Process Biochem.* **2007,** *42,* 1443–1448.

Raghavan, S.; Kristinsson, H. G. ACE Inhibitory Activity of Tilapia Protein Hydrolysate. *Food Chem.* **2009,** *117,* 582–588.

Rajapakse, N.; Jung, W.; Mendis, E.; Moon, S.; Kim, S. K. A Novel Anticoagulant Purified from Fish Protein Hydrolysate Inhibits Factor XIIa and Platelet Aggregation. *Life Sci.* **2005,** *76,* 2607–2619.

Ren, J.; Zhao, M.; Shi, J.; Wang, J.; Jiang, Y.; Cui, C.; Kukuda, Y.; Xue, J. S. Optimization of Antioxidant Peptide Production from Grass Carp Sarcoplasmic Protein Using Response Surface Methodology. *Food Sci. Technol.* **2008,** *41,* 1624–1632.

Sheriff, S. A.; Sundaram, B.; Ramamoorthy, B.; Ponnusamy, P. Synthesis and In Vitro Antioxidant Functions of Protein Hydrolysate from Backbones of *Rastrelliger kanagurta* by Proteolytic Enzymes. *Saudi J. Biol. Sci.* **2014,** *21* (1), 19–26.

Shirai, K.; Ramirez, J. C. Utilization of Fish Processing By-products for Bioactive Compounds. In *Fish Processing – Sustainability and New Opportunitie,* 1st ed.; Hall, G. M. Ed.; Blackwell Publishing Limited: Hoboken, NJ, 2011; pp 236–265.

Sun, Y.; Hayakawa, S.; Ogawa, M.; Naknukool, S.; Guan, Y.; Matsumoto, Y. Evaluation of Angiotensin I-converting Enzyme (ACE) Inhibitory Activities of Hydrolysates Generated from Byproducts of Freshwater Clam. *Food Sci. Biotechnol.* **2011,** *20* (2), 303–310.

Tanuja, S.; Viji, P.; Zynudheen, A. A.; Joshy, C. G. Composition, Functional Properties and Antioxidative Activity of Hydrolysates Prepared from the Frame Meat of Striped Catfish (*Pangasianodon hypophthalmus*). *Egypt. J. Biol.* **2012,** *14,* 27–35.

Thiansilakul, Y.; Benjakul, S.; Shahidi, F. Antioxidative Activity of Protein Hydrolysate from Round Scad Muscle Using Alcalase and Flavourzyme. *J. Food Biochem.* **2007,** *31* (2), 266–287.

Vercruysse, L.; Van-Camp, J.; Smagghies, G. ACE Inhibitory Peptide Derived Enzymatic Hydrolysates of Animal Muscle Protein: A Review. *J. Agric. Food Chem.* **2005,** *53* (21), 8106–8115.

Wong, D. W. S. Oxidation. In *Encyclopedia of Food Science and Technology;* Hui, Y. H., Ed.; John Wiley and Sons Inc.: Hoboken, NJ, 1992; Vol. 3, pp 1965–1970.

Yoon, H. D.; Kim, Y. K., Lim, C; Yeun, S; Lee, M.; Moon, H.; Yoon, N.; Park, H.; Lee, D. ACE-inhibitory Properties of Proteolytic Hydrolysates from Giant Jelly Fish *Nemopilema nomurai. Fish Aquatic. Sci.* **2011,** *14* (3), 174–178.

CHAPTER 8

EFFECT OF OREGANO ESSENTIAL OIL ON THE STABILITY OF MICROENCAPSULATED FISH OIL

JEYAKUMARI A.[1*], ZYNUDHEEN A. A.[2], BINSI P. K.[2], PARVATHY U.[1], and RAVISHANKAR C. N.[2]

¹Mumbai Research Centre of ICAR—Central Institute of Fisheries Technology, Vashi 400703, Navi Mumbai, India

²ICAR—Central Institute of Fisheries Technology, Kochi 682029, India

*Corresponding author. E-mail: jeya131@gmail.com

CONTENTS

Abstract 216
8.1 Introduction 216
8.2 Materials and Methods 217
8.3 Conclusion 222
Keywords 222
References 223

ABSTRACT

Omega-3 polyunsaturated fatty acids (PUFAs) have numerous human health benefits includes reduction of cardiovascular disease risk, rheumatoid arthritis, reduced risk of certain cancer forms and important for brain and nervous tissue developments in infants. Fish oil is an excellent source of omega-3 fatty acids like eicosapentaenoic acid (EPA) and docosahexaenoic acid (DHA), but their addition to foods is limited on account of oxidative rancidity. Hence, fish oil has to be protected against oxidation preferably by microencapsulation. In this study, microencapsulation of fish oil was done by spray drying. Sodium caseinate, bovine gelatin, gum arabic, and maltodextrin were used as encapsulant material. Fish oil and wall material were used at the ratio of 1:2. In order to study the effect of natural antioxidants on the fish oil encapsulates, oregano essential oil was added at 0.50% concentration. Physical, chemical, and oxidative stability of fish oil microencapsulate were analyzed. Microcapsules had a moisture content of 2.56–4.2%. Higher encapsulation efficiency (EE) (65.13%) was observed for microencapsulate contained gum arabic and sodium caseinate as a wall material. Morphological characterization of fish oil encapsulates by scanning electron microscopy (SEM) revealed spherical shape of particles without any cracks. Flow properties of fish oil encapsulate were passable to poor. Oxidative stability of fish oil encapsulates were tested under accelerated temperature of 60°C for 7 days at 24 h interval and it was monitored by determining peroxide value and thiobarbituric acid value. Oxidative stability studies revealed that encapsulate prepared by sodium caseinate and gum arabic with oregano essential oil showed lower TBA (0.58 mg malonaldehyde/kg) value than control (9.92 mg malonaldehyde/kg). Results from the study suggested that oregano essential oil can be used to improve the oxidative stability of fish oil microencapsulates.

8.1   INTRODUCTION

Fish oil represents a functional food ingredient due to the fact that they are excellent dietary sources of the important fatty acids especially polyunsaturated fatty acid (PUFA) like eicosapentaenoic acid (EPA) and docosahexaenoic acid (DHA). Many researchers showed that supplemental fish oil may be beneficial for the healthy function of the heart, brain, and nervous system (Wu et al., 2005). However, fish oil has a strong odor and unless protected it is easily oxidized. The oxidative deterioration of PUFA in fish oil results in a loss of nutritional value and the development of off-flavors (Lee et al., 1999).

The usual approach to minimizing oxidation involves the addition of antioxidants and microencapsulation (Matsuno & Adachi, 1993). Microencapsulation of fish oil produces a dry powder from liquid fish oil, which enables its use for the enrichment of instant foods, formulae, bread, etc., in order to deliver omega-3 fatty acids. Spray drying is commonly used in the food and pharmaceutical industries to transform liquid materials into dried powders (Bhandari et al., 2008) and it has been widely applied to prepare omega-3 PUFA microcapsules. Spray drying offers many advantages over other drying methods, such as freeze drying, extrusion, complex-coacervation, etc., which includes low operational cost, ability to handle heat-sensitive materials, readily available machinery and reliable operation, and ability to control the mean particle size of the powders for spray dried emulsions. Typical wall materials for microencapsulation by spray drying are low molecular weight carbohydrates like maltodextrins or saccharose, milk or soy proteins, gelatin and hydrocolloids like gum arabic or mesquite gum. The physicochemical properties of spray-dried powders are affected by process variables such as inlet air temperature, concentration of carrier agent, feed flow rate, etc. (Chegini et al., 2008). Although encapsulation itself prevents lipid oxidation, additional stabilization with antioxidants is required to ensure maximum protection during processing and subsequent storage of microencapsulated bioactive ingredients. There is an increasing interest in essential oils of herbs and spices as source of natural antioxidants (Baratta et al., 1998; Bakkali et al., 2008). Essential oils and their components are gaining interest because of their relatively safe status, their mode of acceptance by consumers and numerous studies have been carried out on their antimicrobial, antifungal, antioxidant, and radical-scavenging properties (AOAC, 2000). In this background, this study was carried out with the objective to prepare fish oil encapsulates by spray drying with inclusion of essential oil and to evaluate its oxidative stability under accelerated condition.

8.2 MATERIALS AND METHODS

8.2.1 RAW MATERIALS

Fish oil (Seacod, Universal Medicare, Mumbai, India) was used for the preparation of emulsion and encapsulates. Sodium caseinate, gum arabic, bovine gelatin, and maltodextrin were procured from Himedia Laboratory and used as wall material for encapsulation. Oregano essential oil was purchased from Synthite Industries (Cochin, India).

8.2.2 PREPARATION OF EMULSION AND SPRAY DRYING

Four formulations were made for encapsulation, namely (1) Bovine gelatin + maltodextrin; (2) sodium caseinate + gum arabic; (3) Bovine gelatin + maltodextrin + oregano essential oil (0.50%); and (4) sodium caseinate + gum arabic + oregano essential oil (0.50%). Fish oil and wall material were used at the ratio of 1:2. After the dissolution of wall material, fish oil was added and further it was homogenized using a tissue homogenizer (Poly system PT 2100, Kinematica AG) at 25,000 rpm for 5 min and spray dried using a pilot-scale spray dryer (Hemraj Enterprises, Mumbai, India) under the following experimental conditions, namely, inlet temperature 160°C; outlet temperature 80°C; spray flow feed nozzle diameter 3 mm; and air pressure 2.5 bar. Fish oil encapsulates prepared by spray drying were used for further analysis.

8.2.3 CHARACTERIZATION OF MICROENCAPSULATES

8.2.3.1 PARTICLE MORPHOLOGY

Surface morphology of microencapsulates was evaluated by Field emission scanning electron microscope (JEOL-JSM-7600F, Japan). Samples were uniformly spread onto double-sided adhesive carbon tape mounted on SEM stubs, coated with platinum using auto fine coater (JEOL-JFC-1600, Japan) in a vacuum and examine by SEM at 10 kV with magnification of 2000×.

8.2.3.2 MOISTURE CONTENT

Moisture content of fish oil encapsulates was determined by the method of AOAC (2000). The microencapsulates weight loss after oven drying at 105°C until a constant weight was obtained, and the percentage of moisture content was evaluated.

8.2.3.3 ENCAPSULATION EFFICIENCY

EE was derived from the relationship between total oil and surface oil or free oil. Total oil extraction was based on the Soxhlet method (AOAC, 2000). Free oil fraction was extracted according to Sankarikutty et al. (1988). EE was calculated by the following formula:

$$EE (\%) = Total\ oil - Surface\ oil/Total\ oil \times 100.$$

8.2.3.4 FLOW PROPERTIES

Bulk density (ρB) and tapped densities (ρT) of microencapsulates were determined by the method of Chinta et al. (2009). The flow properties of microencapsulates were estimated from Carrs Index and Hausner ratio according to Turchiuli et al. (2005).

8.2.3.5 COLOR ANALYSIS

Color of microencapsulated fish oil was measured using a Hunter-Lab scan XE—Spectrocolorimeter (Hunter Associates Laboratory, Reston, USA) at D-65 illuminant and 10° observer. Results were expressed by Commission Internationale de L "Eclairage"s (CIE) color values [L*(lightness), a*(redness), and b*(yellowness)]. The instrument was calibrated using white and black standard ceramic tiles and the readings were recorded in the inbuilt software.

8.2.4 DETERMINATION OF OXIDATIVE STABILITY

Oxidative stability of fish oil encapsulates stored under accelerated temperature of 60°C was tested up to for 7 days at 24 h interval. Lipid oxidation in microencapsulates was determined by measuring peroxide value (Shantha & Decker, 1994) and thiobarbituric acid reactive substances (TBARS) (McDonald & Hultin, 1987).

8.2.5 RESULTS AND DISCUSSION

8.2.5.1 PHYSICAL PROPERTIES OF MICROENCAPSULATED FISH OIL

Moisture content of the fish oil encapsulates ranged between 2.56% and 4.2%. This is little higher than the moisture content (2.89–3.02%) reported for tuna oil powders by Klaypradit and Huang (2008). It may be due to wall material composition used for the study. The flowability of powder is an important property in handling and processing operations, such as storage, transportation, formulation, mixing, compression, or packaging (Kim et al., 2005). Quality parameters like bulk density, Carr Index, and Hausner ratio have been used to assess powder flowability (Fitzpatric & Ahrne, 2005). Results showed that fish oil encapsulates had higher Carr Index (26.57–36.27) and Hausner ratio (1.36–1.57) which indicated poor flowability of the spray dried powder. Determination of free and encapsulated oil will provide the information of EE. EE is used to assess the quality of the dried microcapsules because it has an effect on oxidation sensitivity and flow powder properties. In this study, EE of fish oil encapsulates ranged from 39.60% to 65.13%. The difference in EE can be attributed to different core and wall material (Jeyakumari et al., 2015; Binsi et al., 2016). SEM analysis of fish oil encapsulates prepared by spray drying showed a spherical shape and wrinkle appearance (Figs. 8.1 and 8.2). It was observed that encapsulates made by spray drying showed higher variation in sizes, indicating that during this process different sizes of droplets were formed. Kolanowski et al. (2006) observed the spherical structure of microencapsulates with different sizes by spray drying of fish oil. Kagami et al. (2003) also reported different degrees of formation of surface indentations for fish oil containing microcapsules produced from protein and dextrin wall materials.

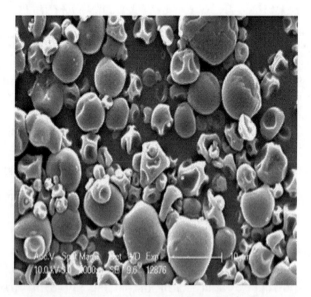

FIGURE 8.1   SEM image of microencapsulated fish oil prepared without oregano essential oil.

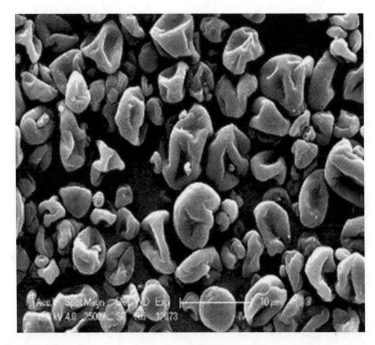

FIGURE 8.2   SEM image of microencapsulated fish oil prepared with oregano essential oil.

Tristimulus color values of L*, a*, and b* are used as indices for the color changes in fish oil encapsulates during storage. This study showed decreased L* value (86.02–78.87) under accelerated storage condition. Jeyakumari et al. (2014) observed similar trend for fish oil encapsulates stored under room temperature.

8.2.5.2   *OXIDATIVE STABILITY OF MICROENCAPSULATED FISH OIL*

Oxidative stability was evaluated by measuring TBARS. TBARS are secondary oxidation products formed from the breakdown of oxidized PUFAs (Shahidi & Wansundara, 2002) TBARS assay is a useful tool in monitoring lipid peroxidation owing to its sensitivity and simplicity. Measuring secondary oxidation products is important in the determination of lipid oxidation in food products for human consumption because they are generally odor active, whereas primary oxidation products are colorless and flavorless. In this study, TBARS values showed increasing trend during storage. However, the levels of secondary lipid oxidation products were lower in the fish oil encapsulates prepared with the addition of oregano essential oil than others. Oxidative stability studies revealed that encapsulate contained sodium caseinate, gum arabic, and oregano essential oil showed lower TBARS value (0.58 mg malonaldehyde/kg) value than control (9.92 mg malonaldehyde/kg). Jeyakumari et al. (2014) observed low TBARS value in fish oil encapsulates prepared with fish gelatin and 0.25% ginger essential oil under room temperature. Lauren et al. (2007) also reported that combination of tocopherol and EDTA was effective in increasing oxidative stability of spray-dried multilayer emulsion. Although oil/fat is encapsulated, the oxidative stability may be due to the presence of surface oil and air inclusion which influences the shelf-life of microencapsulated fish oil (Keogh et al., 2001).

8.3   CONCLUSION

In this study, it can be concluded that fish oil encapsulates prepared by spray drying using gum arabic and sodium caseinate revealed that wall material used in the study were able to encapsulate the fish oil without affecting the emulsion properties as indicated by uniform and spherical structure of microencapsulates and higher EE. Fish oil encapsulates prepared with oregano essential oil showed better oxidative stability of microencapsulated fish oil than control under accelerated condition. Further, encapsulation of fish oil with oregano essential oil revealed the prospect of incorporating essential oils from diverse herbal sources into encapsulates for protecting highly susceptible omega-3 fatty acids from lipid oxidation.

KEYWORDS

- **fish oil**
- **microencapsulation**
- **spray drying**
- **oregano essential oil**
- **stability**

REFERENCES

AOAC. Official Methods of Analysis, 17th ed.; Association of Analytical Chemists: Washington, DC, 2000.

Baratta, M. T.; Dorman, H. J.; Deans, S. G.; Figueiredo, A. C.; Barroso, J.; Ruberto, G. Antimicrobial and Antioxidant Properties of Some Commercial Essential Oils. *Flavour Frag. J.* **1998,** *13,* 235–244.

Bakkali, F.; Averbeck, S.; Averbeck, D.; Idomer, M. Biological Effects of Essential Oils—A review. *Food Chem. Toxicol.* **2008,** *46,* 446–475.

Bhandari, B. R.; Patel, K. C.; Chen, X. D. Spray Drying of Food Materials—Process and Product Characteristics. In Drying Technologies in Food Processing; Chen, X. D., Mujumdar, S. A., Eds.; Blackwell Publishing Ltd.: Oxford, UK: 2008; pp 113–157.

Binsi, P. K.; Natasha, N.; Sarkar, P. C.; Jeyakumari, A.; Muhamed, A.; George, N.; Ravi Shankar, C. N. Structural and Oxidative Stabilization of Spray Dried Fish Oil Microencapsulates with Gum Arabic and Sage Polyphenols: Characterization and Release Kinetics. *Food Chem.* **2016,** *219,* 158–168. Doi.og/10.1016/ j.food chem. 2016-09-126

Chegini, G. R.; Khazaei, J.; Ghobadian, B.; Goudarzi, A. M. Prediction of Process and Product Parameters in an Orange Juice Spray Dryer Using Artificial Neural Networks. *J. Food Eng.* **2008,** *84,* 534–543.

Chinta, D. D.; Graves, R. A.; Pamujula, S.; Praetorius, N.; Bostanian, L. A.; Mandal, T. K. Spray-dried Chitosan as a Direct Compression Tableting Excipient. *Drug Dev. Ind. Pharm.* **2009,** *35,* 43–48.

Fitzpatrick, J. J.; Ahrne, L. Food Powder Handling and Processing: Industry Problems, Knowledge Barriers and Research Opportunities. *Chem. Eng. Process.* **2005,** *44,* 209–214.

Jeyakumari, A.; Kothari, D. C.; Venkateshwarlu, G. Microencapsulation of Fish Oil-milk Based Emulsion by Spray Drying: Impact on Oxidative Stability. *Fish. Technol.* **2014,** *51,* 31–37.

Jeyakumari, A.; Kothari, D. C.; Venkateshwarlu, G. Oxidative Stability of Microencapsulated Fish Oil during Refrigerated Storage. *J. Food Proces Preserv.* **2015,** doi:10.1111/ jfpp.12433

Kagami, Y.; Sugimura, S.; Fujishima, N.; Matsuda, K.; Kometani, T.; Matsumura, Y. Oxidative Stability, Structure, and Physical Characteristics of Microcapsules Formed by Spray Drying of Fish Oil with Protein and Dextrin Wall Materials. *J. Food Sci.* **2003,** *68,* 2248–2255.

Keogh, M. K.; O'kennedy, B. T.; Kelly, J.; Auty, M. A.; Kelly, P. M.; Fureby, A.; Haahr, A. M. Stability to Oxidation of Spray-dried Fish Oil Powder Microencapsulated Using Milk Ingredients. *J. Food Sci.* **2001,** *66,* 217–224.

Kim, E. H. J.; Chen, X. D; Pearce, D. Effect of Surface Composition on the Flow Ability of Industrial Spray-dried Dairy Powders. *Colloids Surf. B.* **2005,** *46,* 182–187.

Kolanowski, W.; Ziolkowski, M.; Weibrodt, J.; Kunz, B.; Laufenberg, G. Microencapsulation of Fish Oil by Spray Drying—Impact on Oxidative Stability. Part 1. Eur. Food Res. Technol. **2006,** 222, 336–342.

Klaypradit, W.; Huang, Y. W. Fish Oil Encapsulation with Chitosan Using Ultrasonic Atomizer. *LWT Food Sci. Technol.* **2008,** *41,* 1133–1139.

Lauren, A.; Shaw, D.; McClements, J.; Decker, E. A. Spray Dried Multilayered Emulsions as a Delivery Method for ω-3 Fatty Acids into Food Systems. *J. Agric. Food Chem.* **2007,** *55,* 3112–3119.

Lee, S. W.; Kim, M. H.; Kim, C. K. Encapsulation of Ethanol by Spray Drying Technique: Effects of Sodium Lauryl Sulfate. *Int. J. Pharm.* **1999,** *187,* 193–198.

Matsuno, O. R.; Adachi, S. Lipid Encapsulation Technology-Techniques and Application to Food. *Trends Food Sci.Technol.* **1993,** *4,* 256–261.

Mcdonald, R. E.; Hultin, H. O. Some Characteristics of the Enzymic Lipid Peroxidation System in the Microsomal Fraction of Flounder Skeletal Muscle. *J. Food Sci.* **1987,** *52,* 15–21.

Sankarikutty, B.; Sreekumar, M. M.; Narayann, C. S.; Mathew, A. G. Studies on Microen-capsulation of Cardamom Oil by Spray Drying Technique. *J. Food Sci. Technol.* **1988,** *25,* 352–356.

Shahidi, F.; Wanasundara, U. N. Chemistry, Nutrition and Biotechnology. In Food Lipids; Akoh, C. C., and Min, D. B., Eds.; Marcel Dekker, Inc.: New York, NY, 2002; pp 465–487.

Shantha, N. C.; Decker, E. A. Rapid, Sensitive, Iron-based Spectrophotometric Methods for Determination of Peroxide Values of Food Lipids. *J. AOAC Inter.* **1994,** *77,* 421–424.

Turchiuli, C.; Fuchs, M.; Bohin, M.; Cuvelier, M. E.; Ordonnaud, C.; Peyrat-Maillard, M. N.; Dumoulina, E. Oil Encapsulation by Spray Drying and Fluidized Bed Agglomeration. *Innov. Food Sci. Emerg. Technol.* **2005,** *6* (1), 29–35.

Wu, K.; Chai, X.; Chen, Y. Microencapsulation of Fish Oil by Simple Coacervation of Hydroxypropyl Methylcellulose. *Chin. J. Chem.* **2005,** *23,* 1569–1572.

CHAPTER 9

SAFETY CONSIDERATIONS OF RAW MEAT: AN INDIAN PERSPECTIVE

VISHVA MARU and VAIJAYANTI V. RANADE[*]

Department of Microbiology, Guru Nanak Khalsa College of Arts, Science and Commerce (Affiliated to University of Mumbai), Nathalal Parekh Marg, Matunga, Mumbai 400019, Maharashtra, India

[*]*Corresponding author. E-mail: vaijayantiranade@gmail.com*

CONTENTS

Abstract 226
9.1   Overview of Indian Livestock Industry 226
9.2   Organized Sector of the Indian Livestock Industry..........227
9.3   Challenges of the Unorganized Sector.............................229
9.4   Conditions of Indian Slaughterhouses for Domestic Supply 234
9.5   Statistical Overview of Indian Livestock Industry 235
9.6   Existing Laws and Regulations in India with Respect to the
       Livestock Industry..236
9.7   Establishing Quality Parameters of Meat in the
       Unorganized Sector...239
Keywords 245
References..245

ABSTRACT

With changing times, livestock industry in India has seen tremendous modification and development. An inherent desire to actively contribute in world trade has led to magnificent growth in Indian beef export industry, placing India on the top notch of beef export. Current scenario of unorganized sector of Indian livestock has been portrayed to highlight various loopholes in the existing system. Several laws and regulations proposed by national and international government agencies have been implemented in organized sector of India for export of good-quality meat. Yet, quality of meat supplied to local population that forms the unorganized sector is inadequate and hence a cause of societal concern. This chapter lays emphasis on biological, chemical, and physical factors that affect meat quality. This would indeed demand for necessary measures that need to be inculcated in the unorganized sector to provide better quality meat from local slaughterhouses.

9.1 OVERVIEW OF INDIAN LIVESTOCK INDUSTRY

Meat, a staple food of majority of the world population, has made livestock a multibillion-dollar industry in the world economy. Shift of diets and food consumption patterns has caused a paradigm shift in world food economy toward livestock products, bringing in the "food revolution" (Delgado et al., 1999).

Past four decades have witnessed a transformation of livestock in India, from a mere tool of supplementary income and nutritious food for the family to a major commercial activity generating required revenue. Changing food habits, rising income of the middle-class Indian, presence of private players, and rising demand of Indian livestock produce in the export market are some of the contributing factors to the growth of this industry. However, there is a wide gap in the quality of meat exported and that sold in the retail market. This wide gap has inadvertently divided the livestock industry in India into organized and unorganized sectors.

The organized sector has few players, but still is the one contributing to Indian export and economy. Production and processing of buffalo meat is the main focus of this sector because there is less demand for buffalo meat in India. Therefore, the organized sector employs stringent quality measures for processing buffalo meat as most of it is exported. However, this is not the case for meat in unorganized sector.

The unorganized sector comprises unprocessed meat sold in the local markets in rural and urban areas. It mainly consists of poultry, goat/sheep meat, and pig meat to a lesser extent. In rural areas, animal husbandry basically acts as a source of additional income and an asset that can be sold at the time of crisis (Randolph et al., 2007). In urban areas, fresh raw meat is sold in slaughterhouses cramped in small establishments and feeding the ever-increasing population of the cities.

Poultry and goat meat is the most consumed food commodity in developing countries like India, where ruminant meat is still considered as a taboo by many. However, the quality of meat supplied in the unorganized sector is not up to the standards laid down by various enforcement agencies.

The focus of this chapter, therefore, is on the unorganized sector of meat industry in India, dealing with relevant problems such as meat quality, hygiene standards, climatic conditions, illiteracy, exploitation of poor, and lack of awareness at all levels. We here, discuss an entire farm to fork approach of meat industry in India. An attempt to include a multidisciplinary perspective has been made with discussion on key issues related to animal science, economics, epidemiology, and public health. Employing a "harm reduction" approach to the issues related to above areas in a public perspective is, therefore, justified.

9.2   ORGANIZED SECTOR OF THE INDIAN LIVESTOCK INDUSTRY

According to data released by US Department of Agriculture and Agriculture and Processed Export Development Authority (APEDA), buffalo meat has emerged as India's top agri-export commodity leaving Basmati rice at second position(http://www.indiantradeportal.in/vs.jsp?lang=0&id=0,10,662,http://time.com/3833931/india-beef-exports-rise-ban-buffalo-meat/).

For over four decades, India is reputed to be a reliable exporter of risk-free, lean, nutritious, and competitively priced meat. This has resulted in consistent compounded growth in export of processed meat. India exports both frozen and fresh chilled meat to countries such as Vietnam, Malaysia, Thailand, Australia, UAE, Saudi Arabia, and Egypt (http://apeda.gov.in/apedawebsite/SubHead_Products/Buffalo_Meat.htm.; http://money.cnn.com/2015/08/05/news/economy/india-beef-exports-buffalo). The private sector has invested significantly and the government has been equally supportive in initiation and progress. This has enabled putting in place various aspects, such as water, labor, and infrastructure, programs on buffalo breeding, veterinary services, and disease control which are very efficient.

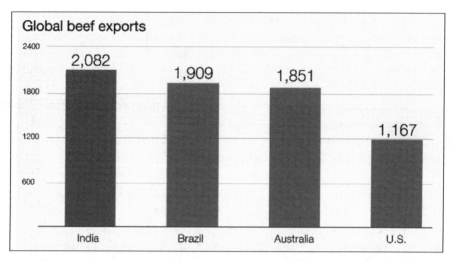

FIGURE 9.1 Global beef exports (adapted from US Department of Agriculture; http://money.cnn.com/2015/08/05/news/economy/india-beef-exports-buffalo).

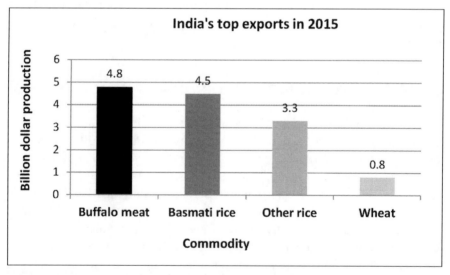

FIGURE 9.2 India's top exports in 2015 (adapted and modified from APEDA exports during 2014–2015; http://money.cnn.com/2015/08/05/news/economy/india-beef-exports-buffalo).

There are about a dozen export-oriented, modern, integrated abattoirs, or meat processing plants registered with the Agricultural and Processed Food Export Development Authority (APEDA).

Some of them include Venky's, Government Bacon, Godrej, Hind Agro Industries, Al Noor Exports, Al Nafees Frozen Food Exports, Frigerico Conserva Allana, Rustom Foods, and Rayban Foods which are all major exporters of meat to different countries (http://www.apeda.gov.in/apedawebsite).

9.3 CHALLENGES OF THE UNORGANIZED SECTOR

The term unorganized sector when used in Indian context is defined by National Commission for Enterprises in the Unorganized Sector, in their Report on Conditions of Work and Promotion of Livelihoods in the Unorganized Sector as "consisting of all unincorporated private enterprises owned by individuals or households engaged in the sale or production of goods and services operated on a proprietary or partnership basis and with less than ten total workers"(www.nceuis.nic.in).

Unorganized sector in food industry at one end has fresh food supplied in the local market as well as partially processed food. Wide variation in food preferences has given rise to a semi-organized sector in urban areas, which prefers processed meat made available in supermarkets over locally sold fresh meat. However, lesser demand for this type of meat has created stagnancy. Even today majority of the meat-eating population of the country belongs to the lower socio-economic class and prefers fresh meat from local butcher shops which are available at a cheaper cost. Hence, the quality of meat supplied by this sector is of great concern as it feeds larger population. Any compromises made in this sector can give rise to deteriorating quality and serious public health consequences.

9.3.1 GENERAL CONCERNS

In developing countries like India, agriculture still holds the prime status and livestock industry is yet to receive its due share. Rearing of animals for milk is the mainstay but obtaining meat from animals is still a subsidiary aspect. Apart from this, following are several other issues that Indian livestock sector is still struggling with:

- *Agriculture/land availability*: Livestock production is the world's largest user of land, either directly through grazing or indirectly through consumption of fodder and feed grains (Thornton, 2010).

Major problem is the availability of land for rearing of animals, due to high cost of land, poultry farms are shifting to outskirts which then increases the transportation cost. Land for agriculture is still a major crisis, at the same time making land available for growth of animal feed is a big problem. Most of the animals are just left to graze on open lands thus they lack adequate nutrition. In case of poultry, the birds are kept in small cages placed in unhygienic environment. Cereal-based feed is the one still preferred for these animals and hence, it increases the burden on agriculture. Traditional mixed farming practices are no longer capable of meeting the demand for animal-based food.

- **Breed of animal**: In the unorganized sector, there is no proper lineage of animal or poultry that is reared for food. Mixed breed of animals and poultry is used based on their availability and health. Animals and poultries are purchased from rural areas where different breeds are kept together. This genetic diversity and failure of genetic selection can be a reason for the occurrence and spread of various zoonotic diseases. Lack of awareness and knowledge related to quality of breed compromises on supplying safe and uniform product to the consumers (Opara, 2003).

- **Economic aspects and reduced returns**: A wide gap in money value from farm to market is another reason why the unorganized sector has not received the kind of attention it deserves. Improper chain of supply from farm to wholesaler to retailer and exploitation of the poor is the reason for failure of livestock in terms of economy. Regulations, laws, etc. for setting up of proper market value for animal-based foods are still not completely implemented. Financial barriers prevent small farmers from intensifying their production.

- **Apathy of enforcement agencies**: No structured system for formulating laws, guidelines, regulations, and failure to implement Indian or Worldwide standards has significantly affected the growth of livestock in India. Lack of political consideration, religious issues, and negligence by private enterprises has caused lack of interest and support from government which further worsens the situation. Hence, there is paucity of enforcement of laws and regulations for the unorganized sector for meat.

- **Consumer preferences**: Lack of education and awareness about the hygiene and sanitation condition of meat product has a direct effect on its quality. In spite of availability of processed meat in the market, there are no buyers as majority of the population cannot afford the cost and consumers fail to understand the difference between raw and processed meat. Ignorance and biased preferences of consumers restrict the growth of livestock industry (Gadekar, 2011).

- **Technical barrier:** Lack of appropriate infrastructure to preserve perishable products affects the negotiating power of small production units particularly if they are distant from consumption centers. Absence of modernized abattoirs, skilled and trained labor, scientific approach toward cutting of animal, etc., all play a part in the failure of meat industry in unorganized sector. Labor intensive approach of unorganized sector increases chances of contamination by handling as opposed to semi-automated processing techniques.

- **Miscellaneous concerns:** Apart from the above-mentioned factors, the challenges faced by the concerned individuals would be many. Small-scale of operation, competition among individuals, local ownership, improper registration, fluctuation in prices, less sophisticated packing, absence of a brand name, unavailability of good storage facilities, and lack of co-operative measures are some other problems faced by the unorganized sector. Employees of enterprises belonging to the unorganized sector have lower job security and poorer chances of growth due to lack of education, skill, and awareness. All these factors are collectively responsible for the failure of unorganized sector.

9.3.2 BIOLOGICAL FACTORS

Animal food is a source of quality nutrients, such as fat, protein, carbohydrates, and minerals. Animal protein, when compared to plant protein in terms of protein digestibility corrected amino acid score (PDCAAS), protein efficiency ratio (PER), and biological value (BV) shows a higher nutritional value (http://ecoursesonline.iasri.res.in/mod/resource/view.php?id=147711). PDCAAS is a method of evaluating the protein quality based on both the amino acid requirements of humans and their ability to digest it. This rating was adopted by the US Food and Drug Administration (FDA) and the Food and Agricultural Organization of the United Nations/ World Health Organization (FAO/WHO) in 1993 and referred to as the best method to determine protein quality. PER is based on the weight gain of a test subject divided by its intake of a particular food protein during the test period. BV is a measure of the proportion of absorbed protein from a food which becomes incorporated into the proteins of the organism's body.

It captures how readily the digested protein can be used in protein synthesis in the cells of the organism. Plant origin proteins are deficient in at least one of more essential amino acids, for example, some cereals in lysine and some legumes in methionine but the animal proteins contain all essential amino acids (http://ecoursesonline.iasri.res.in/mod/resource/view.php?id=147711).

- *Microbial aspect:* Meat gets putrefied due to the action of bacteria, molds, and yeasts. Meat, a rich source of nutrient, with high moisture content and nearly neutral pH provides a conducive environment for growth and proliferation of meat spoilage microorganisms and common food borne pathogens. These organisms may be present in the lymph nodes, bones, and muscle of animals(Dave, 2012). Contamination with spoilage organisms is therefore, almost unavoidable. All these make preservation of meat more difficult than that of most kinds of food. Food spoilage organisms generally affect the organoleptic properties of the food, whereas foodborne pathogens can make the food toxic, in either case, the meat is unfit for consumption. Load of aerobic and anaerobic microflora is important in determining the shelf life of meat. Therefore, if cooling is not prompt and rapid after slaughter, meat may undergo undesirable changes before being processed in some way.

    The greater the bacterial load in the gut of the animal, higher is the chances for invasion of tissues. The state of animal also plays an important role in determining meat quality. If the animal is excited, feverish, or fatigued, bacteria are more likely to enter the tissues, causing rapid bacterial proliferation. Immediate cooling will reduce the rate of invasion of microorganisms to the tissues (Doyle, 2006).

    Contamination of meat with food borne pathogens thus becomes a cause of great concern and an important public health issue. *Salmonella* and *Campylobacter* are of major concern even today. Others include *Clostridium perfringens, Escherichia coli O157:H7,* and *Listeria monocytogenes.* (Sudershan, 2014; Bhandare et al., 2007; Mor-Mur, 2010).

- *Zoonotic diseases:* In developing countries, infectious and parasitic diseases of livestock put constraints on productivity and profitability of livestock production. Infectious diseases, such as Nipah virus, Hendra virus, Avian influenza, Japanese encephalitis, rinderpest, foot-and-mouth disease, contagious bovine pleuropneumonia, and Kyasanur forest disease (KFD) are still major threats to Indian livestock (Hellina, 2015).

- Through increased movements of livestock, livestock products, and people, these diseases also threaten production in industrial countries. Certain livestock diseases (zoonoses) can also affect humans, for example, brucellosis and tuberculosis, and new zoonoses may originate from livestock populations (e.g., Nipah and Avian flu). Emergence of diseases, such as bovine spongiform encephalopathy (BSE) in cattle and its effect on human health (variant Creutzfeldt-Jakob disease) demonstrate potential dangers to livestock (Sekar, 2011). It is estimated that new diseases have been detected at the rate of one a year over the last 30 years.

From a production viewpoint, helminthosis, and tick-borne diseases are particularly important. Helminths (worms), while rarely fatal, can seriously affect productivity and profitability. Although helminths can be effectively controlled, parasite resistance to drugs through the inappropriate use of antihelminthics is a growing problem (Kumar, 2015). Ticks also transmit diseases but the cost of traditional dipping with acaricides for tick control is becoming prohibitive and raises environmental concerns about disposal of the waste chemical. Intensification of livestock production is thus going to face growing constraints both from epidemic and endemic disease agents. Presence of insects and rodents is yet another problem. These animals are generally found in filthy unhygienic conditions. They are vectors of many disease-causing germs and hence their presence can be a threat to the animal and in turn human health. But if proper hygienic conditions are maintained, these problems can be eliminated.

Diseases reduce farm incomes directly and indirectly: directly, by causing considerable losses in production and stock as well as forcing farmers to spend money and labor on their control; and indirectly by the consequent restrictions on exports. Stringent and systematic eradication efforts have eliminated some livestock diseases in developed countries and drastically reduced the incidence of others. However, the maintenance of this situation is becoming more and more difficult with the ever-increasing intensity of international travel and the growing long-distance trade in animals and livestock products. Import of feeds from distant areas, increasing densities of livestock, particularly in peri-urban and urban areas, and shifts in dietary habits have raised concerns about diseases, microbial contamination of food and general food safety. Changes in production systems, changing feeding practices, and the safety of animal feed may increase the risk and change the pattern of disease transmission.

9.3.3   CHEMICAL FACTORS

Antimicrobials are widely employed in livestock sector, irrespective of the health of animal and without proper medical examination. In modern farming systems with high animal density, it is a common practice to supplement feed with sub-therapeutic doses of antimicrobials so as to enhance growth rates (Reig, 2008). Resistance to antimicrobials in farm animal pathogens can be passed on to bacteria of humans through the exchange of genetic material between microorganisms. This constant exposure to antimicrobials thus promotes development of resistant microbes which in turn increases public health costs. Most antimicrobial resistance in human pathogens stems from inappropriate use of these drugs in human medicine, but the use of antimicrobials in the livestock sector plays a contributing role. Growth promotants, drugs to increase body mass, body muscle, etc., are included in animal feed and used indiscriminately with the aim to gain better price value. Presence of heavy metals in animals through food, water, etc., is also of great concern as it indirectly affects the health of humans consuming such food (Andree et al., 2010). However, these malpractices adversely affect the health of animal and indirectly that of the consumer. Growth and steroid hormones that increase feed conversion efficiency are used in many parts of the world, particularly in the beef and pig industry. Though no negative impacts on human health as a result of their correct application have been scientifically proven, a stricter stand on the use of hormones in livestock production may be required.

9.4   CONDITIONS OF INDIAN SLAUGHTERHOUSES FOR DOMESTIC SUPPLY

Existing condition in majority of the traditional slaughterhouses of India feeding the local population is far from satisfactory. Most of these lack basic facilities like water, electricity, ventilation, drainage, ceramic flooring, overhead rails, and waste disposal. Slaughtering of animal on open ground or dirty wooden block with or without further processing or dressing is a common sight in most slaughterhouses. Carcasses get exposed to heavy contamination from dung and soil and the situation is further aggravated by inadequate ante- and post-mortem inspection practices (Guleria, 2015). Quality of meat thus available in these slaughterhouses is unhygienic and harbours high microbial load. Other sources of contamination include water, tools used for slaughtering, storage containers, hygiene, and health of handlers, temperature, environment in which animals are stored and transported, presence of flies, rodents, and insects (http://ecoursesonline.iasri.res.in/mod/resource/view.php?id=147711). Though cooking may kill many of the microorganisms in meat and reduce the load, cross-contamination of the products eventually occurs under the prevailing conditions of meat handling. Enormous quantities of by-products generated are not utilized efficiently and economically. One of the main reasons for such apathy of slaughterhouses is absence of any sort of authoritative control.

9.5   STATISTICAL OVERVIEW OF INDIAN LIVESTOCK INDUSTRY

The livestock business in India is an age-old practice and this food industry is one of the important contributors to economy of rural and semi-urban India. India ranks first in meat export according to the data released by US Department of Agriculture (http://money.cnn.com/2015/08/05/news/economy/india-beef-exports-buffalo). India is the fifth largest producer of eggs and ninth largest producer of poultry meat amongst all countries (http://ecoursesonline.iasri.res.in/mod/resource/view.php?id=147711). According to the Ministry of Food Processing Industries (MOFPI), about 70% of poultry processing is in the organized sector and 30% is in the unorganized sector (http://www.mofpi.nic.in).

The unorganized sector with respect to meat is dominant in southern states in our country with nearly 60–70% total output coming from southern states especially, Andhra Pradesh, Tamil Nadu, and Maharashtra. Other major contributors include Uttar Pradesh, Bihar, West Bengal, Punjab, and Haryana. Uttar Pradesh (UP) has emerged as the biggest exporter of buffalo meat in the organized sector, followed by Punjab and Maharashtra (http://www.thehindu.com/news/national/india-on-top-in-exporting-beef/article7519487.ece).

Rural poultry production constitutes an important component of the agricultural economy in India. Small poultry farmers are capable of more significant contribution to alleviate malnutrition, poverty, and unemployment. The Indian poultry industry is 5000 years old and since then has witnessed a remarkable growth from backyard to industry. India requires both mass production as well as production by masses. Chicken meat contributes about 37% of total production of meat and the growth is expected to increase in the near future (http://articles.economictimes.indiatimes.com/2016-01-19/news/69900476_1_icra-broiler-poultry-industry). This might be due to easy availability, no religious taboos, and the cheapest source of protein.

However, in terms of quality of supply, it is still far behind as there are no stringent enforcement.

9.6   EXISTING LAWS AND REGULATIONS IN INDIA WITH RESPECT TO THE LIVESTOCK INDUSTRY

In response to food safety and animal disease concerns, many countries have encouraged initiatives toward the establishment of guidelines with a scientific approach. Government of India has also initiated several efforts in this direction. Laws and guidelines exist at each step of meat production from farm to processing to market, but the degree of implementation of these is still questionable.

9.6.1   LAWS RELATED TO QUALITY ASPECT OF MEAT

- *National Meat and Poultry Processing Board (NMPPB):* A major initiative of Government of India was launching of the National Meat and Poultry Processing Board on February 19, 2009. The main focus of this apex body is to work as a national hub for addressing all key issues related to meat and poultry processing for the systematic and proper development of this sector. The board serves as a single window service provider for producers/manufacturers and exporters of meat and meat products. It promotes and regulates the meat industry, with the aim to create employment opportunities. It also attempts to help rural farmers to generate more income. The board also aims to help set up quality control laboratories for meat and meat products, promote meat manufacturers to adopt good manufacturing practices (GMP), hazard analysis and critical control points (HACCP) systems, ISO-9001: 2000 standards help industry to create and disseminate data, train workers, and technicians and utilize slaughterhouse waste materials efficiently (http://www.mofpi.nic.in/writereaddata/AnnexureI.pdf).
- *Indian Council of Agricultural Research (ICAR)—National Research Center on Meat, Hyderabad:* The importance of developing Indian meat sector toward an efficient and organized activity to utilize livestock resources to provide meat and by-products of human utility to the growing population needs has been realized. With this challenging backdrop, ICAR has visualized the need for establishing

an exclusive research center on meat. National Research Centre on Meat was initiated at IVRI (Indian Veterinary Research Institute) campus, Izatnagar in the year 1986 and later shifted to Hyderabad in the year 1999. It started functioning as its own campus from April 2007 (http://nrcmeat.org.in/about-us/nrc-on-meat/).

Growth of centralized meat processing units necessitated the need for central authority to ensure meat quality. In 1973, Government of India promulgated an order (Meat Products Order, 1973) to enforce strict quality parameters on the production and processing of meat products under Essential Commodities Act 1955. The order is implemented by Directorate of Marketing and Inspection and it controls production, quality, and distribution of raw and processed meat (http://www.fssai.gov.in/Portals/0/Pdf/MFPO%201973-Amended%20_English.pdf).

- *Agricultural and Processed Food Products Export Development Authority (APEDA):* The APEDA was established by the Government of India under the Agricultural and Processed Food Products Export Development Authority Act passed by the Parliament in December 1985. It encourages export of agro and processed food products including meat and meat products. APEDA is engaged in determining the area of operation and its action plan in developing technical and analytical quality assurance capabilities. It also conducts specific training programs for quality management systems, such as HACCP, total quality management (TQM), and ISO 9002 series and so on. Registration of abattoirs and meat processing plants is done by APEDA followed by inspection of the meat processing plants. During inspection, focus is given to hygiene and sanitary conditions maintained by the plant, ante-mortem and post-mortem inspections, infrastructure, staff hygiene, laboratory facilities, and record maintenance. Registration of the meat processing plant is renewed every year after a detailed plant inspection by the committee (http://apeda.gov.in/apedawebsite/index.html).
- *Bureau of Indian Standards specifications (BIS):* BIS specifications (Certification marks) Act 1952 provides third party assurance for the consumers. Under this system, BIS issues license to food manufacturing units which comply with the specifications laid down in the relevant Indian Standards (http://www.bis.org.in/bis_origin.asp).
- *Codex Alimentarius:* Codex Alimentarius has been established as the benchmark for ensuring food safety and consumer protection. FAO's manual of food quality control and food for export has become the

standard reference in the improvement of quality of food for national and international trade. Continuous surveillance of product by way of sampling and laboratory analysis is essential to ensure that contaminant levels comply with those prescribed by importing countries (http://www.fao.org/fao-who-codexalimentarius/en/).

9.6.2 LAWS RELATED TO EXPORTS

- *All India Meat And Livestock Exporters Association (AIMLEA):* AIMLEA formed in 1972, headquartered in South Mumbai works in coordination with APEDA and the Ministry of Commerce, Government of India for the export of risk free, frozen/chilled buffalo, sheep, goat meat from India (http://www.meat-ims.org/groups/all-india-meat-livestock-exporters-association/).
- *Directorate-general of Foreign Trade (DGFT):* DGFT is attached to Ministry of Commerce and Industry and is headed by Director General of Foreign Trade. The organization is essentially involved in regulation and promotion of foreign trade with the prime focus on increasing safety and quality exports of meat (http://dgft.gov.in/).
- *Export Quality Control and Inspection Act 1963:* This act was promulgated to promote export trade by ensuring exports of international quality products. The Export Inspection Council has been established to ensure compulsory quality control and inspection of various commodities manufactured for export (http://faolex.fao.org/docs/pdf/ind69241.pdf.; http://commerce.nic.in/export_quality_control.htm).
- *Prevention of Food Adulteration Act (PFA Act) 1954:* This is the basic food act which empowers the Central Government to make rules and amend the existing ones. *Central Committee for Food Standards (CCFS)* is the main standardization and advisory body to make the food control system effective in terms of science-based approach to develop standards for food and other implementation aspects of the food regulation (http://www.fssai.gov.in/Portals/0/Pdf/pfa%20acts%20and%20rules.pdf).
- *Central Committee Food Standards (CCFS):* CCFS has been constituted under PFA, to advise the Central and State Governments on matters related to administration of the act (http://admis.hp.nic.in/himpol/Citizen/LawLib/C223.htm).
- *HACCP for food safety:* HACCP is a management system in which food safety is addressed through the analysis and control of

biological, chemical, and physical hazards from raw material production, procurement, and handling, to manufacturing, distribution and
consumption of the finished product (http://www.fda.gov/Food/
GuidanceRegulation/HACCP/).

Thus, the Government of India has laid down standards for meat at each
level of the process which includes standards for abattoir, processing plants,
marketing of various meat products, etc. These guidelines are implemented
very meticulously by the organized sector of India; however, they have failed
at the unorganized sector. The meat supplied by India for export purposes
follows strict regulations and guidelines as enforced by various countries.
The abattoirs of the organized sector follow GMP, HACCP, ISO 9002: 2000
for quality systems, ISO 14001 for Environment management, ISO 18001
for Occupational health and safety management system accreditations, OIE
terrestrial animal health code, etc.

In the unorganized sector, the failure starts at the farm level, continues up
to registration of abattoirs and influences prices at market level. Currently,
unorganized sector is the one feeding nation's population and hence quality
of meat supplied to the market is of utmost importance. If collective respon-
sibility by individuals at each level of the unorganized sector is carried out
in enforcement of laws, then this sector too can contribute to export of meat
and its processed product. There are several laws laid by developed countries for meat and poultry industry, these laws can be modified and applied
for Indian context.

9.7   ESTABLISHING QUALITY PARAMETERS OF MEAT IN THE UNORGANIZED SECTOR

The Indian meat and poultry market stands at a low position with respect
to the global markets. The biggest challenge to it is low availability of
adequate infrastructural facilities, inadequately developed linkages between
R&D; labs and industryand unawareness about hygienic practices. India is
exploring opportunities like setting of mega food parks, modern abattoirs,
state-of-art processing units, providing added incentive to develop existing
projects while taking care of the changing consumption patterns. The integration of development in contemporary technologies, such as electronics,
material science, biotechnology, etc. offers vast scope for rapid improvement and progress and opening of the global markets.

Following are some of the process parameters which when put into place will help improve the meat quality of the unorganized sector.

- **Government efforts:** National Meat and Poultry Processing Board (NMPPB) has played an instrumental role in contributing to the top position of India in beef export. The private sector and government have joined hands for raising the bar of organized sector to International standards. Same kind of intention and efforts are required for upliftment of the unorganized sector. Funds by government and investment from private sector can help in providing necessary infrastructure helpful to create state of art abattoirs.

- **State-of-art abattoirs:** Though the unorganized sector does not generate enough revenue for the country, it is the one feeding the Indian population and hence quality of meat supplied on the platter becomes a matter of concern. Currently, the unorganized sector is in an extremely dire state in all aspects. The need of the hour is to establish law-enforcing bodies from state level reaching to the grass root level. Registration of all abattoirs with issue of valid license should be the foremost step followed by modernization of abattoirs and setting up rural abattoirs. Municipal Corporation can play a significant role in achieving this target. Slaughterhouses or abattoirs need to follow strict scientific design, with the practice of meat hygiene regulations (sanitation, ante-mortem inspection, and post-mortem inspection). Setting up large commercial meat farms is recommended to address traceability issue. Timely inspection by enforcement bodies, stringent laws, and punishment to those who do not abide by law—all such cumulative efforts can transform the face of the unorganized sector to a large extent (Jongwanich, 2009).

- **Education and training:** Education and awareness at various levels should be carried out, starting from farmers, slaughterhouse personnel, retailers, and consumers. Masses should be made aware of the health hazards of eating spoiled or contaminated meat and about the importance of hygiene and sanitation. Prevention of contamination of the carcass and separation of clean and unclean operations is required. Clean operations like dressing of carcass should be separated from unclean operations like cleaning of stomach or guts, so as to prevent contamination of the carcasses. This will ensure GMP. Farmers should be educated about the ill effects of using antibiotics, hormones, drugs, etc., for animals and its adverse effects on humans. Skill development at various levels should be initiated by way of continuous training of personnel.

- *Comparison with white revolution:* The way there are research organizations dedicated to problems of dairy industries, efforts should be made to establish research organizations for livestock industry as well. Funds should be made available to address key issues related to livestock health. Vocational and value added courses aimed at training personnel for job opportunities in livestock sector should be put in place. Bringing in organization and system in this sector can streamline the whole set up and can develop it into a revenue-generating model.
- *Media:* Media is the most influential body which has power to reach every nook and corner of the country. Media should be involved in creating awareness among people, in organizing programs with open forum, where issues and problems can be raised and efforts are made to find suitable solutions. Greater involvement of the masses will help in faster improvement of the unorganized sector.
- *Health of animal and consumers*: Health of consumers is directly depended on the quality of meat made available to them. Hence a great deal of efforts is required at the farm level itself.

1. Selection of good quality breed, preservation, and propagation of true breeds should be put to practice by farmers.
2. Feeding animals with nutritious food sans antibiotics, drugs, etc., should be enforced. Indiscriminate and irrational use of hormones, antibiotics, and chemical growth promotant should be prevented and regular inspection to be carried out.
3. Incorporation of prebiotic and natural growth promotant over antibiotics should be preferred.
4. Hygiene and sanitation condition should be strictly maintained with proper utilization and disposal of waste.
5. Measures to prevent insects, pests, and rodents in farms should be carried out as these act as vectors for various zoonotic diseases.
6. Employing a veterinarian for regular health check-up of animals and timely follow up of vaccination and immunization schedule should be enforced.
7. Surveys should be conducted on regular basis to monitor prevalence of microorganisms/parasites in these animals, in the surrounding and subsequent precautionary measures for their outbreak should be put in place.

8. Possible risks to the public health should be analyzed through thorough scientific studies.

    Epizootic diseases can be brought under control only gradually through intensive, internationally coordinated animal health programs. There has been a major shift away from countrywide eradication programs toward more flexible control strategies, with focus on areas offering highest returns (Rajkovic, 2010). Risk analysis and animal health economics help determine where disease control investment will have the greatest impact and benefit. Because of the large externalities of outbreaks of these diseases, management of their control remains a public sector responsibility. However, many developing countries do not have effective veterinary institutions capable of the task and public funding for disease control has been on the decline over the last decades.

- *Veterinary services:* Compulsory ante-mortem inspection of livestock, post-mortem examination of carcasses, and microbiological testing of the frozen meat should be undertaken by the competent Government Veterinary authorities to ensure the use of only healthy livestock for meat processing. Government of India should provide comprehensive veterinary health services for livestock. Veterinary health care to be provided through facilities like polyclinics, hospitals, dispensaries and first-aid centers (Rajkovic, 2010). Control of infectious and contagious diseases of livestock should be undertaken by systematic vaccination programs. Epidemiological studies should be conducted for monitoring and disease diagnosis. There are laws to restrict movement of livestock in disease outbreak areas and movement regulations of livestock meant for trade which should be implemented when required.
- *Meat quality tests*: Rapid detection methods for pathogens such as PCR, serology, and quick control measures during disease outbreaks should be established (Toldra, 2006). Nutritional profile testing, sensory (appearance, flavor, tenderness, texture, etc.), physical (pH, water activity, etc.), microbiological (total plate count, yeast and mold count, etc.), and chemical (allergens tests, antibiotic residue testing, etc.) are all mentioned by FAO and should be implemented in the unorganized sector (http://www.fao.org/docrep/010/ai407e/ai407e24. htm). PDCAAS, PER, and BV are some of the parameters that can be used to test nutritional value of meat produced in India. Guidelines

related to nutrition profile of meat supplied should be established and implemented. Hygiene or safety implies freedom from harmful microorganisms and any residues. These can be controlled through legislation, proper feeding designs, and strategies, quality management schemes on the farm and procedures in the slaughterhouse and processing plant.

- **Economy:** Once the streamlining of unorganized sector is initiated, it will give rise to lot of opportunities for skilled and trained labors, veterinarians, researchers, entrepreneurs, etc. Proper financial support from government can draw interest of people toward this unexplored sector of the food processing industry. Government should streamline market policies for the unorganized sector in order to prevent exploitation of the farmers in the hands of retailers. A combination of high production and high transaction cost can help improve the condition of the small scale. Credit facilities should be provided to farmers for maintaining livestock. Producers association or cooperatives would enable producers to benefit from economies of scale by reducing transaction costs.

- **Improvement of shelf life:** Apart from GMP, decontamination technologies may be adapted to ensure meat safety and wholesomeness. Decontamination procedures, such as high-pressure washes, use of organic acid rinses, hot water rinses, and antimicrobial chemicals have been attempted to increase the shelf life of meats. Gamma radiation and electron beam are promising non-thermal technologies which are being considered at an industrial level for decontamination of meat products. Combinations of decontamination procedures may be used to improve effectiveness (Aymerich, 2008).

- **Exploiting biotechnology:** Biotechnology offers the promise of solving some of the technical constraints through improved prevention, diagnosis, and treatment of animal disease. Genomics, for example, may well contribute to the development of new generations of vaccines using recombinant antigens to pathological agents. A far wider range of effective, economic vaccines that are easy to use and do not require a cold chain can be expected in the future. The development of cost-effective, robust pen-side diagnostics will enhance the veterinary services offered in developing countries. However, technological advances must be matched by enhanced epidemiological and logistical capacities and by greatly improved coordination of all institutions involved in animal disease control from local up to international level.

1. **HACCP approach**: Process control systems should limit microbial contamination of meat to as low practicable level as possible and prevent subsequent growth to levels that may constitute a hazard. HACCP provides a scientific approach to meet safety and wholesomeness throughout the production, processing, and distribution of fresh meat and the HACCP approach should wherever possible, be utilized with other quality assurance procedures (Ropkins, 2000; Wilhelm, 2011). For example, chilling and freezing of meat are the common methods of meat preservation to extend shelf life. It is considered desirable that the temperature of the meat in any part of the carcass should be reduced to below 10°C within 12 h of slaughter. As a general guide, a deep muscle temperature of 6–7°C should be achieved in 28–36 h. Failure to bring down the internal temperature of carcass will result in rapid multiplication of bacteria deep in the meat resulting in off odors and bone taint.

2. **By-products of meat**: Edible offal or an edible by-product is meat trimmed from the head. Edible fats are obtained during slaughter, such as the cowl fat surrounding the rumen or stomach. Fatty tissue, hide, skin, etc., is obtained from animal. Blood derived from animal can be separated into several fractions that have therapeutic properties one such example is purified bovine serum albumin. Bone meal, meat cum bone meal, meat meal, and tallow are used locally. Irshad et al. (2015) and Jayathilakan et al. (2012) provide a deep insight into the utilization of these by-products and the revenue generated from it.

Henson et al. (1999) give a detailed description of risk assessment and measures to be implemented thereafter. With proper support from policymakers and researchers, the livestock sector can become a driving force for agricultural economy. In order to improve food safety from feed production to the supermarket shelf, basic food quality control systems are evolving into quality assurance systems, and these, in turn, are moving toward TQM systems. The costs of compliance with these systems are high but at the same time, it will reduce the burden on health care sector. To produce wholesome meat, safe to the consumer, it is essential to enforce strict hygienic standards at all stages of meat production. After the success of white and green revolution, it is now the right time to bring in the so-called pink revolution for meat in India and once again exhibit the potential that India has.

KEYWORDS

- livestock industry
- slaughterhouses
- unorganized sector
- zoonosis
- hazard analysis critical control point

REFERENCES

Aymerich, T.; Picouet, P. A.; Monfort, J. M. Decontamination Technologies for Meat Products. *Meat Sci.* **2008,** *78,* 114–129.

Andree, S., et al, Chemical Safety of Meat and Meat Products. *Meat Sci.* **2010,** *86,* 38–48.

Agriculture and Processed Food Products Export Development Authority. http://apeda.gov.in/apedawebsite/index.html (accessed April 15, 2015).

All India Meat Livestock Exporters Association. http://www.meat-ims.org/groups/all-india-meat-livestock-exporters-association/ (accessed April 15, 2015).

Buffalo Meat Now India's Top Agri Export Item, Indian Trade Portal. http://www.indiantradeportal.in/vs.jsp?lang=0&id=0,10,662 (accessed April 15, 2015).

Buffalo Meat. APEDA. http://apeda.gov.in/apedawebsite/SubHead_Products/Buffalo_Meat.htm

Broiler Meat Demand to Grow 5%–7% a Year in Long Term, Says ICRA, The Economic Times, http://articles.economictimes.indiatimes.com/2016-01-19/news/69900476_1_icra-broiler-poultry-industry (accessed April 15, 2015).

Bhandare, S., et al, A Comparison of Microbial Contamination on Sheep/goat Carcasses in a Modern Indian Abattoir and Traditional Meat Shops. *Food Control.* **2007,** *18,* 854–858.

Bureau of Indian Standards, http://www.bis.org.in/bis_origin.asp (accessed April 15, 2015).

Chapter 19. Present Status of Meat, Poultry and Fish Industry in India. http://ecoursesonline.iasri.res.in/mod/resource/view.php?id=147711 (accessed April 15, 2015).

Dave, D.; Ghaly, A. E. Meat Spoilage Mechanisms and Preservation Techniques: A Critical Review. *Am. J. Agric. Biol. Sci.* **2011,** *6,* 486–510.

Doyle, M. P.; Erickson, M. C. Emerging Microbiological Food Safety Issues Related to Meat. *Meat Sci.* **2006,** *74,* 98–112.

Delgado, C., et al. Livestock to 2020—The Next Food Revolution:

Food, Agriculture and the Environment Discussion Paper 28, IFPRI/FAO/ILRI: Washington, DC, 1999.

Directorate General of Foreign Trade. http://dgft.gov.in/ (accessed April 15, 2015).

Detailed Description of all Enforcing Bodies for Meat Industry worldwide. http://www.meat-ims.org/groups (accessed April 15, 2015).

Export Quality Control and Inspection Act 1963. http://faolex.fao.org/docs/pdf/ind69241.pdf (accessed April 15, 2015).

Export Quality Control and Inspection Act 1963. http://commerce.nic.in/export_quality_ control.htm (accessed April 15, 2015).

FAO Corporate Document Repository-simple Test Methods for Meat Products. http://www. fao.org/docrep/010/ai407e/ai407e24.htm (accessed April 15, 2015).

Food Safety and Standards Authority of India. http://www.fssai.gov.in/Portals/0/Pdf/ MFPO%201973-Amended%20_English.pdf (accessed April 15, 2015)

Food and Agriculture Organization of the United Nations. http://www.fao.org/fao-who-codexalimentarius/en/ (accessed April 15, 2015).

Food Safety and Standards Authority of India http://www.fssai.gov.in/Portals/0/Pdf/pfa%20 acts%20and%20rules.pdf (accessed April 15, 2015).

Gadekar, Y. P.; Shinde, A. K. Indian Meat Industry: Opportunities and Challenges, Indian. *Food Ind.* **2011,** *30,* 17–22.

Guleria, P.; Kumari. S.; Khan. A.; Dangi. N. Present Scenario of Indian Meat Industry-A Review. *Int. J. Enhanc. Res. Sci. Technol.Eng.* **2015,** *4,* 251–257.

Global Meat News. http://www.globalmeatnews.com/Regions/Asia/India (accessed April 15, 2015).

Hellina, J.; Krishnab, V.; Erensteine, O.; Boeber, C. India's Poultry Revolution: Implications for its Sustenance and the Global Poultry Trade. *Int. Food Agribus. Manag. Rev.* **2015,** *18,* 151–164.

Henson, S.; Caswell, J. Food Safety Regulation: An Overview of Contemporary Issues. *Food Policy* **1999,** *24,* 589–603.

HACCP. http://www.fda.gov/Food/GuidanceRegulation/HACCP (accessed April 15, 2015).

Holy Cow, India is the World's Largest Beef Exporter, Virginia Harrison, CNN traders, http:// money.cnn.com/2015/08/05/news/economy/india-beef-exports-buffalo/ (accessed April 15, 2015).

India on Top in Exporting Beef, The Hindu.

http://www.thehindu.com/news/national/india-on-top-in-exporting-beef/article7519487.ece (accessed April 15, 2015)

India Stays World's Top Beef Exporter Despite New Bans on Slaughtering Cows, Time. http:// time.com/3833931/india-beef-exports-rise-ban-buffalo-meat/ (accessed April 15, 2015)

Irshad, A.; Sureshkumar, S.; Shalima Shukoor, A.; Sutha, M. Slaughter House by-product Utilization for Sustainable Meat Industry-A Review. *Int. J. Dev. Res.* **2015,** *5,* 4725–4734.

Jongwanich, J. Impact of Food Safety Standards on Processed Food Exports from Developing Countries, No. 154 | April 2009 ADB Economics Working Paper Series No. 154.

Jayathilakan, K.; Sultana, K.; Radhakrishna, K.; Bawa, A. S, Utilization of by Products and Waste Materials from Meat, Poultry and Fish Processing Industries: A Review. *J. Food Sci. Technol.* **2012,** *49,* 278–293.

Kumar, R.; Singh, S. P.; Savalia, C. V. Overview of Emerging Zoonoses in India: Areas of Concern, *J. Trop. Dis.* **2015,** *3,* 1000165.

List of Indian Integrated Abattoirs and Meat Processing Plants Approved by APEDA. http:// www.apeda.gov.in/apedawebsite/Announcements/A_Approved_Indian_Abattoirs_cum_ Meat_Processing_Plants.pdf (accessed April 15, 2015).

Livestock Sector Brief, Food and Agriculture Organization of the United Nations July 2005 http://www.fao.org/ag/againfo/resources/en/publications/sector_briefs/lsb_IND.pdf (accessed April 15, 2015)

Livestock Commodities, FAO Corporate Document Repository, http://www.fao.org/ docrep/005/y4252e/y4252e05b.htm (accessed April 15, 2015).

MOFPI, http://www.mofpi.nic.in/(15th April 2015)

More Slaughter Houses than Milk Units in India, India Today. http://indiatoday.intoday.in/story/there-are-more-slaughterhouses-than-milk-units-in-india/1/578242.html (accessed April 15, 2015).

Mor-Mur, M.; Yuste, J. Emerging Bacterial Pathogens in Meat and Poultry: An Overview. *Food Bioproc. Technol.* **2010,** *3,* 24–35.

National Meat and Poultry Processing Board. http://www.mofpi.nic.in/writereaddata/AnnexureI.pdf

National Research Centre on Meat, Hyderabad. http://nrcmeat.org.in/about-us/nrc-on-meat/ (15th April 2015)

Opara, L. Traceability in Agriculture and Food Supply Chain: A Review of Basic Concepts, Technological Implications, and Future Prospects. *Food Agric. Environ.* **2003,** *1,* 101–106.

Reig, M.; Toldra, F. Veterinary Drug Residues in Meat: Concerns and Rapid Methods for Detection. *Meat Sci.* **2008,** *78,* 60–67.

Rajkovic, A.; Smigic, N.; Devlieghere, F. Contemporary Strategies in Combating Microbial Contamination in Food Chain. *Int. J. Food Microbiol.* **2010,** *141,* 29–42.

Randolph, T. F., et al. Role of Livestock in Human Nutrition and Health for Poverty Reduction in Developing Countries. *J. Anim. Sci.* **2007,** *85,* 2788–2800.

Report on Conditions of Work and Promotion of Livelihoods in the Unorganised Sector 2007. National Commission on Enterprises in the Unorganised Sector. www.nceuis.nic.in (accessed April 15, 2015).

Ropkins, K.; Beck, A. Evaluation of Worldwide Approaches to the Use of HACCP to Control Food Safety. *Trends Food Sci.Technol.* **2000,** *11,* 10–21.

Sudershan, R. V., et al. Foodborne Infections and Intoxications in Hyderabad India. *Epidemiol. Res. Int.* **2014,** *2014,* 5.

Sekar, N.; Shah, N.; Abbas, S.; Kakkar, M. Research Options for Controlling Zoonotic Disease in India, 2010–2015. *PLoS One.* **2011,** *6,* e17120.

The Prevention of Food Adulteration Act, 1954. http://admis.hp.nic.in/himpol/Citizen/LawLib/C223.htm (accessed April 15, 2015).

Thornton, P. Livestock Production: Recent Trends, Future Prospects. *Phil. Trans. R. Soc. B,* **2010,** *365,* 2853–2867.

Toldra, F.; Reig, M. Methods for Rapid Detection of Chemical and Veterinary Drug Residues in Animal Foods. *Trends Food Sci. Technol.* **2006,** *17,* 482–489.

Wilhelm, B. The Effect of Hazard Analysis Critical Control Point Programs on Microbial Contamination of Carcasses in Abattoirs: A Systematic Review of Published Data. *Foodborne Pathog. Dis.* **2011,** *8,* 949–959.

CHAPTER 10

ACRYLAMIDE IN FOOD PRODUCTS: OCCURRENCE, TOXICITY, DETOXIFICATION, AND DETERMINATION

KALYANI Y. GAIKWAD and K. A. ATHMASELVI*

Department of Food Process Engineering, School of Bioengineering, SRM Institute of Science and Technology, Chennai, Tamil Nadu, India

*Corresponding author. E-mail: athmaphd@gmail.com

CONTENTS

Abstract..250
10.1   Introduction..250
10.2   Chemistry...252
10.3   Formation of Acrylamide...254
10.4   Risk Associated with Exposure and Dietary Limits....................258
10.5   Detoxification/Mitigation Strategies..259
10.6   Analytical Methods of Acrylamide Detection 263
Keywords..280
References...280

ABSTRACT

The one of the main causes of wide spectrum of toxic effect is acrylamide. Its presence in human food has made extensive studies researching its formation and level of exposure. It has also induced search into suitable analytical procedures to find its presence in food products. In this chapter, we have discussed about the source of occurrence, its toxicity level in food, determination of procedures to find its presence, and detoxication methods.

10.1 INTRODUCTION

Acrylamide is an industrial chemical produced in large quantities. Basically, it is used for the production of polyacrylamide. In many other fields, polyacrylamide has many applications as flocculent in water treatment, conditioner in soil treatment, as grouting (binding) agent in concrete binder, in paper manufacture, ore processing, and in gel electrophoresis for scientific research.

In 1997, when water reservoir leaked during building of a tunnel in Sweden, the large number of dead fish and paralyzed cattle were found nearby construction site as tunnel walls were containing acrylamide and N-methylolacrylamide. Then during the investigation of researchers, the exposure of acrylamide found in tunnel workers after accidental spillage of grouting agent in their blood as adduct of acrylamide-hemoglobin and some peripheral nerve symptoms found (Hagmar et al., 2005). Long-term exposure to acrylamide also damages the nervous system both in humans and animals to certain extent.

The occurrence of acrylamide graph, the consciousness about its formation in food products, and level of toxicity were studied and first reported by scientists from Swedish National Food Authority and the University of Stockholm in April 2002 (Surdyk et al., 2004; Tareke et al., 2002; Svensson, 2003). On the basis of the animal studies, International Agency for Research on Cancer (IARC) has classified acrylamide as Group 2A, that is, "Probable Human Carcinogen." These reports were verified by several countries and attracted worldwide attention and concern of acrylamide effects. As a result, the United Nations' Food and Agricultural Organization (FAO) and World Health Organization (WHO) formed an Expert Consultation on "Health Implications of Acrylamide in Food" in June 2002.

Thermal processes are very commonly used in food processing to obtain digestible, safe, and shelf-stable food with some quality effects on nature of food. Baking, toasting, frying, pasteurization, and sterilization are some methods of heat treatments. Each treatment has some desired and undesired effect due to chemical reactions like Maillard reaction (MR), caramelization, and lipid oxidation are most common and prominent.

Another important aspect information of acrylamide is that low water content is important for these reactions, as acrylamide not found in boiled foods containing starch. Deep frying or roasting seems to be propitious to the formation of acrylamide (Tsutsumiushi et al., 2004). The performance of current analytical methods used for determination of acrylamide is not adequate for more "difficult" food matrices, such as cocoa, coffee, and high salt flavorings. For intake and risk assessment, the data obtained from current methods is not sufficient and it will need some robust and sensitive methods. Staple foods like bread contain trace amounts (in the low part-per-billion range) also contribute to the overall dietary intake. There are few analytical techniques like gas chromatography (GC)-mass spectrometry (MS) and liquid chromatography with Tandem MS (LC-MS/MS) analysis are described as most useful and authoritative method for acrylamide determination by many researchers including its validation, extraction, and cleanup procedures with optimization.

All this research led to the discovery and research of acrylamide in different food products, resulting in occurrence of many carbohydrate-rich food products. It is majorly formed in foods, which are rich in sugars and free amino acids particularly asparagine when processed at higher temperatures (120–180°C). The first breakthrough in acrylamide research was the simultaneous discovery by several groups that acrylamide is formed from reducing sugars and asparagine in the MR in a complex mechanism (Mottram et al., 2002; Stadler et al., 2002; Weisshaar & Gutsche, 2002; Zyzak, 2002). Later, additional formation mechanisms, for example, from peptides, proteins, and biogenic amines were identified (Buhlert et al., 2006; Claus et al., 2006b; Granvogl et al., 2004; Yaylayan et al., 2004). With this fundamental knowledge, detailed studies to reduce acrylamide in food products were initiated.

Acrylamide contents were much lower in meat (protein rich) products. This review article summarizes the knowledge of acrylamide and related toxicity, its formation, health effects, detoxification, and determination particularly in all kinds of food products (Sorgel et al., 2002; Robarge et al., 2003). The risk assessment of acrylamide evaluated by the Scientific Committee on Toxicity and the Environment (CSTEE) of the European Union (EU) demonstrated that exposure of acrylamide to humans should be kept as low as possible with regard to inherent toxic properties like carcinogenicity, genotoxicity to both somatic and germ cells, neurotoxicity, and reproductive toxicity. During toxicological studies, the acrylamide vapors irritate eyes and skin and cause paralysis of cerebrospinal system and it has been demonstrated that it has carcinogenic properties in animals (Zhang et al., 2005; Johnson et al., 1986; Smith & Oehme, 1991).

Also, some toxicological studies suggested that acrylamide vapors irritate the eyes and skin and cause paralysis of the cerebrospinal system. In 2005, the Swedish National Food Administration announced that foods processed and cooked at high temperatures contain relatively high levels of acrylamide.

10.1.1   PROCESSING CONTAMINANTS

These are the contaminants generated during the processing of foods (e.g., heating, fermentation, etc.). They are absent in raw materials and are formed by chemical reactions between natural and/or added food constituents during processing. Examples are nitrosamine, polycyclic aromatic hydrocarbons (PAH), heterocyclic amines, histamine, acrylamide, furan, benzene, trans fat, monochloropropanediol (MCPD), semicarbazide, 4-hydroxynonenal (4-HNE), and ethyl carbamate. Also, acrylamide can fall under the new concept of food contaminant and is emerging as food contaminant as these have been discovered recently. Some other emerging food contaminants are furan, benzene, perchlorate, perfluorooctanoic acid (PFOA), 3-monochloropropane-1,3-diol (3-MCPD), and 4-hydroxynonenal (4-HNE).

Similarly, 5-hydroxymethylfurfural (HMF) is one of the processing contaminant or toxin, forms as an intermediate in the MR (Ames, 1992) and from dehydration of sugars under acidic conditions (caramelization) during thermal treatments applied to foods (Kroh, 1994). It majorly occurs in coffee (instant and decaffeinated), chicory, malt, barley, honey, etc., with maximum concentration than other food products.

10.2   CHEMISTRY

Acrylamide (CH_2=CH—CO—NH_2; 2-propenamide) is a white crystalline solid with a molecular weight of 71.08 kDa. It has a melting point of $84.5°C \pm 0.3°C$, low vapor pressure of 0.007 mmHg at 25°C, and a high boiling point (136°C at 3.3 kPa/25 mmHg) (Norris, 1967; Ashoor & Zent, 1984; Eriksson, 2005). AA is a small hydrophilic small molecule. It is an odorless solid and its color ranges from colorless to white. Acrylamide is generally formed from the hydration of acrylonitrile with sulfuric acid between 90°C and 100°C or by catalytic hydration using a copper catalyst. It is soluble in a number of polar solvents, for example, acetone, acetonitrile, and water. Acrylamide is susceptible to polymerization during heating, which prevents the determination of boiling point at ambient pressure. At 3.34 kPa (25 mmHg), it boils at 125°C. It is regarded as a thermally unstable compound.

10.2.1 PHYSICAL PARAMETERS OF ACRYLAMIDE

Acrylamide possesses two functional groups, amide group, and the electron-deficient vinylic double bond that makes it readily available for a wide range of reactions, including nucleophilic and Diel–Alder additions and radical reactions (Table 10.1).

TABLE 10.1 Physical Parameters of Acrylamide.

Parameter	Specification
Chemical formula	$C_3H_5\,NO$
Molecular weight	71.08 g/mol
Melting point	84–85°C
Solubility	216 g/100 g water at 30°C
Boiling point	125°C at 3.34 kPa
Vapor pressure	0.007 mmHg at 20°C
Vapor density	(Air = 1) 2.4 at 175°C
Specific gravity	1.1222 kg/dm³ at 30°C

10.2.2 STRUCTURE OF ACRYLAMIDE

The structure of acrylamide has been given in Figure 10.1.

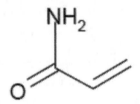

FIGURE 10.1 Structure of acrylamide.

10.3 FORMATION OF ACRYLAMIDE

After the discovery of acrylamide in foods, the major pathway for acrylamide formation in foods is established. The MR with free asparagine as main precursor (Mottram et al., 2002; Stadler et al., 2002; Stadler & Scholz, 2004; Zyzak, 2002), asparagine can thermally decompose by deamination and decarboxylation. But, when a carbonyl source is present, the yield of acrylamide from asparagine is much higher. It shows high concentration of acrylamide detected in foods rich in reducing sugars and free asparagine such as fried potatoes and bakery products (Mottram et al., 2002; Weisshaar & Gutsche, 2002; Yaylayan et al., 2003). Other minor reaction routes for acrylamide formation in foods have been found and described is from acrolein (Yasuhara et al., 2003), from acrylic acid (Yasuhara et al., 2003), and from wheat gluten (Claus et al., 2006a). Finally, acrylamide can be generated by deamination of 3-aminopropionamide (3-APA) (Granvogl & Schieberle, 2006). 3-APA is an intermediate in MR can also form by enzymatic decarboxylation of free asparagine and can yield acrylamide on heating even in the absence of a carbonyl source (Granvogl et al., 2004; Granvogl & Schieberle, 2007).

In April 2009, the European Food Safety Agency reported the results of the monitoring of acrylamide levels in foods on the request of the European Commission. Acrylamide is typically formed during frying, roasting, and baking not typically during boiled or microwaved foods. The highest levels of acrylamide found in fried potato products, bread, bakery wares, and coffee. Products other than mentioned in table like hazelnuts and almonds (Amrein et al., 2005b), olives and foods, which are not subjected to severe heating such as dried fruits (like plums, pears, and apricots) (Amrein et al., 2007).

10.3.1 MECHANISM OF FORMATION OF ACRYLAMIDE IN FOOD

According to the model studies, it is proved that and calculated formation of acrylamide from quantities of precursors. Heating equimolar amounts of asparagine and glucose at 180°C for 30 min resulted in the formation of 368 μmol/mol of acrylamide/mol of asparagine (Stadler et al., 2002). Incorporation of water in same reaction increases the acrylamide content to 960 μmol/mol. Also, temperature dependence study showed that increase in acrylamide from 120°C to 170°C and decreases on further higher temperatures. Mass spectral studies resulted that three C atoms and N atom in acrylamide structure are derived from asparagine.

10.3.1.1   ACRYLAMIDE IN PROCESSED FOODS

It is largely derived from heat-induced reactions of amino groups of free amino acid asparagine and carbonyl group of reducing sugars like glucose during processes like baking and frying. These kinds of foods majorly derived from plant sources like potatoes and cereals (barley, rice, and wheat) and not majorly in animal foods like meat, poultry, and fish. Most common and popular processed foods with higher levels of acrylamide include French fries, potato chips, tortilla chips, bread crust, crisp bread, and various baked and cereal-based products.

10.3.1.2   FORMATION BY MAILLARD REACTION

The first detailed information regarding formation of acrylamide was provided by Zyzak et al. (2003) (Fig. 10.1) and Yaylayan et al. (2003), and confirmed by the hypothesis of Stadler et al. (2002) and Mottram et al. (2002) that it is formed by MR from asparagine and carbonyl precursors like reducing sugars. This conversion of asparagine to acrylamide by thermally induced decarboxylation and deamination (Yaylayan et al., 2003) in the presence of carbohydrates. The very first step in the reaction is formation of Schiff base intermediate as low energy alternative to decarboxylation of the intact Amadori product. The Schiff base intermediate either hydrolyzes to for 3-amino-propionamide (potent acrylamide precursor) (Granvogl et al., 2004) or undergoes 1,2-elimination to direct acrylamide. Still this route of formation of acrylamide not fully elucidated but including this as main pathway, some additional routes also form acrylamide.

10.3.1.3   ALTERNATIVE ROUTES FOR ACRYLAMIDE FORMATION

Other than MR, several mechanisms have been reported. Acrolein and acrylic acid can be formed by dehydration of glycerol mostly when fats are heated at higher temperatures and are also formed by combination of ammonia from the degradation of amino acids (Becalski et al., 2003).

Yaylayan et al. (2004) reported that acrylamide gets formed in meat when heated at higher temperatures from dipeptide carnosine. This peptide hydrolyzes to β-alanine, which reacts with ammonium resulting from the Strecker degradation of amino acids. Because of this reason, very less or no acrylamide is found in meat as it readily forms methyl derivatives can affect with unknown toxicological effects.

In model study of using peptides, for wheat gluten and gluten supplemented wheat bread rolls the key amino acid is protein bound alanine adjacent to an amino acid with a β-H atom. After heating these proteins, an electrocyclic domino reaction leads to the formation of acrylamide and other acid amides such as cinnamic amide from phenylalanine.

10.3.2   MODEL STUDIES OF ACRYLAMIDE FORMATION IN FOODS

According to recent research, the formation mechanism of acrylamide in asparagine-carbohydrate model systems has been clarified nearly. But its actual formation affected by the food matrixes and some important effect parameters on generation of acrylamide based on carbohydrate-asparagine model studies. Some researchers have found such factors that affect formation of acrylamide in food matrixes like potato, cereal, bread, almond, coffee, etc.

10.3.2.1   POTATOES

These tubers contain considerable amounts of acrylamide precursors, that is, free asparagine, glucose, and fructose, which may form higher levels of acrylamide in some potato products. Potatoes of the cultivar Erntestolz stored at 4°C for 15 days showed that an increase in reducing sugar from 80 to 2250 mg/kg (referring to fresh weight). As a consequence, the potential of acrylamide formation at 120°C rose by a factor of 28 (Biedermann et al., 2002b). Currently, French fries the most popular processed food item of potatoes is the primary source of acrylamide. Any modifications performed on raw material will inevitably affect the MR and its products including its organoleptic properties like taste and color. Small-scale and laboratory trials have shown that French fries containing acrylamide amounts less than 100 µg/kg can be prepared (Grob et al., 2003). According to the research, controlling reducing sugars to 0.7 g/kg and frying temperature 170°C reduces acrylamide concentration to 50 µg/kg.

10.3.2.2   CEREAL

Cereal-based foods are one of the key grounds of acrylamide formation after heat treatment. Wheat and rye are two important cereal mixtures, contains two acrylamide precursors, free asparagine, and reducing sugars, Noti et al. (2003) reported 0.15–0.4 g/kg of asparagine in 10 samples of wheat flour. Also, Surdyk et al. (2004) measured asparagine level of 0.17 g/kg in white wheat flour.. Between 5 and 20 min, major losses of asparagine, water, and total reducing sugars were accompanied by large increase in acrylamide, which maximized between 25 and 30 min, followed by a slow liner reduction when cereal matrixes were cooked at 180°C. Acrylamide did not form to a large degree until the moisture contents of the cakes fell below 5%. Possibly, a combined treatment of pH (citric acid) and low levels of added amino acids (soy protein hydrolysate) could result in significant reductions but effect of these treatments on sensory properties remain for evaluation (Cook & Taylor, 2005).

10.3.2.3   BREAD

Other than presence of two precursors, the acrylamide formation occurs in bread due to baking time and temperature. It is studied and reported the dry starch systems, containing freeze-dried rye based flatbread dough's, flatbread and bread, which baked at varying temperatures and time. Maximum formation occurs at approximately 200°C, depending on the system and the baking time, as longer baking time decreased formation of acrylamide. However, acrylamide content increased with both baking time and temperature in bread crust in the interval test. Another possible source of high amount of acrylamide is gingerbread, up to 1000 μg of acrylamide/kg of fresh weight. As for the control of acrylamide in gingerbread, the use of hydrogen carbonate as baking agent, by minimizing free asparagine and avoiding prolonged baking need to be followed.

10.3.2.4   ALMONDS

Almonds also contain appreciable amounts of acrylamide precursors. The free asparagine content reported in the range of 2000–3000 mg/kg (Seron et al., 1998) while glucose and fructose levels found around 500–1300 mg/kg, and sucrose ranged from 2500 to 5300 mg/kg (Ruggeri et al., 1998). Therefore, the concentration of acrylamide in roasted almonds ranged from 260 to 1530 μg/kg (Becalski et al., 2003). The occurrence of acrylamide found less in European origin almonds as they contain significantly less free asparagine compared to U.S. almonds (Amrein et al., 2005b). Also, the reduction in acrylamide was found in roasted almonds, which stored at room temperature (Amrein et al., 2005a). Therefore, it can be assumed that reactive compounds formed during the roasting process responsible for decrease in acrylamide during storage.

10.3.2.5   COFFEE

It is generally roasted at temperatures 220–250°C, as time and speed of roasting, plays very crucial role in sensorial properties of coffee such as color, aroma, and taste. Coffee beans subjected to relatively higher temperature than other foods, which increases the possibility of formation of acrylamide in more than one particular pathway (Stadler & Scholz, 2004). The free asparagine content in green coffee falls under very narrow range, 30–90 mg/100 g. Also, two groups of researchers have found that acrylamide is not stable in commercial coffee stored in its original container (Andrzejewski et al., 2004; Delatour et al., 2004). Losses of 40–60% have been recorded in ground coffees stored in room temperature. In coffee, acrylamide is formed at the very beginning of the roasting step, reaching more than 7 mg/kg, and then declining sharply toward the end of the roasting cycle due to higher rates of elimination. Deeper roasting as a potential choice to reduce acrylamide could generate other undesirable compounds and negatively affect the sensory characteristics of product.

10.4   RISK ASSOCIATED WITH EXPOSURE AND DIETARY LIMITS

The WHO estimates a daily dietary intake of acrylamide in the range of 0.3–2.0 mg/kg body weight for the general population and up to 5.1 mg/kg body weight for the 99th percentile consumers (WHO, 2005).

Dietary intake of acrylamide has been recorded for several populations (Dybing et al., 2005). The average of acrylamide intake ranges from 0.3 to 0.6 ug/kg of body weight per day for adults and for children and adolescent's 0.4–0.6 ug/kg of body weight. As these values obtained accordingly intake of acrylamide-rich food such as French-fries and potato crisps by children and adolescents than adults (Dybing et al., 2005; Wilson et al., 2006).

The food pattern and dietary intake vary from country to country. In general, potato products, coffee, and bakery products are the most important sources. All these source s are rich in acrylamide cannot reduce the burden on consumers intake by reducing formation in one specific food item.

10.5   DETOXIFICATION/MITIGATION STRATEGIES

There are various mitigation strategies to reduce acrylamide formation in food products. Mostly they are focused on potato products and cereal-based products, as these are major contributors to the dietary exposure in most of the population. Similarly, for coffee as one of the major source of acrylamide has limited process options for acrylamide reduction without affecting final quality (CIAA, 2009; Guenther et al., 2007).

10.5.1   STRATEGIES TO REDUCE ACRYLAMIDE IN POTATO PRODUCTS

The mitigation or detoxification steps include change in recipes and formulations, selection of raw materials with low in acrylamide precursors, addition of proteins, cysteine, glycine, or other amino acids than asparagine, acidulants, calcium ions, cyclodextrin, antioxidants (natural or artificial), or its extracts, replacing reducing sugars with non-reducing sugars like sucrose, ammonium bicarbonate with sodium bicarbonate or change in processing conditions, or many various technologies can possibly reduce the formation during frying, baking, or other thermal processes. Also, selection of such potato varieties with low reducing sugars offers promising strategy for acrylamide reduction (De Wilde et al., 2006a), or storage at temperature above 6°C (De Wilde et al., 2005; Grob et al., 2003; Noti et al., 2003). Even increase in nitrogen fertilization increases free asparagines and reduces reducing sugar levels followed by acrylamide (De Wilde et al., 2006b). In addition to this pre and post-harvest strategies may reduce the possible formation of acrylamide. The removal of precursors by blanching or soaking potato slices combined with organic acid treatment (Gama-Baumgartner et al., 2004; Grob et al., 2003), decreases up to 70% of acrylamide. The reduction in pH inhibits formation of Schiff base by favoring protonation of amine group of asparagine. Similarly, the most crucial parameters for acrylamide control in case of potato products are regulation of heat and moisture content. However, with application of suitable maximum temperature, other factors such as load of fries in fryer surface to volume ratio shows significant reduction. A low surface to volume ratio significantly lowers acrylamide levels formed in crust (Surdyk et al., 2004). Instead of all these methods, the most promising method for reduction of acrylamide is use of enzyme asparaginase. This enzyme (L-asparagine amidohydrolase) able to catalyze the hydrolysis of asparagine in aspartic acid and ammonia as it lowers the precursor asparagine. The trial was successfully applied at lab scale for potato (Zyzak et al., 2003) and cereal-based products (Capuano et al., 2008) with great reduction in acrylamide content (85–90%) without affecting sensorial characteristic of product. Also, the addition of di and trivalent cations reduce acrylamide as per proposed literature in case of potato products (Elder, 2005; Elder et al., 2004). Application of vacuum frying reduces formation as lower temperatures can be applied without alteration in sensory properties of chips (Granda et al., 2004). If potato slices are dried before frying results in lower acrylamide as frying time has been reduced (Pedreschi et al., 2007). The addition of fish meat, nearly pure protein, significantly reduced acrylamide (Taeymans et al., 2004), which is probably due to the reaction of previously formed acrylamide with the SH and amino groups of the proteins (Schabacker et al., 2004).

10.5.2 STRATEGIES TO REDUCE ACRYLAMIDE IN BAKERY PRODUCTS

There are many mitigation results proposed by researchers, which limited up to laboratory scale. The practical use of these studies as per their claim of reduction in acrylamide not yet started in processing of food at industrial level. Also, there are some risks associated with these some mitigation strategies. For example, prolong yeast fermentation reduces acrylamide concentration in bread but increases the level of 3-monochloropropandiol (3-MCPD), one of the neo-formed contaminant (Hamlet et al., 2005).

10.5.3 STRATEGIES TO REDUCE ACRYLAMIDE IN CEREAL PRODUCTS

10.5.3.1 IMPACT OF RAW MATERIAL

The impact of variety, harvest year, fertilization, and storage conditions on acrylamide formation in potatoes studied already (De Wilde et al., 2005, 2006a, 2006b; Grob et al., 2003; Noti et al., 2003), and data on cereal products are limited. As farmers cannot influence climatic conditions, fertilization is the key factor in crop production. N-fertilization positively influenced acrylamide formation in potato products (De Wilde et al., 2006b) and inverse effect has been reported in bakery products (Claus et al., 2006b). The authors reported significantly increased amounts of asparagine in wheat grains with higher nitrogen dosage. When a zero level of fertilizer was compared with the highest dosage, the level of acrylamide in bread increased fourfold from 10.6 to 55.6 µg/kg. In contrast to potatoes, reducing sugars in wheat were not affected by fertilization. As N-fertilization is the pre-requisite to increase crop yields and flour quality but shows elevated amounts seen, therefore, minimization of nitrogen fertilization will be helpful. As flour containing higher amounts of dietary fiber and ash are highly valued from a nutritional point of view, but they cause to increase acrylamide level in bread. Unfavorable weather conditions can also result in sprouting, a severe problem in cereal production. As result of this, protease activities get almost doubled (Claus et al., 2006b), results in higher asparaginase levels in flour from sprouted wheat grains. Therefore, flours from sprouted wheat or rye should not be used for bakery products, even after blended with other flours.

10.5.3.2   IMPACT OF TECHNOLOGY

The use of sucrose instead of invert sugar syrup in wheat cracker reduces acrylamide by 60%. Similar effects observed in in gingerbread. The impact of other baking agents on acrylamide formation was studied. Enzymes containing bakery improvers, especially amylase, amyloglucosidase, and up to some extent protease are widely used in bread and bread roll production which might have an impact on acrylamide. In all preparations, amylase enhanced activity used and thereby reducing sugars, no impact on asparagine content was observed. Consequently, acrylamide remain unchanged by using enzyme containing bakery improvers. Accordingly, there is a significant effect on use of NaCl on acrylamide formation in model systems. Addition of 1% NaCl reduced formation by 40% and even slight increase NaCl amounts supported in acrylamide reduction. It is explained by inhibition of formation of a Schiff base between reducing sugars and asparagine (Gokemen & Senyuva, 2007). Ingredients other than salt, flour, and baking agents may also affect the acrylamide content of bakery products. The impact of roasting on acrylamide formation in almonds, hazelnuts, sesame, and poppy seeds is studied. Exception of hazelnut, all other typical baking ingredients significantly increased as these levels depend on the surface-to-volume ratio. When optimized flour was used in combination with a changed temperature profile, acrylamide reduced by almost 50%. Heat regulation and final moisture content are the most important factors for the formation of acrylamide (Stadler, 2006). Several studies carried out on acrylamide formation by time and temperature during baking (Biedermann, 2003). The results of research done by Surdyk et al. (2004), Taeymans et al. (2004), and Haase et al. (2003) showed that prolonged heating at lower temperatures is a suitable measure to acrylamide minimization during baking up to 30% in rye bread. The breads baked in forced air circulation oven resulted in lower levels of acrylamide levels than breads baked in convection oven as, in forced circulation of air faster and intense drying occurs in bread crust. Low moisture content enhances acrylamide formation during the MR and deck ovens are advantageous to reduce formation in bakery products. Analogous to oven type, extruder technology is also used in breakfast cereal production, which is crucial for acrylamide formation. The two different extrusion technologies, Pellet-to-flaking extrusion cooking (PFEC), and direct expansion extrusion cooking (DEEC) used. The higher moisture content of puffed breakfast cereals, reduced while processing by intensive drying and tempering applied in order to equilibrate humidity to approximately 10%, which improves the shelf-life. Compared DEEC with PFEC, higher thermal input in DEEC gives rise to acrylamide levels more than PFEC process. But by some other authors (CIAA, 2009; Stadler, 2006), reported that acrylamide levels in puff extruded breakfast cereals are lower end of the scale, usually below 100 µg/kg. The extrusion cooker gelatinizes starch but hardly develops Maillard products, as water in cereals evaporated at the end of extruder and no roasting occurs. Other than these temperature and machineries used for processing of cereal products, process handling also plays important role in acrylamide formation.

According to previous studies (Claus et al., 2007; Fredriksson et al., 2004), time of fermentation also has strong impact on acrylamide levels. In these studies, fermenting yeast consumed high amounts of free asparagine, around 60% (Claus et al., 2007) and 90% (Fredriksson et al., 2004) decrease, respectively, in the limiting precursor in cereal products was reported. Acrylamide decrease occurred in first hour of fermentation and subsequently remained constant, therefore, prolongation of at least 1 h was found to be sufficient for acrylamide reduction in industrial bread production. If the fermentation exceeds by 3 h, Fredriksson et al. (2004) suggested that it was unsuitable as gluten levels decreases and subsequent flattening of bread and bread rolls occurs. Lacto fermentation, as it is used in sourdough preparation, has also been assessed for reduction of acrylamide. That was reported in 75% of reduction in crisp bread when fermentation carried out using lactic acid bacteria (NCIMB 40450). This effect is due to the reduced pH (3.7 compared to 6.0 in the control) rather than consumption of asparagine by the bacteria. These results were confirmed by Fredriksson et al.

(2004), who reported a decrease in asparagine after 72 h fermentation with spontaneous sour dough.

10.5.3.3 IMPACT OF ADDITIVES

Some basic modification in formulation, process technology, and management affect the sensory and some other qualities of food. The additives, such as consumable acids, amino acids, and cations have recently been gaining attention. Efficiently some consumable acids minimize acrylamide in bakery products. Citric acid added in baked corn chips, the linear decrease seen with increasing concentration. At both concentrations (0.1% and 0.2%) no negative effects on sensory properties were observed. Similar effects were reported for lactic, tartaric, citric, and hydrochloric acids in semi-finished biscuits and cracker models (Taeymans et al., 2004; Graf, 2006). The application of asparaginase is one more strategy for acrylamide reduction. As in presence of acids, asparagine is hydrolyzed to aspartic acid, thus inhibits acrylamide generation in the MR (Table 10.2).

10.6 ANALYTICAL METHODS OF ACRYLAMIDE DETECTION

From the analytical point of view, initiatives have been taken at national levels to coordinate the activities in different countries. From the food industry, private sector, and official food control laboratories in Germany, they have good example of proficiency test, which was organized by the German Federal Institute for Risk Assessment (BfR) in 2002.

There are various chromatographic methods used to determine various structural groups present during this process. The majority is classical methods based on HPLC or GC techniques, such as GC, MS, and LC-MS/MS analysis are both acknowledged as the most validated, useful, and authoritative methods for the detection of acrylamide determination and other chromatographic methods are introduced for detection. These methods include extraction; clean up steps, which are important for the detection. However, the food matrices are very complex, thus simple chromatographic methods do not suffice for the analysis of acrylamide in those heat-treated food with trace amount (Riediker and Stadler, 2003). Rosén and Hellenäs reported first that analysis of acrylamide in different heat-treated food using the isotope dilution LC-MS technique. They developed MS method for direct detection acrylamide which couple with acrylamide group and help to

TABLE 10.2 Summary of Studies on the Reduction of Acrylamide and Major Contributions from Various Laboratory Colleagues by the CIAA "Toolbox" Approach.

Food matrix	Reducing method	CIAA "toolbox" parameters*	Specification of experience level**	Reducing effect	Reference
Potato tubers	Moderate temperature storage	Final preparation: product storage	Industrial scale: storing at low temperature may increase acrylamide concentration	The acrylamide level in potato chips made from tubers stored at 2°C was higher than ten times of that stored at 20°C	Chuda et al. (2003)
Potato chips	Blanching or soaking	Processing: Pre-treatment	Lab and pilot scale: blanching or soaking may effectively reduce acrylamide	A reduction of the sugar content by blanching or soaking decreased the acrylamide conc. By about 60%	Haase et al. (2003)
Fried and baked corn chips and French fries	Lowering pH by citric acid	Recipe: pH	Lab, pilot, and industrial scale: application of two foods used as ingredient assessed and provided marginal improvement	The 0.2% CA treatments induced 82.2 and 72.6% inhibition of acrylamide formation in two food matrixes, respectively. Dipping potato cuts in 1% and 2% CA solutions for 1 h before frying showed 73.1% and 79.7% inhibition of acrylamide formation in French fries	
Fried or roasted potatoes	Suitable cultivar and storage temperature	Final preparation: product storage	Industrial scale: storing at 4°C may increase acrylamide content	Potatoes, which may be used for roasting and frying, should contain less than 1 g/kg fresh weight of reducing sugar. This can be easily fulfiled with the most important potato cultivars grown but presupposes that potatoes are no longer stored at 4°C	Biedermann-Brem et al. (2003)
French fries	Suitable cultivar and storage temperature	Final preparation: product storage	Industrial scale: storing below about 10°C should be avoided	French fries with 40–70 µg/kg acrylamide were consistently produced. The raw potato should be of a suitable cultivar and storage at temperatures below about 10°C must be avoided.	Grob et al. (2003)

TABLE 10.2 *(Continued)*

Food matrix	Reducing method	CIAA "toolbox" parameters*	Specification of experience level**	Reducing effect	Reference
Bread	Fermentation	Processing: fermentation	Lab, pilot, and industrial scale: extension of fermentation time in bread may be an option to lower acrylamide levels	Compared with short fermentation time, longer fermentation reduced acrylamide content in bread made with whole grain wheat 87%. For bread made with rye bran, the corresponding reduction was 77%	Fredriksson et al. (2004)
Gingerbread	The use of sodium hydrogen carbonate	Recipe: ammonium bicarbonate	Pilot and industrial scale: industrial feasibility of reducing ammonium bicarbonate in recipes for finished products has yet to be established, and could lead to significant	Such method reduced the acrylamide level by >60%. Significant reduction of acrylamide can also be achieved by minimizing free asparagine and avoiding prolonged baking	
French fries	Lactic acid fermentation	Processing: fermentation	Lab scale: extension of fermentation time in French fries may be an option to lower acrylamide levels	Lactic acid fermentation of non-blanched potato rods for 45 and 120 min reduced 48% and 71% of acrylamide in French fries. By blanching potato rods before fermentation, reductions of acrylamide after 45 min and 120 min were 79% and 94%, respectively	

*The CIAA "Toolbox" parameters include the agronomical (sugars and asparagine), recipe (ammonium bicarbonate, pH, minor ingredients, dilution, and rework), processing (fermentation, thermal input, and pre-treatment), and final preparation (color endpoint, texture/flavor, and storage/shelf-life/consumer preparation).

**This content specifies the level of experience available for a proposed mitigation study.

detect easily. The choice of LC-MS due to hydrophilic properties of acryl-amide and MS/MS for high degree of verification as transition of elements also can be detected. Most of the published analytical methods based on MS as determinative technique, coupled with chromatographic step either by LC or GC with or without derivatization. LC-MS/MS and GC-MS appeared to be the most useful and authoritative method for acrylamide detection.

10.6.1   GC-BASED METHODS

10.6.1.1   GC-MS WITH BROMINATION

10.6.1.1.1   Extraction and Derivatization of Acrylamide

Bromination of acrylamide produces more volatile compounds, which increases the selectivity of the determination. Some derivatization approaches are laborious and time consuming. Adding prepared bromination solution containing potassium bromide and hydrogen bromide carries these out and bromine to either pretreated or raw aqueous extracts (Castle et al., 1991; Castle, 1993; Ahn et al., 2002; Tareke et al., 2002; Ono et al., 2003). While using different derivatization reagent consists of potassium bromide and sodium bromate in acidic medium some improvements achieved (Nemoto et al., 2002). Necessity of additional pretreatments depends on the matrix of food sample. Matrices, such as tomatoes, mushrooms do not require specific cleanup before derivation but some more general methodology applicable for protein (e.g., meat) and carbohydrate (e.g., potato crisps, bread, or potato chips) rich matrices (Tareke et al., 2000, 2002).

Bromination is frequently carried out overnight at temperatures slightly above freezing point of water. Castle (2003) reported that there is the large in the reaction kinetics of bromination reaction of acrylamide and meth-acrylamide at workshop on "Analytical Methods for Acrylamide Determina-tion if Food." Also, it was stated that isotopically labeled internal standards reduced reaction time from overnight to 1 h. These methods proposed in accordance with Japanese scientists (Nemoto et al., 2002; Ono et al., 2003).

10.6.1.1.2   Clean-up of the Derivative

The excess of the bromine is removed after reaction by titration with sodium thiosulfate solution (0.7–1 M) until color of that reaction mixture changes to light yellow. This brominated acrylamide is less polar as compared to original compound; therefore, it gets better solubility in non-organic solvents. Most of the time ethyl acetate or mixture of ethyl acetate and cyclohexane are used for the extraction of analyte from aqueous phase. This phase separation is mostly carried out by centrifugation. Older articles published on this separation (Castle et al., 1991; Castle, 1993), which are now modified and clean up carried out by fractionation of the organic extracts on silica gel cartridges. To avoid in change of silica activity because of ethyl acetate it should be dried (absorbs up to about 10% of water) or replaced with cyclohexane. A similar precaution was taken with the application of florisil as an adsorbent (Nemoto et al., 2002). In that case, analyte and internal standard were eluted with mixture of acetone and hexane. Alternatively, gel permeation chromatography was performed as final sample clean up (Tareke et al., 2000). Recent studies described drying of extract by adding Na_2SO_4 (Ahn et al., 2002; Tareke et al., 2002). The removal of residual water also eliminates or reduces its interference from water-soluble co-extractants. Even before injection of sample into GC, the solvent volume has to be reduced to 30–200 μL to reach the limits of detection in the range 1–5 μL kg^{-1} (Castle et al., 1991; Castle, 1993; Tareke et al., 2000, 2002).

10.6.1.1.3 Analyte Separation and Quantification

It was performed on standard GC capillary columns of middle to high polarity with a length of 30 m and internal diameter of 0.25 mm (standard in GC-MS). The 1–2 μL sample was injected in split-less mode, with initial oven temperature settings in the range of 60–85°C and heating rate most commonly as 15°C min^{-1}, and final oven temperature about 250°C. The original method (Castle et al., 1991) using GC-MS with bromination was with the addition of methacrylamide standard as an internal standard to the homogenized sample, which produces the derivative 2,3-dibromo-2-methyl-propionamide. Methacrylamide (Castle, 1993) were derivatized separately and added to sample directly before the final solvent volume adjustments. In these cases, the methacrylamide acts as a chromatographic internal standard, which can monitor potential changes in performance and quantification of the instruments achieved by standard addition. Castle (1993) reported that 2,3-dibromopropionamide might eliminate hydrogen bromide during injection or chromatographic separation. Others (Andrawes et al., 1987; Takata & Okamoto, 1991; Nemoto et al., 2002) used dehydrobromination for the acrylamide analysis.

10.6.1.2   GC-MS WITHOUT DERIVATIZATION

Most of the GC determinations of acrylamide carried out in food toxicology labs are based on derivatization with bromine. In addition, few methodologies rose which omit this derivatization step and measure acrylamide directly after extraction and cleanup. The sample preparation and measurement procedures differ accordingly from the GC methods with bromination.

10.6.1.2.1   Extraction and Clean-up of Acrylamide

The most primarily used extraction solvents are water or mixtures of water and organic solvents, such as n-propanol or 2-butanone (SQTS, 2003). After studies, it was found that use of dry n-proponal as extractant decreased acrylamide recovery drastically. Overall, 68–75.4% recovery was reported for methods, which applies pure methanol for extraction of baked goods (Tateo & Bononi, 2003). Soxhlet extraction of acrylamide from potato crisps with methanol as solvent was recently published (Pedersen & Olsson, 2003). The extracted amount of acrylamide, which had been reported for other extraction methods, including one drawback is it, takes long extraction time of 10 days. A special aspect of the extraction procedure is swelling of the matrix in order to obtain better access of the extraction solvents to potentially absorbed or enclosed acrylamide with the side effect of providing some time for the development of matrix/internal standard interactions. This is the reason; the homogenized sample is mixed with water and internal standard solution and kept at the prespecified temperatures for 10–20 min. Depending on the matrix, swelling yielded an increase in analyte recovery up to 100fold (Biedermann et al., 2002a). Hot water (60–80°C) was used to enhance extraction; even for some sample types, it gives satisfactory results at room temperatures. Also, samples treated in ultrasonic bath (30 min at 60°C) (Schaller, 2003) increased recovery rate, with approximate 20–30 min extraction time. Some problem found with high viscosity foods, which solved by addition of small amounts of amylase to the extraction mixture (SQTS, 2003). In case of high fatty foods, defatting step has to be included in sample preparation. It can be done by removal of fat fraction by extraction with hexane after swelling and extraction (SQTS, 2003) or by fractionation of the aqueous phase on graphitized carbon cartridges (Tareke et al., 2000, 2002). Other methods include phase separation first by centrifugation followed by removal of water fraction by azeotropic distillation (Biedermann et al., 2002a). In other case, defatting can also be accomplished by extraction of sample residue containing the analyte with an n-hexane/aceto-nitrile mixture (Biedermann et al., 2002a). Acrylamide is easily separated from fat due to the immiscibility of the two solvents. The one more different approach of defatting of the sample with n-hexane in Soxhlet extractor and removal of n-hexane residues from the sample under vacuum followed by extraction of acrylamide with methanol under stirring and sonication (Tateo & Bononi, 2003).

10.6.1.2.2 Analyte Separation and Quantification

The high polarity of non-derivatized acrylamide, those samples mostly injected on column in to the GC. Few articles (Biedermann et al., 2002a; Tateo & Bononi, 2003; WEJ, 2003) suggest the use of splitless injection as an alternative without any drawbacks or advantages. For analyte separation, columns with polar phases, for example, polyethylene glycol used. Temperature does not differ much, which used in case of derivatization. The major drawback of GC analysis without derivatization is the lack of characteristic ions in the mass spectrum of underivatized acrylamide. In electron ionization mode, the major fragment ions at m/z 71 and 55, respectively. These ions are also used for quantification. The co-extracted substances such as maltol or heptanoic acid produces same fragmentation pattern, which may interfere in results (Biedermann et al., 2002a).

Chemical ionization using methane as reagent gas gives increase of selectivity. Quantification carried out by addition of different internal standards, ranging from propionamide to isotopically labeled acrylamide. Level of detection (LOD) was reported to be below 10–5 mg kg^{-1}. Similarly, acrylamide can also be determined by positive chemical ionization with ammonia as reagent gas (WEJ, 2003) and tandem mass spectrometric detection of daughter ions released from the single charged molecular ion adduct. Compared to chemical ionization method, which is mentioned above, it is possible to reduce the LOD by GC-MS/MS to 1–2 μg kg^{-1} (Table 10.3).

10.6.2 LC-BASED METHODS

10.6.2.1 EXTRACTION

In most of the LC methods, water is used at room temperature according to researchers (Tareke et al., 2000; Rosén & Hellenäs, 2002; Becalski et al., 2003).

TABLE 10.3 GC-based Methods for the Determination of Acrylamide in Food Products.

Ref.	Matrix	Extraction	Internal standard	Derivatization	Clean-up	Column	GC parameter	Detection	MS parameter
Tareke et al. (2000)	Fried food	10 g sample mixed with 100 mL water	N,N-dimethylacrylamide	Bromination (7.5 g KBr, HBr till pH 1–3, bromine water), 4°C 18 h	Filtration of extract through glass fiber (Satorious) filter purification on Carbograph 4 column (1000 mg carbon) (LARA S.r.l, Rome Italy), addition of IS, bromination, titration with sodium thiosulphate solution (1 M), 15 g Na_2SO_4 added, extraction with 2×10 mL ethyl acetate (EtAc)/hexane = ¼, filtration of combined extract through 0.45 µm SRP25 PTFE filter (Satorious Mini-Sart), GPC on Bio beads S-X3 gel (400 × 25 mm i.d. column) (Bio-Rad), eluent:ethyl acetate/cyclohexane = 1/1, fraction from 80 to 200 mL collected and evaporated to 100 µL	HP PAS 1701, 25 m × 0.32 mm i.d., 0.25 µm film thickness (SGE)	65°C held for 1 min then 15°C min⁻¹ to 250°C held for 10 min, Inj.: 2 µL splitless	GC/MS LOD: 5 µg kg⁻¹, recovery: 98%, WR = 5–500 µg kg⁻¹	Analyte: m/z = 106, 108, 150, 152 IS: m/z = 178, 180
Tateo and Bononi (2003)	Baked food	Finely ground sample defatted with n-hexane in a Soxhlet extractor, residual solvent removed under vacuum, 10 g defatted sample extracted with 50 mL methanol for 15 min under stirring, followed by shaking in an ultrasonic bath for 1 min, centrifugation of the mixture for 5 min at 2500 rpm	External standardization		Reduction of the volume of the methanol phase to less than 2 mL by means of a rotary evaporator, transference of sample into a graduated vial and made up with methanol to 2 mL	Supelcowax.30 m × 0.25 mm i.d., 0.2 µm film thickness	60°C for 1 min then 10°C min⁻¹ to 240°C	GC/MS LOD: 2 µg kg⁻¹ LOQ: 10 µg kg⁻¹ recovery: 68–75.4%	AA: m/z = 27, 55, 71

TABLE 10.3 (Continued)

Ref.	Matrix	Extraction	Internal standard	Derivatization	Clean-up	Column	GC parameter	Detection	MS parameter
Ahn et al. (2002)	Toasted bread, fried potato chips	10 g homogenized sample extracted with 100 mL hot water, 2 h in water bath at 80°C ± 2°C	Methacrylamide	According to Castle (1993)	Filtration of the rapidly cooled water extract through glass fiber filter (Sartorius), IS added to 10–15 g extract, 15 mL bromination solution added, after reaction titration with sodium thiosulfate solution (1 M), followed by extraction on shaker with 8 mL EtAc for 15 min, centrifugation (5 min at 2000 rpm). Drying of 4 mL extract over 0.5 g Na_2SO_4 for 15 min, organic solution decanted and evaporated to < 1 mL, reconstitute with EtAc to 1 mL	DB 17, 30 m × 0.25 mm i.d., 0.25 μm film thickness (J and W)	85°C held for 1 min, then 25°C min^{-1} to 175°C held for 6 min then heated at 40°C min^{-1} to 250°C, held for 5 min, Inj: 1 μL splitless	GC/MS	Analyte: m/z = 106, 108, 150, 152 IS: m/z = 120, 122
Bieder-mann et al. (2002a)	All types	15 g sample homogenized with 75 mL water, addition of IS1 (1 μL g^{-1} sample), mixing, equilibration 30 min at 70°C, 10 g homogenate mixed with 40 mL 1-propanol, centrifugation of about 12 mL mixture	IS1: methacrylamide + [D_3]-acrylamide (500 mg L^{-1}) IS2: butyramide		10 mL supernatant added by 15 droplets of vegetable oil, evaporation of solvent at 50Torr and 60–70°C, addition of 3 mL acetonitrile (can) and 20 mL hexane to the residue, second extraction of can with another 5 m hexane, transference of about 1.5 mL can-phase into autosampler vial, addition of butyramide standard (IS2)	40 cm guard column 0.53 mm i.d. deactivated with Carbowax 20 M, Carbowax 20 M, 10 m × 0.25 mm, i.d., 0.4 μm film thickness	70°C held for 1 min, then 15°C min^{-1} to 220°C, held for 2 min, Inj: 1 μL on column	GC/MS with chemical ionization, reagent gas methane, LOD: 10–20 μg kg^{-1} recovery of IS1: 20–90%	

TABLE 10.3 *(Continued)*

Ref.	Matrix	Extraction	Internal standard	Derivatization	Clean-up	Column	GC parameter	Detection	MS parameter
Biedermann et al. (2002b)	All types	20 g sample addition of 200 mL warm water (60°C), addition of IS (1 µg g^{-1}), mixing, equilibration 20 min, centrifugation of 80 mL homogenate for 20 min at 4000 rpm	propionamide (1 g L^{-1})		Removal of fat layer and transference of 25 mL into another 100 mL centrifuge glass, addition of 20 g (NH$_4$)$_2$SO4 and 25 mL 2-butanone, shaking for 1 min, centrifugation 10 min at 4000 rpm, 20 mL supernatant transferred into pear-shaped flask, solvent evaporated to dryness (50°C at 250 mbar), dry residue reconstituted in 1 mL can and extracted with 3 mL hexane, hexane phase discarded, can phase injected	1 m guard column 0.32 mm i.d. deactivated with diphenyltetra-methyldisilazane, BGB-FFAP, 30 m × 0.32 mm i.d., 0.25 µm film thickness	110°C, then 10°C min^{-1} to 230°C and with 50°C min^{-1} to 250°C, held for 1 min, Inj: 1–2 µL on column or splitless	GC/MS, LOD: 50–100 µg kg^{-1} recovery (mean): 94–101%	m/z = 42–80
Wiertz-Eggert-Jorissen GmbH (2003	Baby food	10 g homogenized sample added by IS (0.1 µg g^{-1}sample), extraction with 50 mL water in an ultrasonic bath for 30 min at 60°C	[D$_3$]-acrylamide		At first addition of Carrez I and Carrez II and 30 mL n-hexane to the aqueous solution, afterward, centrifugation at 45,000 g for 10 min, saturation of the aqueous phase with NaCl and extraction with 50 mL EtAc (twice), the organic phase are combined, dried over Na$_2$SO$_4$ and evaporate to 1 mL	DB Wax, 30 m × 0.25 mm i.d., 0.25 µm film thickness	70°C held for 1 min then 20°C min^{-1} to 230°C, held for 10 min Inj: 2 µL splitless	GC-CI-MS/ MS Reagent gas: ammonia LOD: 1–2 µg kg^{-1} recovery: 80–100%	AA: m/z = 89 > 55, IS: m/z = 92 > 75

TABLE 10.3 *(Continued)*

Ref.	Matrix	Extraction	Internal standard	Derivatiza-tion	Clean-up	Column	GC parameter	Detection	MS parameter
Castle (1993)	Mushrooms	50 g sample, 200 g water, homogenization with Ultra Turrax (1 min), pour off foam, 20 mL subsample centrifuged (15 min at 3500 rpm), supernatant derivatized	IS: 2,3- dibromo-2-methypropionamide (0.5 µg mL⁻¹)	15 mL bromination reagent (200 g KBr, 10 mL HBr, 160 mL bromine water, with water to 1000 mL), overnight at 4°C	Titration with thiosulfate solution (0.7 M), extraction with EtAc (1 × 8 mL), centrifugation (2000 rpm) drying of EtAc over Na₂SO₄, evaporation of the extract to 30 µL (hot block 40°C), addition of 200 µL hexane, evaporation to dryness, reconstitution in 20% EtAc/hexane, fractionation over preconditioned Bond-Elutsilics gel column (500 mg), elution: 3 mL EtAc/hexane = 3/7, discard first 2 mL, collect next 2 mL, addition of 50 µL IS, evaporation to 30 µL	DB 17, 30 m × 0.25 mm i.d., 0.2 µm film thickness (J & W)	65–250°C at 15°C min⁻¹ Inj: 1 µL splitless	GC/MS LOD: 1 µg kg⁻¹	Analyte: m/z = 106, 108, 150, 152 IS: m/z = 120,122

Some reported as heating or sonicating during extraction leads to generate large amounts of fine particles that can saturate the solid phase extraction (SPE) columns used in further cleanup steps (FDA, 2003). Pre-heated water to 80°C has been reported as without any operational problems during clean up (Ahn et al., 2002). Some laboratories took part in the proficiency testing organized by the BfR used a mixture of water and acetone as extractant (Fauhl, 2003). Different laboratories used different mechanical methods for initial extraction step in the analysis, for example, shaking at high speed on horizontal shaker (Becalski et al., 2003), use of a rotating shaker (FDA, 2003), occasional swirling (Ahn et al., 2002; Takatsuki et al., 2003), and mixing with a blender or mixing on a vortex. Extraction in an ultrasonic bath (SQTS, 2003) has also been proposed, as its particle size during extraction as side effects may generate problem during clean up as mentioned earlier.

Also, the defatting step also included by some laboratories before or in combination with the extraction step. Using hexane, toluene, or cyclo-hexane carries it out. New approach of using accelerated solvent extraction (ASE) device has been done. But, there was no such improvement found in chromatographic separations when samples were defatted with hexane before aqueous extraction. ASE approach also used by other laboratories and 60–95% recovery difference was found as this could be due to difference in composition of extraction solvents and matrices to be analyzed.

After extraction before clean-up procedures, the aqueous phase was centrifuged, as centrifugation conditions vary according to laboratories. To control the recoveries and keep track of possible losses occurring during the whole sample pretreatment (extraction and clean up), an internal standard was added to the food extractant mixture. In a similar way to the GC-MS methods, isotopically labeled [$^{13}C_3$]-acrylamide (Tareke et al., 2002; Becalski et al., 2003), [D_3]-acrylamide (Ahn et al. 2002; Rosén & Hellenas, 2002; Becalski et al., 2003; Ono et al., 2003) and [$^{13}C_1$]-acrylamide (Takatsuki et al., 2003) were used in most of the procedures.

10.6.2.2 CLEAN-UP

Most clean up procedures consisted of the combinations of several SPEs (Tables 10.3 and 10.4). One approach was to combine Oasis HBL (Waters, Milford, MA, USA) and Bond Elut-Accucat (mixed mode: C_8, SAX, and SCX) (Varian, Palo Alto, CA, USA) cartridges. Becalski et al. (2003) used a combination of three different cartridges: Oasis MAX (mixed-mode anion exchange) (Waters), Oasis MCX (mixed-mode cation exchange) and

TABLE 10.4 HPLC-based Methods for the Determination of Acrylamide in Food Products.

Ref.	Matrix	Extraction	Internal standard	Clean-up	Liquid Chromatography Detection			MS or MS/MS parameters
					Column	HPLC parameter	Detector +LOD	
Tareke et al. (2002)	Potato crisps, meat, bread	10 g homogenized sample +100 mL water +1mL IS (1 µg mL⁻¹ in water) Mixture centrifuged in 12 mL Pyrex tubes Supernatant centrifuged (10 min at 14,000 rpm) in two Eppendorf tubes (1.5 mL/tube)	[¹³C₃]-acrylamide (CIL, Andover, MA)	SPE on Isolute Multi-Mode 300 mg (International Sorbent Technology), activated with 1 mL acetonitrile and washed with water (2 + 2 mL) The two fractions obtained after the extraction are recombined (3 mL) and passed through the SPE column First 1 mL of the filtrate is discarded, the rest is passed through a syringe filter of 0.45 µm (Sartorius Minisart Hydrophilic). 500 µL are filtered on a Microcon YM-3 (Millipore) in a centrifuge at 14,000 rpm for 10 min or at least until 200 µL are passed	Hypercarb 50 × 2.1 mm min⁻¹, 5 µm (Thermo Hypersil)	Mobile phase + gradient: 0.2 mL min⁻¹ water (6.1 min) Inj: 20 µL Column temperature: room temperature	MS/MS DL: 10 µg kg⁻¹ WR: 10–5000 µg kg⁻¹ Recovery: 91–112%	Ionization mode: positive desolvation temperature: 350°C Source temperature: 125°C desolvation gas flow: 2111 h⁻¹ Collision gas pressure: 2.5 mbar (Argon) Cone voltage: 50 V Capillary voltage: 3.2 kV MRM: Dwell time: 0.3 s AA: 72 > 54 (collision energy:16 eV) [¹³C₃]-acrylamide:75 > 58 (collision energy:11 eV)
Becalski et al. (2003)	Potato chips, potato crisps, cereals, bread, coffee	16 g ground sample + 80 mL water + 16 µL IS (250 µg mL⁻¹) + 10 mL dichloromethane mixture was shaken horizontally for 15 min mixture centrifuged for 2 h at 15,000 rpm (24,000 g) at 4°C. Approximately 10 mL of top layer (water) is transferred to a 5 kDa centrifuge filter and centrifuged at 3500 rpm (4000 g) for 4 h or longer at 4°C	[¹³C₁]-acrylamide or [D₃]-acrylamide (CIL, Andover, MA)	5 mL of the filtrate passed through Oasis MAX (waters) cartridge connected in tandem with Oasis MCX cartridge. The elute of the tandem Oasis MAX/MCX loaded onto the preconditioned ENVI-carb cartridge. First, 1 mL discarded and remaining fraction collected (f1), cartridge washed with 1 mL water (f2) and with 1.5 mL 10% methanol in water (f3). Fractions f1, f2, and f3 analyzed by LC-MS/MS	Hypercarb 150 × 2.1 mm, 5 µm (Thermo Hypersil)	Mobile phase + gradient: 0.175 mL min⁻¹ 15% methanol in 1 m aqueous ammonium formate (isocratic). Inj: 5–10 µL Column temperature: 28°C Autosampler temperature: 10°C	MS/MS LOD: ≈6 µg kg⁻¹ Recovery: 84–100%	Ionization mode: positive Desolvation temperature: 250°C Source temperature: 120°C Desolvation gas flow: 525 l h⁻¹ Cone gas flow: 50 l h⁻¹ Collision gas pressure: 2.6 × 10⁻³ mbar (Argon) Ion energies: 1.0 V for both quadrupoles MRM: Dwell time: 0.3 s Cone voltage: 34 V Interchannel delay: 0.05–0.1s AA: 72 > 55 (Collision energy: 11 eV) 72 > 54(collision energy: 11 eV) 72 > 44(collision energy: 14 eV) 72 > 27(collision energy:16 eV) [¹³C₁]-acrylamide: 75 > 58(collision energy:11 eV)

TABLE 10.4 (Continued)

Ref.	Matrix	Extraction	Internal standard	Clean-up	Liquid Chromatography Detection			
					Column	HPLC parameter	Detector +LOD	MS or MS/MS parameters
Ahn et al. (2002)	Crispbread, po-tato crisps, toasted bread	10 g homogenized sample extracted with 100 mL hot water (80°C ± 2°C) for 2 h occasional swirling rapid cooling vacuum filtration through sintered-glass filter partition of extract for duplicate analysis and one spiking experiment	Methacryl-amide (Sigma, Poole, UK) [13C1]-acryl-amide (QMX, Thaxted, UK) 2,3,3-[D3]-acrylamide (Polymer source, Inc., Dorval, Canada)	4 mL extract passed through Isolute M-M cartridge (300 mg) (Jones Chromatography, Hengoed, UK) First 2 mL discarded, next 2 mL collected and filtered through 0.45 μm	Different columns tested Finally Primisphere C18-HC, 250 × 3.2 mm, 5 μm, 110 Å (Phenomenex)	Mobile phase: 2.1% can + 0.1% acetic acid in water, 0.5 mL min⁻¹ Inj: 50 μL	MS/MS WR: 0.01–100 mg l⁻¹ LOD: not specified	Ionization mode: positive MRM: AA: 72 > 55 [13C1]-acrylamide: not specified 2,3,3-[D3]-acrylamide: not specified
Ono et al. (2003)	Different kinds	To 50 g homogenized sample, IS and 300–400 mL water added and homogenized, Centrifugation 20 min at 48,000 g, supernatant is frozen, melted, and centrifuged again (10 min at 21,700 g)	[D3]-acrylamide	0.5–2 mL supernatant are fraction-ated on mixed-mode SPE cartridges (500 mg). Elution solvent is water. The first 1 mL is discarded, and fractions 2–7, 0.5 mL each) are collected, centrifuged (10 min at 21,700 g), subsequent filtration of the supernatant through 0.22 μm syringe filters. The final step is centrifugal filtration with a cut-off of 3000 Da (50 min at 14,000 g)	Atlantis dC18, 150 × 2.1 mm, 3 μm (Waters)	Mobile phase: 0.1 mL min⁻¹ 10% metha-nol, isocratic Column temperature: 40°C Inj: 2 μL	LOD: 0.2 ng mL⁻¹ LOQ: 0.8 ng mL⁻¹	MS: Ionization: electrospray, Positive mode SRM: AA: 72 > 55 [D3]-acrylamide: 75 > 58

ENVI-Carb (graphitized carbon) (Supelco, Bellefonte, PA, USA). A similar combination of SPE cartridges consisting of Bond Elut C18, Bond ElutJr-PSA (anion exchange) and Bond ElutAccucat (all Varian) was chosen for clean-up of samples, which were measured by LC-MS with column switching (Takatsuki et al., 2003). Höfler et al. (2002) reported that liquid-liquid extraction and SPE did not lead to any significant improvement in analysis. Therefore, filtering through 0.22 μm nylon filters was the only treatment for sample after extraction before HPLC. Completely different approach was used by other laboratories, which took part in proficiency testing of German. First acetonitrile was added to the aqueous extract, subsequently, 500 μL Carrez I ($K_4[Fe(CN)_6]3H_2O$) and Carrez II ($ZnSO_4.7H_2O$) were pipetted onto the sample to obtain a clear solution.

10.6.2.3   LC SEPARATION AND DETECTION

Most of the laboratories used reverse phase chromatography for acrylamide and most frequent column has been used were Hypercarb (5 μm) (Tareke et al., 2000; Rosén & Hellenäs, 2002; Becalski et al., 2003). Also, different columns in reverse phase have been compared are Atlantis dC_{18} (Ono et al., 2003), μ-Bondapack C_{18} and Acqua C_{18}, Vydac 201SP54, Primi sphere C_{18}-HC, Hypercarb, YMC-Pack ODS-AQ, Hypersil, HyPURITY Elite, and Hichrom HIPRP were reported (Ahn et al., 2002). From the comparison the Primi sphere, C_{18}-HC was finally chosen as it provides sufficient retention of acrylamide to minimize matrix interference while allowing a 10 min chromatographic cycle time.

LC-MS/MS, working in multiple reaction monitoring mode (MRM), in which transition from a precursor ion to a product ion is monitored with high selectivity. MRM means transition of precursor ion, which is separated in first Quadra pole, to a product ion, generated by collision with argon in the second Quadra pole, is monitored in third Quadra pole. The transition 72>55 was usually selected to quantify acrylamide as it shows relatively high intensity (Tareke et al., 2000; Ahn et al., 2002; Rosén & Hellenäs, 2002; Becalski et al., 2003). For the detection of isotopically labeled acrylamide used as internal standard, monitored at 75>58 transition for [D_3]- and [$^{13}C_3$]-acrylamide and 73>56 for (Placeholder1) (Abramsson-Zetterberg et al., 2005; Ahn et al., 2002; Ames, 1992; Amrein et al., 2005b; Andrawes et al., 1987; Amrein et al., 2007; Ashoor & Zent, 1984; Becalski et al., 2003; Biedermann et al., 2002a; Buhlert et al., 2006; Gertz, 2002; Capuano et al., 2008; Castle et al., 1991; Castle, 2003; Castle, 1993; Claus et al., 2006; Zhong, 2004;

Andrzejewski et al., 2004; De Wilde et al., 2006b; Doerge, 2005; Dybing et al., 2005; Tareke et al., 2002; Elder, 2005; Elder et al., 2004; Eriksson, 2005; Fauhl, 2003; Tateo & Bononi, 2003; Gama-Baumagartner et al., 2004; Granda et al., 2004; Granvogl et al., 2004)[^{13}C$_1$]-acrylamide. Although, interference occurs with peaks showing identical retention time to acrylamide and deuterated acrylamide observed. This problem could be solved by increasing pH of (Granvogl & Schieberle, 2006; Granvogl & Schieberle, 2007) solution from which acrylamide was extracted into organic solvent (e.g., the ASE device used during the extraction) (Swiss Federal Office of Public Health, 2002). According to Becalski et al. (2003) existence of early eluting compounds interferes in transition 72>55. To eliminate this problem, increase in column length from 100 to 150 mm and by applying Isolute Multi-Mode cartridges during sample preparation. Further chromatographic conditions with optimum parameters used for MS/MS and UV detectors given in tables.

10.6.3   HPLC-BASED METHODS FOR THE DETERMINATION OF ACRYLAMIDE IN FOOD PRODUCTS

The limited number of methods has been published for determination of acrylamide in different food products, despite large number of different matrices observed where acrylamide occurs. Out of these only few reports indicate rigorous and systematic determination approach of the method, which have been performed according to international guidelines like ISO 17025, CEN, and Eurachem (Table 10.4).

10.6.3.1   OFFICIAL AND VALIDATED METHODS

The details of few validated methods for acrylamide determination by corresponding research institutes were described as follow:

10.6.3.1.1   US Food and Drug Administration Center for Food Safety and Applied Nutrition

In 1 g of homogenized sample, 9 mL of water and 1 mL of internal standard [^{13}C$_3$] acrylamide (200 ng/mL in 0.1 % formic acid) were added. The mixture was shaken on rotating shaker for 10 min and centrifuged at 9000 rpm for

30 min. A 5 mL aliquot of supernatant was placed in a Maxi-Spin filter tube, 0.45 μm PVDF (Alltech Associates, IL, USA), and centrifuged at 9000 rpm for 4 min. An Oasis HLB 6 mL SPE cartridge (200 mg; Waters, Miliford, MA, USA) was conditioned with 5 mL of methanol and 5 mL of water. A 2 mL aliquot of the extract was loaded onto the Oasis SPE cartridge, 2 mL of water was added, and the eluent was collected. A Bond Elute-Accucat (mixed mode C_8, SAX, and SCX) 3 mL SPE cartridge (Varian, Palo Alto, CA, USA) was conditioned with 3 mL of methanol and 3 mL of water. The first two portions of the eluent from the previous steps were discarded and the rest was passed through the second SPE cartridge.

LC-MS/MS (ESI+) using Aqua C18 column (250 × 2 mm, 5 μm; Phenomenex, Torrance, CA, USA) analyzed the samples. The HPLC parameters had the following settings: mobile phase, 0.5% methanol in aqueous 0.1% acetic acid: 0.2 mL/min; injection, 20 μL; column temperature, 26°C. The electrospray source had the following settings (with nitrogen): probe temperature, 240°C; source temperature, 120°C; collision gas pressure, 1 Torr.

Acrylamide was identified by MRM mode. The precursor ion $[M+H]^+ = 27$ was fragmented and product ions $[H_2C=CHC=O]^+ = 55$ and $[H_2C=CH]^+ = 27$ were monitored. The ion m/z 55 was used for quantification. Monitored product ions for the internal standard were $[^{13}C_3H_3O]^+ = 58$ and $[^{13}C_2H_3]^+ = 29$ from precursor ion $[M+H]^+ = 75$. The collision energy was 19 eV.

10.6.3.2.2   Swiss Federal Office of Public Health

The homogenized sample of 5 g was spiked with a deuterated acrylamide solution (50 μL, corresponding to ca. 5 μg $[^2H_3]$ acrylamide) and well mixed to 4 g of hydro matrix. The mixture was placed into a 22 mL PLE cartridge filled previously with 0.5 g of hydro matrix material. The material was subjected to a two-step extraction procedure with an ASE-200 instrument from Bionex (Sunnyvale, CA, USA). The same was first degreased three times during 5 min with hexane (PLE experimental set-up: 40°C, 100 bar static, 100% eluent). The residual hexane was flushed with nitrogen for 5 min and the sample was extracted three times 20 min (85/15% acetonitrile/water, 40°C, 100 bar static, 150% eluent). The complete extraction program lasted ca. 1.5 h per sample. A portion (20 mL) of the obtained extract (ca. 45 mL) was concentrated to 1–2 mL in a rotative evaporator at 40°C. The extract was taken in 15 mL water, basified b addition of 1 mL buffer solution (0.1 M trisodium phosphate in water) and shaken 1 min in an ultrasonic bath.

The mixture was poured onto a 20 mL Chemelut cartridge (Varian) with the 5 mL of water used to rinse the flask. After a delay of 15 min, the extraction was carried on with 100 mL of ethyl acetate. The eluent was concentrated to a final volume of ca. 0.5 mL in a rotative evaporator and transferred into an autosampler vial. Turbid solutions were filtered additionally through a syringe nylon filter (Titan Nylon Promax, 0.45 µm, Scientific Resources, Eatontown, USA).

LC-MS/MS (ESI+) using Hypercarb column (125 mm × 2.1 mm, 5 m; Thermo Electron, San Jose, CA, USA) analyzed the samples. The HPLC parameters had the following settings: mobile phase+ gradient, 100% A for 5 min (A: 0.01 M formic acid in water); flow rate, 0.2 mL/min; injection, 10 µL; column temperature, 20°C. The electrospray source had the following settings (with nitrogen): source temperature, 350°C.

Acrylamide was identified by MRM mode. The precursor ion m/z 72 was fragmented, and product ion m/z 55 was monitored and used for quantification. Monitored product ion for the internal standard was m/z 58 from precursor ion m/z 75.

KEYWORDS

- **acrylamide**
- **potato**
- **cereal**
- **Maillard reaction**
- **processed food**
- **detoxification**
- **bakery products**

REFERENCES

Abramsson-Zetterberg, L.; Wong, J.; Ilbäck, N. G. Acrylamide Tissue Distribution and Genotoxic Effects in a Common Viral Infection in Mice. *Toxicology* **2005,** *211,* 70–76.

Ahn, J. S.; Castle, L.; Clarke, D. B.; Lloyd, A. S.; Philo, M. R.; Speck, D. R. Verification of Findings of Acrylamide in Heated Foods. *Food Addit. Contam.* **2002,** *19,* 1116–1124.

Ames, J. M. The Maillard Reaction. In *Biochemistry of Food Proteins;* Hudson B. J. F., Ed.; Elsevier: London, 1992; pp 99–153.

Amrein, T. M.; Andres, L.; Schönbächler, B.; Conde-Petit, B.; Escher, F.; Amadò, R. Acrylamide in Almond Products. *Eur. Food Res. Technol.* **2005a,** *221,* 14–18.

Amrein, T. M.; Andres, L.; Escher, F.; Amadò, R. Occurrence of Acrylamide in Selected Foods and Mitigation Options. *Food Addit. Contam.* **2007,** *24,* 13–25.

Amrein, T. M.; Lukac, H.; Andres, L.; Perren, R.; Escher, F.; Amadò, R. Acrylamide in Roasted Almonds and Hazelnuts. *J. Agric. Food Chem.* **2005b,** *53,* 7819–7825.

Andrawes, F.; Greenhouse, S.; Draney, D. Chemistry of Acrylamide Bromination for Trace Analysis by Gas Chromatography and Gas Chromatography-Mass Spectrometry. *J. Chromatogr.* **1987,** *399,* 269–275.

Andrzejewski, D.; Roach, J. A. G.; Gay, M. L.; Musser, S. M. Analysis of Coffee for the Presence of Acrylamide by LC-MS/MS. *J. Agric. Food Chem.* **2004,** *52,* 1996–2002.

Ashoor, S. H.; Zent, J. B. Maillard Browning of Common Amino Acids and Sugars. *J. Food Sci.* **1984,** *49,* 1206–1207.

Becalski, A.; Lau, B. P. Y.; Lewis, D.; Seaman, S. W. Acrylamide in Foods: Occurrence, Sources and Modelling. *J. Agric. Food Chem.* **2003,** *51,* 802–808.

Biedermann, M.; Biedermann-Brem, S.; Noti, A.; Grob, K.; Egli, P.; Mändli, H. Two GC-MS Methods for the Analysis of Acrylamide in Foodstuffs. *Mitt. Lebensm. Hyg.* **2002a,** *93,* 638–652. [http:// www.klzh.ch/downloads/acrylamid_1.pdf]

Biedermann, M.; Biedermann-Brem, S.; Noti, A.; Grob, K.; Egli, P.; Mändli, H. Experiments on Acrylamide Formation and Possibilities to Decrease the Potential of Acrylamide Formation Potatoes. *Mitt. Lebensm. Hyg.* **2002b,** *93,* 668–687.

Biedermann, M.; Biedermann-Brem, S.; Noti, A.; Grob, K.; Egli, P.; Mändli H. Two GC-MS Methods for the Analysis of Acrylamide in Foods. *Mitt. Lebensm. Hyg.* **2003,** *93,* 638–652.

Buhlert, J.; Carle, R.; Majer, Z.; Spitzner, D. Thermal Degradation of Peptides and Formation of Acrylamide. *Lett. Org. Chem.* **2006,** *3,* 356–357.

Capuano, E.; Ferrigno, A.; Acampa, I.; Ait-Ameur, L.; Fogliano, V. Characterization of Maillard Reaction in Bread Crisps. *Eur. Food Res. Technol.* **2008,** *228,* 311–319.

Castle, L.; Campos, M. J.; Gilbert, J. Determination of Acrylamide Monomer in Hydroponically Grown Tomato Fruits by Capillary Gas Chromatography-mass Spectrometry. *J. Sci. Food Agric.* **1991,** *54,* 549–555.

Castle, L. Determination of Acrylamide Monomer in Mushrooms Grown on Polyacrylamide Gel. *J. Agric. Food Chem.* **1993,** *41,* 1261–1263.

Castle, L. Determination of Acrylamide in Food: GC-MS Bromination Method. Presentation at the Workshop 'Analytical Methods for Acrylamide Determination in Food', Oud-Turnhout, Belgium, 2003.

Chuda, Y.; Ono, H.; Yada, H.; Ohara-Takada, A.; Matsuura-Endo, C.; Mori, M. Effects of Physiological Changes in Potato tubers (*Solanum tuberosum* L.) After Low Temperature Storage on the Level of Acrylamide Formed in Potato Chips. *Biosci. Biotechnol. Biochem.* 2003, *67* (5), 1188–90.

CIAA e Confederation of the Food and Drink Industries of the EU. CIAA Acrylamide "Toolbox." REV 12. Feb 16, 2009. www.ciaa.be/documents/acrylamide.

Claus, A.; Weisz, G. M.; Schieber, A.; Carle, R. Pyrolytic Acrylamide Formation from Purified Wheat Gluten and Gluten-supplemented Wheat Bread Rolls. *Mol. Nutr. Food Res.* **2006a,** *49,* 87–93.

Claus, A.; Schreiter, P.; Weber, A.; Graeff, S.; Hermann, W.; Claupein, W.; Schieber, A.; Carle, R. Influence of Agronomic Factors and Extraction Rate on the Acrylamide Contents in Yeast-leavened Breads. *J. Agric. Food Chem.* **2006b,** *54,* 8968–8976.

Cook, D. J.; Taylor, A. J. On-line MS/MS Monitoring of Acrylamide Generation in Potato- and Cereal-based Systems. *J. Agric. Food Chem.* **2005,** *53,* 8926–8933.

Delatour, T.; Périsset, A.; Goldmann, T.; Riediker, S.; Stadler, R. H. Improved Sample Preparation to Determine Acrylamide in Difficult Matrixes Such as Chocolate Powder, Cocoa, and Coffee by Liquid Chromatography Tandem Mass Spectrometry. *J. Agric. Food Chem.* **2004,** *52,* 4625–4631.

De Wilde, T.; De Meulenaer, B.; Mestdagh, F.; Govaret, Y.; Vanderburie, S.; Ooghe, W.; Fraselle, S.; Demeulemeester, K.; Van Peteghem, C.; Calus, A.; Degroodt, J. M.; Verhé, R. Influence of Fertilization on Acrylamide Formation during Frying of Potatoes Harvested in 2003. *J. Agric. Food Chem.* **2006b,** *54,* 404–408.

De Wilde, T.; De Meulenaer, B.; Mestdagh, F.; Govaert, Y.; Vandeburie, S.; Ooghe, W.; Fraselle, S.; Demuelemeester, K.; Van Petegheim, C.; Calus, A.; Degroodt, J. M.; Verhé, R. Influence of Storage Practices on Acrylamide Formation during Potato Frying. *J. Agric. Food Chem.* **2005,** *53,* 6550–6557.

De Wilde, T.; De Meulenaer, B.; Mestdagh, F.; Govaert, Y.; Ooghe, W.; Fraselle, S.; Demeulemeester, K.; Van Peteghem, C.; Calus, A.; Degroodt, J. M.; Verhé, R. Selection Criteria for Potato Tubers to Minimize Acrylamide Formation during Frying. *J. Agric. Food Chem.* **2006a,** *54,* 2199–2205.

Doerge, D. R.; Young, J. F.; McDaniel, L. P.; Twaddle, N. C.; Churchwell, M. I. Toxicokinetics of Acrylamide and Glycidamide in B6C3F1 Mice. *Toxicol. Appl. Pharmacol.* **2005,** *202,* 258–267.

Dybing, E.; Farmer, P. B.; Andersen, M.; Fennell, T. R.; Lalljie, S. P.; Muller, D. J.; Olin, S.; Petersen, B. J.; Schlatter, J.; Scholz, G.; Scimeca, J. A.; Slimani, N.; Törnqvist, M.; Tuijtelaars, S.; Verger, P. Human Exposure and Internal Dose Assessments of Acrylamide in Food. *Food Chem Toxicol.* **2005,** *43,* 365–410.

Elder, V. A. Method for Enhancing Acrylamide Decomposition. U.S. Patent 20050118322, 2005.

Elder, V. A.; Fulcher, J. G.; Leung, H.; Topor, M.G. Method for Reducing Acrylamide in Thermally Processed Foods. U. S. Patent 20040058045, 2004.

Eriksson, S. Acrylamide in Food Products: Identification, Formation and Analytical Methodology. Ph.D. Thesis, Department of Environmental Chemistry, Stockholm University, Stockholm, Sweden, 2005.

Fauhl, C. *Proficiency Testing Acrylamide,* Presentation at the Workshop "Analytical Methods for Acrylamide Determination in Food." Oud-Turnhout, Belgium, 2003.

Food and Drug Administration (FDA). Draft: Detection and Quantitation of Acrylamide in Foods. February 24, **2003**. http://vm.cfsan.fda. gov/dms/acrylamide.html.

Fredriksson, H. J.; Tallving, J.; Rosén, P.; Åman. Fermentation Reduces Free Asparagine in Dough and Acrylamide Content in Bread. *Cereal Chem.* **2004,** *81,* 650–653.

Gama-Baumgartner, F.; Grob, K.; Biederman, M. Citric Acid to Reduce Acrylamide Formation in French Fries and Roasted Potatoes? *Mitt. Lebens. Hyg.* **2004,** *95,* 110–117.

Gertz, C.; Klostermann, S. Analysis of Acrylamide and Mechanism of Its Formation in Deep-fried Products. *Eur. J. Lipid Sci. Technol.* **2002,** *104,* 762–771.

Granda, C.; Moreira, R. G.; Tichy, S. E. Reduction of Acrylamide Formation in Potato Chips by Low-temperature Vacuum Frying. *J. Food Sci.* **2004,** *69,* 405–411.

Granvogl, M.; Jezussek, M.; Koehler, P.; Schieberle, P. Quantification of 3-amminopropionammide in Potatoes: A Minor but Potent Precursor in Acrylamide Formation. *J. Agric. Food Chem.* **2004,** *52,* 4751–4757.

Granvogl, M.; Schieberle, P. Thermally Generated 3-amminopropio-nammide as a Transient Intermediate in the Formation of Acrylamide. *J. Agric. Food Chem.* **2006,** *54,* 5933–5938.

Grob, K.; Biedermann, M.; Biedermann-Brem, S.; Noti, A.; Imhof, D.; Amrein, T.; Pfefferle, A.; Bazzocco, D. French Fries with Less than 100 µg/kg Acrylamide. Collaboration between Cooks and Analysts. *Eur. Food Res. Technol.* **2003,** *217,* 185–194.

Granvogl, M.; Schieberle, P. Quantification of 3-aminopropionamide in Cocoa, Coffee and Cereal Products. Correlation with Acrilamide Concentrations Determined by an Improved Clean-up Method for Complex Matrices. *Eur. Food Res. Technol.* **2007,** *225,* 857–863.

Guenther, H.; Anklam, E.; Wenzl, T.; Stadler, R. H. Acrylamide in Coffee: Review of Progress in Analysis, Formation and Level Reduction. *Food Addit. Contam.* **2007,** *24* (1), 60–71.

Hagmar, L.; Wirfält, E.; Paulsoon, B.; Törnqvist, M. Differences in Haemoglobin Adduct Levels of Acrylamide in the General Population with Respect to Dietary Intake, Smoking Habits and Gender. *Mutat. Res.* **2005,** *580,* 157–165.

Hamlet, C. G.; Baxter, D. E.; Sadd, P. A.; Slaiding, I.; Liang, L.; Muller, R.; Yaratne, S. M.; Booer, C. RHM Technology Report C014 for the Food Standards Agency, Project Codes C03032 and C03026: Exploiting Process Factors to Reduce Acrylamide in Cereal-based Foods. RHM Technology Ltd.: High Wycombe, UK, 2005.

Höfler, F.; Maurer, R.; Cavalli, S. Schnelle Analyse von Acrylamid in Lebensmitteln mit ASE und LC/MS. *GIT Labor Fachz.* **2002,** *48,* 968–970.

Johnson, K. A.; Gorzinski, S. J.; Bodner, K. M.; Campbell, R. A.; Wolf, C. H.; Friedman, M. A.; Mast, R. W. Chronic Toxicity and Oncogenicity Study on Acrylamide Incorporated in the Drinking Water of Fischer 344 Rats. *Toxicol. Appl. Pharmacol.* **1986,** *85,* 154–168.

Kroh, L. W. Caramelisation in Food and Beverages. *Food Chem.* **1994,** *51,* 373–379.

Mottram, D. S.; Wedzicha, B. L.; Dodson, A. T. Acrylamide is Formed in the Maillard Reaction. *Nature* **2002,** *419,* 448–449.

Nemoto, S.; Takatsuki, S.; Sasaki, K.; Maitani, T. Determination of Acrylamide in Foods by GC/MS Using 13C-labeled Acrylamide as an Internal Standard. *J. Food Hyg. Soc. Jpn.* **2002,** *43,* 371–376.

Norris, M. V. Acrylamide. In *Encyclopedia of Industrial Chemical Analysis;* Dee Snell F., Hilton C. L., Eds.; Interscience: New York (NY), 1967; Vol. 4, pp 160–168.

Noti, A.; Biedermann-Brem, S.; Biedermann, M.; Grob, K.; Albisser, P.; Realini, P. Storage of Potatoes at Low Temperature Should be Avoided to Prevent Increased Acrylamide Formation during Frying or Roasting. *Mitt. Lebensm. Hyg.* **2003,** *94,* 167–180.

Ono, H.; Chuda, Y.; Ohnishi-Kameyama, M,; Yada, H.; Ishizaka, M.; Kobayashi, H.; Yoshida, M. Analysis of Acrylamide by LC-MS/MS and GC-MS in Processed Japanese Foods. *Food Addit. Contam.* **2003,** *20,* 215–220.

Pedersen, J. R.; Olsson, J. O.; Soxhlet Extraction of Acrylamide from Potato Chips. *Analyst.* **2003,** *128,* 332–334.

Pedreschi, F.; León, J.; Mery, D.; Moyano, P.; Pedreschi, R.; Kaack, K.; Granby, K. Color Development and Acrylamide Content of Pre-dried Potato Chips. *J. Food Eng.* **2007,** *79,* 786–793.

Riediker, S.; Stadler, R. H. Analysis of Acrylamide in Food by Isotope-dilution Liquid *Chromatography* Coupled with Electrospray Ionization Tandem Mass Spectrometry. *J. Chromatogr. A.* **2003,** *1020,* 121–130.

Robarge, T.; Phillips, E.; Conoley, M. *LC-GC Europe, The Applications Book;* Thermo Electron Corporation Press: San Jose, CA, 2003, p 2.

Rosén, J.; Hellenäs, K. E. Analysis of Acrylamide in Cooked Foods by Liquid Chromatography Tandem Mass Spectrometry. *Analyst.* **2002,** *127,* 880–882.

Ruggeri, S.; Cappelloni, M.; Gambelli, L.; Nicoli, S.; Carnovale, E. Chemical Composition and Nutritive Value of Nuts Grown in Italy. *Ital. J. Food Sci.* **1998**, *10*, 243–252.

Schabacker, J.; Schwend, T.; Wink, M. Reduction of Acrylamide Uptake by Dietary Proteins in a Caco-2 Gut Model. *J. Agric. Food Chem.* **2004**, *52*, 4021–4025.

Schaller, U. In *Experiences with Acrylamide Determination—View from a Retailer's Laboratory.* Presentation at the Workshop 'Analytical Methods for Acrylamide Determination in Food', Oud-Turnhout, Belgium, 2003.

Seron, L. H.; Poveda, E. G.; Moya, M. S. P.; Carratalá, M. L. M.; Berenguer-Navarro, V.; Grané-Teruel, N. Characterisation of 19 Almond Cultivars on the Basis of Their Free Amino Acids Composition. *Food Chem.* **1998**, *61*, 455–459.

Smith, E. A.; Oehme, F. W. Acrylamide and Polyacrylamide: A Review of Production, Use, Environmental Fate and Neurotoxicity. *Rev. Environ. Health* **1991**, *9*, 215–228.

Sörgel, F.; Weissenbacher, R.; Kinzing-Schippers, M.; Hofmann, A.; Illauera, M.; Skott, A.; Landersdorfer, C. Acrylamide: Increased Concentrations in Homemade Food and First Evidence of Its Variable Absorption from Food, Variable Metabolism and Placental and Breast Milk Transfer in Humans. *Chemotherapy* **2002**, *48*, 267–274.

Stadler, R. H. The Formation of Acrylamide in Cereal Products and Coffee. In *Acrylamide and Other Hazardous Compounds in Heat-treated Foods;* Skog, K., Alexander, J., Eds.; Woodhead Publishing: Cambridge, 2006; pp 23–40.

Stadler, R. H.; Blank, I.; Varga, N.; Robert, F.; Hau, J.; Guy, P. A.; Robert, M. C.; Riediker, S. Acrylamide from Maillard Reaction Products. *Nature* **2002**, *419*, 449–450.

Stadler, R. H.; Scholz, G. Acrylamide: An Update on Current Knowledge in Analysis, Levels in Food, Mechanisms of Formation, and Potential Strategies of Control. *Nutr. Rev.* **2004**, *62*, 449–467.

Sumner, S. C. J.; MacNeela, J. P.; Fennell, T. R. Characterization and Quantitation of Urinary Metabolites of 1,2,3-C-13 Acrylamide in Rats and Mice Using C-13 Nuclear Magnetic Resonance Spectroscopy. *Chem. Res. Toxicol.* **1992**, *5*, 81–89.

Surdyk, N.; Rosén, J.; Andersson, R.; Åman, P. Effects of Asparagine, Fructose and Baking Conditions on Acrylamide Content in Yeast-leavened Wheat Bread. *J. Agric. Food Chem.* **2004**, *52*, 2047–2051.

Svensson, K.; Abramsson, L.; Becker, W.; Glynn, A.; Hellenas, K. E.; Lind, Y.; Rosen. J. Dietary Intake of Acrylamide in Sweden. *Food Chem. Toxicol.* **2003**, *41*, 1581–1586.

Swiss Federal Office of Public Health. Determination of Acrylamide in Food. . http://www. bag. admin.ch/verbrau/aktuell/ d/AA_methode.pdf (accessed August 10, *2002*)

Swiss Quality Testing Service (SQTS). Untersuchungs MethodezurBestimmung von Acrylamidmit GC/MS. Standard Operation Procedure (SOP) Received from SQTS. Doetokon, Switzerland, 2003.

Takata, K.; Okamoto, T. A Method for Determination of Acrylamide in Environmental Samples by Gas Chromatography Using Bromination-dehydrobromination. *Kankyo Kagaku.* **1991**, *1*, 559–565.

Taeymans, D.; Wood, J.; Ashby, P.; Blank, I.; Studer, A.; Stadler, R. H.; Gondé, P.; Van Eijck, P.; Lalljie, S.; Lingnert, H.; Lindblom, M.; Matissek, R.; Müller, D.; Tallmadge, D.; O'Brien, J.; Thompson, S.; Silvian, D.; Whitmore, T. A Review of Acrylamide: An Industry Perspective on Research, Analysis, Formation, and Control. *Crit. Rev. Food Sci. Nutr.* **2004**, *44*, 323–347.

Tateo, E.; Bononi, M. A GC/MS Method for the Routine Determination of Acrylamide in Food. *Ital. J. Food Sci.* **2003**, *15*, 149–151.

Takatsuki, S.; Nemoto, S.; Sasaki, K.; Maitani, T. Determination of Acrylamide in Processed Foods by LC/MS Using Column Switching. *J. Food Hyg. Soc. Jpn.* **2003**, *44*, 89–95.

Tareke, E.; Rydberg, P.; Karlsson, P.; Eriksson, S.; Tornqvist, M. Acrylamide: A Cooking Carcinogen? *Chem. Res. Toxicol.* **2000**, *13*, 517–522.

Tareke, E.; Rydberg, P.; Karlsson, P.; Eriksson, S.; Törnqvist, M. Analysis of Acrylamide, a Carcinogen Formed in Heated Foodstuffs. *J. Agric. Food Chem.* **2002**, *50*, 4998–5006.

Tsutsumiushi, K.; Hibino, M.; Kambe, M.; Oishi, K.; Okada, M.; Miwa, J.; Taniguchi, H. Application of Ion-trap LC/MS/MS for Determination of Acrylamide in Processed Foods. *Shokuhin Eiseigaku Zasshi.* **2004**, *45*, 95–99.

Weisshaar, R.; Gutsche, B. Formation of Acrylamide in Heated Potato Products-model Experiments Pointing to Asparagine as Precursor. *Deut. Lebensm. Randsch.* **2002**, *98*, 397–400.

WHO. Summary Report of the Sixty-fourth Meeting of the Joint FAO/WHO Expert Committee on Food Additive (JECFA). The ILSI Press International Life Sciences Institute, Rome, Italy, 2005; pp 1–47.

Wiertz-Eggert-Jorissen (WEJ) GmbH. Standard Operation Procedure for the Determination of Acrylamide in Baby Food. Standard Operation Procedure (SOP). WEJ GmbH: Hamburg: Germany, **2003**.

Wilson, K. M.; Rimm, E. B.; Thompson, K. M.; Mucci, L. A. Dietary Acrylamide and Cancer Risk in Humans: A Review. *J. Verbrauch. Lebensm.* **2006**, *1*, 19–27.

Yaylayan, V. A.; Wnorowski, A.; Perez-Locas, C. Why Asparagine Needs Carbohydrates to Generate Acrylamide. *J. Agric. Food Chem.* **2003**, *51*, 1753–1757.

Yasuhara, A.; Tanaka, Y.; Hengel, M.; Shibamoto, T. Cromatographic Investigation on Acrylamide Formation in Browning Model Systems. *J. Agric. Food Chem.* **2003**, *51*, 3999–4003.

Zhang, Y.; Zhang, G.; Zhang, Y. Occurrence and Analytical Methods of Acrylamide in Heat-treated Foods: Review and Recent Developments. *J. Chromatogr. A.* **2005**, *1075*, 1–21.

Zhong, C. Y.; Chen, D. Z.; Xi, X. L. *Guangdong Chem. Eng.* **2004**, *46*, 15.

Zyzak, D. V.; Sanders, R. A.; Stojanovic, M.; Tallmadge, D. H.; Eberhardt, B. L.; Ewald, D. K.; Gruber, D. C.; Morsch, T. R.; Strothers, M. A.; Rizzi, G. P.; Villagran, M. D. Acrylamide Formation Mechanism in Heated Foods. *J. Agric. Food Chem.* **2003**, *51*, 4782–4787.

Zyzak, D. V. Acrylamide Formation Mechanism in Heated Food. JIFSAN/NCFST Acrylamide in Food Workshop. http://www.jifsan.umd.edu/Acrylamide/acrylamide_workshop.html, 2002

FIGURE 1.1 *H. erinaceus* fresh fruit bodies grown in tropical climate of Malaysia.

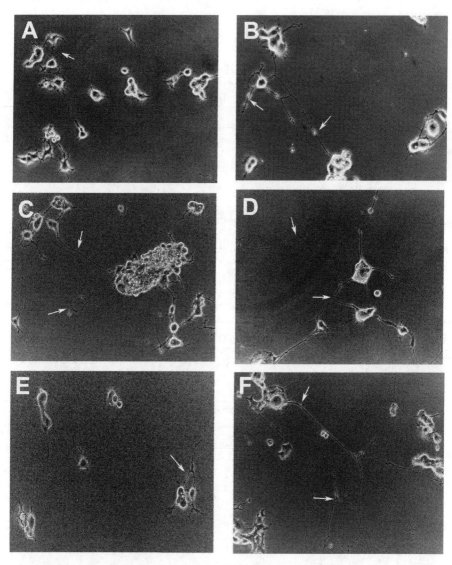

FIGURE 1.2   The morphology of neural hybrid clone NG108-15 cells treated with 0.2% (v/v) aqueous extracts of fruit bodies and mycelium of *H. erinaceus* after 72 h of incubation at $37 \pm 2°C$ in a 5% CO_2 humidified incubator ×100. NG108-15 cells without extract or treated with NGF (20 ng/mL) were used as negative and positive controls, respectively. Arrows indicate neurites. (A) Negative control without extract. (B) Positive control: NGF (20 ng/mL). (C) Fresh fruit bodies aqueous extract 0.2% (v/v). (D) Freeze-dried fruit bodies aqueous extract 0.2% (v/v). (E) Oven-dried fruit bodies aqueous extract 0.2% (v/v). (F) Mycelial aqueous extract 0.2% (v/v). (A) and (E): Cells attached strongly and exhibited short cellular extensions, but these have not elongated adequately to be scored as neurites. (B–D) and (F): Cells show an exuberant outgrowth of long, diverse, beaded, and branching neurites.

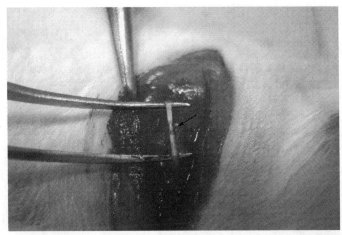

FIGURE 1.3   A crush injury was created using a fine watchmaker forceps no. 4 for 10 s on the peroneal nerve at 10 mm from extensor digitorum longus muscle and complete crush was confirmed by the presence of a translucent band across the nerve as indicated by an arrow. All operations were performed on right limb and the left limb served as an uninjured control. (Reprinted from International Journal of Medicinal Mushrooms, Vol 14, Issue 5, K.-H. Wong, M. Naidu, P. David, R. Bakar & V. Sabaratna, Neuroregenerative Potential of Lion's Mane Mushroom, Hericium erinaceus (Bull.: Fr.) Pers. (Higher Basidiomycetes), in the Treatment of Peripheral Nerve Injury (Review), pp. 427-446, © 2012, with permission from Begell House, Inc.)

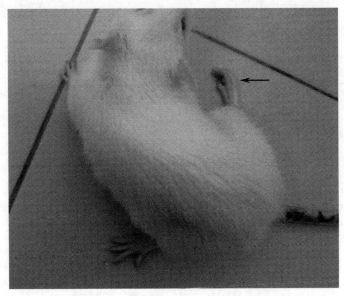

FIGURE 1.4   Gait changes associated with peroneal nerve injury–joint contracture, making measurement impossible because the rat walks on the dorsum of the affected foot. Arrow indicates the operated limb.

FIGURE 1.5   Hot plate apparatus comprises of an open-ended cylindrical space with a floor of heated plate. Skin of the plantar surface of the foot was stimulated by gently placing the rat on the heated surface. Thermal nociception was evaluated by observing the withdrawal reflex latency (WRL) of the operated limb in response to heat stimulation (as indicated by an arrow) by a built-in *timer* activated by an external foot switch. (Reprinted from Wong, Kah-Hui et al . Restoration of sensory dysfunction following peripheral nerve injury by the polysaccharide from culinary and medicinal mushroom, Hericium erinaceus (Bull.: Fr.) Pers. through its neuroregenerative action. **Food Sci. Technol (Campinas)**, Campinas , v. 35, n. 4, p. 712-721, Dec. 2015 . Available from http://www.scielo.br/scielo.php?script=sci_arttext&pid=S0101-20612015000400712&lng=en&nrm=iso https://creativecommons.org/licenses/by/4.0/deed.en)

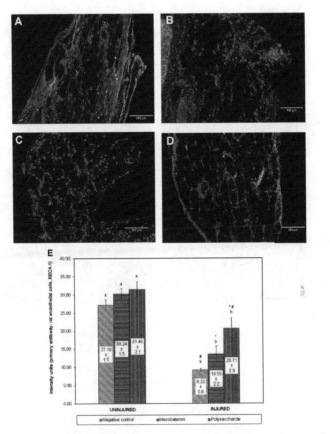

FIGURE 1.7 Distribution of RECA-1-positive microvessels in the peroneal nerve. Fluorescence imaging staining for RECA-1 as green fluorescent lines. DNA was stained as blue fluorescence. (A) Uninjured nerve (contralateral side). (B) Distal to the injury site of injured nerve in negative control group. (C) Distal to the injury site of injured nerve in positive control group. (D) Distal to the injury site of injured nerve in polysaccharide group. Scale bar = 100 μm. (E) RECA-1 levels in the peroneal nerve as measured by intensity of immunoreactivity. Each bar represents 24 sections of DRG from 6 animals per group. Asterisks (*) indicate significant differences ($p < 0.05$, DMRT) in values for different groups compared to the negative control within a same uninjured/injured category. Hash signs (#) indicate significant differences ($p < 0.05$, DMRT) in values for different groups compared to mecobalamin within a same uninjured/injured category. The same alphabet indicates no significant differences and different alphabets indicate significant differences ($p < 0.05$, ANOVA) between different categories in the same group. (Reprinted from Wong, Kah-Hui et al . Restoration of sensory dysfunction following peripheral nerve injury by the polysaccharide from culinary and medicinal mushroom, Hericium erinaceus (Bull.: Fr.) Pers. through its neuroregenerative action. **Food Sci. Technol (Campinas)**, Campinas , v. 35, n. 4, p. 712-721, Dec. 2015 . Available from http://www.scielo.br/scielo.php?script=sci_ arttext&pid=S0101-20612015000400712&lng=en&nrm=iso   https://creativecommons.org/ licenses/by/4.0/deed.en)

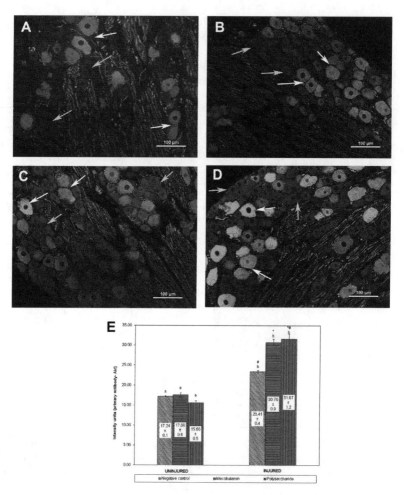

FIGURE 1.8   Akt activation in DRG neurons after crush injury. Double immunofluorescence staining between Akt (orange) and NF-200 (green) in contralateral and ipsilateral DRG. DNA was stained as blue fluorescence. (A) DRG from uninjured nerve (contralateral side). (B) DRG from injured nerve in negative control group. (C) DRG from injured nerve in positive control group. (D) DRG from injured nerve in polysaccharide group. Akt did not co-localize with large neurons. White arrows indicate large neurons, whereas yellow arrows indicate small neurons. Scale bar = 100 μm. (E) Akt levels in the DRG as measured by intensity of immunoreactivity. Each bar represents 24 sections of DRG from six animals per group. Means with different letters in uninjured/injured category are significantly different ($p < 0.05$). (Reprinted from Wong, Kah-Hui et al . Restoration of sensory dysfunction following peripheral nerve injury by the polysaccharide from culinary and medicinal mushroom, Hericium erinaceus (Bull.: Fr.) Pers. through its neuroregenerative action. **Food Sci. Technol (Campinas)**, Campinas , v. 35, n. 4, p. 712-721, Dec. 2015 . Available from http://www. scielo.br/scielo.php?script=sci_arttext&pid=S0101-20612015000400712&lng=en&nrm=iso https://creativecommons.org/licenses/by/4.0/deed.en)

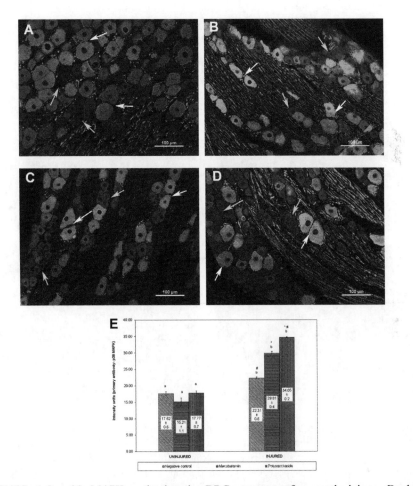

FIGURE 1.9 p38 MAPK activation in DRG neurons after crush injury. Double immunofluorescence staining between p38 MAPK (orange) and NF-200 (green) in contralateral and ipsilateral DRG. DNA was stained as blue fluorescence. (A) DRG from uninjured nerve (contralateral side). (B) DRG from injured nerve in negative control group. (C) DRG from injured nerve in positive control group. (D) DRG from injured nerve in polysaccharide group. p38 MAPK did not co-localize with large neurons. White arrows indicate large neurons, whereas yellow arrows indicate small neurons. Scale bar = 100 μm. (E) p38 MAPK levels in the DRG as measured by intensity of immunoreactivity. Each bar represents 24 sections of DRG from six animals per group. Means with different letters in uninjured/injured category are significantly different ($p < 0.05$). (Reprinted from Wong, Kah-Hui et al . Restoration of sensory dysfunction following peripheral nerve injury by the polysaccharide from culinary and medicinal mushroom, Hericium erinaceus (Bull.: Fr.) Pers. through its neuroregenerative action. **Food Sci. Technol (Campinas)**, Campinas , v. 35, n. 4, p. 712-721, Dec. 2015 . Available from http://www.scielo.br/scielo.php?script=sci_ arttext&pid=S0101-20612015000400712&lng=en&nrm=iso https://creativecommons.org/ licenses/by/4.0/deed.en)

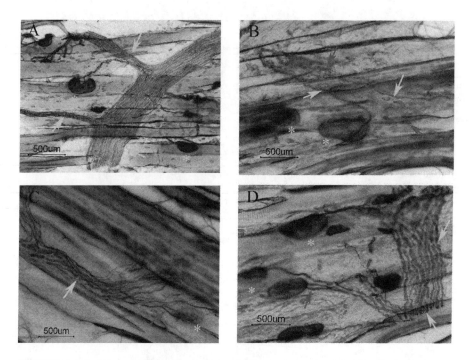

FIGURE 1.10 The morphology of silver-cholinesterase-stained longitudinal section of EDL muscle in normal unoperated limb and operated limb after 14 days of peroneal nerve crush injury. Yellow arrows indicate the axons. Violet arrows indicate the degenerating axons. *Red arrows* indicate polyneuronal innervation. Asterisks indicate the motor endplates. Scale bar = 500 μm. (A) Normal unoperated limb. Axon bundles are clear and compact. (B) Operated limb in negative control group—distilled water (10 mL/kg body weight/day). Wallerian degeneration can be detected. Degenerated axons are being phagocytosed by the cooperative action of denervated Schwann cells and infiltrating macrophages. (C) Operated limb in positive control group—mecobalamin (130 μg/kg body weight/day). Loose axon bundles indicate regeneration process is ongoing. (D) Operated limb in polysaccharide group—*H. erinaceus* fresh fruit bodies (10 mL/kg body weight/day). Axon bundles are more compact and regeneration process is more advanced compared to positive control group. (Reprinted from International Journal of Medicinal Mushrooms, Vol 14, Issue 5, K.-H. Wong, M. Naidu, P. David, R. Bakar & V. Sabaratna, Neuroregenerative Potential of Lion's Mane Mushroom, Hericium erinaceus (Bull.: Fr.) Pers. (Higher Basidiomycetes), in the Treatment of Peripheral Nerve Injury (Review), pp. 427-446, © 2012, with permission from Begell House, Inc.)

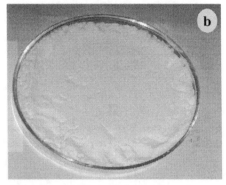

FIGURE 5.1 (a) Culinary banana at matured edible stage and (b) starch from culinary banana.

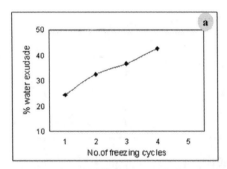
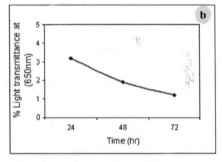

FIGURE 5.2 (a) Freeze-thaw stability of culinary banana starch and (b) Paste-clarity of culinary banana starch.

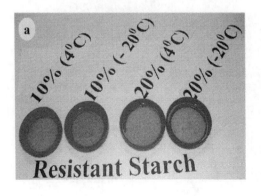
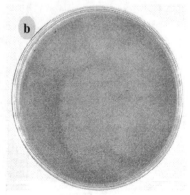

FIGURE 5.8 Resistant starch (RS) obtained by (a) Hydrothermal process and (b) Enzyme debranching.

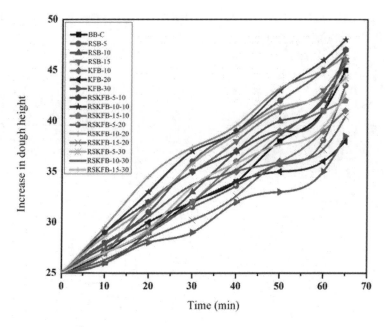

FIGURE 5.14   Proofing of dough prior to baking.

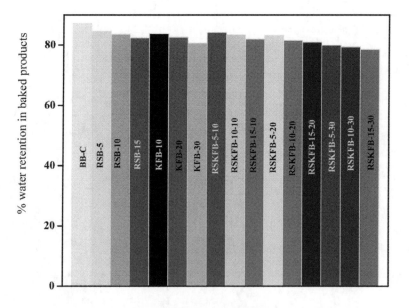

Brown bread samples

FIGURE 5.15   Percentage water retention in baked products.

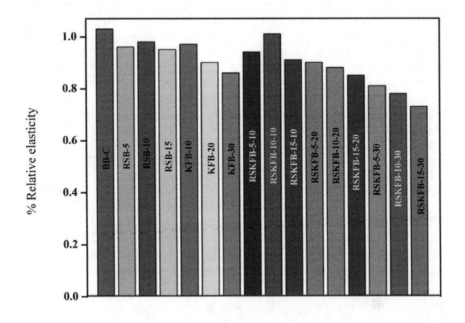

FIGURE 5.16   Relative elasticity of brown bread samples.

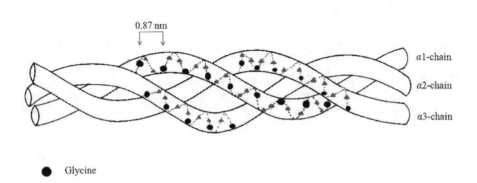

FIGURE 6.2   Illustrative drawing of the collagen triple helix structure and the arrangement of amino acids (Gly-X-Y) in the collagen molecule (Modified from Nimni and Harkness, 1988).

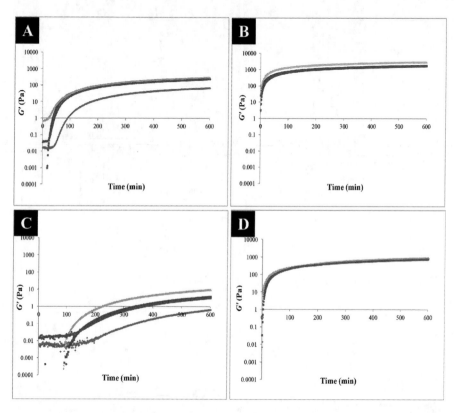

FIGURE 6.10   Elastic modulus, G', as a function of gelation time for 5 % (w/w) of (A) dory, (B) hoki, (C) ling, and (D) salmon gelatins containing various concentration levels of TGase. Concentrations of TGase were: control (■), TGase 1.6 mg per g (■), TGase 3.33 mg per g (■), and TGase 5 mg per g (■).

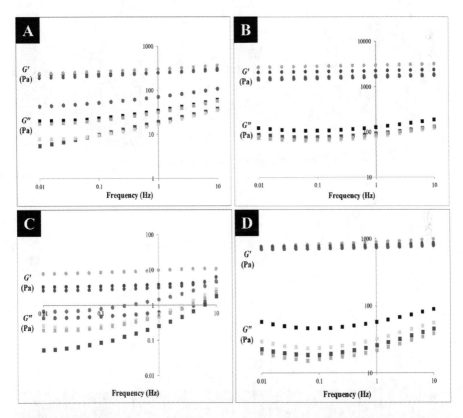

FIGURE 6.11 Elastic modulus, *G'* (circle symbols) and loss modulus, *G"* (square symbols) as a function of frequency for 5 % (w/w) of (A) dory, (B) hoki, (C) ling, and (D) salmon gelatins containing various concentration levels of TGase. Gelatin concentration was 5 % (w/w) and TGase concentrations were: Control (•,■), TGase 1.6 mg per g (•,■), TGase 3.33 mg per g (•,■), and TGase 5 mg per g •,■). Please note that the Y-axis scales are different for the different fish gelatins. This was done to allow the reader to see the differences between the different curves of the same fish gelatin.

FIGURE 11.2   Preparing fresh-cut pommecythere (*Spondias cytherea*).

FIGURE 11.3   Preparing fresh-cut carambola (*Averrhoa carambola*).

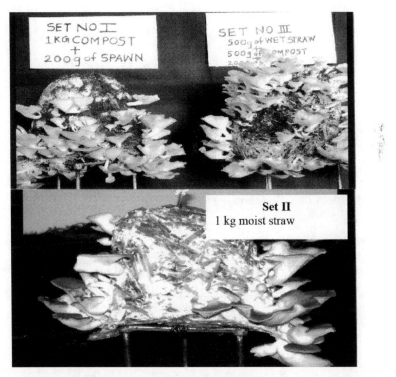

FIGURE 20.1   Fruiting bodies (sporophores) growing on compact mass of three different sets of substrates of Set I (top left), Set III (top right), and Set II (bottom).

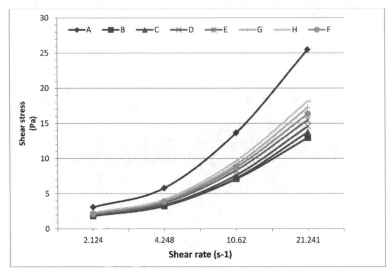

FIGURE 21.2   Shear stress vs. shear rate relationship of different custard formulations as per Table 22.1.

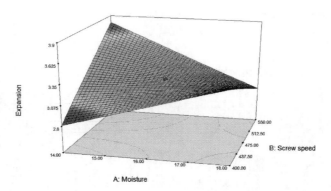

FIGURE 22.1   Effect of moisture and screw speed on the ER.

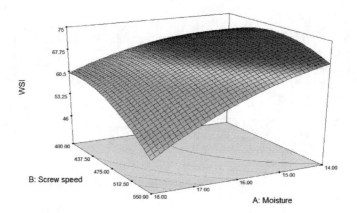

FIGURE 22.2   Effect of moisture and screw speed on the WSI.

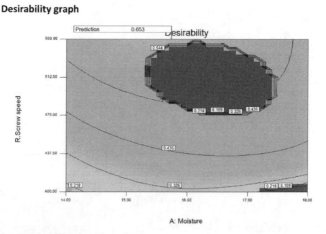

FIGURE 22.3   Desirability function graph.

PART III

Innovative Methods in Processing and Preservation of Food and Bioactive Components

CHAPTER 11

IMPROVING QUALITY OF READY-TO-EAT, MINIMALLY PROCESSED PRODUCE USING NON-THERMAL POSTHARVEST TECHNOLOGIES

ROHANIE MAHARAJ*

Biosciences, Agriculture and Food Technologies (BAFT), EC IAF Campus, The University of Trinidad and Tobago, Caroni North Bank Road, Centeno, Piarco, Trinidad & Tobago

E-mail: rohanie.maharaj@utt.edu.tt

CONTENTS

Abstract 290
11.1   Introduction 290
11.2   Ozone, Ultrasound, and Other Chemical Treatments for Product Decontamination 294
11.3   Modified Atmosphere Packaging (MAP) 297
11.4   Pressure Treatment and Dense Phase Carbon Dioxide (DPCD)...302
11.5   Food Irradiation 303
11.6   Emerging Technologies: Light Emitting-Diodes (LED), Laser, Pulsed Light (PL) and Pulsed Electric Field (PEF), and High Power Ultrasound 308
11.7   Conclusion 312
Keywords 313
References 313

ABSTRACT

Consumer demands for high-quality and safe fresh-cut and ready-to-eat food products, which are free from contaminants and pathogens, require novel, and complementary food preservation technologies, which will ensure integrity and wholesomeness of the foods. This chapter will examine recent advances in the application of primarily non-thermal postharvest technologies for the preservation of such crops. Novel technologies such as irradiation, pulsed light, pulsed electric field, ultrasound, and high-pressure processing can be used to produce safe foods with a high nutritional value similar to its fresh state. These technologies have become an attractive option to preserve fruits and vegetables, creating high-quality products, for both domestic and export markets. However, the challenge facing manufacturers is balancing both the economic and quality costs in incorporating these advanced methods into their processing operations.

11.1   INTRODUCTION

Fresh fruits and vegetables are highly perishable due to their active metabolism during the postharvest stage. Minimally processed fruits and vegetables are those prepared for convenient consumption and distribution to the consumer in a fresh-like state. Minimal processing refers to process steps, such as trimming, peeling, washing, slicing, or shredding before packaging, which makes the product ready-to-eat by the consumer. The International Fresh-Cut Produce Association (IFPA) defines fresh-cut products as fruits or vegetables that have been trimmed and/or peeled and or/cut into 100% usable product that is bagged or pre-packaged to offer consumers high nutrition, convenience, and flavor while still maintaining freshness (Rico et al., 2007). Postharvest operations for ready-to-eat, fresh-cut, or minimally processed produce are outlined in Figures 11.1–11.3.

The consumption of minimally processed, ready-to-eat fresh fruits and vegetables have grown rapidly in recent decades as a result of changes in consumer perceptions, health benefits of desired phytonutrients, and the desire to preserve organoleptic properties as a result of minimal processing. Such products are expected to offer food safety and preservation, while at the same time maintaining fresh-like characteristics (Allende et al., 2006). Importation of fresh exotic tropical fruits, particularly for markets in temperate countries where these products are not readily available, requires a consistently high quality and reliable supply of crops from producing countries.

These crops will need to satisfy regulatory requirements to ensure safety and quality. Currently, the majority of these tropical fruits are consumed fresh in the producing countries, due to high perishability of the harvested product (Romano et al., 2016) and limited postharvest technologies.

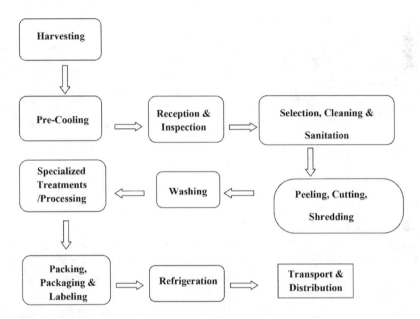

FIGURE 11.1 Postharvest operations for fresh-cut produce.

FIGURE 11.2 **(See color insert.)** Preparing fresh-cut pommecythere (*Spondias cytherea*).

FIGURE 11.3 **(See color insert.)** Preparing fresh-cut carambola (*Averrhoa carambola*).

Processing of crops causes biochemical changes, microbial degradation, and injury to tissues, which results in a reduction of their shelf life particularly in tropical countries where temperature and humidity conditions can be high and favor rapid microbial growth. Minimal processing (including peeling, cutting, slicing, shredding, or trimming operations) of fruits and vegetables are generally more perishable than whole intact crops. Mechanical injury during processing results in cellular delocalization of enzymes and their substrates, leading to biochemical deteriorations, such as enzymatic browning, off-flavors, and texture breakdown as well as increased respiration rate and ethylene synthesis. Enzymatic browning, caused mainly by the action of polyphenol oxidase (PPO), is a major factor limiting shelf life of minimally processed fruits (Gómez et al., 2010). While blanching or use of steam has been used in the food industry for thermal inactivation of undesirable enzymes, these also are responsible for undesirable changes including changes in texture and color (Gómez et al., 2010). With respect to controlling postharvest storage rots, the use of chemical agents has its limitations. Chlorine solutions are the most common sanitizer used for fruits and vegetables particularly in the fresh-cut industry and generally, the reported levels of free chlorine for washing vegetables range from 0 to 30 ppm (Bilek & Turantaş, 2013). However, the association of chlorine with the possible formation of carcinogenic chlorinated compounds in water such as, trihalomethanes has called into question, the use of chlorine in food processing (Ribeiro et al., 2012). Similarly, the use of fumigants for insect disinfestation after harvest can cause health hazards to consumers and pose a threat to the environment. Fungal growth increases the risk of spoilage and some undesirable changes in crops like taste. The application of fungicide on the surface of produce after washing is quite common. In minimally processed fruits and vegetables, such treatments are applied alone or in combination, such as impregnation in food grade waxes or edible coating layers (Llano et al., 2016). Their industrial application may be sometimes limited by regulations or, most frequently, by the awareness of consumers to food additives and concern toward human health.

Temperature management and humidity control are critical factors that influence the quality of perishable food crops after harvest. Various methods in combination with these are used to control insect and microbial growth as well as maintain the quality and prolong the shelf life of fresh-cut produce. These methods include ozone decontamination, modified, or controlled atmosphere-packaging, irradiation, alone, or in different combinations. Hurdle technology which is the combination of different preservation techniques and its synergistic effects as a preservation strategy will be covered in detail in another chapter. Chemical compounds, such as antioxidants and antimicrobials, either from natural or synthetic origin, have been broadly used to improve shelf life of fresh-cut produce along with good practices in storage and transportation through the use of refrigerated trucks (Rico et al., 2007). The use of combined treatments, such as ascorbic acid and calcium chloride or the use of anti-browning agents with physical methods, such as heat treatments or refrigeration in combination with controlled atmosphere have been found to be effective in reducing browning in fruits and vegetables. Other non-thermal treatments used for fresh-cut and ready-to-eat produce include treatment with high pressure and electric field pulses (Garcia & Barrett, 2002).

Concerns about safety of the food supply chain along with the impact of agricultural chemicals on the environment are increasing public and scientific demand in the development of non-chemical preservation treatments (Yahia, 2016). Such treatments can add value for growers and processors by tapping into health-oriented markets. Consumer demands for high quality and safe fresh-cut and ready-to-eat food products, which are free from contaminants and pathogens, require novel and complementary food preservation technologies, which will ensure integrity and wholesomeness of the foods. The objective of this chapter will focus on a review of some recent research focusing primarily on the use of non-thermal technologies for fresh-cut, ready-to-eat produce that may improve quality and shelf life of these products.

11.2   OZONE, ULTRASOUND, AND OTHER CHEMICAL TREATMENTS FOR PRODUCT DECONTAMINATION

Water use is critical in several postharvest unit operations that are carried out in packinghouse facilities and in many cases recycling of water is a common practice. Postharvest wash water must be constantly monitored and sanitized. A common practice in the preparation of fresh-cut products is rinsing the peeled and/or cut product in cold water, which helps lower the temperature in addition to removing cellular exudates released during this process. The washing methods can reduce the microbial load of the product; however, if the washing treatment has not been applied properly, this step can cause cross-contamination (Bilek &Turantaş, 2013). Allende et al. (2006) reported on the use of immersion therapy which involves cutting the fruit while it is submerged in water which prevents movement of fluids while the product is being cut and prevents enzymatic degradation. This is used in combination with acidic electrolyzed water produced by electrolysis of an aqueous sodium chlorite solution as a disinfectant for minimally processed vegetable products. Similarly, UV-C light and ozone have been used to reduce browning and injury in fresh-cut produce. Moisture content is another factor that influences the deteriorative reactions and can affect the quality of crops. De-watering of rinsed products can control decay and is usually done commercially through centrifugation or can also be achieved with forced air (Garcia & Barrett, 2002).

11.2.1   OZONE

Ozone is a strong oxidizing agent that acts on carbon residues dissolved in the washing water as well as on the product surface, thus rendering it effective against microorganisms (Huyskens-Keil et al., 2011). When compared to chlorine, ozone has a greater effect on certain microorganisms and rapidly decomposes to oxygen, leaving no residues. However, a higher corrosiveness and initial capital cost for generator are the main disadvantages compared to the use of chlorine (Rico et al., 2007). Due to its efficacy at low concentrations, short contact times, and the absence of detectable residues in or on treated food, makes it an effective sanitizer for water and washing processes. Ozone at concentrations of 0.15–5.0 ppm has been shown to inhibit the growth of spoilage bacteria and yeasts (Rawson et al., 2011). In the food industry, due to its efficient bactericidal and fungicidal properties, ozone is applied to several food products, such as tomato, strawberry, grape, plum, carrot, apple, strawberry, lettuce, and cantaloupe. As a postharvest technology, ozone has significantly improved quality and storage life of broccoli, cucumber, and persimmon and its inhibitory effect on undesired changes of anthocyanins, phenols, ascorbic acids in some strawberry cultivars, and similarly of carbohydrates, carotenoids, and phenolic compounds in tomato fruits.

The effects of ozone on physiology and quality of fruits vary according to chemical composition of the food material, ozone dose, mode of application, and time (Rawson et al., 2011). Application of low ozone is thought to enhance plant secondary metabolites, such as phenolic compounds. Studies have shown an ozone-induced delay in fruit softening for strawberry and tomato and an inhibition of undesired textural related cell wall modifications in green asparagus. However, negative effects have been reported in carrots and, enzyme inactivation associated with the loss of crispiness in lettuce (Huyskens-Keil et al., 2011). Ozone has been declared in many countries to have potential use for food processing and declared in the United States (US) as Generally Recognized as Safe (GRAS) (Rico et al., 2007).

11.2.2   ULTRASOUND

Ultrasonic processing can be adapted to the wash water decontamination process for fruits and vegetables. The application of ultrasound, a non-thermal technology contributes to improved microbial safety and prolongs shelf life, especially in foods with heat-sensitive, nutritional, sensory, and functional characteristics as noted by Bilek & Turantaş (2013). Ultrasonic fields consist of pressure waves at high amplitude with a frequency of 20 kHz or more and generally, ultrasound equipment uses frequencies from 20 kHz to 10 MHz. Higher-power ultrasound at lower frequencies (20–100 kHz), is referred to as "power ultrasound" and has the ability to cause cavitation, which has uses in food processing to inactivate microorganisms. Cavitation enhances the mechanical removal of attached or entrapped bacteria on the surfaces of fresh produce by displacing or loosening particles through a shearing or scrubbing action (Allende et al., 2006). Power ultrasound has been used to speed up the convective drying of several foodstuffs and has emerged as an alternative to conventional thermal treatments for pasteurization and sterilization of food products. The ultrasonic waves generate alternating expansions and contractions when traveling across a medium, and advantages include reduced processing time, higher throughput, and lower energy consumption (Rawson et al., 2011). A major advantage of ultrasound over other techniques in the food industry is that sound waves are generally considered safe, non-toxic, and environmentally friendly. The combination of ultrasound with some non-thermal and/or physical–biological methods constitutes an attractive approach to enhance microbial inactivation and elimination. Additionally, from the standpoint of consumer demand, ultrasound, and physical-biological combined processes show a potential for further investigation and application in a plant scale and dependent on this, ultrasound technology could have a wide range of current and future applications in the food industry. Most published data indicate that the antimicrobial efficiency of ultrasound is relatively low in some conditions and only under special situations could ultrasound become an actual and effective alternative to the decontamination process (Bilek &Turantaş, 2013).

11.2.3 ELECTROLYZED WATER (EW)

Electrolyzed water (EW), also known as electrolyzed oxidizing water, is conventionally generated by electrolysis of aqueous sodium chloride to produce an electrolyzed basic aqueous solution at the cathode and an electrolyzed acidic solution at the anode as reported by Rico et al. (2007). Acidic EW (pH 2.1–4.5) has a strong bactericidal effect against pathogens and spoilage organisms, more effective than chlorine due to a high oxidation-reduction potential (ORP). It has shown higher effectiveness in reducing viable aerobes than ozone on whole lettuce although at the expense of produce quality when used on fresh-cut vegetables (Rico et al., 2007).

11.2.4 CHEMICAL TREATMENTS

Minimally processed fresh fruits and vegetables must be subjected to chilled storage to prolong shelf life as preparation steps, such as scrubbing peeling, slicing, and shredding of such produce renders them susceptible to microbial spoilage, desiccation, browning, bleaching, textural changes, and developments of off-flavor and off-odor (Garcia & Barrett, 2002). Enzymatic browning can be controlled through the use of physical and chemical treatments and in most cases, both are employed. Physical treatments include temperature reduction and/or oxygen, use of modified atmosphere packaging or edible coatings or irradiation treatments while chemical treatments include post-cut dipping in anti-browning agents (Table 11.1). The use of natural preservatives may be effective to retain quality of minimally processed produce by having an antimicrobial effect, inhibiting spoilage, or avoiding oxidative processes (Rico et al., 2007). Certain antioxidant treatments including ascorbic acid dips or naturally occurring thiol-compounds are commercially used to delay the development of signs of browning and discoloration on the cut surface of fresh-cut produce. Calcium salts have been widely used as firming agents in the fruit and vegetable industry for both whole and fresh-cut commodities. On the other hand, calcium treatments have been widely applied in combination with ascorbic acid and thiol-compounds, such as cysteine, N-acetylcysteine, and reduced glutathione to prevent enzymatic browning and maintain firmness of fruits (Llano et al., 2016).

TABLE 11.1   Examples of Anti-browning Agents (Garcia & Barrett, 2002).

Chemical treatment/anti-browning agent	Example
Acidulants	Malic acid
Reducing agent and thiol-containing compounds	Ascorbic acid, Cysteine
Chelating agent	EDTA and Sporix
Complexing agent	Cyclodextrins
Enzyme inhibitors	4-hexylresorcinol
Other anti-browning agents	Sodium chloride, calcium treatments (calcium ascorbate/NatureSeal™), honey, proteases, and aromatic carboxylic acids

11.3   MODIFIED ATMOSPHERE PACKAGING (MAP)

Modified or controlled atmosphere is usually used as a supplement to temperature and humidity control for prolonging the shelf life of fresh fruits and vegetables including fresh-cut produce. The modified atmosphere can be achieved passively where the package is sealed under normal air conditions or actively where the package is flushed with a gas mixture before closure (Rico et al., 2007). For packaged fruits and vegetables, a modified atmosphere can be achieved by vacuum packaging (VP), gas flushing, or controlled permeability of the packaging to control the oxidation reaction by reducing O_2 concentration in the storage atmosphere. Essentially, MAP can be achieved by replacing air in the package with a single gas or mixture of gases where the proportion of each component is fixed when the mixture is introduced. Nitrogen filling for dry or intermediate moisture foods or CO_2 filling, which acts as a biostat for fresh fruits and vegetables can also be used to prevent anaerobic microbial growth and lipid oxidation. The atmosphere concentrations recommended for preservation depend on the product (Rico et al., 2007). Fresh-cut produce is generally more tolerant to higher CO_2 concentrations than intact products because the resistance to diffusion is smaller. Successful applications of MAP have been reported for fresh-cut pineapple, apples, kiwifruit, honeydew, banana, and mango as reported by Corbo et al. (2010).

Controlled atmosphere packaging involves changing the gaseous composition at which the products are exposed. Usually, the packaging is flushed with an inert gas, such as nitrogen followed by injecting gases with specified O_2 and CO_2 concentrations. The concentration of the filling gases (O_2, CO_2, and N_2) should be arranged depending on the nature of the product, packaging materials, and storage temperature. According to Vigneault et al. (2012), using modified or controlled atmospheres to enhance the quality of horticultural produce involves one of the following types of treatments: short-term (pulse to minutes), medium-term (an hour to days), or long-term (weeks to a year). It is thought that gas treatments should produce more consistent results compared to radiation treatments since they are uniformly applied to the whole surface of the fresh produce.

Commodity-specific packaging is commercially available for both fresh and minimally processed products (Ozturk et al., 2016; Gontard & Guillaume, 2010). MAP uses specialized polymers or packaging film materials which are designed with differential permeability to facilitate the required gas composition in the packaging and are tailored to the specific demands of the food and fresh-cut products. Some reported results as noted by Corbo et al. (2010) are listed in Table 11.2. Development of biodegradable packaging materials can replace conventional synthetic materials made from essentially four basic polymers: polyvinyl chloride (PVC), polyethylene terephthalate (PET), polypropylene (PP), and polyethylene (PE). Edible coatings are applied directly on the food surface are designed to create a modified atmosphere. The edible coatings can be furnished with active compounds, such as antimicrobials or antioxidants to obtain additional desired effects. Such materials and coatings may protect very powerfully against microbial spoilage and loss of intrinsic product quality, resulting in a prolonged shelf life (Ozturk et al., 2016). However, there is still a major concern about product safety associated with the use of MAP primarily as a result of conditions that encourage slower growing pathogenic bacteria or spoilage organisms which are suppressed under these conditions (Allende et al., 2006).

TABLE 11.2   Some Beneficial Effects of MAP for Fresh-cut Fruits (Corbo et al., 2010).

Examples of fresh-cut fruits packages	MAP storage conditions	Benefits/shelf life
Conference pears	Plastic bags of permeability 15 cm^3 O_2/m^2/bar/24 h, initial atmospheres 0 kPa O_2	3 weeks under refrigeration
Smooth cayenne pineapple pieces	10% CO_2 combined with a maximum 8% O_2	4 days at 10°C 2 weeks at 0°C
Smooth cayenne pineapple pieces	2–5% CO_2 combined with a maximum 12–15% O_2	14 days at 10°C
Cantaloupe cubes	Film-sealed containers flushed with 4% O_2 and 10% CO_2	Maintained saleable quality at 5°C

11.3.1   EDIBLE POLYMERIC FILMS AND COATINGS

An edible coating or film could be defined as primary packaging made from edible components where a thin layer of material can be coated to a food or formed into a film and used as a food wrap without changing the original ingredients or the processing method (Pascall & Lin, 2013). Edible films can be distinguished from coatings based on the method of manufacture and application to the food product. Generally, films are dried preform thin materials (50–250 µm) that are used to wrap product or make pouches and bags. While edible coatings are also thin layers of edible material, the key difference is that they are applied as a liquid of varying viscosity to the outer surface of the product by spraying, dipping, brushing, or other appropriate methods (Pascall & Lin, 2013).

The use of edible coatings and films in the food industry is of great interest due to their potential to increase shelf life of foods (Genevois et al., 2016). These films can act as a carrier for the incorporation of food additives as antimicrobials, antioxidants, flavors, and colors, which can improve the antimicrobial protection, appearance, texture, and taste during storage in fruits and vegetables (Genevois et al., 2016). According to Pascall and Lin (2013), advantages and improvements of such films and coatings include:

- gas and moisture barriers and mechanical properties;
- improving appearance and sensory perceptions;
- convenience;
- microbial protection;
- increasing shelf life;

- health benefits when nutrients are incorporated within the film matrix; and
- edible films being biodegradable.

Edible films and coatings can either be prepared from lipids, polysaccharides, resins or proteins, or a combination of these. Some of the polysaccharides that are suitable for use as edible films and coatings include chitosan, starches, pectins, alginates, and cellulose derivatives. Improved stability and growth inhibition effect of Microorganisms, were reported when chitosan-based coating was applied to the surface of strawberry, carrot, mango, cantaloupe, pineapple, and mushroom (Genevois et al., 2016). These authors also reported that the application of a cassava starch-based film, containing potassium sorbate, reduced significantly the activity of microorganisms in samples of pumpkin. Edible coatings may be improved by the addition of plasticizers, surfactants, and emulsifiers (Garcia & Barrett, 2002). The selection of edible coatings is important due to the hydrophilic nature of cut surfaces of some fresh-cut produce. Lipid components offer important water-barrier characteristics to some coatings; however, they may give a waxy or gummy mouth-feel to the product. On the other hand, hydrophilic polymers have poor moisture barrier properties and do not aid in reducing water loss of coated products. Emulsion coatings containing mixed components seem to have a better performance as noted by Garcia and Barrett (2002).

Bioactive compounds containing nutrients, such as calcium and vitamin E have been incorporated in the use of films and coatings. Ready-to-eat pumpkin fortified with iron and ascorbic acid in a starch-based coating improved product stability (Genevois et al., 2016).

11.3.1.1 MICRO-ENCAPSULATION AND THE USE OF NATURAL COMPOUNDS

For fresh-cut produce, research has been conducted on natural antimicrobial agents that are compatible with the chemical properties of post-cut dipping solutions. Naturally occurring antimicrobials include compounds derived from biological materials including:

- animal-derived enzymes, proteins, and small peptides;
- plant-derived secondary metabolites (phytoalexins, phenolics, and essential oils); and
- microorganism-based bacteriocins (nisin and pediocin).

Many plant extracts have been shown to possess antimicrobial activities against a wide range of microorganisms related to food spoilage and safety. Incorporating vanillin (12 mM) in the post-cut dipping solution (NatureSeal) of apple slices could inhibit the microbial growth during the 19-day post-cut storage by 37% and 66%, respectively, in two cultivars of apple slices (Rupasinghe et al., 2006). Whey permeates (WP), a by-product of the cheese industry when incorporated in fresh-cut vegetables showed good antimicrobial activity (Corbo et al., 2010).

According to Mohammadi et al. (2016), natural compounds, such as essential oils (EOs), particularly with high percentage of thymol and carvacrol have been reported to be beneficial in controlling postharvest decay in fruits, in addition to possessing antioxidant, antibacterial, and antifungal properties. The United States Food and Drug Administration (FDA), indicates that such compounds are Generally Recognized as Safe (GRAS) food additive. Limitations of these EOs include, instability and low bioavailability since its bioactive compounds are chemically reactive species, which can easily degrade under abiotic stress factors (e.g., heat, oxygen, light, pressure, moisture, and pH). In this regard, the nano-/micro-encapsulation of EOs allow protection of their sensitive bioactive compounds from unfavorable environmental conditions and enhanced their bioactivity during food processing and storage. In the case of antimicrobial agents, encapsulation can increase the concentration of bioactive compounds in food areas where microorganisms are preferentially located (e.g., in water-rich phases or at solid-liquid interfaces). Over the last few years, chitosan (CS) has been used as a matrix for encapsulating different active compounds because of its advantageous biological properties, such as biodegradability, biocompatibility, and nontoxicity. While microencapsulation systems may guarantee protection of hydrophobic antimicrobial compounds against evaporation or degradation, in general, they do not affect antimicrobial activity. In contrast, novel nano-encapsulation systems, due to the subcellular size and high surface area to volume ratio, increase the passive cellular absorption mechanisms and penetrate areas (intracellular and extracellular areas) that may be inaccessible to other drug delivery systems, thus increasing antimicrobial and antioxidative activity. In this regard, it was previously reported that antimicrobial effects of thymol have been improved by preparation of CSNPs, while the results of one study have revealed that the CSNPs of carvacrol, showed different degrees of growth inhibition (Mohammadi et al., 2016).

Possible downsides to the commercial development of natural compounds as antimicrobials for controlling postharvest diseases of fruits and vegetables as noted by Ippolito and Nigro (2005) include:

- expensive procedures for extracting the active compounds;
- the need for chemical standardization, stability, and quality control;
- extended studies on toxicological aspects for specific compounds;
- production at competitive costs with existing pesticides;
- the lack of studies on development of resistance;
- difficulties in registration as pesticides;
- efficacy sometimes not consistent and acceptable unless in complexly integrated approaches;
- restricted market confined to postharvest environment; and
- scarce interest by companies in testing for botanicals since it is still unclear whether proprietary claims can be made.

11.4   PRESSURE TREATMENT AND DENSE PHASE CARBON DIOXIDE (DPCD)

11.4.1   PRESSURE TREATMENT

Pressure treatment is a physical non-thermal treatment involving the application of a pressure other than normal atmospheric pressure to the commodity to extend shelf life. It is applied instantaneously and uniformly around each single product or throughout an entire mass of food, independently of its size, shape, or composition (Vigneault et al., 2012). Pressure treatments can be divided into two-man categories: low and high-pressure treatments. Low-pressure treatment (0–1 MPa) can be hypobaric or hyperbaric and it can be applied to fresh produce while high-pressure treatments (100–1200 MPa) are generally applied to processed food and not fresh produce as it causes irreversible damage to the cell structure of fresh fruits and vegetables (Goyette et al., 2012). Literature indicates that high-pressure processing preserves the nutritional value of processed food products. The advantage of this treatment is that pressure at a given position and time is the same in all directions, transmitted uniformly and immediately through the pressure transferring medium and independent of geometry (Rawson et al., 2011).

Hypobaric or sub-atmospheric pressure treatment involving low-pressure changes (0–0.5 MPa absolute pressures) can rapidly remove heat, reduce oxygen levels, and expel harmful gases released over time during storage and thus increase shelf life of crops. Pressures of less than 0.1 MPa tend to increase shelf life; however, desiccation is a limiting factor due to a reduction in the water vapor partial pressure as in the case of mango using a pressure below 0.007 MPa. Usually, water is sprayed on the produce to resolve the problem of insufficient relative humidity causing desiccation during hypobaric storage. Beneficial effects have been reported in several crops, such as mango, asparagus, cucumber, and apple (Vigneault et al., 2012).

Postharvest hyperbaric treatment consists of subjecting the crops to a pressurized environment ranging from 0.1 to 1 MPa in which the proportion of each gas in the air is normally maintained. In a study described by Goyette et al. (2012), breaker stage tomato fruits subjected to hyperbaric pressures ranging from 5 to 7 atmospheres for a duration of 10 days at 13°C were found to delay ripening.

11.4.2 DENSE PHASE CARBON DIOXIDE (DPCD)

According to a recent review by Rawson et al. (2011), dense phase carbon dioxide processing (DPCD or DP-CO$_2$), is a collective term for liquid CO$_2$ and supercritical CO$_2$ or high-pressurized carbon dioxide (HPCD). It is a non-thermal alternative to heat pasteurization for liquid foods, that utilizes pressure (<90 MPa) in combination with carbon dioxide (CO$_2$) to destroy microorganisms as a means of food preservation. DPCD or supercritical CO$_2$ denotes phases of matter that remain fluid, yet are dense with respect to gaseous CO$_2$. Applications of DPCD to the horticultural industry included treatment of whole fruits, such as strawberry, honeydew melon, and cucumber to inhibit mold growth but some limitations include severe tissue damage in some fruits even at low pressures. There are a limited number of studies available in the literature regarding the effect of DPCD on bioactivity of exotic fruits. However, Rawson et al. (2011) reported on the effects of DPCD (30.6 MPa, 8% CO$_2$ and 6.8 min, 35°C) treatments on total antioxidant activity, and phenolic content of guava puree, where total antioxidant activity (DPPH) in organic fractions was not significantly changed by DPCD treatments. Similar results were reported for phenolic content.

11.5 FOOD IRRADIATION

UV light and gamma irradiation are the main irradiation methods used in food industry and this treatment involves exposure of raw or processed foods to ionizing or non-ionizing radiation in order to extend shelf life. The ionizing radiation source could be high-energy electrons, X-rays (machine generated), or gamma rays which emit radiation in closed chambers (from cobalt-60 or cesium-137), while the non-ionizing radiation is electromagnetic radiation that does not carry enough energy/quanta to ionize atoms or molecules, represented mainly by ultraviolet rays (UV-A, UV-B, and UV-C), visible light, microwaves, and infrared (Rawson et al., 2011).

11.5.1　GAMMA IRRADIATION

Gamma irradiation is a clean and efficient physical technology for food decontamination and is an effective means of processing and preserving food products as it is applicable to fresh, frozen, cooked produce, and packaged foods (Tawema et al., 2016). It has been recognized as an alternative to methyl bromide for treating fresh and dried agricultural products to overcome quarantine barriers in international trade. Although gamma rays, high energy electrons, and X-rays all have similar effects, gamma rays are most commonly used in food irradiation because of their ability to deeply penetrate pallet loads of food, however, the choice of irradiation method will depend on the material needing to be treated (IFST, 2006). For treating surfaces or thin layers of food, beta particles (i.e., electrons) are the preferred source of energy, while for bulk products gamma rays are used. The dose of absorbed radiation is called the gray (Gy), and 1 gray is equivalent to 1 joule of ionizing energy per kilogram. Irradiation has been used in the US, Canada, and the EU since the 1950s. In 1986, the US FDA approved the use of gamma radiation treatments of up to 1 kGy (100 krad) on fruits and vegetables while in 2008, the FDA amended the regulations to provide for the safe use of ionizing radiation for control of food-borne pathogens, at a dose of up to 4 kGy for fresh iceberg lettuce and fresh spinach. Gamma irradiation has shown as a promising method to prevent deterioration as well as to control pathogenic microorganisms and insect disinfestation contamination in fruits and vegetables, and for improving nutritional attributes and shelf life (Lacroix & Vigneault, 2007). The doses required for sterilization of most insects is below 0.50 kGy. Irradiation doses below 1 kGy can help to inhibit germination (onions, potatoes, etc.), to delay ripening and senescence, and to control several insects on fruits and vegetables. However, the doses required for inactivation of pathogens can cause tissue damage and changes in food sensory properties. Therefore, it is recommended to use gamma irradiation in combination with other methods to reduce the irradiated doses, while maintaining adequate antimicrobial effectiveness (Lacroix & Vigneault,

2007). Research shows that gamma irradiation treatments either increase or decrease the antioxidant content of foods, which is dependent on the dose delivered, exposure time, and the raw material used. Gamma irradiation treatment has been found to extend the storage life of peach and enhance its antioxidant content, similarly with midrib and non-midrib leaf tissues of Romaine and Iceberg lettuce. According to Rico et al. (2007), irradiation of minimally processed carrots improved their color and flavor but impaired texture while doses of up to 0.5 kGy did not impair quality of minimally processed lettuce but higher doses did. Consumption of foods irradiated at doses up to 10 kGy has been considered safe by the World Health Organization (WHO), Food and Agriculture Organization (FAO), and the International Atomic Energy Agency (IAEA). A dose of up to 7 kGy is considered acceptable for fish and other marine products. Gamma irradiation using a dose of ≥7 kGy could be effective in reducing murine norovirus-1 (MNV-1) in traditional Korean half-dried seafood products during storage at 10°C for 7 days (Kang et al., 2016).

Irradiated produce sold in retail packages must be labeled with the radura symbol and contain the statement, "treated with radiation (or irradiation)." For irradiated food not sold in retail packages, the radura symbol and statement must appear on either the individual item, the bulk container or a sign at the point of purchase (Yahia, 2016). Symptoms of irradiation stress on fruits include accelerated softening, uneven ripening, and surface damage (Yahia, 2016). Consumer concern is still associated with the safety of gamma-irradiated produce. Another drawback of this technology, is that irradiation is considered a capital-intensive technology, as capital costs include the radiation source, hardware, and radiation shields among other items.

11.5.2 ULTRAVIOLET RADIATION (UV-C)

The use of artificial non-ionizing UV-C is a promising non-thermal process in postharvest technology. UV-C has been extensively used for many years in the disinfection of equipment, glassware, and air by the food and medical industries. The UV portion of the electromagnetic spectrum ranges from approximately 100 to 400 nm, but the optimal germicidal effectiveness of UV-C irradiation is obtained at a wavelength of 254 nm which is the most efficient wavelength for damaging deoxyribonucleic acid (DNA) as noted by Civello et al. (2006). The UV-C light acts as a microbial agent directly due to DNA damage and indirectly due to the induction of resistance mechanisms in different crops against pathogens. It is mainly used as a surface treatment because it penetrates only 5–30 μ of the tissue and UV light used as disinfection method in the food industry, can have a greater germicidal efficacy than that of chlorine, hydrogen peroxide, or ozone. Comparing the limited research on the impact of ozone on quality parameters and mechanisms of action in fruits and vegetables during storage, there is an abundance of research on the effects of UV-C and its beneficial effect on keeping the integrity and freshness of fruits and vegetables as described by several authors (Charles & Arul, 2007; Civello et al., 2006; Ribeiro et al., 2012; Turtoi, 2013; Maharaj, 2015). UV-C irradiation is a postharvest technology which can be adjunct to refrigeration for delaying postharvest ripening, senescence, and decay in different fruit and vegetable species (Maharaj, 2015). Exposure to abiotic UV-C radiation stress is well known to have deleterious effects on plant tissues however, low levels may stimulate beneficial responses of plants, a phenomenon is known as hormesis (Charles & Arul, 2007). The use of UV-C hormesis to improve the postharvest quality of fresh fruits and vegetables has been the subject of numerous research activities during the last two decades, with some early applications in the eighties on the control of postharvest pathogens. The treatments have been reported to be effective in reducing decay caused by most common postharvest pathogens in several products, such as apple, peach, grapefruit, strawberry, pepper, mango, grape, and tomato among others (Civello et al., 2006). UV-C has also been used to modulate ripening and senescence in several UV-C treated produce. The results obtained are quite variable depending on the commodity, stage of maturity, harvest season, and dose applied. Exposure to UV-C also induces bioactive compounds and phytochemicals which can retard ripening and senescence and delay storage life of some horticultural crops (Maharaj, 2015). Thus UV-C effects are associated not only with its germicidal properties but physiological modifications of the crop in response to abiotic stress (Allende et al., 2006; Civello et al., 2006; Maharaj, 2015). There are several advantages in the use of UV-C technology as such as:

- it leaves no residues after treatment, including moisture residues;
- it does not involve complex expensive equipment;
- it requires low maintenance and has low cost;
- it is simpler and more economical to use than ionizing radiation;
- it can be applied as a standalone technology or combined with other postharvest technologies; and
- it generally lacks regulatory restrictions.

UV dose has the preferred units of kJm^{-2} with the beneficial effects from UV-C radiation result from the use of very low UV doses ranging from 0.125 to 9 kJm^{-2} (Vigneault et al., 2012; Charles & Arul, 2007) and the time scale for the induction of such effects is generally measured over hours or even days. Application of doses below 1 kJm^{-2} helped to prevent the development of postharvest diseases and to slow fruit deterioration and ripening. Application of UV-C doses around 5 kJm^{-2} resulted in a reduction of more than 1 log of the microbial population on fresh-cut watermelon, while doses above 10 kJm^{-2} were too high and affected fruit quality manifested as surface browning of the fruit (Vigneault et al., 2012; Maharaj et al., 1999). The weakness of this technique is its low penetrating power and its limited effectiveness on irregular surfaces. Also, some food compounds can absorb the applied wavelengths and thus decrease the treatment effectiveness (Tawema et al., 2016).

With respect to fresh-cut produce similar to whole produce, UV-C is used primarily to destroy pathogens, delay senescence, and increase shelf life of produce. UV-C dose of 10 kJm^{-2} with mode of application in both the inner and outer side of the pericarp of fresh-cut red ripe bell peppers combined with refrigeration, exhibited reduced soft rots, respiration, and softening without impairing other quality attributes, such as sugar, acids, color or antioxidant capacity, and extended the shelf life of such red peppers (Rodoni et al., 2015). Microbial activity was reduced after treatment of zucchini squash slices, processed lettuce, watermelon cubes, and in cut pineapple (Civello et al., 2006). Fresh-cut mango treated with UV-C had increased antioxidant capacity mainly influenced by total phenols and flavonoids which increased with exposure time (González-Aguilar et al., 2007). A limiting factor relates to the dose-response relationship, in that higher UV-C doses induced higher CO_2 production, and resulted in a firmer lettuce, while in slices of zucchini squash, the highest dose used to reduce microbial activity, increased respiration rate, and caused a slight reddish-brown discoloration on the surface of the produce. When combined with other preservation and postharvest techniques, for example, UV-C with high CO_2, modified atmosphere packaging for fresh-cut produce, improved results to manage both postharvest diseases and reduce senescence were noted (Civello et al., 2006). Proposing this technology for commercial applications, requires research in the development of engineering systems, ensuring better control and more uniform application (Vigneault et al., 2012).

11.6   EMERGING TECHNOLOGIES: LIGHT-EMITTING-DIODES (LED), LASER, PULSED LIGHT (PL) AND PULSED ELECTRIC FIELD (PEF), AND HIGH POWER ULTRASOUND

11.6.1   LIGHT EMITTING-DIODES (LED)

Illumination with light-emitting-diodes (LED) has become a more available technology and a more economic and energetically efficient way for light treatments on vegetables. Light from LEDs provides a narrow output spectrum that allows determined wavelengths to match more specifically to plant photoreceptors. For photosynthesis, the absorption quantum yield curve only has two broad maxima, centered at 620 nm (red) and near 450 nm (blue), with a shoulder at 670 nm. The white phosphor types LEDs have peaks of emission from 450 to 500 nm and from 550 to 600 nm. In the case of postharvest treatments, the addition of green or red LED light has been reported to improve the quality of broccoli florets maintaining higher quantity of chlorophylls (Hasperué et al., 2016). In a study reported by Liu et al. (2009), mature-green tomato when treated with short bursts of red light (610–750 nm) and stored for up to 21 days at 12–14°C showed an increased concentration of lycopene in tomato exocarp, after 4 days, where a 9-fold increase between days 4 and 21 was noted compared to 3.5-fold increase for control tomato over the same period.

Blue light has a variety of important roles in plants, such as photomorphogenesis, stomatal control, phototropism, development of photosynthetic apparatus, and chlorophyll biosynthesis among others. Moreover, short-duration blue LED treatments before harvest significantly increased shoot tissue carotenoid and glucosinolates content in sprouting broccoli. Hasperué et al. (2016) reported that the white-blue LED treatment on entire broccoli heads during storage at 5°C or 22°C was effective in delaying senescence and extending the product shelf life. The combination of white-blue LED light sources is a clean and cheap technology to improve the postharvest shelf life of broccoli heads.

11.6.2   LASER

According to the literature, near-infrared (NIR) spectroscopy is a non-destructive method that, despite the high cost and the lack of spatial resolution, has been successfully used to detect numerous attributes in horticultural products, including textural characteristics. Hyperspectral imaging in the visible light spectrum was found to be a promising alternative optical method to quantify local color uniformity of pickling cucumbers due to the integration of spatial information with spectroscopic data. The advance in technology, unfortunately, makes NIR spectroscopy even more complex and costly. The use of light scattering analysis for food and agricultural applications has gained considerable attention in recent years as a cost-effective approach for non-destructive analyses. Particularly, backscattering analysis of laser light in the Vis/NIR spectrum has been shown to be capable of monitoring various physicochemical changes in fruit products. Light scattering analysis has also been tested as a method to obtain information on drying processes in a non-destructive way, in golden-colored products which are highly susceptible to browning, such as banana, apple, and yellow bell pepper as reported by Romano et al. (2016).

11.6.3   PULSED LIGHT (PL)

Pulsed light (PL) is an emerging postharvest process developed initially as a non-thermal sterilization technology for the superficial decontamination of surfaces of fresh fruits and vegetables (Llano et al., 2016; Lopes et al., 2016). This technology, which may be an alternative to continuous UV light treatment, involves a wide broad-spectrum light in the wavelengths from about 100–1100 nm with energy emitted in the UV ranging from 15% to 50% (Lopes et al., 2016). According to the US FDA web site, PL is a method of food preservation that involves the use of intense and short duration pulses of broad-spectrum "white light." The spectrum of light for pulsed light treatment includes wavelengths in the UV to the near-infrared region. The material to be treated is exposed to a least 1 pulse of light having an energy density in the range of about 0.01–50 Jcm^{-2} at the surface. A wavelength distribution such that at least 70% of the electromagnetic energy is within the range from 170 to 2600 nm is used. The material to be sterilized is exposed to at least 1 pulse of light (typically 1–20 flashes per s) with a duration range from 1 µs to 0.1s. For most applications, a few flashes applied in a fraction of a second, provide a high level of microbial inactivation.

During a PL treatment, high-frequency electrical energy accumulated in a high power capacitor is released over an inert gas (e.g., xenon) generating intermittent and intense pulses of light, of short duration lasting for milliseconds (Llano et al., 2016). The intensity of the light pulses as well as their wide range of wavelengths, starts a cascade of photo-thermal and photochemical processes on the surface tissue of the fruit. Previous studies demonstrated the applicability of PL treatments for the decontamination of fresh-cut products, such as watermelon, different apple cultivars, avocado, or mushroom as reported by Llano et al. (2016). Published studies demonstrated that fresh-cut "Kent" mangoes responded to postharvest PL application by maintaining their phytochemical levels; however, there was an increment in browning, associated with polyphenoloxidase activity and darkening of samples. PL also induced darkening of fresh-cut "Golden Delicious" apples, reduced microbial contaminants of "Climberley" tomatoes without compromising its nutritional value; however, it did induce a loss of visual quality. PL treatment influenced fresh-cut "Abrusen" watermelon physiology through an increment in fruit respiratory rate and decreased ethylene production (Lopes et al., 2016).

French scientists have evaluated the impact of pulsed light technology on external appearance (firmness and color) and on nutritional quality (carotenoids, ascorbic acid, phenols, dry matter, phenylalanine ammonia lyase (PAL) activity, PPO activity, and proteins in the fresh-cut mango "Kent" as reported by Charles et al. (2013). For the study, mangoes were peeled, diced, and then the dices were washed in water with sodium hypochlorite, drained, treated by PL, placed in glass jars, and stored at 6°C for 7 days. PL treatments, were carried out using an automatic flashlamp system (Mulieribus, Claranor) composed of eight lamps situated all around the sample with a total fluence of 8 Jcm^{-2}. Cubes of mangoes were treated by four successive pulses, for a total fluence of 8 Jcm^{-2} (one pulse has a total fluence of 2 Jcm^{-2}, light wavelength ranged from UV-C to IR). The scientists found that the PL treatment, preserved the tissue firmness, color, and carotenoid content, but increased the PPO activity after 3 days of storage and maintained the PAL activity. As regards the nutritional quality, the PL treatment, preserved phenols and total ascorbic acid contents like the control treatment. The scientists concluded that PL technology could be used to preserve fresh-cut mango quality, however, they noted that: (1) the necessity of further studies on the effect of pulsed light on specific processes occurring in the fruit and (2) the legislation on pulsed light technology should be improved. They also indicated that the European Union has not approved the PL technology as yet, but approves specific food and food ingredients treated with pulsed light (Charles et al., 2013). However, it is well known that light can have a negative effect on the quality of the fresh-cut products, leading to the degradation of various compounds, such as those significantly contributing to the antioxidant properties of fruit. The potential for using PL as a commercial postharvest technology depends on determining the best conditions, such as fluence and penetrability of light, which may impart desirable effects on fruit quality without impairing the fresh-like attributes of treated produce.

11.6.4   PULSED ELECTRIC FIELD (PEF)

Pulsed electric field (PEF) can be used in many food-related applications, such as preservation and enhancing the heat and/or mass transfer based processes (Wiktor et al., 2015). PEF treatment is the application of very short, high voltage pulses to a food material placed between two or more electrodes. PEF treatment using higher intensity fields (15–40 kVcm^{-1}, 5–100 pulses, 40–700 µs, and 1.1–100 Hz) are more effective toward microbial inactivation while low and medium intensity fields (0.6–2.6 Vcm^{-1}, 5–100 pulses, 10^{-4}–10^{-2} s, 1 Hz) have been successfully used for enhancing mass transfer in solid foods (Rawson et al., 2011). The treatment lasts milli or microseconds and does not cause high-temperature elevation if the dose of energy is not very high. Depending on the electric field parameters, the shape, and material properties, the phenomenon of electroporation manifested as local structural changes and cell membrane breakdowns can be reversible or irreversible. According to Evrendilek (2016), processing of fruit juices with the application of PEFs is one of the non-thermal alternative technologies with the great potential for microbial and enzyme inactivation, shelf-life extension, and preservation of nutritional and sensory properties, and aroma compounds. In apple extract treated with PEF, antioxidant properties were higher by 33% in comparison to the untreated extract. Electrical field strengths aimed at increasing extraction yields from metabolites from plant foods are generally in the range of 1–10 kVcm^{-1}). At these field strengths, microbial inactivation is negligible for most species and the extent of cell poration tends to be reversible (Rawson et al., 2011). Bioactive compounds, such as total phenolic content in apple tissue and total carotenoid content in carrot tissue increased up to 85.8% and 11.3%, respectively, with PEF however, intensities greater than 1.85 kVcm^{-1} was found to reduce extractability of both carotenoids and polyphenolic compounds (Wiktor et al., 2015).

11.6.5   HIGH POWER ULTRASOUND

Ultrasound is defined as sound waves having frequency that exceeds the hearing limit of the human ear (approximately 20 kHz). The propagation of ultrasound through a material induces compressions and decompressions of the medium particles, which imparts a high amount of energy. Ultrasound is one of the emerging technologies developed to minimize processing cost and time as well as maximize quality and food safety. High power ultrasound with frequency higher than 20 kHz has mechanical, chemical, and/or biochemical effects, which are used to modify the physicochemical properties and enhance the quality of various food systems during processing. It can be applied using sonication baths or ultrasonic immersion probes with different lengths, diameters, and tip geometries depending on applications. These effects have positive effects in food processing, such as improvement in mass transfer, food preservation, etc. High power (high energy and high intensity) ultrasound operates at frequencies between 20 and 500 kHz and intensities higher than 1 Wcm^{-2} which are disruptive and induce effects on the physical, mechanical, or chemical (biochemical) properties of food (Jambrak, 2013).

11.7 CONCLUSION

Consumer demand for high quality, fresh, nutritious, safe, and conveniently prepared food items has increased dramatically in recent years and has led to the need for adequate technologies to preserve these food products and reduce postharvest losses. Minimally processed foods have a limited shelf life due to their high perishability as a result of mechanical, physiological, or pathological factors. The improvement of the safety and quality of minimally processed foods is of major interest to both the consumer and the food industry. There is a need to provide novel technologies that can ensure the delivery of high-quality products with high levels of the desired phytonutrients and are primarily based on physical methods that reduce the growth of microorganisms, provide milder conditions and less intense heating while at the same time inducing chemical modifications in cell. In the past decades, research has been done to develop alternatives to traditional chemical methods to control postharvest diseases and extend shelf life. The development of physical non-thermal treatments for the preservation of crops has been described in this chapter, however, most of the alternative techniques and technologies have not gained widespread acceptance by the food and fresh-cut industry. Clearly, there is need for more research effort to better understand underlying mechanisms of action with different types of produce, as well as the many engineering considerations, such as uniformity of dose, dose-response relationship, and ease of control in the selection of an appropriate treatment. The use of a combination of such technologies is also warranted. One of the challenges with using such postharvest technologies either single or in combination will ultimately depend on the transfer of laboratory research using relatively small-scale devices and technology transfer into commercial opportunities. Apart from determining which one or more technologies are the most appropriate, economic considerations for potential commercial-scale applications must be considered.

KEYWORDS

- **ready-to-eat**
- **minimally processed crops**
- **non-thermal postharvest technologies**

REFERENCES

Allende, A.; Tomás-Barberán, F. A.; Gil, M. I. Minimal Processing for Healthy Traditional Foods. *Trends Food Sci. Technol.* **2006,** *17,* 513–519.

Bilek, S. E.; Turantaş, F. Decontamination Efficiency of High Power Ultrasound in the Fruit and Vegetable Industry, A Review. *Int. J. Food Microbiol.* **2013,** *166,* 155–162.

Charles, M. T.; Arul, J. UV Treatment of Fresh Fruits and Vegetables for Improved Quality: A Status Report. *Stewart Postharvest Rev.* **2007,** *3* (6), 1–8.

Charles, F.; Vidal, V.; Olive, F.; Filgueiras, H.; Sallanon, H. Pulsed Light Treatment as New Method to Maintain Physical and Nutritional Quality of Fresh-cut Mangoes. *Innov. Food Sci. Emerg. Technol.* **2013,** *18,* 190–195.

Civello, P. M.; Vicente, A. R.; Martínez, G. A. UV-C Technology to Control Postharvest Diseases of Fruits and Vegetables. In *Recent Advances in Alternative Postharvest Technologies to Control Fungal Diseases in Fruits and Vegetables*; Troncoso-Rojas, R., Tiznado-Hernández, M. E., Gonzalez-Leon, A., Eds.; Transworld Research Network: Trivandrum, India, 2006; Vol. 37, p 2.

Corbo, M. R.; Speranza, B.; Campaniello, D.; D'Amato, D.; Sinigaglia, M. Fresh-cut Fruits Preservation: Current Status and Emerging Technologies. In *Current Research, Technology and Education Topics in Applied Microbiology and Microbial Biotechnology;* Méndez-Vilas, A., Ed.; Formatex Research Center: Badajoz, 2010; pp 1143–1154.

Evrendilek, G. A. Change Regime of Aroma Active Compounds in Response to Pulsed Electric Field Treatment Time, Sour Cherry Juice Apricot and Peach Nectars, and Physical and Sensory Properties. *Innov. Food Sci. Emerg. Technol.* **2016,** *33,* 195–205.

Garcia, E.; Barrett, D. M. Preservative Treatments for Fresh-cut Fruits and Vegetables. In *Fresh-cut Fruits and Vegetables: Science, Technology and Market;* Lamikanra, O., Ed.; CRC Press: Boca Raton, FL, 2002; pp 267–303.

Genevois, C. E.; de Escalada Pla, M. F.; Flores, S. K. Application of Edible Coatings to Improve Global Quality of Fortified Pumpkin. *Innov. Food Sci. Emerg. Technol.* **2016,** *33,* 506–514.

Gómez, P. L.; Alzamora, S. M.; Castro, M. A.; Salvatori, D. M. Effect of Ultraviolet-C Light Dose on Quality of Cut-apple: Microorganism, Color and Compression Behavior. *J. Food Eng.* **2010,** *98,* 60–70.

Gontard, N.; Guillaume, C. Packaging and the Shelf Life of Fruits and Vegetables. In *Food Packaging and Shelf Life a Practical Guide*; Robertson, G. L., Ed.; CRC Press: Boca Raton, FL, 2010; pp 297–315.

González–Aguilar, G. A.; Villega-Ochroa, M. A.; Martinez-Telez, M. A.; Gardea A. A.; Ayala–Zavala, J. F. Improving Antioxidant Capacity of Fresh Cut Mangoes Treated with UV-C. *J. Food Sci.* **2007,** *36,* S197–S202.

Goyette, B.; Vigneault, C.; Charles, M. T.; Raghavan, V. G. S. Effect of Hyperbaric Treatments on the Quality Attributes of Tomato. *Can. J. Plant Sci.* **2012,** *92,* 541–551.

Hasperué, J. H.; Luciano Guardianelli, L.; Rodoni, L. M.; Alicia, R.; Chaves, A. R.; Martínez, G. A. Continuous White-blue LED Light Exposition Delays Postharvest Senescence of Broccoli. *Lebensm. Wiss. Technol. Food Sci. Technol.* **2016,** *65,* 495–502.

Huyskens-Keill, S.; Hassenberg, K.; Herppich, W. B. Impact of Postharvest UV-C and Ozone Treatment on Textural Properties of White Asparagus (*Asparagus officinalis* L.). *J. Appl. Bot. Food Qual.* **2011,** *84,* 229–234.

IFST. *The Use of Irradiation for Food Quality and Safety*; Institute of Food Science and Technology (IFST) Information Statement: London, 2006; pp 1–19.

Ippolito, A.; Nigro, F. Natural Antimicrobials in Postharvest Storage of Fresh Fruits and Vegetables. In *Improving the Safety of Fresh Fruit and Vegetables;* Wim Jongen, A., Ed.; CRC Press, Woodhead Publishing Ltd.: Cambridge, England, 2005; pp 513–541.

Jambrak, A. Application of High Power Ultrasound and Microwave in Food Processing Extraction. *J. Food Process. Technol.* **2013,** *4* (1), e113.

Lacroix, M.; Vigneault, C. Irradiation Treatment for Improving Fruit and Vegetable Quality. *Stewart Postharvest Rev.* **2007,** *3* (7), 1–8.

Liu, L. H.; Zabaras, D.; Bennett, L. E.; Aguas, P.; Woonton, B. W. Effects of UV-C, Red Light and Sun Light on the Carotenoid Content and Physical Qualities of Tomatoes During Postharvest Storage. *Food Chem.* **2009,** *115,* 495–500.

Llano, K. R. A.; Marsellés-Fontanet, A. R.; Martín-Belloso, O.; Soliva-Fortuny, R. Impact of Pulsed Light Treatments on Antioxidant Characteristics and Quality Attributes of Fresh-cut Apples. *Innov. Food Sci. Emerg. Technol.* **2016,** *33,* 206–215.

Lopes, M. M. A.; Silva, E. O.; Canuto, K. M.; Lorena, M. A.; Silva, L. M. A.; Gallão, M. I.; Urban, L.; Fernando Ayala-Zavala, J. F.; Miranda, M. R. A. Low Fluence Pulsed Light Enhanced Phytochemical Content and Antioxidant Potential of 'Tommy Atkins' Mango Peel and Pulp. *Innov. Food Sci. Emerg. Technol.* **2016,** *33,* 216–224.

Maharaj, R. Effects of Abiotic Stress (UV-C) Induced Activation of Phytochemicals on the Postharvest Quality of Horticultural Crops, In *Phytochemicals-Isolation, Characterization and Role in Human Health;* Rao, V., Rao, L. G., Eds.; InTech: Croatia, 2015; pp 221–244.

Maharaj, R.; Arul, J.; Nadeau, P. Effect of Photochemical Treatment in the Preservation of Fresh Tomato (*Lycopersicon esculentum cv. Capello*) by Delaying Senescence. *Postharvest Biol. Technol.* **1999,** *15* (1), 13–23.

Mohammadi, A.; Hashemi, M.; Hosseini, S. M. Postharvest Treatment of Nanochitosan-based Coating Loaded with *Zataria multiflora* Essential Oil Improves Antioxidant Activity and Extends Shelf-life of Cucumber. *Innov. Food Sci. Emerg. Technol.* **2016,** *33,* 580–588.

Ozturk, I.; Sagdic, O.; Yalcin, H.; Capar, T. D.; Asyali, M. H. The Effects of Packaging Type on the Quality Characteristics of Fresh Raw Pistachios (*Pistacia vera* L.) during the Storage. *LWT Food Sci. Technol.* **2016,** *65,* 457–463.

Pascall, M. A.; Lin, S. The Application of Edible Polymeric Films and Coatings in the Food Industry. *J. Food Process. Technol.* **2013,** *4* (2), 1–2.

Rawson, A.; Patras, A.; Tiwari, B. K.; Noci, F.; Koutchma, T.; Brunton, N. Effect of Thermal and Non Thermal Processing Technologies on the Bioactive Content of Exotic Fruits and Their Products: Review of Recent Advances. *Food Res. Int.* **2011,** *44,* 1875–1887.

Ribeiro, C.; Canada, J.; Alvarenga, B. Prospects of UV Radiation for Application in Postharvest Technology. *Emir. J. Food Agric.* **2012,** *24* (6), 586–597.

Rico, D.; Martín-Diana, A. B.; Barat, J. M.; Barry-Ryan, C. Extending and Measuring the Quality of Fresh-cut Fruit and Vegetables: A Review. *Trends Food Sci. Technol.* **2007,** *18* (7), 373–386.

Rodoni, L. M.; Zaro, M. J.; Hasperué, J. H.; Concellón, A.; Vicente, A. R. UV-C Treatments Extend the Shelf Life of Fresh-cut Peppers by Delaying Pectin Solubilization and Inducing Local Accumulation of Phenolics. *LWT Food Sci. Technol.* **2015,** *63,* 408–414.

Romano, G.; Nagle, M.; Müller, J. Two-parameter Lorentzian Distribution for Monitoring Physical Parameters of Golden Colored Fruits during Drying by Application of Laser Light in the Vis/NIR Spectrum. *Innov. Food Sci. Emerg. Technol.* **2016,** *33,* 498–505.

Rupasinghe, H. P. V.; Boulter-Bitzer, J.; Ahn, T.; Odumeru, J. A. Vanillin Inhibits Pathogenic and Spoilage Microorganisms in Vitro and Aerobic Microbial Growth in Fresh-cut Apples. *Food Res. Int.* **2006,** *39,* 575–580.

Santacatalina, J. V.; Guerrero, M. E.; Garcia-Perez, J. V.; Mulet, A.; Carceln, J. A. Ultrasonically Assisted Low-temperature Drying of Desalted Codfish. *LWT Food Sci. Technol.* **2016,** *65,* 444–450.

Tawema, P.; Han, J.; Dang Vu, K.; Salmieri, S.; Lacroix, M. Antimicrobial Effects of Combined UV-C or Gamma Radiation with Natural Antimicrobial Formulations Against *Listeria monocytogenes, Escherichia coli* O157: H7, and Total Yeasts/Molds in Fresh Cut Cauliflower. *LWT Food Sci. Technol.* **2016,** *65,* 451–456.

Turtoi, M. Ultraviolet Light Treatment of Fresh Fruits and Vegetables Surface: A Review. *J. Agroalimentary Process. Technol.* **2013,** *19* (3), 325–337.

Vigneault, C.; Leblanc, D. L.; Goyette, B.; Jenni, S. Invited Review: Engineering Aspects of Physical Treatments to Increase Fruit and Vegetable Phytochemical Content. *Can. J. Plant Sci.* **2012,** *92,* 373–397.

Wiktor, A.; Sledz, M.; Nowacka, M.; Rybak, K.; Chudoba, T.; Lojkowski, W.; Witrowa-Rajchert, D. The Impact of Pulsed Electric Field Treatment on Selective Bioactive Compound Content and Color of Plant Tissue. *Innov. Food Sci. Emerg. Technol.* **2015,** *30,* 69–78.

Yahia, E. M. *Irradiation of Produce: What for, How and is it Safe?* World Food Logistics Organization (WFLO) Scientific Advisory Council Report: Alexandria, VA, 2016.

CHAPTER 12

RADIATION PROCESSING: AN EMERGING POST HARVEST PRESERVATION METHOD FOR IMPROVING FOOD SAFETY AND QUALITY

SUMIT GUPTA and PRASAD S. VARIYAR*

Food Technology Division, Bhabha Atomic Research Centre, Mumbai 400085, India

Corresponding author. E-mail: prasadpsv@rediffmail.com

CONTENTS

Abstract 318

12.1 Introduction 318

12.2 History of Food Irradiation 321

12.3 The Process 323

12.4 Applications of Food Irradiation 329

12.5 Wholesomeness of Irradiated Foods 335

12.6 Detection of Irradiated Food 337

12.7 International Status 341

12.8 National Status 342

12.9 Conclusion and Future Goals 344

Keywords 346

References 346

ABSTRACT

Among the currently practiced methods for preservation of food, treatment by ionizing radiation is a highly promising non-thermal technique that can aid in extending shelf-life and maintain food quality without affecting sensory properties. The process involves exposing food to a carefully controlled amount of ionizing radiation that includes gamma rays emitted from radionuclide (cobalt 60 and cesium 137), X-rays ($\leq$ 7.5 MeV), and the high energy ($\leq$ 10 MeV) electrons generated by machine sources. The technology can be used for disinfestation of food grains and pulses, inhibition of sprouting in bulbs and tubers, extending shelf-life, and ensuring microbiological safety. It can also aid in overcoming quarantine barriers to international trade as it is an effective alternative to fumigation that is being phased out due to adverse effects on environment and human health. The process can also be used for hygienization and sterilization of non-food items including cut-flowers, pet food, cattle feed, aqua feed, Ayurvedic herbs, and medicines and packaging materials. This chapter discusses the effects of radiation treatment on the microbial status and chemical and nutritional quality as well as the methods currently available for identification of such treated foods. The legislation enacted at both the national and international level that regulate trade in irradiated foods, the current status of trade in such treated foods and novel applications of this technology in the area of food preservation are detailed. The future goals toward wider commercialization of the technology are also presented.

12.1   INTRODUCTION

Consumption of contaminated food has been implicated in widespread health problems and consequent reduced economic productivity (WHO, 1984). Although reliable statistics on foodborne diseases are not available from most of the countries, a recent study by WHO (WHO, 2015) has reported that 550 million people worldwide fall ill yearly due to foodborne illness and 230,000 deaths occur annually. Situation of foodborne illness is more worrying in children under the age of five years. Two hundred and twenty million children under the age of five years suffer from foodborne illness every year which results in over 96,000 deaths. Although the United States food supply chain is considered to be safest in world, yet according to United States Food and Drug Administration (USFDA) there are over 48 million cases of foodborne illnesses in the US resulting in an estimated 128,000 hospitalizations and over 3000 deaths (USFDA, 2016).

In the developed countries like the United States, diseases, such as Salmonellosis and Campylobacteriosis are the major causes of foodborne illnesses (Buzby & Roberts, 1997). In contrast, much wider range of food-borne diseases including parasitical diseases is prevalent in developing nations (Todd, 1996). According to WHO (2015), major causes of diarrheal diseases globally are Norovirus, *Campylobacter*, Nontyphoidal *Salmonella*, and *Escherichia coli*. Apart from causing foodborne illnesses, contamination of food products with pathogenic microorganisms has severe economic implications. Microbial contaminations can lead to food shipment rejections or even total ban on food imports from specific countries by importing nations (Molins et al., 2001). In addition, pathogenic contaminations also result in food recalls leading to massive losses to food processing industry in form of product losses and decreased consumer confidence.

Apart from pathogenic contaminations, food losses and wastage are another major concern for global food supply chain. Food losses refer to the decrease in edible food mass specifically meant for human consumption throughout the supply chain. Food losses take place at production, post-harvest, and processing stages in the food supply chain. Food losses occurring at the end of the food chain (retail and final consumption) are rather called "food waste," which relates to behavior of retailers and consumers (Parfitt et al., 2010). According to Food and Agriculture Organization (Gustavsson et al.,, 2011), one-third of food produced for human consumption is lost or wasted annually, amounting to about 1.3 billion tons. Food losses occur throughout the supply chain that is, from initial agricultural production to final consumer. In developed countries significant amount of food is wasted at consumption stage, meaning it is discarded even when it is suitable for human consumption while in developing nations most of the food is lost during early and middle stages of food supply chain (Gustavsson et al.,, 2011). The same report estimates that in case of cereals, pulses, and oilseeds up to 20–30% of produce are wasted while in case of fruits, vegetables, roots, and tubers higher amount of wastage up to 50% was noted.

The global population is presently over 7.2 billion and according to UN estimates (UN, 2015) is expected to rise beyond 9 billion in 2030. Besides, rapid urbanization and increasing incomes also lead to diverse food habits and demands. Food security is now prime concern around the world because, without adequate food availability, natural disasters, and emergencies can lead to local or regional food riots resulting in economic and political instability (Pillai & Shyanfar, 2015). In view of above, there is significant need for adoption of food processing technologies which can prevent pathogenic microbial contamination and simultaneously extend shelf-life of food products thus reducing wastage.

Presently, thermal processing, canning, freezing, refrigeration, and use of chemical preservatives are widely practiced for preventing microbial spoilage and extension of shelf-life (Mostafavi et al., 2010). Thermal processing is well suited and widely applied for processing of liquid and semi-solid foods but it cannot be applied efficiently to solid foods and dry ingredients (Farkas, 1998). In addition, it is not possible to process fresh fruits and vegetables and nuts using heat treatment. Chemical fumigation technology (using sulfur dioxide, potassium nitrate, or methyl bromide) is widely used to preserve crops after harvest (Pillai & Shayanfar, 2015). However, this technology has been banned by health authorities in many countries due to concerns regarding their carcinogenic and ozone-depleting properties. Therefore, there is a need for food preservation technology which can ensure microbial safety and extend shelf-life, while maintaining high nutritional and sensory quality without or with minimal chemical preservatives. Food irradiation is one such technology which can fulfill all these requirements.

Food irradiation is a non-thermal technology which involves exposing food to carefully controlled amounts of ionizing radiations generated from either natural radionuclides (cobalt 60 and cesium 137) or from machine (X-rays ($\leq$7.5 MeV), high energy ($\leq$10 MeV) electrons) sources (Farkas et al., 2014). The amount of radiation dose given to a food material is denoted by Gray (Gy) which is defined as the energy in Joules absorbed per unit mass (kg) of the irradiated product.

$$1 \text{ Gy} = 1 \text{ J/kg}, 1000 \text{ Gy} = 1 \text{ kGy} = 238.9 \text{ cal kg}^{-1} = 6.2418 \times 10^{18} \text{ eV kg}^{-1}$$

Dose rate of a source is defined as the absorbed dose per unit time (Gy s^{-1}, kGy min^{-1}, or kGy h^{-1}).

The technology can be used for disinfestation of food grains and pulses, inhibition of sprouting in bulbs and tubers, extending shelf-life, and ensuring microbiological safety. It can also aid in overcoming quarantine barriers to international trade as it is an effective alternative to fumigation that is being phased out due to adverse effects on environment and human health. The process can also be used for hygienization and sterilization of non-food items including cut flowers, pet food, cattle feed, aqua feed, ayurvedic herbs and medicines, and packaging materials.

This chapter details history, applications, effects of radiation processing, detection methods of radiation foods, national, and international status and some recent applications.

12.2 HISTORY OF FOOD IRRADIATION

Concept of irradiating foodstuffs to bring about beneficial changes is almost a century old. First patent was issued to J. Appleby and A. J. Banks in 1905 in the United Kingdom for using ionizing radiation "to bring about an improvement in the conditions of foodstuffs" and in "their general keeping quality" (Diehl, 1995). These inventors proposed using alpha, beta, and gamma radiations from radioactive radium or other sources to enhance keeping quality of cereal products. They also highlighted the advantage that radiation processing could eliminate use of chemical compounds for food preservation. However, the idea could not take off because radiation sources were not available in sufficient quantities at that time to scale up the process commercially. Subsequently, two patents were granted in the United States between 1915 and 1925 for development of X-ray equipment which could be used for treatment of food products, both for vegetable and animal origin, furs, woolens, books, feathers, and for inactivation of trichinae in pork (Gillett, 1918; Schwartz, 1917). However, first commercial application of food irradiation was to kill eggs, larvae, and adults of tobacco beetle in cigars using water-cooled X-ray machines. Although, the process was successful it was soon replaced by chemical fumigations due to unreliability of X-ray equipment manufactured at that time (Diehl, 1995). Publication by Brasch and Huber (1947) regarding sterilization of meat and some other products by using pulsed electron beam renewed interest in food irradiation. At the same time studies at MIT concluded that neutron radiation is unsuitable for food irradiation applications because it can induce radioactivity in food while alpha and ultraviolet light are inappropriate because of their low penetration. X-ray could not be used because of low power X-ray machines available at that time. Therefore, it was suggested that accelerated electrons are most appropriate for food irradiation applications. Studies using gamma rays were not carried out due to the fact that suitable radioisotopes were not available in sufficiently large quantities. United States Atomic Energy Commission (USAEC) started a coordinated research program in 1950 on use of ionizing radiation for food preservation using spent fuel from irradiators as gamma source. But this was soon abandoned due to limitations of dosimetry protocols for spent fuel source (Diehl, 1995).

Further research in food irradiation was pursued by US army from 1953 to 1960 on both high and low dose irradiation. After 1960, the US army concentrated mainly on developing radiation sterilized meat products to substitute for canned military rations. Positive results from these researchers encouraged similar efforts in other countries like United

Kingdom, France, Germany, Netherlands, and several other European countries. First international symposium on food irradiation organized by IAEA in 1966 in Karlsruhe, Germany for reviewing progress made in research laboratories that saw participation from 28 countries (Farkas & Mohacsi-Farkas, 2011).

Despite the promising results obtained from around the world regarding usefulness of food irradiation for microbial control and shelf-life extension, health authorities around the world were reluctant to give clearances for irradiated foods. Due to its uniqueness, association with radiation, lack of application history, scant data on wholesomeness (toxicological, microbial safety, and nutritional adequacy), and safety for human consumption the technology was widely questioned. These were recognized as major obstacles for commercialization of food irradiation. To address these concerns International Project in Field of Food Irradiation (IFIP) was formed in 1970 by partnership of 24 countries with specific aim of sponsoring a worldwide research program on wholesomeness of irradiated foods (Diehl, 1995).

The joint FAO/IAEA/WHO expert committee on the Wholesome of Irradiated Food (JECFI) after carefully examining wholesomeness data obtained through IFIP framework and research carried out in various laboratories concluded that "the irradiation of any food commodity to an overall average dose of 10 kGy presents no toxicological hazard, hence, toxicological testing of food so treated is no longer required." The committee also proposed that irradiation of food up to an overall average dose of 10 kGy introduces no specific nutritional or microbiological problems (WHO, 1981). In 1980, JECFI also considered need of higher doses (>10 kGy) for certain applications but due to unavailability of sufficient data at that time could not provide any recommendations on toxicology and wholesomeness of foods irradiated beyond 10 kGy. Nevertheless, in 1997 joint FAO/IAEA/ WHO study group after carefully examining data regarding safety of food irradiated in dose range of 25–60 kGy concluded that such foods are safe as well as nutritionally adequate (WHO, 1999).

Based on the JECFI recommendations the FAO/WHO Codex Alimentarius programme developed a "Codex General Standard for Irradiated Foods" and a "Recommended International Code of Practice for the Operation of Radiation Facilities Used for the Treatment of Foods." These documents have been revised later (Codex, 2003a, 2003b) and became widely adopted by the UN member states. Due to these efforts, food irradiation facilities were granted clearances in several parts of the world. IAEA maintains a complete database of all the food irradiation facilities installed worldwide on its nucleus database (https://nucleus.iaea.org/fitf/Default.aspx). It can be noted from these databases that presently radiation processing is practiced in 60 countries for over 100 food items.

In spite of these encouraging data, it is true that the implementation of the food irradiation process is seriously hampered, particularly in the European region due to opposing attitudes of certain activist groups forming "public opinion," inducing unfunded fears by misinformation, and unwillingness of legislators and industrial stakeholders to act (Farkas & Mohacsi-Farkas, 2011).

12.3 THE PROCESS

Food irradiators are facilities where radiation processing is carried out. The irradiator is composed of a bunker, in which products are exposed to a source of ionizing radiation to desired doses. Radiation source can either be gamma-ray sources, such as cobalt-60 or machine sources, such as electron beam or X-rays.

12.3.1 RADIATION SOURCES

The Codex Alimentarius General Standard for Irradiated Foods (Codex, 2003a) describes the following sources of ionizing radiations for the radiation processing of food products:

(a) Gamma radiation from ^{137}Cs or ^{60}Co;
(b) Accelerated electrons (forming electron beams) with a maximum energy of 10 MeV;
(c) X-rays with a maximum energy of 5 MeV.

In USA radiation processing with X-rays of up to 7.5 MeV is permitted (IAEA, 2015).

Characteristics of different types of radiations are summarized in Table 12.1.

Gamma rays and X-rays have higher penetration power of up to 40–50 cm in product depending upon density whereas electrons can penetrate only up to depth of 5 cm. Therefore, gamma and X-rays result in more uniform dose distribution in product as compared to electron beam. But electron beam can provide very high dose rate in the order of kGy/s and, therefore, truckload of product can be processed in few hours using e-beam while it can take few days in gamma irradiators.

TABLE 12.1   Different Types of Ionizing Radiation, Their Penetration, and Characterization.

	Isotope sources ^{137}Cs or ^{60}Co	Machine sources	
Type of radiation	Gamma rays	X-rays	Electron beam
Composed of	Photons	Photons	Electrons
Electric charge	None	None	Yes (−1)
Penetration	Very good	Very good	Limited
Emission of radiation	Cannot be switched off	Can be switched off	Can be switched off
	Isotropic and direction cannot be controlled	Unidirectional and direction can be controlled	Unidirectional and direction can be controlled
Dose rate (orders of magnitude)	kGy/h	kGy/min	kGy/s
Type of products that can be treated	Products of low and medium density can be treated in cartons, drums or pallets	Products of low and medium density can be treated in cartons, drums or pallets	Suitable for products of low density only
Overall process	Since emission cannot be stopped so non-stop (24/7) operation to optimize source usage	Emission can be stopped. Large quantities of products can be treated in short time	Emission can be stopped. Large quantities of products can be treated in short time

12.3.2   GAMMA IRRADIATOR

Radionuclide used in gamma irradiators is either caesium-137 (^{137}Cs) or cobalt-60 (^{60}Co). Cobalt-60 is by far the most common source of gamma irradiation and ^{137}Cs is now only found in laboratory facilities. The advantage of ^{60}Co is that it is present in a water-insoluble form and thus presents very little risk of environmental contamination. The half-life of ^{60}Co is 1925.2 days, which means that at the end of this period the activity is half the initial activity. ^{60}Co disintegrates with emission of gamma radiation (1.17 and 1.33 MeV) and beta radiation (0.31 MeV) to stable nickel.

Strength of the source used in gamma irradiator is measured by activity. Generally, activity of source is measured using SI unit the Becquerel (Bq), which is the number of radioactive decays per second or in Curies (Ci), which is the activity of one gram of ^{226}Ra. The use of curies, which is not an SI unit, still prevails in the radiation processing industry. The relationship between the two units is:

$$1 \ Ci = 3.7 \times 10^{10} \ Bq$$

Due to the fact that radionuclides decay continuously, their activity decreases in time-dependent manner. Hence, the activity value needs to be associated with a date. The initial value of the activity given in the source certificates established by the source supplier is to be used as a reference. Following equation describes relation between initial and final activity:

$$A = A_0 \exp(-\ln 2 \times d/t_{1/2}),$$

where A = final activity

A_0 = initial activity

d = number of days

$t_{1/2}$ = half-life in days.

Schematic diagram of a typical food irradiation plant or facility is shown in Figure 12.1. Radionuclides (^{60}Co source) are contained within sealed metallic tubes called "pencils" that are placed on a metallic rack referred to as the source (Fig. 12.2). The activity contained in each pencil, the number and dimensions of the pencils and the shape and size of the rack have an effect on the dose distribution of the products. These elements make the source geometry, which is characterized and documented. Each pencil has a unique identification number, and its activity and exact position on the source rack are also recorded. Depending on the respective heights of the source and of the products, gamma irradiators fall into two basic categories: overlapping source or overlapping product (Fig. 12.2). Generally, gamma irradiators having overlapping product design have absorption of higher proportion of emitted radiation.

As shown in Figure 12.1, commercial food irradiation plant consists of a concrete shielded radiation chamber. The thickness of the concrete shield is usually about 1.518 m to prevent gamma radiation from coming out of radiation chamber. Radiation source is stored underwater pool having a depth of about 6 m (Diehl, 1995). For carrying out radiation treatment food materials are loaded into metallic carriers and moved into radiation chamber using a conveyor system. Source rack is lifted from water storage tank to start radiation processing. The carriers move around the ^{60}Co rack and turn around their own axis, which ensures that contents are irradiated equally from both the sides. The radiation dose absorbed by the material contained in the carrier is determined by the strength of the cobalt source, that is, by the number of Curies or Becquerel, and by the dwell time of the carrier in the irradiation position (Fig. 12.1). In most of the commercial irradiators source strength is fixed and cannot be changed easily, therefore, the length of time each carrier spends in irradiation chamber will decide the amount of dose received by the product. When products travel in a single file in a symmetric pattern on each side of the source the irradiator is known as single pass irradiator. Whereas in multi-pass irradiators products travel in multilayers on each side of source (Fig. 12.2). In single pass irradiators, the radiation having crossed the products is wasted in the concrete walls of the irradiation chamber. However, there is little attenuation effect between the products being irradiated. In multiple pass irradiators, more of the energy emitted by the source is absorbed by the products. However, the dose received at a given position by a given product depends on the quantity and density of other products between this product and the source.

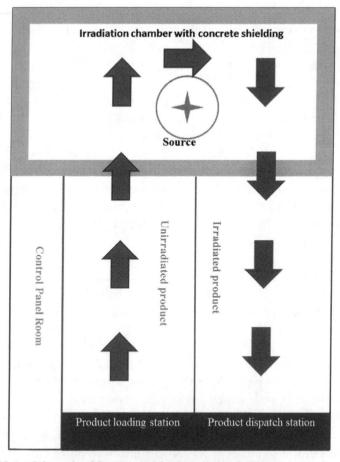

FIGURE 12.1   Schematic of Co-60-based food irradiation facility.

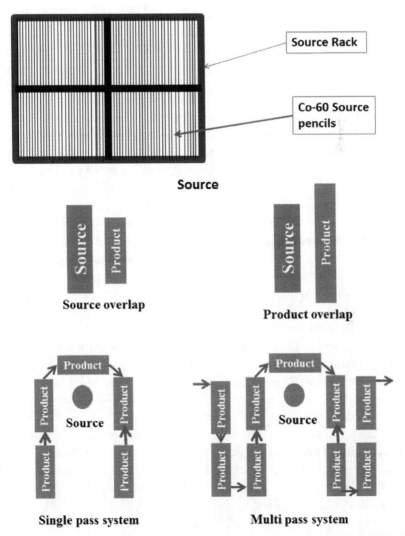

FIGURE 12.2   Diagram of Co-60 source rack and various configurations of food irradiation facility.

Gamma irradiators work on the fail-safe principle. Any abnormal occurrence (e.g., power outage, jammed conveyor, fire alarm, or timer failure) causes the source to return to the safe storage position and the irradiation stops. The source rack goes back to the bottom of the storage pool by gravity.

12.3.3   ELECTRON BEAM AND X-RAY IRRADIATORS

Although earlier experiments in food irradiation were carried out using X-rays, X-ray machines have never been used for commercial food irradiation apart from laboratory scale experiments (Diehl, 1995). Nevertheless, the current state of the science of food irradiation is the electron beam (eBeam) technology. This technology involves the use of compact equipment, termed linear accelerators, to generate high energy electrons which are ionizing and are used to achieve the desired end result (Pillai & Shyanfar, 2015). The effective penetration range of an electron beam depends on its energy level. Generally, the depth of penetration of an electron beam in most foodstuffs is 5 mm/MeV. Low-energy (up to 300 keV) and medium-energy (300 keV to 1 MeV) electron beam can penetrate up to a thickness of few millimeters only; therefore, they are mainly employed for the sterilization of films or sheets of packaging material. High-energy electron beam accelerators produce electrons with energies above 1 MeV. For purposes of food irradiation, 10 MeV is the upper limit. A 10 MeV electron beam can thus be used for irradiation of thicknesses up to 5 cm if irradiated from one side, or 10 cm if irradiated from two sides (Diehl, 1995).

Two different kinds of accelerator designs are used for generating high energy electron beams for food irradiation applications. (1) Direct current accelerators and (2) Radiofrequency linear accelerator (LINAC). Electrons are accelerated to speeds up to 99.9% of the speed of light in evacuated tube. In DC accelerators, electrons are accelerated by application of high voltage across the terminals in the tube. Electrons emitted from an electron source, a heated wire or an "electron gun," are pushed away from the negative end of the tube and attracted by the positive end. The higher the potential difference, the higher the speed attained by the electrons. But in DC accelerators it is very difficult to achieve energies greater than 5 MeV, whereas in LINACS higher energy (up to 10 MeV) electron beams can be easily produced. In a LINAC, pulses or bunches of electrons produced at the thermionic cathode are accelerated in an evacuated tube by driving RF electromagnetic fields along the tube. Electron beams produced by LINAC are monoenergetic but pulsed rather than continuous. In LINAC the beams produced are electron pulse of a few microseconds duration, generally followed by a dead time of a few milliseconds (Clemmons et al., 2015).

The high energy electron beam produced by accelerators has a diameter of only a few millimeters or centimeters. Therefore, to uniformly provide radiation dose to food products electron beam after coming out from accelerator tube is rhythmically deflected by a scanning magnet attached to the end of the accelerator tube. The beam is moved to and fro like a pendulum with a frequency of 100–200 Hz. To achieve uniform dose distribution in material electron beam is scanned continuously and product moved in perpendicular direction to the scanning line of the electron beam (Fig. 12.3).

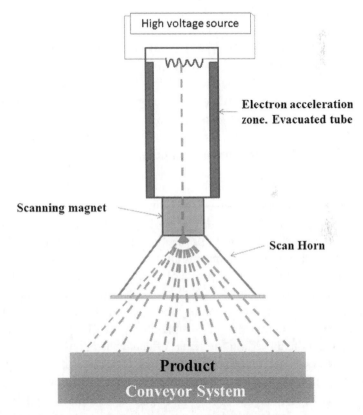

FIGURE 12.3   Schematic of electron beam food irradiation facility.

12.4   APPLICATIONS OF FOOD IRRADIATION

Irradiation in one of the many physical processes used for shelf-life extension of foods but it has certain distinct advantages over other technologies as summarized by Roberts (2014).

    a.   Radiation processing is a versatile technology, that is, it can be used for microbial safety, shelf-life extension (food security), and for quarantine application thus enhancing international trade.

b. Extremely effective as it has a broad spectrum effect against microbes including bacteria and fungi, insects and pests.
c. Irradiation is a cold process and therefore, advantageous to several products.
d. Radiation used for treating foods has high penetration, which helps in treating foods in their final packaging without affecting the process efficiency. Treating pallet loads is also possible.
e. Solid and raw foods which are not possible to process by thermal treatment can also be readily processed using radiation treatment.
f. Radiation processing is a residue-free technology. No harmful chemical residues are left in foods processed with radiation technology.
g. Food irradiation process is simple to control as dose delivered will only be dependent on source activity and conveyor speed.
h. Food treated with radiation can be immediately distributed in food supply chain.

Food and agricultural products may be irradiated for phytosanitary, sanitary, or physiological purposes. Phytosanitary measures aim at improving the health of plants and include preventing the introduction or spread of regulated pests. It includes increasing mortality, prevents successful development, that is, preventing emergence of adults and inducing sterility in the targeted pests. It also includes microbial inactivation. The treatment may be performed in the importing country or prior to export, in which case the regulatory requirements of the importing country apply. Sanitary measures include applications based on the lethal effects of irradiation on (1) microorganisms which cause foodborne diseases, reduce shelf-life, or cause food spoilage thus rendering food unsuitable for human consumption; (2) parasites, such as helminths that infect meat or protozoa that infect fresh cut vegetables; and (3) insect disinfestation to prevent post-harvest losses. Apart from phytosanitary and sanitary applications radiation processing can be carried out for physiological effects on plant products. Applications based on physiological effects are (1) sprout inhibition (in the case on bulbs and tubers), (2) delayed senescence (fresh fruits and vegetables), and (3) delayed ripening in case of fruits.

Although radiation processing of foods has several applications broadly they can be classified into three categories, that is, low dose applications (where dose required is < 1 kGy), medium dose applications (dose delivered between 1 and 10 kGy) and high dose applications (dose required is beyond 10 kGy). Major applications in all the three categories are summarized in Table 12.2.

TABLE 12.2 Applications of Food Irradiation.

Dose range	Effects	Examples
Low dose (up to 1 kGy)	Sprout inhibition	Onions, potatoes
	Delay in ripening	Fruits
	Quarantine treatment	Fruits, fresh produce
	Insect disinfestation	Cereals, legumes, and pulses
	Pest disinfestation	Fresh produce, dried products
	Parasite inactivation	Pork (*Trichinella*)
Medium dose (1–10 kGy)	Reduce spoilage organisms thereby extending shelf-life	Fruits and vegetables
	Eliminating vegetative pathogenic micro-organisms	Meats, dried fish, meat products and spices
High dose (beyond 10 kGy)	Reduce micro-organisms to the point of sterility	Hospital diets, Spices, emergency rations, foods for severely immune-compromised patients, foods for astronauts

12.4.1 LOW DOSE APPLICATIONS (<1 KGY)

12. 4.1.1 SPROUT INHIBITION

Almost 20–40% of total post-harvest storage losses (50–60%) in onions are due to sprouting (Majid et al., 2015). Prevention of these losses during storage of several months is necessary to maintain year-long supply of commodities, such as onions, potatoes, garlic, and yams to consumers. To inhibit sprouting, refrigeration during storage, or the application of various chemicals, such as hydrazide (pre-harvest) and isopropyl chlorocarbonate (post-harvest) are presently employed. However, refrigeration is expensive particularly so in the tropical and sub-tropical zones of the world. Chemical treatments although relatively cheap and efficient have serious health concerns due to the fact that they leave harmful residues after treatment (Mostafavi et al., 2010). There-fore, in view of the above reasons, radiation can be a suitable alternative for preventing damages due to sprouting. Radiation processing in the range of 50–150 Gy had shown to result in sprout inhibition, reduced rotting and weight loss in potatoes, garlic, onions, sweet potato, turnip, carrot, sugar beet, and yams (Thomas et al., 1975; Kader, 1986).

12.4.1.2 INSECT DISINFESTATIONS

Insect pests inflict damage on stored products mainly by direct feeding. Some species feed on the endosperm causing loss of weight and quality, while other species feed on the germ, resulting in poor seed germination and less viability (Malek & Parveen, 1989; Santos et al., 1990). Thus, insect infestation reduces value of grains for marketing, consumption, and further reduces their viability for planting. In addition to direct consumption of the product, insect pests contaminate their feeding media through excretion, molting, dead bodies, and their own existence in the product, which is not commercially desirable. Damage done by insect pests encourages infection with bacterial and fungal diseases through transmission of their spores. Fumigants, such as ethylene dibromide or ethylene oxide are highly effective and were widely used for control of insect pests. However, these chemicals have been banned or stringently restricted in most countries for health and environmental reasons. Other processing methods, such as heat and cold treatments although capable of insect disinfestations could degrade the sensory quality of the produce (Mostafavi et al., 2010). Radiation processing has, therefore, been suggested as an alternative to fumigation. Radiation treatment can prevent losses caused by insects in stored grains, pulses, flour, cereals, coffee beans, dried fruits, nuts, and dried fish (Landgraf et al., 2006). Required radiation dose for insect disinfestation is in the range of 150–700 Gy. A dose level of 250 Gy can be effective for quarantine treatment of fruit flies, whereas a dose of 500 Gy can control all stages of most pests (; Miller, 2005).

12.4.2 MEDIUM-DOSE APPLICATIONS (1–10 KGY)

12.4.2.1 FOOD-BORNE PATHOGENS

Meat, eggs, and dairy products are major sources of foodborne illnesses. The most serious microbial pathogens are *E. coli, Salmonella, Campylobacter and Listeria*. Tapeworm is another organism of concern for beef. Radiation treatment in dose range of 1–3 kGy can eliminate all these pathogenic microorganisms (Molins et al., 2001).

12.4.2.2 SHELF-LIFE EXTENSION

Radiation dose required for controlling microbial pathogens (1–3 kGy), apart from controlling pathogenic bacteria can also extend shelf-life of treated foods. Shelf-life extension is mainly due to reduction in population of spoilage bacteria, molds, and yeasts. Shelf-life of meat products, such as chicken can be extended by few weeks while shelf-life extension of few days (10–20), can be achieved in case of fruits and vegetables using radiation treatment (Mostafavi et al., 2010). Similarly, radiation dose in range of 0.25–1 kGy can result in delayed ripening in case of climacteric fruits, such as mango, banana, and papaya when irradiated before onset of ripening (Lacroix & Ouattara, 2000).

12.4.2.3   MICROBIAL DECONTAMINATION

Plant materials from which spices are derived are generally contaminated by microorganisms from the soil, windblown dust, and by bird droppings. Due to these contaminations, spices can have microorganism population densities exceeding 10^6 organisms per gram of material (Farag et al., 1995). If these spices are used in manufacture of food products for which the manufacturing process does not include a sterilizing step, these organisms can cause rapid food spoilage and can lead to foodborne illness. Since fumigation is now banned in several parts of the world; radiation treatment is the method of choice for spice manufacturers. Elimination of bacteria, mold spores, and insects without negative impact on chemical or sensory properties can be achieved using doses between 5 and 10 kGy (; IAEA, 2015; Koopmans & Duizer, 2004).

12.4.3   HIGH DOSE (>10 KGY) APPLICATIONS

Foods can be commercially sterilized as in canning by using radiation treatment in the dose range of 25–45 kGy (WHO, 1999). But not all foods are amenable to such radiation sterilization procedures (Diehl, 2002). Fresh fruits and vegetables will deteriorate, while, products with excellent quality can be obtained from meat, poultry, and seafood by employing high dose of radiation. Generally, products subjected to high dose radiation are preheated to inactivate enzymes, vacuum packed to exclude oxygen and irradiated at low temperatures (-20–$40°C$) thereby preventing off-flavors resulting from lipid oxidation. These products can be stored at room temperature almost indefinitely. Although additional procedures and high dose radiation add to costs of the final product, these sterilized products are useful for immune-compromised patients, in natural disasters, military rations, and for a special group, such as astronauts (Pietranera et al., 2003; Kumar et al., 2016; Diehl, 1995).

12.4.4 OTHER APPLICATIONS

Apart from these applications recent research in the area of food irradiation has shown some very interesting applications of radiation processing, such as enhancing aroma quality, antioxidant status, and shelf-life extension of minimally processed products. Some of these recent applications are summarized in Table 12.3.

TABLE 12.3 Some Recent Applications of Food Radiation.

Product	Dose range (kGy)	Application	References
Spices, such as nutmeg	5 kGy	Enhanced aroma notes	Ananthakumar et al. (2006)
Soybean	0.5–5 kGy	Radiation-induced enhanced antioxidant activity	Variyar et al. (2004)
Guar gum	1–25 kGy	Partially hydrolyzed guar gum for flavor encapsulation and as soluble type of dietary fiber	Sarkar et al. (2012) and Gupta et al. (2015b)
Guar gum based biodegradable films	500 Gy	Improved mechanical properties of films	Saurabh et al. (2013)
French beans, ash gourd, pumpkin, cabbage, and cauliflower	up to 2 kGy	Minimally processed products with extended shelf-life Inhibition of browning	Gupta et al. (2012), Tripathi et al. (2013), Tripathi et al. (2014), Banerjee et al. (2016a), Vaishnav et al. (2015)
Cabbage	Up to 2 kGy	Enhanced nutraceutical value	Banerjee et al. (2016b)
Grapes	Up to 2 kGy	Wine with improved antioxidant status, higher red color and improved flavor	Gupta et al. (2015a)
Ambient storage meal for calamity victims	15 kGy	Stuffed baked food having long shelf-life at ambient storage temperatures developed for calamity victims	Kumar et al. (2016)

12.5 WHOLESOMENESS OF IRRADIATED FOODS

12.5.1 MICROBIAL QUALITY

Several concerns have been raised regarding microbial safety of radiation processing (Diehl, 1995), such as: (1) Could the selective effect of radiation on microflora result in survival of more pathogenic population as compared to non-pathogenic leading to higher risk of food-borne illness, (2) Whether radiation will induce mutations leading to conversion of nonpathogenic organisms to pathogenic organisms or less virulent strains to more virulent strains and could radiation stimulate toxin formation in non-toxin producing bacteria or molds, (3) Can repeat sub lethal dose of radiation lead to increased radiation resistance in microorganisms, (4) Whether radiation can change diagnostic characteristic of microbe thus making its detection difficult and finally, and (5) Is radiation only a cosmetic treatment. Years of research worldwide in different laboratories have provided answers to all these questions and helped in dispelling all doubts regarding food irradiation.

Most foodborne illness-causing microorganisms are more sensitive to radiation treatment when compared to food spoilage causing species. A dose of 5 kGy can reduce population of most food pathogens, such as *Salmonella species*, *Staphylococcus aureus*, *Shigella*, *Campylobacter*, *Listeria*, and *E. coli* by up to 6 log cycles (Ingram & Farkas, 1977; El Zawahry et al., 1979). Therefore, there will be no selective advantage of these pathogenic microorganisms over spoilage organisms due to radiation processing. Regarding, threat of radiation-induced mutations causing more virulence in microorganisms is concerned, no reports on observance of such a phenomenon have been found in literature (Ingram & Farkas, 1977). Generally, it is observed that microorganisms become more sensitive to heat, drying, and antimicrobial agents due to radiation processing. Moreover, because of this reduced fitness, survivors of sublethal radiation treatment do not compete well with the non-irradiated organisms present in practical situations and disappear unnoticed (Farkas & Andrassy, 1985; Farkas, 1988). Also, as shown by several studies radiation did not increase the rate of production of microbial toxins, such as aflatoxins, rather a decreased content of toxins was reported in irradiated foods (Diehl, 1995). With regards to the probability of increased radio-resistance of microorganism, there are reports demonstrating up to two-fold increase in radiation resistance of some species by repeated cycles of sublethal radiation treatment followed by growth in nutrient-rich media under appropriate temperature (Parisi & Antoine, 1974). There are also several other reports in other species wherein no increase in radiation resistance has been demonstrated (Diehl, 1995). Nevertheless, it can be concluded that production of radiation resistant microbes will require highly specific laboratory conditions, such as repeated radiation and culturing which never exists in industrial processing conditions. Therefore, concerns regarding the production of radiation resistant pathogenic microorganisms are highly unfounded (Maxcy, 1983). Another concern about microbial safety of irradiated food is that radiation can change diagnostic characteristics of pathogens making their detection difficult. But several investigations in this regard have proved that diagnostic characteristics of microorganisms do no change due to radiation processing. Lastly, regarding radiation as a cosmetic treatment, it can be said that spoilage of food products occurs with a change in their sensory qualities. There are distinct changes in visual quality, odor, and taste of product with the spoilage. Radiation processing can reduce the number of spoilage microorganisms in food products but it cannot change its sensory properties. Therefore, radiation processing cannot be used for making bad or spoiled products look good.

12.5.2 NUTRITIONAL QUALITY OF IRRADIATED FOODS

12.5.2.1 MACRONUTRIENTS

Carbohydrates, lipids, and proteins are major components of food providing energy and building blocks for growth and maintenance of the body. Several animal feeding and human trials have successfully shown that radiation processing even with very high doses of 10–50 kGy do not result in any changes in metabolizable energy, nitrogen balance, and coefficients of digestibility (Read et al., 1961; Kraybill, 1958). Although chemical analysis did show some radiation-induced changes in carbohydrates, proteins, and lipids, these changes were found to be very small and non-specific even at very high doses of 50 kGy (Diehl, 1995). With regards to proteins several animal and human trials have clearly shown that biological value and digestibility of food proteins are not adversely affected by radiation processing (Reber et al., 1968a; Reber et al., 1968b; Metta & Johnson, 1956).

12.5.2.2   MICRONUTRIENTS

Radiation-induced deterioration in the content of vitamins has been widely reported by carrying out irradiation of pure solutions of vitamins. However, it is usually observed that radiation-induced degradation in significantly higher in pure solutions as compared to food products. It is due to the mutual protection of various food components on each other. Nevertheless, radiation can cause degradation in vitamin content in meat and fresh fruits and vegetables. Further, sensitivity of vitamins to radiation processing also depends upon composition of food, processing, and storage conditions. Generally, irradiation under vacuum and low-temperature results in significantly lesser vitamin losses as compared to irradiation in presence of oxygen and at room temperature. Different vitamins have different sensitivity toward radiation treatment. A detailed discussion is beyond the scope of this chapter. An excellent detailed review of effects of radiation on vitamins has been published (Diehl, 1995; Kilcast, 1994). Lastly, it can be concluded that potential nutritional losses due to radiation processing are not different from losses in other processing treatments. Traditional processing methods, such as heating and drying may cause higher nutritional losses that irradiation.

12.6   DETECTION OF IRRADIATED FOOD

Although irradiated food is safe and wholesome, correct, and comprehensive information is sought by the consumer to make a choice between irradiated and non-irradiated food. Analytical methods have, therefore, been developed to directly detect in the product whether or not it has been treated with ionizing radiation. Availability of such methods would help authorities to regulate marketing of irradiated foods, accelerate approval of additional food irradiation applications, and enhance international trade in irradiated food. Detection methods to be developed should be highly sensitive and must focus on minute changes as there are no major chemical, physical, or sensory changes in irradiated foods. Techniques developed so far can be classified into three broad categories, namely, (1) physical (2) chemical, and (3) biological. Table 12.4 summarizes the detection methods available till date for identifying irradiated food.

TABLE 12.4   Detection Methods for Irradiated Foods.

(a) Physical	1) Electron spin resonance (ESR) or electron paramagnetic resonance (EPR)
	2) Impedance
	3) Near infra-red reflectance
	4) Nucleation temperatures (NT)
	5) Photostimulated luminescence (PSL)
	6) Thermoluminescence
	7) Viscometry
(b) Chemical	Cyclobutanones
	Gas evolution (e.g., H_2, CO, and possibly H_2S and NH_3)
	Hydrocarbons
	Peroxides
	Proteins
	o-Tyrosine (ortho-hydroxy phenylalanine)
	Thymine glycol
(c) Biological	DEFT (direct epifluorescent filter techniques)/APC (aerobic plate count)
	DNA (bacterial count, micro-gel electrophoresis/comet assay, mitochondrial linear vs. circular vs. supercoiled)
	Germination
	Immunological/enzyme-linked immunosorbent assay (ELISA) for cyclobutanone, dihydrothymidine, DNA, proteins, and thymidine glycol.
	LAL (Limulus amoebocyte lysate test)

12.6.1   PHYSICAL METHODS

12.6.1.1   ELECTRON SPIN RESONANCE

Among the physical methods, electron spin resonance (ESR) is a widely accepted technique based on radiation specific stable radicals formed in solid and dry components of food (bones, shells, etc.). The technique is non-destructive, specific, highly sensitive, and rapid. Cost of the equipment and special technical skills required to operate, however, restricts its use. The method has been successfully applied to the detection of irradiated foods containing bones (beef, chicken, and fish), in low moisture foods containing crystalline cellulose (pistachio nutshells, paprika powder, and fresh and frozen berries) and in dehydrated fruits containing crystalline sugar (mango, papaya, figs, and raisins) (Chauhan et al., 2009). ESR can also detect radiation-induced radicals in packaging materials offering an indirect method of detection.

12.6.1.2 LUMINESCENCE METHODS

Thermoluminescence (TL) is a method whereby heat is applied to release trapped energy deposited in crystalline lattice during irradiation. The intensity of energy thus released in the form of light is then measured. The method has a high degree of selectivity and sensitivity and no reference standard is needed. An unequivocal classification of irradiated and non-irradiated samples is possible. However, differing amounts and types of minerals exhibit varying TL intensities thereby requiring normalization of TL intensities and response. The technique is quite laborious since minerals have to prepared free of organic matter and stringent quality assurance procedures are needed to avoid possible dust contamination within the laboratory. Standards now exist for spices, herbs, and shrimps and hold promise for application to dehydrated fruits and vegetables and shellfish.

Photo-simulated luminescence (PSL) uses light rather than heat to release stored energy. It obviates the need to isolate minerals and thus acts as an effective alternative to TL in terms of sensitivity, speed, and efficiency.

12.6.1.3 OTHER PHYSICAL METHODS

Methods based on cell membrane damage, such as measurement of changes in viscosity and electrical impedance are the other physical methods worth mentioning. Viscosity measurement is simple, quick, and cheap technique based on reduction in viscosity of solutions due to break down of polymers during gamma-irradiation. Although batch to batch variation is quite high, the method seems to be quiet valuable for screening purposes particularly in high starch foods, such as black pepper.

12.6.2 CHEMICAL METHODS

Food irradiation does not bring about significant changes in major nutrients (proteins, lipids, and carbohydrates) of food. In addition, majority of the chemical changes induced by irradiation closely resembles those formed during processing of food by other methods. Detection of chemical changes is further complicated by variation in composition of similar samples due to differences in breed/variety, handling, storage practices, and agro-climatic conditions.

12.6.2.1 DETECTION OF RADIATION-INDUCED HYDROCARBONS

The most important among chemical methods is based on the detection of quantitative increase with dose in characteristic hydrocarbons formed from lipids during exposure to ionizing radiation using gas-liquid chromatography (GLC). The method is applicable to all lipid-containing foods including those with low-fat content. Most food labs are equipped with this instrument. The method has been successfully validated on raw chicken, pork, and beef as well as on Camembert cheese, avocado, papaya, and mango (Chauhan et al., 2009).

12.6.2.2 DETECTION OF 2-ALKYLCYCLOBUTANONES

Another method based on radiolysis of lipids is the detection of 2-alkylcyclobutanones (2-ACB's) using gas chromatography-mass spectrometry (GC/MS). 2-ACB's are considered to be radiation specific with no evidence of it being detected either in non-irradiated or in cooked lipid-containing food. This eliminates the need for non-irradiated samples to be used for comparison making the method highly promising. The method can be successfully applied to all fat-containing foods (including those with low-fat content) and has been validated on raw chicken, pork, and liquid whole egg. Fluorescent labeling of cyclobutanones and their detection by thin-layer chromatography (TLC), high-performance liquid chromatography (HPLC), and enzyme-linked immune assay (ELISA) techniques have recently been developed to detect 2-alkylcyclobutanones.

Other methods, such as determination of chemical changes in DNA and estimation of o-tyrosine formed during exposure of foods to ionizing radiation offer good potential as a detection method.

12.6.2.3 BIOLOGICAL METHODS

Among the biological methods, direct epifluorescent filter technique (DEFT)/aerobic plate count (APC) that measures the difference between dead and live microorganisms although not specific for irradiated foods serves as a useful, simple, and fast primary screening method.

Methods based on damage to DNA (DNA hybridization techniques, microgel electrophoresis/comet assay, and mitochondrial DNA strand breaks), ELISA methods and measurement of the rate of seed germination are some of the other methods currently under active investigation.

In summary, although there are several promising techniques to screen and detect a few irradiated foods no one technique is likely to be applicable to all food materials. Food control laboratories in various countries routinely employ some of the methods, such as hydrocarbon and 2-ACB's for lipid-containing foods, ESR for bone-containing food, and TL and PSL for foods containing silicate minerals. It is expected that newer and better techniques will emerge in near future.

12.7 INTERNATIONAL STATUS

Independent research carried out worldwide for the last five decades by several countries including Canada, Denmark, France, India, Sweden, UK, European Union, and US FDA has convincingly demonstrated the safety of irradiated food and confirmed that such processed foods are safe for human consumption. Several international bodies including the World Health Organization (WHO), the Food and Agriculture Organization (FAO), the International Atomic Energy Agency (IAEA), and Codex Alimentarius have also approved radiation technology as a safe food preservation technique. There has been a considerable increase in international trade in radiation processed foods in the last two decades. The quantity of various food groups irradiated worldwide in 2015 is shown in Table 12.5.

TABLE 12.5 Quantity of Food Items Irradiated Worldwide in 2015.

Food items and purpose	Quantity irradiated worldwide (tons)
Disinfection of spices and dry vegetables	186,000
Disinfection of grains and fruits	82,000
Disinfection of meat and fish	33,000
Sprout inhibition of garlic and potato	88,000
Other food items	17,000

According to a recent report on trade in irradiated food products, the quantity of the irradiated foods increased by 100,000 tons in Asia and 10,000 tons in the US in 2010 compared with 2005 (Kume & Todoriki, 2013). A corresponding increase in the quantity of fruits and agricultural produce irradiated for phytosanitary purposes also increased during this period.

12.8   NATIONAL STATUS

India's first pilot plant radiation processing facility The Food Package Irradiator was commissioned in 1967. The facility could process commodities right from requiring very low dose (0.06–0.1 kGy), for example, onion and potato to products requiring very high doses (10–30 kGy), such as spices. Approval for formulating a policy that governs food irradiation in India was sought from Indian government based on large-scale studies in this facility, Government of India in 1987 set up a National Monitoring Agency (NMA) to oversee commercial application of radiation processing technology. Atomic Energy (Control of Irradiation of Food) rules were notified in 1991 that was later amended in 1996. Prevention of Food Adulteration Act (1954) was amended in 1994 and irradiation of onion, potato, and spices for domestic market was approved by Government of India. In April 1998 and in May 2001, additional items were approved. The Directorate General of Health Services (PFA) issued a draft notification in 2007 for generic or food class-wise approval of radiation processing. Ministry of Agriculture, Government of India in February 2004, included irradiation as a quarantine measure by amending the plant protection and quarantine regulations. The primary legislation that regulates food irradiation, "The Atomic Energy (Control of Irradiation of Food) rule 1996," was also amended and the notification issued in June 2012. The Ministry of Health and Family Welfare also issued an amended draft notification relating to the Prevention of Food Adulteration Act (Fifth Amendment) rules, 1994.

Approval of radiation processing based on generic class, Atomic Energy (Radiation processing of Food and Allied Products) rules 2012 can aid in increasing the product range and economic viability of irradiation plants by providing year-long availability of feedstock. Under this amendment allied products, such as dried foods of animal origin and their products, ethnic foods, military rations, space foods, ready-to-eat/ready-to-cook minimally processed foods have also been separately listed. Class wise segregation of food products and doses approved for various applications is provided in Tables 12.6 and 12.7.

KRUSHAK facility at Lasalgaon, Nashik district, Maharashtra state, India is the major facility in catering to irradiation of fresh horticultural produce. A total of 1880 tons of mango has been irradiated and exported from this facility from its inception in 2007 till 2015. Phytosanitary treatment of other products, such as litchi and pomegranate has also been demonstrated. Processing spices and dry ingredients for microbial decontamination has been routinely carried out at the Radiation Processing Plant, BRIT,

TABLE 12.6 Atomic Energy (Radiation Processing of Food and Allied Products) Rules 2012 Schedule – I. Classes of Food Products and Dose Limits for Radiation Processing.

Class	Food	Purpose	Dose limit (kiloGray)	
			Minimum	Maximum
Class 1	Bulbs, stem and root tubers, and rhizomes	Inhibit sprouting	0.02	0.2
Class 2	Fresh fruits and vegetables (other than Class 1)	Delay ripening	0.2	1.0
		Insect disinfestation	0.2	1.0
		Shelf-life extension	1.0	2.5
		Quarantine application	0.1	1.0
Class 3	Cereals and their milled products, pulses and their milled products, nuts, oilseeds, and dried fruits and their products	Insect disinfestation	0.25	1.0
		Reduction of microbial load	1.5	5.0
Class 4	Fish, aquaculture, seafood and their products (fresh or frozen), and crustaceans	Elimination of pathogenic microorganisms	1.0	7.0
		Shelf-life extension	1.0	3.0
		Control of human parasites	0.3	2.0
Class 5	Meat and meat products including poultry (fresh and frozen) and eggs	Elimination of pathogenic microorganisms	1.0	7.0
		Shelf-life extension	1.0	3.0
		Control of human parasites	0.3	2.0
Class 6	Dry vegetables, seasonings, spices, condiments, dry herbs and their products, tea, coffee, cocoa, and plant products	Microbial decontamination	6.0	14.0
		Insect disinfestation	0.3	1.0
Class 7	Dried foods of animal origin and their products	Insect disinfestation	0.3	1.0
		Control of molds	1.0	3.0
		Elimination of pathogenic microorganisms	2.0	7.0
Class 8	Ethnic foods, military rations, space foods, ready-to-eat, ready-to-cook/minimally processed foods	Quarantine application	0.25	1
		Reduction of microorganisms	2	10
		Sterilization	5	25

Vashi, Navi Mumbai that functions under the administrative control of the Department of Atomic Energy, Government of India.

TABLE 12.7 Atomic Energy (Radiation Processing of Food and Allied Products) Rules 2012 Schedule – II. Dose Limits for Radiation Processing of Allied Products.

Sr. No.	Allied product	Purpose	Dose limits (kiloGray) Minimum	Maximum
1	Animal food and feed	Insect disinfestation	0.25	1.0
		Microbial decontamination	5.0	10.0
2	Ayurvedic herbs and their products, and medicines	Insect disinfestations	0.25	1.0
		Microbial decontamination	5.0	10.0
		Sterilization	10	25
3	Packaging materials for food/allied products	Microbial decontamination	5.0	10.0
		Sterilization	10	25
4	Food additives	Insect disinfestations	0.25	1.0
		Microbial decontamination	5.0	10.0
		Sterilization	10	25
5	Health foods, dietary supplements, and nutraceuticals	Insect disinfestation	0.25	1.0
		Microbial decontamination	5.0	10.0
		Sterilization	10	25
6	Bodycare and cleansing products	Microbial decontamination	5.0	10.0
		Sterilization	10	25
7	Cut flowers	Quarantine application	0.25	1.0
		Shelf-life extension	0.25	1.0

Since its inception in 2007, the facility has processed approx. 30,000 tons of these products fetching revenue of US $ 2.0 million. A total of 12 irradiation plants have also now been set up in the private sector.

12.9 CONCLUSION AND FUTURE GOALS

In the earlier periods, substantial progress has been achieved on research and development and semi-commercial/pilot-scale processing and storage of irradiated foods leading to technology transfer activities, process control procedures, and evaluation of public acceptance. Market survey and actual retail sales demonstrate that consumers are willing to pay a premium for food safety. However, availability of irradiated foods in the retail marketplace is highly limited limiting consumer choice in the marketplace. Further, lack of acceptance of irradiated food mainly due to misconceptions and irrational fear of nuclear-related technologies and confusion by people to differentiate between irradiated foods from radioactive foods has hampered commercialization of irradiated food.

It is, therefore, necessary to overcome consumer's fear of buying irradiated food and device strategies so that they begin to see the advantages of having this option available. There is also a greater need for understanding of the potential advantages of irradiation by both consumers and entrepreneurs so that they will be in a position to evaluate the process in a rational manner. The key to the future expansion of the technology on a larger scale and to stimulate investment lies in increasing the understanding and interest of the business community.

Business community at large emphasizes that the key issue in selecting products on their shelves that are derived from new technologies is identifying and evidencing not the novel technology itself but customer benefit arising from the technology. The focus is thus on what the technology does and not how it does it. As far as novel food technologies are concerned, entrepreneurs consider entering the market early but not first, as the pioneers typically face a greater risk of failure. The clear preference is to have the novel technology become mainstreamed by leading brands and sticking with what they have or what is well-known.

Building strong partnerships between government and private industry in order to expand the scope of the technology from the research laboratory to the market-place should be the major focus in the coming years. Commercial exploitation of the technology in order to bring about trade in irradiated food products on a wider scale should thus include both strengthening public acceptance and market development. As consumers and the food control authorities become increasingly cautious on the question of food safety, food irradiation technology is expected to have wider commercial acceptance.

Food irradiation either as a stand-alone technology or in combination with others is effective to resolve technical problems in trade of many food and agricultural products, a prerequisite to gain a higher share of world trade demands compliance with international food standards. In accordance with the guidelines framed by the International Consultative Group on Food Irradiation (ICGFI), for harmonization of regulation internationally, several countries have authorized food irradiation by groups/classes of food. Profitable products will have to be continuously identified, international markets opened and consumer education strengthened for successful commercialization for which regional cooperation in this regard is necessary. Irradiation as a phytosanitary treatment for quarantine of horticultural produce to meet export market requirements can boost exports to international markets for which ethnic fruits with potential for export need to be identified.

In view of the public concerns regarding the use of cobalt-60 and issues related to handling and transportation of radioactive materials, other types of ionizing radiations, mainly electron beam, and X-ray sources need to be considered. Electron beam facilities not only provide advantage over cobalt-60 plants in terms of safety but also allow processing of large volumes of materials, such as fresh fruits and vegetables as well as frozen foods. Thus feasibility of establishing more machine sources particularly electron beam facilities for irradiation besides encouraging use of gamma radiation for processing agricultural and food commodities should be looked into. In the era of economies becoming global, food products must meet high standards of quality, and quarantine in order to move across borders. Irradiation is an important tool in the fight to prevent the spread of deleterious insects and microorganisms. Further, increasing demand for food supplies to serve the needs of ever-increasing population and for improving the diets of the millions suffering from malnutrition, food irradiation technology could make a useful contribution to help alleviate world food problems.

KEYWORDS

- hygienization
- food irradiation
- radiations
- gamma rays
- pathogens

REFERENCES

Ananthakumar, A.; Variyar, P. S.; Sharma, A. Estimation of Aroma Glycosides of Nutmeg and Their Changes during Radiation Processing. *J. Chromatogr. A* **2006**, *1108*, 252–257.

Banerjee, A.; Chatterjee, S.; Variyar, P. S.; Sharma, A. Shelf Life Extension of Minimally Processed Ready-to-cook (RTC) Cabbage by Gamma Irradiation. *J. Food Sci. Technol.* **2016a**, *53*, 233–244.

Banerjee, A.; Rai, A. N.; Penna, S.; Variyar, P. S. Aliphatic Glucosinolate Synthesis and Gene Expression Changes in Gamma-irradiated Cabbage. *Food Chem.* **2016b**, *209*, 99–103.

Brasch, A.; Huber, W. Ultrashort Application Time of Penetrating Electrons: A Tool for Sterilization and Preservation of Food in the Raw State. *Science* **1947,** *105,* 112–117.

Buzby, J. C.; Roberts, T. Economic Costs and Trade Impacts of Microbial Foodborne illness. *World Health Stat. Q* **1996,** *50* (1–2), 57–66.

Chauhan, S. K.; Kumar, R.; Nadanasabapathy, S.; Bawa, A. S. Detection Methods for Irradiated Foods. *Compr. Rev. Food Sci. Food Saf.* **2009,** *8* (1), 4–16.

Clemmons, H. E.; Clemmons, E. J.; Brown, E. J. Electron Beam Processing Technology for Food processing. In *Electron Beam Pasteurization and Complementary Food Processing Technologies;* Pillai, S. D., Shyanfar, S., Eds.; Woodhead Publishing: Cambridge, 2015; pp 11–25.

Diehl, J. F. Food Irradiation—Past, Present and Future. *Radiat. Phys. Chem.* **2002,** *63,* 211–215.

Diehl, J. F. *Safety of Irradiated Foods,* 2nd ed.; Marcel-Dekker: New York, 1995.

El Zawahry, Y. A.; Rowley, D. B. Radiation Resistance and Injuring of *Yersinia enterocolitica. Appl. Environ. Microbiol.* **1979,** *37,* 50–54.

Farag, S. E. D. A.; Aziz, N. H.; Attia, E. S. A. Effect of Irradiation on the Microbiological Status and Flavouring Materials of Selected Spices. *Z. Lebensm. Unters. Forsch.* **1995,** *201,* 283–288.

Farkas, J. *Irradiation of Dry Food Ingredients;* CRC Press: UK, 1988.

Farkas, J. Irradiation as a Method for Decontaminating Food: A Review. *Int. J. Food Microbiol.,* **1998,** *44* (3), 189–204.

Farkas, J.; Andrássy, E. Increased Sensitivity of Surviving Bacterial Spores in Irradiated spices. In *Fundamental and Applied Aspects of Bacterial Spores*; Dring, G. J., Ellar, D. J., Gould, G. W., Eds.; Academic Press: London, 1985; pp 397–407.

Farkas, J.; Mohácsi-Farkas, C. History and Future of Food Irradiation. *Trends Food Sci. Technol.* **2011,** *22,* 121–126.

Food Borne Illnesses: What You Need to Know. United States Food and Drug Administration (USFDA). http://www.fda.gov/Food/FoodborneIllnessContaminants/FoodborneIllnesses NeedToKnow/default.htm (Accessed on 26/05/2016).

Food Technologies: Food Irradiation. *Encyclopedia of Food Safety*; Elsevier: Amsterdam, 2014; Vol. 3, pp 178–186.

General Standard for Irradiated Foods; CODEX STAN 106–1983, REV.1–2003; CodexAlimentarius Commission: Rome, 2003a.

Gillett, D. C. Apparatus for Preserving Organic Materials by the Use of X-rays. US Patent 1,275,417, August 13, 1918.

Gupta, S.; Chatterjee, S.; Vaishnav, J.; Kumar, V.; Variyar, P. S.; Sharma, A. Hurdle Technology for Shelf Stable Minimally Processed French Beans (*Phaseolus vulgaris*): A Response Surface Methodology Approach. *LWT Food Sci. Technol.* **2012,** *48,* 182–189.

Gupta, S.; Padole, R.; Variyar, P. S.; Sharma, A. Influence of Radiation Processing of Grapes on Wine Quality. *Radiat. Phys. Chem.* **2015a,** *111,* 46–56.

Gupta, S.; Saurabh, C. K.; Variyar, P. S.; Sharma, A. Comparative Analysis of Dietary Fiber Activities of Enzymatic and Gamma Depolymerized Guar Gum. *Food Hydrocoll.* **2015b,** *48,* 149–154.

Gustavsson, J; Cederberg, C; Sonesson, U; Otterdijk, R. V.; Meybeck, A. Global Food Losses and Food Waste–Extent, Causes and Prevention; Food and Agricultural Organization of the United States; Rome, 2011.

High-dose Irradiation: Wholesomeness of Food Irradiated with Doses Above 10 kGy; World Health Organization Technical Report Series 890; Geneva, 1999.

Ingram, M.; Farkas, J. Microbiology of Foods Pasteurised by Ionising Radiation. *Acta Aliment.* **1977,** *6,* 123–184.

Kader, A. A. Potential Applications of Ionizing Radiation in Postharvest Handling of Fresh Fruits and Vegetables. *Food Technol.* **1986,** *40,* 117–121.

Kilcast, D. Effect of Irradiation on Vitamins. *Food Chem.* **1994,** *49,* 157–164.

Koopmans, M; Duizer, E. Foodborne Viruses: An Emerging Problem. *J. Food Microbiol.* **2004,** *90,* 23–41.

Kraybill, H. F. Nutritional and Biochemical Aspects of Food Preserved by Ionizing Radiation. *J. Home Econ.* **1958,** *50,* 695–700.

Kumar, S.; Saxena, S.; Verma, J.; Gautam, S. Development of Ambient Storable Meal for Calamity Victims and Other Targets Employing Radiation Processing and Evaluation of its Nutritional, Organoleptic, and Safety Parameters. *LWT Food Sci. Technol.* **2016,** *69,* 409–416.

Lacroix, M.; Ouattara, B. Combined Industrial Processes with Irradiation to Assure Innocuity and Preservation of Food Products—A Review. *Food Res. Int.* **2000,** *33,* 719–724.

Landgraf, M.; Gaularte, L.; Martins, C.; Cestari, A.; Nunes, T.; Aragon, L.; Destro, M.; Behrens, J.; Vizeu, D.; Hutzler, B. Use of Irradiation to Improve the Microbiological Safety of Minimally Processed Fruits and Vegetables; IAEA-TECDOC-1530: 41–59; Vienna, 2006.

Majid, I.; Dhatt, A. S.; Sharma, S.; Nayik, G. A.; Nanda, V. Effect of Sprouting on Physicochemical, Antioxidant and Flavonoid Profile of Onion Varieties. *Int. J. Food Sci. Tech.* **2016,** *51,* 317–324.

Malek, M.; Parveen, B. Effect of Insects Infestation on the Weight Loss and Viability of Stored BE Paddy. *Bangladesh J. Zool.* **1989,** *17,* 83–85.

Manual of Good Practice in Food Irradiation. Sanitary, Phytosanitary and Other Applications; IAEA Technical Report Series 481; Vienna, 2015.

Maxcy, R. B. Significance of Residual Organisms in Foods After Substerilizing Doses of Gamma Radiation: A Review. *J. Food Safety.* **1983,** *5,* 203–211.

Metta, V. C.; Johnson, B. C. The Effect of Radiation Sterilization on the Nutritive Value of Foods: I. The Biological Value of Milk and Beef Proteins. *J. Nutr.* **1956,** *59,* 479–490.

Miller, R. D. *Electronic Irradiation of Foods: An Introduction to the Technology;* Springer: New York, 2005.

Mohácsi-Farkas, C.; Nyirő-Fekete, B.; Daood, H.; Dalmadi, I.; Kiskó, G. Improving Microbiological Safety and Maintaining Sensory and Nutritional Quality of Pre-cut Tomato and Carrot by Gamma Irradiation. *Radiat. Phys. Chem.* **2014,** *99,* 79–85.

Molins, R. A.; Motarjemi, Y.; Käferstein, F. K. Irradiation: A Critical Control Point in Ensuring the Microbiological Safety of Raw Foods. *Food Control.* **2001,** *12,* 347–356.

Mostafavi, H. A.; Fathollahi, H.; Motamedi, F.; Mirmajlessi, S. M. Food Irradiation: Applications, Public Acceptance and Global Trade. *Afr. J. Biotechnol.* **2010,** *9,* 2826–2833.

Parfitt, J.; Barthel, M.; Macnaughton, S. Food Waste within Food Supply Chains: Quantification and Potential for Change to 2050. *Philos. Trans. R. Soc. Lon. B: Biol. Sci.* **2010,** *365,* 3065–3081.

Parisi, A.; Antoine, A. D. Increased Radiation Resistance of Vegetative *Bacillus pumilus*. *Appl. Microbiol.* **1974,** *28,* 41–46.

Pietranera, M. A.; Narvaiz, P.; Horak, C.; Kairiyama, E. Irradiated Icecreams for Immunosuppressed Patients. *Radiat. Phys. Chem.* **2003,** *66,* 357–365.

Pillai, S. D.; Shyanfar, S. *Electron Beam Pasteurization and Complementary Food Processing Technologies;* Woodhead Publishing: Cambridge, 2015.

Read, M. S.; Kraybill, H. F.; Worth, W. S.; Thompson, S. W.; Isaac, G. J.; Witt, N. F. Successive Generation Rat-feeding Studies with a Composite Diet of Gamma-Irradiated Foods. *Toxicol. Appl. Pharmacol.* **1961,** *3,* 153–173.

Reber, E. F.; Bert, M. H. Protein Quality of Irradiated Shrimp. *J. Am. Diet. Assoc.* **1968a,** *53,* 41–42.

Reber, E. F.; Bert, M. H.; Rust, E. M.; Kuo, E. Biological Evaluation of Protein Quality of Radiation-pasteurized Haddock, Flounder and Crab. *J. Food Sci.* **1968b,** *33,* 335–337.

Recommended International Code of Practice for Radiation Processing of Food. CAC/RCP 19–1979, Rev. 1–2003; Codex Alimentarius Commission: Rome, 2003b.

Roberts, P. B. Food Irradiation is Safe: Half a Century of Studies. *Radiat. Phys. Chem.* **2014,** *105,* 78–82.

Santos, J. P.; Maia, J. D. G.; Cruz, I. Damage to Germination of Seed Corn Caused by Maize Weevil (*Sitophilus zeamais*) and Angoumois Grain Moth (*Sitotrogacerealella*). *Pesq. Agropec. Bras.* **1990,** *25,* 1687–1692.

Sarkar, S.; Gupta, S.; Variyar, P. S.; Sharma, A.; Singhal, R. S. Irradiation Depolymerized Guar Gum as Partial Replacement of Gum Arabic for Microencapsulation of Mint Oil. *Carbohyd. Polym.* **2012,** *90,* 1685–1694.

Saurabh, C. K.; Gupta, S.; Bahadur, J.; Mazumder, S.; Variyar, P. S.; Sharma, A. Radiation Dose Dependent Change in Physiochemical, Mechanical and Barrier Properties of Guar Gum Based Films. *Carbohyd. Polym.* **2013,** *98,* 1610–1617.

Schwartz, B. Effects of X-rays on Trichina. *Jour. Exp. Zool.* **1917,** *22,* 575–576.

The Role of Food Safety in Health and Development. World Health Organization Technical Report Series No. 705: Geneva, 1984.

Thomas, P.; Srirangarajan, A. N.; Limaye, S. P. Studies on Sprout Inhibition of Onions by Gamma Irradiation—I. Influence of Time Interval between Harvest and Irradiation, Radiation Dose and Environmental Conditions on Sprouting. *Radiat. Bot.* **1975,** *15,* 215–222.

Todd, E. C. Epidemiology of Foodborne Diseases: A Worldwide Review. *World Health Statist. Q.* **1996,** *50* (1–2), 30–50.

Tripathi, J.; Chatterjee, S.; Vaishnav, J.; Variyar, P. S.; Sharma, A. Gamma Irradiation Increases Storability and Shelf Life of Minimally Processed Ready-to-cook (RTC) Ash Gourd (*Benincasahispida*) Cubes. *Postharvest Biol. Technol.* **2013,** *76,* 17–25.

Tripathi, J.; Gupta, S.; Mishra, P. K.; Variyar, P. S.; Sharma, A. Optimization of Radiation Dose and Quality Parameters for Development of Ready-to-cook (RTC) Pumpkin Cubes Using a Statistical Approach. *Innov. Food Sci. Emerg. Technol.* **2014,** *26,* 248–256.

Vaishnav, J.; Adiani, V.; Variyar, P. S. Radiation Processing for Enhancing Shelf Life and Quality Characteristics of Minimally Processed Ready-to-cook (RTC) Cauliflower (*Brassica oleracea*). *Food Packag. Shelf Life* **2015,** *5,* 50–55.

Variyar, P. S.; Limaye, A.; Sharma, A. Radiation-induced Enhancement of Antioxidant Contents of Soybean (Glycine max Merrill). *J. Agric. Food Chem.* **2004,** *52,* 3385–3388.

WHO. *Estimates of the Global Burden of Food Borne Diseases. Foodborne Disease Burden Epidemiology Reference Group 2007–2015*; World Health Organization: Geneva, 2015.

Wholesomeness of Irradiated Food. World Health Organization Technical Report Series 659; World Health Organization: Geneva, 1981.

World Population Prospects. The 2015; Department of Economic and Social Affairs, Population Division; United Nations, 2015. http://esa.un.org/unpd/wpp/

CHAPTER 13

FOODS PRESERVED WITH HURDLE TECHNOLOGY

ANURAG SINGH*, ASHUTOSH UPADHYAY, and ANKUR OJHA

Department of Food Science and Technology, National Institute of Food Technology, Entrepreneurship and Management, Sonepat, India

Corresponding author. E-mail: anurag.niftem@gmail.com

CONTENTS

Abstract 352

13.1   Introduction 352

13.2   Food Products Preserved with HT 355

13.3   Conclusion 370

Keywords 371

References 371

ABSTRACT

Food preservation by any single preservation technique requires the use of extreme conditions of that particular technique. This causes an adverse effect on food quality. If more than one preservation parameters (hurdles) are used in combination, the good quality food product can be obtained without compromising the safety. The product preserved with Hurdle Technology is more natural and fresh like. Not only the conventional methods of preservation like thermal treatment, low temperature, drying, chemical preservatives, etc. but also the novel preservation techniques like biopreservation, bacteriocins, HHP, and PEF can also be used in combination. Many food products have been successfully preserved by various researchers using the concept of Hurdle Technology. In this chapter, different food products those have been preserved using this concept are discussed.

13.1   INTRODUCTION

13.1.1   HURDLE TECHNOLOGY

"Hurdle technology" (HT) or "Combination technology" uses two or more preservation parameters in combination at an optimum level so that a maximum lethality can be obtained against microorganisms without compromising with the nutritional and sensory qualities of food products. As the extreme of any single preservative parameter is not being used, the damage to the quality parameters of the food is at minimum level. A wide range of preservation techniques are available, for example, freezing, blanching, pasteurizing, and canning but the spoilage and poisoning of foods by microorganisms is a problem that is not yet under adequate control. Moreover, today's consumer demands for more natural and fresh-like foods, that requires the use of only mild preservation techniques. Hence, for the benefit of food manufacturers and to fulfill the requirement of the consumer, we need an improved mild preservation technology that gives a fresh-like, but stable and safe food (Leistner & Gorris, 1995).

The concept of HT is not new. From ancient time various traditional products are being preserved using this technology with inherent empirical hurdles. The knowledge regarding the principles behind the technology was not known hence it was being used more empirically. The focus was to produce a safe, stable, nutritious, tasty, and economical food. Various terms are used for HT, such as combined methods, combined processes, combination preservation, combination techniques, or barrier technology. In this technology, it is suggested that if a combination of existing and novel preservation techniques are used, a series of preservative factors (hurdles) can be created that any microorganism present should not be able to overcome. This hurdle effect was first highlighted by Leistner (1978).

The effect of combination on stability and safety of foods was known from many centuries but the concept has not been used consciously on industrial basis. Today we are well aware of the major preservative factors such as temperature, pH, water activity (a_w), redox potential (Eh), competitive flora, and their interactive effect, hence the intelligent use of these hurdles can be seen. Intelligent use of the combination of these hurdles in food product assures an extended shelf-life and safety of foods (Leistner, 1994).

Hurdle concept has various advantages such as (1) it avoids the severe use of one hurdle for preservation, (2) it gives the synergistic effect of combination, and (3) our past experiences (i.e., tradition or culture) derive many of the hurdles.

The well understood hurdle concept is now ready to be applied for use in a wide range of food products including fruits and vegetables, bakery products, dairy products, fish, and so on with many novel preservative factors such as gas packaging, bio-conservation, bacteriocins, ultrahigh pressure treatment, etc. HT is a crucial mild preservation concept of foods that controls microbial spoilage and food poisoning, leaving desired-fermentation processes unaffected (Chirife et al., 1991; El-Khateib et al., 1987; Gould, 1988).

13.1.2 TYPES OF HURDLES

There is a huge range of the hurdles available those could be used for food preservation. More than 60 potential hurdles are known for this purpose. The most important hurdles are: (1) high temperature (heating/cooking/pasteurization/sterilization); (2) low temperature (chilling/freezing); (3) lowering the a_w; (4) acidity (pH); (5) Eh; (6) the use of chemical preservatives (e.g., nitrite, sorbate, and sulfite); and (7) the use of microorganisms those suppress the growth of harmful microbes (competitive microorganisms), such as lactic acid bacteria (Leistner, 2007). Recently several novel preservative techniques (e.g., gas packaging, biopreservation, bacteriocins, ultrahigh pressure treatment, edible coatings, etc.), have gained popularity to be used in combination with other traditional preservative factors (hurdles).

Depending on the intensity of the hurdle, both positive and negative effect on foods can be observed. For example, use of low temperature (chilling) below the critical limit of any food can lead to "chilling injury" whereas moderate chilling extends the shelf-life of the food as it retards microbial growth. Similarly, lowering the pH in fermented sausage inhibits the growth of pathogenic bacteria but lowering beyond the required limit can also impair the taste. Therefore, a balanced intensity of any hurdle should be used for food preservation. Too small intensity of hurdle should be strengthened and if it is affecting the food quality adversely, it should be lowered. This adjustment will provide a food that would be safe as well as will have a good quality by the use of hurdles in optimal range (Leistner, 1994).

13.1.3   MECHANISM OF FOOD PRESERVATION BY HT

The whole mechanism of preservation of food by using the concept of HT is comprised of various responses those are given by any microorganism. The whole phenomenon can be understood by following:

Homeostasis

Tendency of any organism to maintain its internal status is known as "Homeostasis." The homeostasis of microorganisms plays a key role in food preservation. If any of the hurdles used in food disturbs the homeostasis of microorganisms, they will not be able to multiply and will remain constant in number or will die before the re-establishment of homeostasis. Hence, a temporary or permanent homeostasis can lead to food preservation (Leistner, 2000).

Metabolic exhaustion

Another important phenomenon for food preservation is "Metabolic Exhaustion" of microorganisms. As a response to the hurdles applied to foods, microorganisms try to repair their homeostasis, use up all their energy for this, and become metabolically exhausted. This leads to an auto-sterilization of such foods. It means, if a food is preserved with concept of hurdle-technology and is microbiologically stable it will become safer during ambient storage condition. The microbes can respond better to the hurdles at ambient temperature than at refrigeration and become metabolically exhausted (Leistner, 2000).

Stress reactions

Some bacteria generate stress shock proteins when they are starved or exposed to various hurdles. These stress shock proteins help microorganisms to cope with the stresses. These stress proteins may affect the preservation, if only one hurdle has been applied. On the other hand, if we use more than one hurdle, different stresses are received by the microorganism at the same time and the activation of genes for the synthesis of several stress shock proteins becomes difficult. A huge amount of energy is consumed to synthesize a number of stress shock proteins due to simultaneous exposure to different stresses and it leads to the metabolic exhaustion of the microorganism (Leistner, 2000).

Multi-target preservation

If different hurdles are used simultaneously, a combined effect could be achieved as various changes within the microbial cell (e.g., cell membrane, DNA, enzyme systems, pH, a_w, and Eh) can be targeted. These changes lead to the disturbed homeostasis of the microorganisms present. In this case, the replenishment of homeostasis and activation of stress shock proteins becomes more difficult. Therefore, simultaneous application of different hurdles in a particular food would lead to optimal microbial stability (Leistner, 2000).

13.2 FOOD PRODUCTS PRESERVED WITH HT

Currently, a number of food products are being developed using multi hurdle concept. As there was no scientific data available in ancient time for the use of these hurdles, the empirical experimentation is being done to optimize the combination and the levels of these hurdles for various foods. It is a challenge for the food technologists/scientists to develop a stable and safe food using multi-hurdle approach. Some of the food products preserved with hurdle concept are discussed in this chapter.

13.2.1 FRUIT AND VEGETABLE PRODUCTS

Blanching coupled with the chemical dip was found effective for the extension of shelf-life of pineapple slices. Alzamora et al. (1993) blanched the slices in saturated vapor for 2 min, cooled the slices in water, and then immersed in glucose syrup containing 150 mg/kg sodium bisulfite and 1000 mg/kg potassium sorbate. The pineapple slices thus preserved were found acceptable up to 120 days of storage. The similar effect was observed for papaya slices by Lopez et al. (1995). Blanching was done for 3 min in saturated vapor, using a concentrated syrup, the a_w was reduced to 0.98 by osmosis, pH was adjusted using citric acid and potassium sorbate and sodium sulfite were added. These slices were found acceptable even after 3 months when stored at 25°C.

Banana puree was preserved for 120 days using the hurdle concept by Guerrero et al. (1994). For this, a_w, pH, heat treatment, and chemical preservatives were used as hurdles. The a_w of the puree was adjusted to 0.97and pH was maintained at 3.4 using 250 ppm of ascorbic acid. Mild heat treatment was applied along with 100 ppm of potassium sorbate and 400 ppm of sodium bisulfate.

Lowering the a_w and pH along with the in pack pasteurization preserved the mango and pineapple pieces. Jayaraman et al. (1998) decreased the a_w using concentrated sucrose syrup, pH was decreased using citric acid and in pack pasteurization was done. Both the fruit pieces were acceptable for more than 6 months at ambient temperature.

Pineapple, mango, and papaya chunks were preserved using a combination of hurdles (pH, mild heat treatment, preservatives, and packaging material) by Vijayanand et al. (2001). The pineapple and mango chunks were blanched in a syrup at 85°C for 5 min. After blanching, the chunks were dipped in a syrup containing 340 mg of potassium metabisulfite/kg and 413 mg sodium benzoate/kg for 8 h. The chunks were then packed in 150 gauge polypropylene pouches. These chunks were found acceptable in terms of sensory and microbiological quality up to a period of 30 days at 27°C and 60 days at 2°C, whereas papaya chunks were treated in the same way but with the increased levels of preservatives. Overall, 680 mg potassium metabisulfite/kg and 826 mg sodium benzoate/kg was used for dip. These chunks were found stable up to 90 days at 2°C and ambient temperature.

The effect of new technology of pulsed electric fields (PEF) coupled with moderate heating and antimicrobial agents was studied by Ziwei et al. (2003) to extend the shelf-life of freshly-squeezed, unclarified apple juice. The microbial count was decreased when the pulse number and the temperature wer increased at constant field strength. There was no added reduction in microbial count observed when nisin (100 U/mL apple juice) or a mixture of nisin and lysozyme (27.5 U nisin/mL and 690 U lysozyme/mL) were used with PEF with heating up to 46°C. Increase in temperature from 44°C to 52°C in the absence of PEF was not significant. The reduction in vitamin C levels in apple juice samples was not seen as a result of these treatments.

High hydrostatic pressure combined with natural additives was applied for preservation of tomato purees by Plaza et al. (2003). The changes in various quality parameters, namely, color, viscosity, enzyme activity, total protein content, total microbial counts, and counts of yeasts and other fungi were analyzed. The experiments were conducted at constant temperature and holding time of high-pressure treatments as 25°C and 15 min, respectively. Response surface methodology was applied to optimize the results. Combined treatments at high values for pressure and additive concentration significantly improved the enzymatic inactivation. The total microbial count was significantly reduced by four logarithmic units at a pressure 400 MPa.

The combined effect of reduction in a_w, reduction in pH, heating, and use of chemical preservative was studied for assai fruit (*Euterpe oleracea*; acai; a South American palm) pulp by Alexandre et al. (2004). An aliquot of 0.5% citric acid was used to lower the pH of pulp to 3.5, whereas sucrose solutions of varying concentration were used for a_w reduction. The pulp was heated at 82.5°C for 1 min and potassium sorbate was added in varying concentration. All the experimental samples were stored at 25°C in the absence of light for 5 months. The pulp samples treated with 25% and 40% sucrose and 0.75% and 0.15% sorbate, respectively, were found stable for 5 months with good overall acceptability.

Lee (2004) resulted that heat, acetic acid, and salt are the factors those contribute toward the preservation of pickled fruits and vegetables. Minimally processed coconut gratings were preserved up to 4 weeks when treated with a combination of acidulants, humectants, chemical preservatives, antioxidants, a mild heat treatment, and use of good packaging material (Gamlath et al. 2004). The changes in physicochemical and sensory quality of coconut gratings were evaluated. Coconut gratings packed in laminated polythene packages and added with 3% NaCl, 0.3% citric acid, 0.009% sodium citrate, and 0.02% butylated hydroxyanisole were found stable for 4 weeks at 5°C.

Soliva-Fortuny et al. (2004) preserved avocado puree using the combined effect of various hurdles. Storage temperature, pH, a_w, modified atmosphere, and the chemical preservatives were selected as the hurdles for the study. The samples were stored for 4 months at refrigerated condition as well as at ambient temperature. Sorbic acid influenced microbiological stability most significantly during storage. An aliquot of 300 mg/kg sorbic acid can control growth of yeasts and fungi in the puree. Purees could also be preserved for 4 months at 4°C after vacuum packaging without the addition of sorbic acid. Addition of maltose to reduce a_w gave better results for stability of the product.

Knol khol (kohlrabi), a spice-based high moisture product, and cauliflower were stabilized by using a combination preservation technique by Vibhakara et al. (2005). The hurdles used were blanching in 4% brine solution with other additives like citric acid (acidulant) and potassium metabisulfite (anti browning agent), treatment with spices, packaging in polypropylene pouches and irradiation. Final product packaged in polypropylene pouches was irradiated @ 3, 5, and 8 kGy doses. The products were found microbiologically stable during storage up to 4 months at ambient temperature. At higher doses, it was observed that the browning in products got intensified and the products became softer after 4 months. The microbial count of the product decreased from an initial value of 10^4–10^1 cfu/g whereas the *Aspergillus flavus* was absent after the irradiation. Knol khol and cauliflower products were successfully preserved for 4 months at ambient temp when irradiated at 5 kGy dose.

Use of blanching coupled with other additives was done for preservation of cauliflowers by Barwal et al. (2005). Salt, potassium metabisulfite, and citric acid were used as additives. The changes in physicochemical, sensory properties and microbiological qualities were studied during storage. The cauliflower samples dipped in salt solution (10% and 15%) containing 0.2% KMS were found acceptable during storage.

Horchata (a traditional low-acid, vegetable beverage) was preserved using PFF treatment along with varying inoculums size and substrate conditions by Selma et al. (2006). The combined effect of these hurdles was observed on the inactivation and recovery of *Listeria monocytogenes* in horchata. The recovery of sub-lethally injured cells was slowed down due to the combined effect of PEF, low temperature (5°C), and low inoculum level but at 12°C or 16°C, this elongation of the lag phases was not achieved. Therefore, for prevention of *L. monocytogenes* development in low-acid products, PEF should be combined with low refrigeration temperature.

Vibhakara et al. (2006) preserved grated carrots using the combination of drying, decrease in pH, decrease in a_w and use of chemical preservatives. Overall, 3.0% sodium chloride was used to reduce a_w to 0.94, pH was reduced below 4.5 using a 0.9% citric acid solution, and antimicrobial agents (1.0% sodium citrate and 0.2% sodium benzoate) were used. The grated carrot, infused with these additives, was partially dehydrated. The partially dehydrated grated carrots (moisture 66.2%) were found acceptable and microbiologically safe for more than six months at ambient temperature and the good retention of carotenoid (up to 82.5%) was observed.

The fungal spoilage of naturally fermented black olives was prevented by using a combination of potassium sorbate, low pH, and NaCl by Yigit and

Korikluoglu (2007). They used different concentration of potassium sorbate (100–1000 mg/L) and NaCl (0%, 3.5%, 5%, 7.5%, and 10%) at different pH values (4.5, 5, 5.5, 6, and 6.5) in the study. The effects of these combinations were studied for common spoilage fungi (*Alternaria alternata*, *Aspergillus niger*, *Fusarium semitectum*, and *Penicillium roqueforti*). All the fungi showed various response to hurdles. *A. alternata* was found to be the most sensitive fungus whereas *P. roqueforti* was the most resistant to all hurdle factors. *A. alternata* and *F. semitectum* were completely inhibited at the lowest inhibitory levels of hurdles (100 mg/L potassium sorbate, 3.5% NaCl, and pH 5). Similarly, a combination of 300 mg/L potassium sorbate with 10% NaCl and 400 mg/L potassium sorbate with 7.5% NaCl were effective against *A. niger* and *P. roqueforti*, respectively, for complete inhibition at pH 5. It was concluded by the scientists that potassium sorbate and NaCl can be suitably used for the inhibition of fungi in fermented olives.

Freezing process was replaced with HT for preservation of mango pulp by Moraes (2010). The a_w of the pulp was lowered down using a glucose syrup and sucrose (25% of pulp weight), pH of pulp was adjusted to 3% and 0.1% potassium sorbate was added as a preservative. The changes in the physicochemical and microbiological quality of hurdle-preserved sample were compared with the frozen pulp sample. Hurdle preserved samples were found better with minimal change in the quality parameters.

Benhura (2012) compared the changes in mango pulp preserved with chemical preservatives and untreated pulp. Sodium metabisulfite, potassium sorbate, and citric acid addition followed by pasteurization at 80–90°C was done in treated samples. One sample was kept under control (untreated). The changes in total soluble solids (TSSs), pH, color, and odor were recorded for 21 days. The changes in treated samples were found minimal as compared to untreated samples. Unpasteurized pulp retained the yellow color while heat-treated samples turned yellowish brown. Mold colonies and stale odors were detected in control samples. Un-pasteurized samples had a pungent odor. Pasteurized samples had a cooked mango odor which was not astringent. Sodium metabisulfite, citric acid, potassium sorbate, and pasteurization increased the shelf-life and stabilized the sensory properties of the pulp.

To ensure the continuous supply of seasonal jackfruit, HT was applied in pre-cut form of tender jackfruit (Pallan, 2012). In-pack sterilization, use of food additives, mild heat treatment, and gamma irradiation were applied as hurdles. The treated samples were then stored at room and sub-room temperature for further studies. The results show that the product can remain excellent for over one year even after the next harvest of the tender fruit.

Shelf-stable intermediate moisture (IM) carrot shreds were developed by Chaturvedi (2013) using "HT." Combination of the factors like drying by two methods—infrared drying (IR)/tray drying (TD) to reduce a_w to 0.6, pre-treatment and packaging was used as hurdles. The product was stored in 400 gauge polyethene and treated with low doses of gamma radiation (0.5 kGy) as a major hurdle and observed for shelf-life stability at ambient conditions (30°C and 65% RH). Vegetables dried by infrared and treated with gamma radiation were found to be stable up to 6 months without substantial loss of flavor, taste, color, and texture than the other treatments. Better rehydration potential, appearance, and nutrient retention were observed in carrot shreds dried by IR as compared to tray dried IM carrot shreds. The product was microbiologically safe throughout the study.

Carrot was preserved by Sinha et al. (2013) also using hurdle treatment. Fresh carrot was blanched at 100°C for 60 s, followed by dipping into 0.25% potassium metabisulfite for 10 min. Different concentrations and combinations of preservatives were used to dip the steeped and blanched carrots, namely, P_0 (control sample-fresh without treatment), P_1 (35°Brix syrup + 8% salt + 500 ppm potassium metabisulfite), P_2 (35°Brix syrup + 10% salt + 400 ppm potassium metabisulfite), P_3 (35°Brix syrup + 12% salt + 300 ppm potassium metabisulfite), P_4 (25°Brix syrup + 8% salt + 500 ppm potassium metabisulfite + 100 ppm sodium benzoate), P_5 (25°Brix syrup + 10% salt + 400 ppm potassium metabisulfite + 200 ppm sodium benzoate), and P_6 (25°Brix syrup + 12% salt + 400 ppm potassium metabisulfite + 300 ppm sodium benzoate). Aseptic packaged carrot was stored at two different temperatures T_1 (30–37°C) and T_2 (5–7°C) up to 180 days. Three treatments P_6/T_1, P_6/T_2, and P_5/T_2 were found microbial safe for 180 days of storage period. Among these three, treatment P_6/T_2 was scored lowest in physical and highest in sensory and nutritional evaluation. So best hurdle treatment for preservation of carrot till 180 days of storage period was (25°Brix syrup + 12% salt + 400 ppm potassium metabisulfite + 300 ppm sodium benzoate) at 5–7°C.

Aloe vera gel coating and calcium chloride spray, separately and in the combination (HT), were applied by Chauhan (2014) to increase the shelf-life of mango stored at 15°C ± 1°C and 85% RH. 60 days shelf-life was recorded with aloe vera gel and calcium chloride, separately but in combination of these two, 90 days shelf-life period was noticed. The changes in fruit firmness, weight loss, skin color, microbial counts, TSSs, and total titratable acidity (TTA) were evaluated. Application of 1%, 5%, and 10% aloe vera gel showed significant inhibition of the total microbial count; however, aloe vera gel was analyzed to be inhibitory in action when applied in combination of calcium chloride.

Sujatha et al. (2014a) developed shelf-stable IM pineapple (*Ananas comosus*) using gamma irradiation. The most important hurdles used were reduction of a_w by osmosis, IR, 400 gauge polyethylene bags along with mild dose of irradiation (R) (1 kGy). The shelf-life of IM pineapple subjected to infrared drying and radiation (IRR) and non-radiated IR was evaluated at ambient ($34 \pm 2°C$ and 65% RH) temperature. The sample subjected to IRR was found stable even after 4 months whereas only IR sample spoiled within 3 months at ambient temperature. No significant ($p > 0.05$) changes in a_w, moisture, and reducing sugars were observed in IRR. Total sugars, TSSs, and acidity increased in both the treatments significantly ($p > 0.05$). Ascorbic acid decreased significantly as storage period increased. IRR treated sample retained vitamin "C" up to 65.3%. The combination of hurdles including osmotic dehydration, IR, and gamma radiation dose of 1 kGy, successfully reduced the microbial load and showed high product sensory, microbial nutritional quality, and storability.

Combined effect of blanching, osmotic dehydration, and preservatives on the quality and stability of minimally processed papaya was observed by Ankita (2014). Different sucrose solution (50, 55, 60, 65, and 70% w/v) was employed at mild temperature (25°C) for osmotic dehydration of papaya. The changes in physicochemical characteristics (TSS, a_w, and ascorbic acid), microbial quality, and sensory quality characteristics were measured at regular intervals throughout the storage period. A decrease in a_w, ascorbic acid (g/kg), firmness (N), color (L^* value), and sensory characteristics were observed during the 30 days storage period. The papaya slices were found acceptable up to 30 days under refrigeration.

Winter melon fruit was processed into puree (WMP) using HT treatment and compared with traditional thermal treatment (TT) to see the effect on ascorbic acid content (AAC) and total phenolic content (TPC) stored at different temperatures for 6 months of storage (Abdullah et al., 2012). HT WMP was adjusted to pH 3 using citric acid while for TT WMP was prepared by heating at 84.5°C for 94 s. Both treated WMP were stored at three different temperatures (25°C , 5°C , and −20°C) and were analyzed at every 1-month interval for 6 months. A gradual decrease in AAC and TPC were observed during storage for both treatments. These changes were more pronounced at 25°C storage for both treatments; however, TT showed high reduction of AAC and TPC throughout 6 months of storage as compared to HT. Overall, WMP for HT stored at 5°C and −20°C shows lowest reduction for AAC and TPC for the 6 months of storage.

Shelf-stable "intermediate moisture" cabbage and cauliflower were developed by Sujatha et al. (2014b). HT was applied using a mild heat treatment, addition of 1% potassium meta-bi-sulfite as antimicrobial agent, partial dehydration to lower a$_w$ using two methods—IR and TD. The samples were packed in 400 gauge polyethylene and treated with low doses of gamma radiation as major hurdle. The shelf-life (physical, chemical, and pathological) stability of the samples was observed at ambient conditions (30°C and 65% RH). Improved rehydration potential, appearance, and maximum nutrient retention up to 43.1–44.6% of vitamin C and maximum shelf-life of 5–7 months was observed in IR vegetables treated with gamma radiation at 0.75–1.0 kGy. No significant changes were observed for color, taste, flavor, texture, and overall acceptability during storage period. The microbial growth was controlled throughout the study resulting in shelf-stable IM vegetables. Among the four treatments studied, IR with radiation dose of 0.75 kGy for cauliflower and 1.0 kGy for cabbage was found to be best in obtaining high-quality IM products with optimum sensory, microbial, nutritional quality, and storability.

Sankhla et al. (2012) and Rawat and Pokhriyal (2014) used thermal treatment, chemical preservatives, packaging, and irradiation as a combination of hurdles for preservation of sugarcane juice. Different time–temperature combination for thermal treatment (80°C for 10 min and 80°C for 20 min) and chemical treatments of 150 ppm KMS and 0.05% citric acid were tried in various combinations. Glass bottles, polyethylene *terephthalate* (PET) bottles, and low-density polyethylene pouches (LDPE) were used for packing the samples. Irradiation of packed samples was done at 0.25, 0.5, and 1.0 kGy dose. The changes in quality parameters of the stored samples were analyzed for 90 days. Microbial count decreased significantly during storage. Organoleptic properties of the juice after irradiation and packaging material did not change statistically but changes in sensory scores were shown during storage. The response of glass and PET was better than the LDPE pouches to extend the shelf-life of the juice. The sample pasteurized at 80°C for 10 min + chemical treatments (KMS@ 150 ppm and citric acid @ 0.05%) + sterilization at 80°C for 20 min and irradiated with 1.0 kGy dose was found acceptable for 60 days when stored at room temperature and at low temperature, it was acceptable up to 90 days.

13.2.2   MEAT, POULTRY, AND FISH PRODUCTS

Aguilera et al. (1992) increased the shelf-life of minced pelagic fish from 3 to 15 days at 15°C by the addition of 6% salt and 0.2% sorbate at pH 5.7

along with the heat treatment (10 min, 80°C). Ready-to-serve, shelf-stable, and microbiologically IM spiced mutton and spiced chicken products were developed by Kanatt et al. (2002). Drying and grilling type applications were done to reduce the a_w of products to 0.80. The products were vacuum packaged and subjected to radiation processing at 0–10 kGy. The irradiated products showed reduced total viable counts and number of *Staphylococcus* species, whereas samples without treatment showed visible fungal growth within 2 months. The product was found acceptable with good sensory qualities up to 9 months at ambient temperature when irradiated at 10 kGy.

Santos et al. (2002) found that the HT was able to develop sausages stable at room temperature. The fine paste was made by adding small meat pieces and other raw materials like pork, fat, soybean derivatives, wheat flour, starch, water, salts, seasonings with the a_w of 0.95 and then filled in artificial impermeable casing. The heating of sausages was carried out at 75°C or 80°C till the temperature at the center of the product is reached the same and then it was maintained at room temperature for up to 3 months. The sausages were analyzed on the basis of physicochemical, microbiological, and sensory parameters. The use of thermal treatment at 80°C in combination with use of the artificial casings yielded a product with a shelf-life of 3 months. The sensory properties, except color, had no changes after storage.

Color stability of ground beef trimmings was studied by Jimenez-Villarreal et al. (2003) after applying various hurdles. The trimmings are treated with different antimicrobials such as 0.5% cetylpyridinium chloride, 10% trisodium phosphate (CT), 200 ppm chlorine dioxide and 2% lactic acid in different combinations. Trimmings were minced and packaged. The change in color of packed trimmings was observed during simulated retail display for 0–7 days at 2°C. According to the results of the study, the antimicrobial treatments had a bad effect on the color of beef mince during retail display.

The chicken patties and fry were preserved by Narahari (2005) for 12 weeks at ambient temperature with the application of multi-HT. The hurdles used were use of healthy chicken, hygienic processing, and handling as per HACCP procedures, washing the carcass in chlorine dioxide and organic acids and mixing with spices, herbs, and GRAS preservatives. Microwave along with grill, hot air oven along with grill or deep-frying in oil were the three different cooking methods applied. The products were either vacuum or modified atmosphere packed (MAP) after cooking; followed by storage at room temperature up to 12 weeks as it is or after gamma irradiation at 3 kGy units. The sensory evaluation, thiobarbituric acid (TBA), and tyrosine values (TVs) were carried out initially and also at biweekly intervals until 12 weeks. The chicken patties and fry prepared using all the combinations were acceptable up to 8 weeks. The shelf-life is retained for the 12 weeks of the samples prepared using MAP, irradiation, and microwave-grill combo methods after 8 weeks. The total bacterial, *Coliform, Salmonella,* and fungal loads were observed of the samples of the products which were found within the safety limits even up to 12 weeks storage. The preservation of the ready-to-eat (RTE) poultry patties and fry at room temperature up to 12 weeks by a combination of common hurdles up to mixing with herbs followed by Microwave-Grill combo (MW+G), MAP, and gamma irradiation was found to be an ideal method.

The HT was also applied in standardization of RTE chicken curry by Rathod et al. (2005). Boneless chicken pieces were maintained by using a mixture of 2% common salt, 0.2% citric acid, 0.05% potassium sorbate, 0.002% butylated hydroxyanisole, and 2.5% water (fresh meat weight basis) for 20 h at 4°C ± 1°C. The dehydration of pieces was done in a cross flow cabinet drier after steam cooking. The evaluation of the processed product for physicochemical, microbiological, and sensorial attributes was done after storing at 37°C ± 2°C in polyethylene pouches. There was a marginal increase in pH (5.57–5.60) and decrease in moisture (38.38–38.25%), protein (29.35–29.06%), and fat contents (27.97–27.77%) during storage up to 6 days. The peroxide value gradually increased from 5.17 to 42.54 meq O_2/kg, FFA values increased from 0.94% to 3.33% (oleic acid) and TV from 20.76 to 79.68 mg/dg and also TPC was slightly increased (3.31–3.41 $\log_{10}$ cfu/g). During storage, there was also an increase in yeast and mold counts gradually from 1.38 to 2.20 log cfu/g. The product was accepted and there was no marked quality deterioration during the storage period of 6 days.

The effect generated by the HT to extend the shelf-life of fresh sausages was investigated by Cepero et al. (2006). Two samples were stored in hung in bar and vacuum packaged conditions at a temperature of 4–8°C at 93% ± 5% of RH. The analysis of physicochemical, sensory, and microbiological quality was done during the storage. According to results, the shelf-life of the Cuban fresh sausage was increased from 4 to 32 days as the vacuum packaging decreased the oxidation–reduction potential with a combined effect of potassium sorbate as a preservative and low temperature (4–8°C) during storage. The hurdles like packaging, gamma irradiation, and reduced a_w were applied in the preparation of safe and shelf-stable natural casing from lamb intestines by Chawla et al. (2006). The a_w of casings was reduced to 0.80 ± 0.02 by common salt and gamma-irradiation @ 5 and 10 kGy was applied. Polyethylene bags were used for packaging. The controlled/non-irradiated samples showed high total viable counts (10^6 CFU/g) and various microbes such as aerobic spores (10^3 CFU/g), spores of *Clostridia* (10^3 CFU/g), *Staphylococci*

(10^4 CFU/g), and *coliforms* (10^2 CFU/g). A dose of 5 kGy reduced *Staphylococci* and *coliforms* total viable counts by three log cycles and spore counts by two log cycles. The use of 10 kGy dose was found devoid of any viable microbes. The growth of the microbes in natural casings was prevented by low a_w of the product during storage at room temperature. The radiation processing did not affect the sausages acceptability and mechanical strength.

Kanatt et al. (2006) developed a shelf-stable, RTE shrimps with the help of hurdles. The reduced a_w (0.85 ± 0.02), packaging, and gamma irradiation (2.5 kGy) were used as hurdles to cooked marinated shrimps. As the irradiation dose increases the total viable count changes along with the change in *Staphylococcus* species. There was mold growth observed within 15 days of storage at ambient temperature ($25°C \pm 3°C$) in non-irradiated samples whereas irradiated products showed very less change in textural properties and sensory qualities. The RTE shrimps thus prepared were microbially safe and organoleptically acceptable when stored at room temperature up to 2 months.

The effect of HT for inactivation of *Escherichia coli* O157:H7 (EHEC) in ground beef was studied by Podolak et al. (2006). Use of packaging, thermal treatment, and hydrodynamic pressure processing (HDP) was applied on the product as hurdles. The ground beef was inoculated with a six strain cocktail of *E. coli* O157:H7 (3027–93, 3055–93, C7927, 43888, C9490, and green fluorescent protein-expressing *E. coli* O157:H7 B6-914) at three different concentrations (10^3, 10^4, and 10^6 cfu/g). The inoculated ground beef samples were wrapped in polyethylene wrap, vacuum packaged into multilayer barrier bags, heat shrunk and treated with HDP. The EHEC has the initial concentrations of 1.29×10^3, 2.88×10^4, and 2.19×10^6 cfu/g. The reduction in EHEC populations was ($p < 0.05$) to 9.12×10^2, 2.40×10^4, and 1.91×10^6 cfu/g, respectively, after HDP treatment. The use of these hurdles was found effective against the microbes.

The assessment of HT to prevent microbial growth in meat products was done by Roedel and Scheuer (2007). The investigation of effects of various combinations of temperature (17–27°C in 1°C steps), a_w (0.990, 0.985, and 0.980) and $NaNO_2$ (0, 200, or 400 mg/L) on growth of *E. coli* was observed at an initial inoculums level of 1×10^3 or 1×10^6 cfu/mL. These three hurdles affected the growth in a very complex interaction. A variety of pattern was exhibited by the microbial growth. The hypothesis of the effect of a common additive between different hurdle factors was correct in no case. There is a consideration for these results in relation to predictive microbiology.

The preparation of shelf-stable buffalo meat chunks using an infusion solution containing humectants such as glycerol 6.0%, sodium chloride

6.0%, and propylene glycerol 1.0% along with sodium nitrite 0.01% and sorbic acid 0.2% was carried out by Malik and Sharma (2011). The vacuum packaging of chunks was done in a multilayered nylon barrier laminates and stored at 30°C ± 3°C for 6 weeks. There was a significant effect on the a_w during storage whereas the moisture content and the shear force value were not significantly affected. There was a decrease in the pH, residual nitrite, total hem pigments, and protein solubility while there was an increase in the TBA, free fatty acids (FFAs), and soluble hydroxyproline. Overall mean TBA value for storage was 0.98 mg malonaldehyde/kg. Eventually, the microbial counts of the product were TPC log 4.18, *Staphylococcus aureus* log 3.50, yeasts and molds log 1.89 and anaerobic counts log 2.61 and sensory rating was maintained between good to very good. The shelf-life of meat chunks remained six weeks at temperature (30°C ± 3°C).

Naga Mallika and Prabhakar (2011) applied vacuum packaging and heat treatment to preserve the pork sausage samples. There was a significantly ($p < 0.05$) better sensory scores and lower microbial counts along with low-fat oxidation and protein denaturation in vacuum packaged samples. There was a partial reduction of microbial counts by application of heat treatment; however, the sensory quality decreased along with enhancement of fat oxidation and protein denaturation. At the end of the storage period of sausages, TBA reactive substance values, TVs, free amino acid values, FFA values, and plate counts were lowered by the combination of hurdles significantly.

Anestis and Labropoulos (2014) suggested that HT can produce salami-type fermented sausages, stable at ambient temperature for extended periods. The microbial growth was affected by the use of a combination of hurdles which is important in different stages of the ripening process. The two important hurdles include salt and nitrite in the stage of the ripening process of salami, which inhibit many of the bacteria present in the meat batter. At the same time, other bacteria multiply by using oxygen and thereby causing a drop in Eh, which inhibits aerobic organisms and favors the selection of lactic acid bacteria. Acidification of the product is caused by the lactic acid bacteria by flourishing and reducing the pH. The numbers of lactic acid bacteria are decreased along with the depletion of nitrate content during long ripening by increasing Eh and pH. The main hurdle is the decrease in a_w in long-ripened sausage.

13.2.3 DAIRY PRODUCTS

The shelf-life of paneer, a very perishable dairy product, can be extended by HT. The freshness of paneer remains intact only for a day at room temperature and 3 days if refrigerated (Bhattacharya et al., 1971). If a combination of mild heat treatment, minor reduction in a_w (0.95), and acidification (pH 5.0) is used, the shelf life of paneer can be extended up to 14 days at 30°C. The shelf-life of paneer was extended by reducing the a_w using 1% each of sodium chloride, sucrose, and glycerol by Rao and Patil (1992).

Use of brine and hydrogen peroxide solution (0.2% v/v) was done by Sachdeva and Singh (1990) to extend the shelf-life of paneer cubes of small size (1.0 × 0.25 × 0.5 inches). Paneer cubes were dipped in 5% brine, acidified brine (5% NaCl, pH 5.5) and hydrogen peroxide solution with or without delvocid (0.5%, w/v). The shelf-life of the samples was found to be 22, 20, 32, and 22 days, respectively, compared to 10 days for control at 8–10°C. There was a good diffusion of solution in the smaller size paneer pieces for a longer shelf-life.

Hossain (1994) optimized microbial stability and sensorial quality of dudh churpi using several hurdles like heating, acid coagulation, addition of sugar and sorbate, smoking, and drying and packaging in a closed container. Dudh churpi is a product of Himalayan region (Bhutan, Sikkim, and Darjeeling) made from yak or cow's milk and is shelf-stable for several months without refrigeration. The texture (elasticity) plays an important role in dudh churpi as people living at high altitude chew it as "energy tablets."

Rao and Patil (1999) used a_w, lowered pH, additive (potassium sorbarte), and thermal treatment as hurdles to develop RTE paneer curry. The hurdle-preserved product was found to be more stable during storage than only heat sterilized product under similar storage conditions. Jayaraj and Patil (2001) also studied the effect of same hurdles, namely, a_w, pH, heat treatment, and potassium sorbate on hardness and chewiness of fried paneer, immediately after processing and during storage. The hardness and chewiness of the product decreased as the result of thermal treatment whereas decrease in a_w and pH increased hardness and chewiness. There was a synergistic effect of these parameters.

The addition of fresh ginger along with other spices as hurdles in milk increased the shelf-life of the curd rice (traditional dairy product popular in South India) up to 7 days at 37°C and 12 days at refrigerated storage (Balasubramanyam et al., 2004). Usually, the shelf-life of curd rice is 24 h at 30°C. The acidity and a_w of fresh curd rice were 0.54 % and 0.994, respectively.

Inactivation of Gram-negative bacteria in skim milk and banana juice using egg white lysozyme (HEWL) and bacteriophage lambda lysozyme (LaL) in combination with high pressure (HP) was studied by Nakimbugwe et al. (2006). HEWL alone was not that effective on all bacteria in both milk and banana juice than the combination of HP treatment with LaL. Under the experimental conditions used, LaL was more effective in banana juice than in milk. The HP food preservation may use LaL as an extra hurdle based on the results.

The effect of conventional cardboard boxes, modified atmosphere, and vacuum packaging techniques was studied by Panjagari et al. (2007) on the sensory, physicochemical, textural, biochemical, and microbiological quality of brown peda during storage for 40 days at 30°C. The caramelized color and highly cooked flavor describe the brown peda, a traditional Indian khoa based confection. According to the results, vacuum packaged samples had the less rate of loss of most quality attributes in comparison with control and modified atmosphere-packaged samples. The brown peda could be best preserved up to 40 days at room temperature (30°C ± 1°C) without appreciable quality loss.

13.2.4   BAKERY PRODUCTS

The combined effect of reduced a_w, reduced pH and preservatives on fungal growth on fermented bakery products were studied by Guynot et al. (2005). Preservatives (sodium benzoate, potassium sorbate, and calcium propionate) in a varying concentration 0–0.3% at pH (4.5–5.5) and a_w (0.80–0.90) were studied against three common spoilage fungi (*Penicillium corylophilum, Eurotium* spp., and *Aspergillus* spp.). Potassium sorbate (0.3%) was found the most effective preservative at all the a_w levels studied whereas other two preservatives at the same concentration were only effective at low a_w values. Results suggest that potassium sorbate could be used to prevent xerophilic fungal spoilage of slightly acidic bakery products.

Nasr et al. (2005) tested various chemical and biological food preservatives against a resistant strain of *Bacillus cereus,* isolated from pizza cheese containing sodium citrate as a preservative. Citric acid, benzoic acid, sorbic acid, propionic acid, potassium sorbate, sodium benzoate, and nisin were studied in different combinations to get the synergistic effect of these preservatives. The result shows that the use of combination is giving the better results than the individual preservative.

Jafari and Emam-Djomeh (2007) reduced nitrite content in hot dogs using HT without sacrificing product safety and quality. Humectants were used to adjust the a_w of the hot dog to 0.95. The pH of the sample was adjusted to 5.4 using Glocono-delta-lactone but the sample had least acceptance at this pH. The sample was heated at a temperature of 80°C ± 1°C for an hour to obtain the internal temperature of 75°C and subsequently cooled to around

5–6°C within 40–45 min. Cooled sausages were stored at chilled temperature (3–10 °C). The total aerobic counts in hurdle treated hot dogs (with 50 ppm nitrite) decreased, compared to the control (with 120 ppm nitrite), whereas *Clostridium perfringens* counts and *Clostridium botulinum* detection were the same ($p < 0.05$) in both hurdle treated and control samples. The hurdle treated sample and control had the same overall acceptability.

13.2.5 MISCELLANEOUS FOODS

Arya et al. (1998) preserved suji halwa for 6 months by hurdle conception 0.25% potassium sorbate, 0.5% calcium propionate, and 0.25% sodium benzoate were added and heating was done in hermetically sealed pouches at 95°C for 2 h. Yano et al. (2006) used spices and herbs along with temperature and nutrient level to kill a foodborne pathogen, *Vibrio parahaemolyticus*. The spices used were basil, clove, garlic, horseradish, marjoram, oregano, rosemary, thyme, etc. These results suggested that use of spices, herbs, and temperature as hurdles can be used for protecting seafood from the risk of contamination by *V. parahaemolyticus*.

Aqueous extracts of *Zingiber officinale* (ginger) and *Piper guineense* (West African black pepper) and heat were studied as hurdles against three fungi (*A. flavus, A. niger* and an unidentified yeast) isolated from a sorrel drink (zobo) by Ilondu and Iloh (2007). The fungal growth was reduced by spice extracts. The fungal biomass was decreased by an increase in the spice concentration in the beverage when compared with the control. The fungal growth was also reduced by a combination of both spice extracts at different ratios. The growth also reduced when heated at 100°C for different periods of times (min). The growth was also inhibited by other combinations of heat treatment with the combined spice treatment.

Braide (2012) extended the shelf-life of zobo beverage by using some commonly used chemical preservatives (acetic acid and sodium benzoate), natural plant extracts (clove, garlic, ginger, and lime) and pasteurization. Total counts and characterization of microorganisms during storage at ambient temperature for 14 days were analyzed. Five bacterial species, namely, *S. aureus, Micrococcus luteus, Micrococcus roseus, Bacillus subtilis,* and *Enterococcus faecalis* were predominant isolates. The isolation of two fungal species, namely, *Saccharomyces cerevisiae* and *Rhizopus stolonifer* was done. The predominant organisms were *B. subtilis* and *S. cerevisiae* throughout the storage. After an increasing for 2 days the microbial load decreased drastically as the effects of the preservative became evident, except for the control sample. Antimicrobial activities of chemical preservatives were most effective against bacteria and fungi species, followed by the natural plant extracts samples. Temporary preservative effect was observed for pasteurization. The chemically treated samples with preservatives remained organoleptically acceptable for 14 days. The preservatives and exhaustion of nutrients reduced the microbial population of zobo drink.

The production of tortellini, an Italian pasta product can be done by using HT (Anestis & Labropoulos, 2014). During storage, the principal hurdles employed for processing, namely, reduced a_w, and mild heating were effective in addition to a modified atmosphere or ethanol vapor in the package and mild chilling of the product. However, ethanol inhibited microbial growth molds and *micrococci*.

There was an attempt by Nisha et al. (2007) to preserve idli batter by HT, using preservatives and temperature as two hurdles. A thick fermented batter of rice and black Gram splits are used to prepare idli, a steamed pudding. Chemical preservatives (potassium metabisulfite, sodium benzoate, and potassium sorbate), biochemical preservative (nisin, a microbial polypeptide) and the combination of these two were used at temperatures of 28–30°C (room temperature storage), 4–8°C (refrigerated storage), and −18°C (frozen storage). This batter was analyzed for acidity, pH, and microbial count which was preserved with various preservatives. The quality parameters, such as bulk density, texture, color, and sensory quality were analyzed for the idli made from the preserved batter. This idli batter was found acceptable for 10 days at room temperature, 30 days at refrigerated storage, and 45 days at frozen storage preserved with a combination of 7.5 ppm nisin and 2000 ppm K-Sorbate.

13.3 CONCLUSION

The HT concept can provide a safe, nutritious, and fresh like food to the consumer. The hurdles when applied in combination prevent the microbial spoilage and food poisoning but do not affect the desired fermentation process. The optimal preservation of food can be done using a multi-target approach as the hurdles in combination exhibit the synergistic effect. The food preserved with this combined technology is safe and better in organoleptic as well as nutritional quality parameters as compared to the foods preserved by conventional methods. That is why this technology is gaining popularity in industrialized as well as in developing countries. Many foods have been successfully preserved using this technology but still a lot of research is needed to develop the protocols for other food products. We can say that the HT is the future of food preservation.

KEYWORDS

- **hurdles**
- **food quality**
- **conventional methods**
- **novel techniques**
- **HHP**
- **PEF**

REFERENCES

Abdullah, N.; Syida, W. S.; Kamarudin, W.; Samicho, Z.; Aziman, N.; Zulkifli, K. S. *In Comparison of Hurdle Treatment and Thermal Treatment on Ascorbic Acid and Total Phenolic Content of Winter Melon Puree Stored at Different Temperature;* International Conference on Environment, Chemistry and Biology, IPCBEE: Singapore, 2012; Vol. 49.

Aguilera, J. M.; Francke, A.; Figueroa, G.; Bornarat, C. Cifuentes, H. Preservation of Minced Pelagic Fish by Combined Methods. *Int. J. Food Sci. Technol.* **1992,** *27* (2), 171–177.

Alexandre, D.; Cunha, R. L.; Hubinger, M. D. Preservation of the Assai Pulp through the Application of Obstacles. *Cienc. Tecnol. Aliment.* **2004,** *24* (1), 114–119.

Alzamora, S. M.; Tapia, M. S.; Welli, A. A. J. Application of Combined Methods Technology in Minimally Processed Fruits. *Food Res. Int.* **1993,** 26, 125–130.

Anestis, S.; Labropoulos, A. The "hurdles" Technology in Food Processing. *Abstract book, International Food Congress Novel Approaches in Food Industry*; Kuşadasi, Turkey, 2014.

Ankita; Singh, R.; Nayansi. Effect of Hurdle Technology on the Quality and Stability of Minimally Processed Papaya. *Int. J. Sci. Res.* **2014,** *3* (8), 173–176.

Arya, S. S.; Rudramma; Arya, S. Preservation of Suji Halwa in Ready to Eat form by Hurdle Technology. *Beverage Food World* **1998,** *25* (1), 36.

Balasubramanyam, B. V.; Kulkarni, S.; Ghosh, B. C.; Rao, K. J. *Application of Hurdle Technology for Large Scale Production of Curd Rice.* Annual Report 2003 & 04, National Dairy Research Institute (Southern Campus): Bangalore, 2004.

Barwal, V. S.; Rakesh, S.; Rajinder, S. Preservation of Cauliflower by Hurdle Technology. *J. Food Sci. Technol.* **2005,** *42,* 26–31.

Benhura, C.; Rukuni, T.; Kadema, C.; Mubvakure, B.; Nazare, R.; Gombiro, P. E.; Tokwe, B.; Matangi, E.; Madzima, A. Preservation of Mango Pulp of Fruit from Rusitu Valley, Chimanimani in Zimbabwe. *Pak. J. Food Sci.* **2012,** *22* (4), 191–196.

Bhattacharya, D. C.; Mathur, O. N.; Srinivasan, M. R; Samlik, O. Studies on the Method of Production and Shelf Life of Paneer. *J. Food Sci. Technol.* **1971,** *8* (5), 117–120.

Braide, W.; Oranusi, S.; Peter-Ikechukwu, A. I. Perspectives in the Hurdle Techniques in the Preservation of a Nonalcoholic Beverage, Zobo. *Afr. Food Sci. Technol.* **2012,** *3* (2), 46–52.

Cepero, Y.; Beldarrain, T.; Campos, A.; Santos, R.; Bruselas, A.; Vergara, N. Use of the Hurdle Effect in a Conservation of Fresh Sausage. *Cienc. Tecnol. Aliment.* **2006,** *16* (2), 7–13.

Chaturvedi, A.; Sujatha, V.; Ramesh, C.; Babu, D. J. Development of Shelf Stable Intermediate Moisture Carrot (*Daucus carota*) Shreds Using Radiation as Hurdle Technology. *Int. Food Res. J.* **2013,** *20* (2), 775–781.

Chauhan, S.; Gupta, K. C.; Agrawal, A. New Approach of Hurdle Technology to Preserve Mango Fruit with the Application of Aloe Vera Gel and Calcium Chloride. *Int. J. Curr. Microbiol. App. Sci.* **2014,** *3* (5), 926–934.

Chawla, S. P.; Chander, R.; Sharma, A. Safe and Shelf-stable Natural Casing Using Hurdle Technology. *Food Control* **2006,** *17* (2), 127–131.

Chirife, J.; Favetto, G.; Ballesteros, S.; Kitic, D. Mummification in Ancient Egypt: An Old Example of Tissue Preservation by Hurdle Technology. *Lebensm. Wiss. Technol.* **1991,** *24,* 9–11.

El-Khateib, T.; Schmidt, U.; Leistner, L. Mikrobiologische Stabilitat von Tiirkischer Pastirrna. *Fleischwirtschaft* **1987,** *67,* 101–105.

Gamlath, G. G. S.; Dassanayaka, L. L. S. K.; Gunatilake, K. D. P. P. Preservation of Fresh Coconut Grating by Hurdle Technique. *Food Australia* **2004,** *56* (4), 140–142.

Gould, G. W. Interference with Homeostasis-Food. In *Homeostatic Mechanisms in Microorganisms*; Whittenbury, R., Gould, G. W., Banks, J. G., Board, R. G., Eds.; FEM Symposium; 1988; Vol. 44, pp 220–228.

Guerrero, S.; Alzamora; Gerschenson, L. N. Development of a Shelf Stable Banana Puree by Combined Factors: Microbial Stability. *J. Food Prot.* **1994,** *57,* 902–907.

Guynot, M. E.; Ramos, A. J.; Sanchis, V.; Marin, S. Study of Benzoate, Propionate, and Sorbate Salts as Mould Spoilage Inhibitors on Intermediate Moisture Bakery Products of Low pH (4.5–5.5). *Int. J. Food Microbiol.* **2005,** *101* (2), 161–168.

Hossain, S. K. A. Technological Innovation in Manufacturing Dudh Churpi. Ph.D. Thesis, University of North Bengal, Siliguri, India, 1994; pp122.

Ilondu, E. M.; Iloh, C. A. N. Inhibition of Three Fungal Isolates from Sorrel Drink (Zobo) Using Hurdle Techniques. *World J. Agric. Sci.* **2007,** *3* (3), 339–343.

Jafari, M.; Emam-Djome, Z. Reducing Nitrite Content in Hot Dogs by Hurdle Technology. *Food Control.* **2007,** *18,* 1488–1493.

Jayaraj Rao, K.; Patil, G. R. A Study on the Effect of Different "Hurdles" on the Rheological Properties of Fried Paneer by Response Surface Methodology. *J. Food Sci. Technol.* **2001,** *38* (3), 207–212.

Jayaraman, K. S.; Vibhakara, H. S.; Ramanuja, M. N.; Gupta, D. K. S. Development of Shelf Stable Fruit Slices for Ambient Storage Using Hurdle Technology. *Process. Food Ind.* **1998,** *12,* 9–10.

Jimenez-Villarreal, J. R.; Pohlman, F. W.; Johnson, Z. B.; Brown, A. H., Jr. Lipid. Instrumental Color and Sensory Characteristics of Ground Beef Produced Using Trisodium Phosphate, Cetylpypiridinium Chloride, Chlorine Dioxide or Lactic Acid as Multiple Antimicrobial Interventions. *Meat Sci.* **2003,** *65* (2), 885–891.

Kanatt, S. R.; Chawla, S. P.; Chander, R.; Sharma, A. Development of Shelf-stable, Ready-to-eat (RTE) Shrimps (*Penaeus indicus*) Using γ-radiation as One of the Hurdles. *LWT Food Sci. Technol.* **2006,** *39* (6), 621–626.

Kanatt, S. R.; Chawla, S. P.; Chander, R.; Bongirwar, D. R. Shelf-stable and Safe Intermediate-moisture Meat Products Using Hurdle Technology. *J. Food Prot.* **2002,** *65* (10), 1628–1631.

Lee, S. Y. Microbial Safety of Pickled Fruits and Vegetables and Hurdle Technology. *Int. J. Food Saf.* **2004,** *4,* 21–32.

Leistner, L.; Gorris, L. G. M. Food Preservation by Hurdle Technology. *Trends Food Sci. Technol.* **1995,** *6,* 41–45.

Leistner, L. Combined Methods for Food Preservation. In *Handbook of Food Preservation;* Rahman, M. S., Ed.; CRC: Boca Raton, FL, 2007; pp 867–894.

Leistner, L. Further Developments in the Utilization of Hurdle Technology for Food Preservation. *J. Food Eng.* **1994,** *22,* 421–432.

Leistner, L. Hurdle Technology in the Design of Minimally Processed Foods. In *Design of Minimal Processing Technologies for Fruits and Vegetables*; Alzamora, S. M., Tapia, M. S., Lo´pez-Malo, A., Eds.; Aspen Publishers: Gaithersburg, MA, 2000.

Leistner, L. *Food Quality and Nutrition*; Downy, W. K., Ed.; Applied Science Publishers: London, 1978; pp 553–557.

Lopez-Malo, A.; Palou, E. W.; Corte, P.; Argaiz, A. Shelf Stable High Moisture Papaya Minimally Processed by Combined Methods. *Food Res. Int.* **1995,** *27,* 545–553.

Malik, A. H.; Sharma, B. D. Use of Hurdle Techniques to Maintain the Quality of Vacuum Packed Buffalo Meat during Ambient Storage Temperatures. *Afr. J. Food Sci.* **2011,** *5* (11), 626–636.

Moraes, I. C. F.; Sampaio, R. M.; Queiroz, N. M.; De Salles, S. M.; Paschoaleti, C. C.; Perez, V. H. Mango Pulp (*Mangifera Indica* L.) Preserved by Hurdle Technology: Physicochemical, Microbiologic and Rheologic Characterization. *J. Food Process Preserv.* **2010,** *35,* 610–614.

Naga Mallika, E.; Prabhakar, K. Effect of Hurdles on the Quality of Low Fat Pork Sausages in Storage. *Vetscan.* **2011,** *6* (2), 98.

Nakimbugwe, D.; Masschalck, B.; Anim, G.; Michiels, C. W. Inactivation of Gram-negative Bacteria in Milk and Banana Juice by Hen Egg White and Lambda Lysozyme under High Hydrostatic Pressure. *Int. J. Food Microbiol.* **2006,** *112* (1), 19–25.

Narahari, D.; Suba, S. In Quality of Poultry Fast Foods Prepared by Multi–Hurdle Technology and Stored at Room Temperature, XVII European Symposium on the Quality of Poultry Meat Doorwerth, World's Poultry Science Association (WPSA): Beekbergen, The Netherlands, 2005; pp 70–75.

Nasr, A.; Kermanshahi, R. K.; Nahvi, A. Study on the Hurdle Effect of Some Organic and Chemical Food Preservatives on a Resistance of *Bacillus cereus* sp. *Iran. Food Sci. Technol. Res. J.* **2005,** *1* (2), 11–21.

Nisha, P.; Ananthanarayan, L.; Sabnis, R. W. In *Preservation of Idli Batter: A Hurdle Approach*, The 233rd ACS National Meeting: Chicago, IL, March 25–29, 2007.

Pallan, N. P.; Variyar, P. S.; Surendranathan, K. K. Preservation of Tender Jackfruit (*Artocarpus heterophyllus* Lam.) with Hurdle Technology, Short Communication. *Acta Biol. Indica.* **2012,** *1* (2), 238–241.

Panjagari, N. R.; Londhe, G. K.; Pal, D. Effect of Packaging Techniques on Self-life of Brown Peda, a Milk Based Confection. *J. Food Sci. Technol.* **2007,** *47,* 117–125.

Plaza, L.; Munoz, M.; de Ancos, B.; Cano, M. P. Effect of Combined Treatments of High Pressure, Citric Acid and Sodium Chloride on Quality Parameters of Tomato Puree. *Eur. Food Res. Technol.* **2003,** *216* (6), 514–519.

Podolak, R.; Solomon, M. B.; Patel, J. R.; Liu, M. N. Effect of Hydrodynamic Pressure Processing on the Survival of *Escherichia coli* O157:H7 in Ground Beef. *Innov. Food Sci. Emerg. Technol.* **2006,** *7* (1–2), 28–31.

Rao, K. J.; Patil, G. R. Development of Ready to Eat Paneer Curry by Hurdle Technology. *J Food Sci. Technol.* **1999,** *36,* 37–41.

Rao, K. J.; Patil, G. R. Water Activity Lowering Ability of Some Humectants in Paneer. *Indian J. Dairy Biosci.* **1992,** *10,* 121–122.

Rathod, K. S.; Zanjad, P. N.; Ambadkar, R. K. Shelf Life of Hurdle Preserved Ready to Eat Chicken Curry Stored at 37±2°C Temperature. *J. Vet. Public Health.* **2005,** *3* (2), 105–110.

Rawat, K.; Pokhriyal, S. Preservation of Sugarcane Juice Using Hurdle Technology. *Int. J. Sci. Eng. Technol.* **2014,** *3* (12), 1455–1458.

Roedel, W.; Scheuer, R. Recent Results on the Hurdle Technology. Measuring of Combined Hurdles. *Fleischwirtschaft* **2007,** *87* (9), 111–115.

Sachdeva, S.; Singh, S. Shelf-life of Paneer Affected by Antimicrobial Agents, Part I, Effect on Sensory Characteristics. *Indian J. Dairy Sci.* **1990,** *43,* 60–63.

Sankhla, S.; Chaturvedi, A.; Kuna, A.; Dhanlakshmi. K. Preservation of Sugarcane Juice Using Hurdle Technology. *Sugar Tech.* **2012,** *14* (1), 26–39. DOI:10.1007/s12355-011-0127-8.

Santos, R.; de la Mella, R. M.; Ramos, M.; Valladares, C.; Garcia, A.; Casals, C.; Cordova, A. Preservation of a Sausage at Room Temperature. *Alimentaria* **2002,** *331,* 21–25.

Selma, M. V.; Salmeron, M. C.; Valero, M.; Fernandez, P. S. Efficacy of Pulsed Electric Fields for *Listeria monocytogenes* Inactivation and Control in Horchata. *J. Food Saf.* **2006,** *26* (2), 137–149.

Sinha, J.; Gupta, E.; Tripathi, P.; Chandra, R. Impact of Hurdle Technology on Preservation of Carrot. *Int. J. Res. Rev. Pharm. Appl. Sci.* **2013,** *3* (4), 476–487.

Soliva-Fortuny, R. C.; Elez-Martinez, P.; Sebastian-Caldero, M.; Martin-Belloso, O. Effect of Combined Methods of Preservation on the Naturally Occurring Microflora of Avocado Puree. *Food Control* **2004,** *15* (1), 11–17.

Sujatha, V.; Chaturvedi, A.; Manjula, K. Storage Stability of Intermediate Moisture Cauliflower *Brassica oleracea,* Var, Botrytis Cabbage *Brassica oleracea,* Var, Capitata Using Radiation as Hurdle Technology. *J. Food Technol. Res.* **2014b,** *1* (2), 60–72.

Sujatha, V.; Chaturvedi, A.; Manjula, K. Effect of Radiation in Combined Method as a Hurdle in Development of Shelf Stable Intermediate Moisture Pineapple (*Ananas comosus*). *Experiment* **2014a,** *25* (3), 1736–1746.

Vibhakara, H. S.; Manjunatha, S. S.; Radhika, M.; Das Gupta, D. K.; Bawa, A. S. Effect of Gamma-irradiation in Combination Preservation Technique for Stabilizing High Moisture Spice Based Vegetables. *J. Food Sci. Technol.* **2005,** *42* (5), 434–438.

Vibhakara, H. S. J.; Das Gupta, D. K.; Jayaraman, K. S.; Mohun, M. S. Development of a High-moisture Shelf-stable Grated Carrot Product Using Hurdle Technology. *J. Food Process Preserv.* **2006,** *30* (2), 134–144.

Vijayanand, P.; Nair, K. K. S.; Narasimham, P. Preservation of Pineapple, Mango and Papaya Chunks by Hurdle Technology. *J. Food Sci. Technol.* **2001,** *38* (1), 26–31.

Yano, Y.; Satomi, M.; Oikawa, H. Antimicrobial Effect of Spices and Herbs on *Vibrio parahaemolyticus*. *Int. J. Food Microbiol.* **2006,** *111* (1), 6–11.

Yigit, A.; Korikluoglu, M. The Effect of Potassium Sorbate, NaCl and pH on the Growth of Food Spoilage Fungi. *Ann. Microbiol.* **2007,** *57* (2), 209–215.

Ziwei, L.; Mittal, G. S.; Griffiths, M. W. Pasteurization of Unclarified Apple Juice Using Low Energy Pulsed Electric Field. *Appl. Biotechnol. Food Sci. Policy* **2003,** *1* (1), 55–61.

CHAPTER 14

RADIO FREQUENCY APPLICATIONS IN FOOD PROCESSING

V. K. SHIBY*, AISHA TABASSUM, and M. C. PANDEY

Defence Food Research Laboratory, Siddharthanagar, Mysore 570011, India

Corresponding author. E-mail: shibyk@gmail.com

CONTENTS

Abstract 376
14.1   Introduction 376
14.2   Fundamental Principles of RF Heating 379
14.3   Mechanisms of RF and Critical Process Parameters 380
14.4   Penetration Depth 383
14.5   Equipment Design 390
14.6   Conclusion 394
14.7   Research Needs and Future Trends 394
Keywords 395
References 395

ABSTRACT

Radio frequency (RF) heating is an advanced and emerging technology for food application because of its higher penetration depth, heat distribution, and low energy requirement. RF can heat large and thick products better than microwave (MW) because of its higher penetration depth. As an electroheat technology, the unit energy costs of an RF system will be higher than an equivalent conventional heating system. RF heating is related to the dielectric properties of the foodstuffs being heated. Absorption and distribution of electromagnetic energy can be explained by the dielectric properties. When foods are placed between the applicator electrodes, there is complex electrical impedance introduced into the RF electrical field. Other properties like specific heat capacity and electrical conductivity of food material will also have an influence on the magnitude of the temperature rise obtained. RF heating has various commercial applications in the food industry. At present, some developments are being applied in the baking industry. Several studies have been done to investigate the effect of RF heating on meat products. Cooking and/or pasteurization of only a few selected meat products have been investigated. This chapter discusses major technological aspects and current applications of RF heating, equipment design, critical process parameters, and major challenges in using RF heating for food processing and preservation.

14.1   INTRODUCTION

The emergence of novel thermal technologies allows producing high-quality products with improvements in terms of heating efficiency and, consequently, in energy savings. These are locally clean processes, more environment-friendly, having less environmental impact than the traditional ones. Hence, these technologies increasingly attract the attention of food processors once they can provide value-added food products with improved quality and a reduced environmental footprint while reducing processing costs. Electroheating methods can be broadly classified as (1) direct electro heating where electrical current is applied directly to the food (e.g., ohmic heating, OH) and (2) indirect electroheating where the electrical energy is first converted to electromagnetic radiation which subsequently generates heat within a product (e.g., microwave, MW or radio frequency, RF). In recent years, there has been an increased interest in the area of RF heating, as evidenced by the increasing number of publications in this area.

RF heating is an advanced and emerging technology for food application because of its higher penetration depth, heat distribution, and low energy requirement (Sona & Mahendran, 2013). While RF energy heats the product rapidly, it also distributes the heat evenly through the product. In other words, heat is generated within the product, so that better temperature distribution is expected compared to conventional heating methods. RF heating is also referred to as dielectric heating and high-frequency heating. RF heating is similar to MW heating and is a result of application of electromagnetic waves to generate heat at regulated frequencies. The heat is generated by a product due to the friction generated by the molecular rotation of polar molecules within the product. The RF portion of the electromagnetic spectrum lies in the range of 3 kHz–1 MHz and 1–300 MHz. Although the frequency range for RF is very wide, there are only three frequencies available for industrial, scientific and medical applications. These frequencies are 13.56, 27.12, and 40.68 MHz. Since wavelength and frequencies are inversely proportional, wavelengths of RF are longer than that of MW (have higher frequencies: 915 MHz, 2450 MHz, 5.8 GHz, and 24.124 GHz). These results in higher penetration depths in RF compared to MW. In other words, thicker samples would be heated without surface overcooking in the RF oven thanks to deep electromagnetic power penetration (Marra et al. 2009). Since energy of RF and MW are not able to ionize biological molecules in a material, they are regarded as non-ionizing radiations. RF can heat large and thick products better than MW because of its higher penetration depth.

As an electroheat technology, the unit energy costs of an RF system will be higher than an equivalent conventional heating system. Nevertheless, when factors, such as increased energy efficiency and increased throughput are taken into account, the total energy cost will be comparable to (or even less than) a conventional system. Moreover, its reduced cost and higher penetration depth are advantages over the MW (Ohlsson, 1999). The power tubes used to generate MW energy are expensive and prone to failure. Radio frequencies have some advantages over MWs but are not as well developed.

Power density is directly proportional to the frequency. Given that the electric field is limited to avoid the occurrence of an electrical breakdown, then the power density will be much higher at MW than radio frequencies. The main consequences of this are that RF systems are usually significantly larger than MW heating systems, with the same power rating, and that faster heating rates can often be achieved with an MW system.

RF cooking is a modern heating method that heats food directly through conversion of electrical energy to heat, which occurs within the food itself

(Rowley, 2001). RF cooks products volumetrically with the product forming a dielectric between two electrodes acting as capacitor plates. These electrodes are alternatively charged from positive to negative several million times in a second (e.g., 27 MHz) and as a result, the polar molecules in the product are constantly realigned causing internal friction to occur and lead to the production of heat. The heat is generated within the product due to molecular friction resulting from oscillating molecules and ions caused by the applied alternating electric field (Piyasena et al., 2003). The energy and heat are absorbed directly by the food; therefore, the cooking not only saves time but also energy. Since, the 1940s RF energy has been applied in the food industry for numerous applications, such as baking, heating, thawing, or pasteurization (Piyasena et al., 2003). Several studies have shown that cooking with RF energy reduces cooking time, lowers juice losses, keeps acceptable color and texture, and increases shelf life (Casals et al., 2010; Guo et al., 2006; Houben et al., 1991; Laycock et al., 2003; Orsat et al., 2001). A major limitation of the available RF technologies, however, is the low number of efficient electromagnetic modes achievable in a tight band of frequencies, which may cause inconsistent heating profiles. This may result in hot and cold spots within the cooked foods, a weakness that possibly would impact food safety. Arcing, an electrostatic discharge as charges jump from one point to another point via dielectric breakdown (Parker et al., 2004), is the main problem that might occur during RF processing. Despite these limitations, it was shown that cooking with RF energy reduced the level of microbial contamination in foodstuff and in this way improved food quality, increased shelf life and probably also increased the food safety (Al-Holy et al., 2004b; Casals et al., 2010; Guo et al., 2006; Houben et al., 1991; Laycock et al., 2003; Orsat et al., 2001; Uemura, 2010). Bacterial destruction using RF has been mostly understood as related to heat generation on the substrate (Fung & Cunningham, 1980; Vela & Wu, 1979). RF technology has been explored in various food processing operations, such as pasteurization and sterilization (Bengtsson et al., 1970; Houben et al., 1991; Luechapattanaporn et al., 2005; Byrne et al., 2010) and insect disinfestations in various agricultural commodities, such as fresh fruits (Birla et al., 2004, 2005; Wang et al., 2006; Tiwari et al., 2008) and legumes (Wang et al., 2010).

Comparison of food quality in different sterilization processes when food was exposed to approximately same sterilization value (F_0) showed that cooked values (relative thermal effect on food quality) of RF heated products are generally half of that of conventional retort heating (Wang et al., 2003). Awuah et al. (2005) used a 2 kW, 27.12 MHz, RF applicator in order to find best conditions to inactivate surrogates of both *Listeria* and *Escherichia coli* cells in milk under continuous laminar flow conditions. The impact of RF treatment on microbial inactivation in orange and apple juices and in apple cider was assessed by Geveke & Brunkhorst (2004, 2008) and by Geveke et al. (2007). The used RF treatments were applied across frequency a range (which varied from a minimum of 15 kHz to a maximum of 41 kHz which varied with the product type) and were relatively mild, increasing product temperatures to a maximum of 65°C at the outlet. The main focus of their work was on microbial inactivation and microbial quality of beverages processed using this system. Under their experimental conditions, they demonstrated the potential of RF toward the inactivation of *E. coli* K12 by up to 3, 3.3, and 4.8 log cycles in apple juice, orange juice, and apple cider, respectively.

This chapter will consider (1) major technological aspects and current applications of RF heating, (2) equipment design and critical process parameters, and (3) major challenges in using RF heating in food processing and preservation.

14.2 FUNDAMENTAL PRINCIPLES OF RF HEATING

Electroheating technologies differ in terms of their methods of application. In MW heating, waves (generated by a magnetron) pass via a waveguide into an oven cavity in which they essentially bounce around off the metal walls of the cavity interior impinging on the product from many directions. In OH, the product is placed in direct contact with a pair of electrodes through which generally a low frequency (traditionally 50 or 60 Hz) alternating current is passed into the food product. Low-frequency alternating current is used in OH because the cyclic change in current direction helps to prevent electrolysis although higher frequencies (e.g., 10 or 4 kHz) have been shown by Samaranayake et al. (2005) to further reduce these electrochemical reactions. RF heating also involves the use of electrodes (with the product being placed either midway between or on top of one of a pair of electrodes) between which a high frequency directional electrical field is generated by high power electrical valves which transfer energy to the electrodes by a transmission line. However, RF heating does not have any requirement for direct contact between the product and electrodes as RF waves will penetrate through conventional cardboard or plastic packaging. The RF portion of the electromagnetic spectrum occupies a region between 1 and 300 MHz, although the main frequencies used for industrial heating lie in the range 10–50 MHz (Tang et al., 2005). These frequencies are included in the industrial, scientific, and medical (ISM) frequencies with 27.12 MHz being the most commonly used band for RF heating, 2450 MHz is used for catering/consumer MW heating and 915 MHz is used for applications, such as industrial MW processing and tempering. RF energy is generated by a triode valve and is applied to material via a pair of electrodes (Rowley, 2001). In the parallel plate RF system, one of these electrodes is grounded which sets up a capacitor to store electric energy. The target material to be heated is placed between but not touching the parallel electrodes. It must be noted that while the use of parallel plate electrodes (or "through-field" applicators) is the most commonly used electrode configuration for heating thicker materials, two other configuration types are included in Jones and Rowley (1997). These are "fringe-field" applicators (which consist of a series of bar, rod, or narrow plate electrodes which are most suited for heating or drying thin layers (< 10 mm)) or "staggered through field" applicators (consisting of rod or tube-shaped electrodes staggered on either side of a belt which are used for heating products of intermediate thickness).

14.3   MECHANISMS OF RF AND CRITICAL PROCESS PARAMETERS

RF heating is related to the dielectric properties of the foodstuffs being heated. Absorption and distribution of electromagnetic energy can be explained by the dielectric properties. When foods are placed between the applicator electrodes, there is complex electrical impedance introduced into the RF electrical field. Dielectric heating is due to energy absorption by a lossy dielectric when it is placed in a high-frequency electrical field. When an alternating electrical field is applied to a food, one phenomenon that occurs is the movement of positive ions in the material toward negative regions of the electric field and the movement of negative ions toward positive regions of the field (Buffler, 1993). This movement of ions in this fashion is often referred to as ionic depolarization and is essentially resistance heating as found in OH. Heating occurs because this field is not static, with polarity continually changing at high-frequencies (e.g., 27.12 MHz for RF or 2450 MHz for MW), which is in contrast to OH where the field polarity changes at much lower frequencies (i.e., 50 Hz in Europe or 60 Hz in the United States). However, regardless of the frequency, the continued reversal of polarity in the electrical field leads to the oscillation of ions forward and backward in the product with the net effect of internal generation of heat within the product by friction (Buffler, 1993). It is generally accepted that ionic depolarization tends to be the dominant heating mechanism at the lower frequencies encountered in the RF range. Both ionic depolarization and dipole rotation can both be dominant loss mechanisms at frequencies relevant to MW heating (i.e., 400–3000 MHz) depending upon the moisture and salt content within a product (Tang, 2005). In RF range dissolved ions are more important for heat generation than the water dipoles in which they are dissolved.

The dielectric properties of food materials can be divided into two parts known as the permeability and permittivity. Permeability values for foodstuffs are generally similar to that of free space and as a result are not believed to contribute to heating.

However, the permittivity (ε), which determines the dielectric constant (ε') and the loss factor (ε'') influences RF heating. The (ε') and (ε''), which are the real and imaginary parts, respectively, of ε, is given by equation

$$\varepsilon = \varepsilon' - j\varepsilon'' \tag{14.1}$$

ε' is a characteristic of any material and is a measure of the capacity of a material to absorb, transmit, and reflect energy from the electric portion of the electrical field and is a constant for a material at a given frequency under constant conditions. The ε'' is a measure of the polarizing effect from applied electric field (i.e., how easily the medium is polarized). ε'' measures the amount of energy that is lost from the electrical field, which is related to how the energy from a field is absorbed and converted to heat by a material passing through it. A material with a low ε'' will absorb less energy and could be expected to heat poorly in an electrical field due to its greater transparency to electromagnetic energy (Decareau, 1985). Another descriptive dielectric parameter is the loss tangent (or the dissipation factor) of the material. Zhuang et al. (2007) reported the dielectric constant and loss factor of the chicken breast meat as 100 and 510, respectively, at 26 MHz at 25°C. Whereas, the dielectric constant and loss factor of the marinated (15%) chicken breast meat were reported as 105.4 and 967.8, respectively, at 27.12 MHz at 20°C (Lee et al. 2008a). The ratio of dielectric loss factor and dielectric constant of a material expresses the loss tangent, and it equals to the tangent of dielectric loss angle or electric loss tangent (tanδ). Hence

$$\tan\delta = \varepsilon'' / \varepsilon'.$$

14.3.1   FACTORS AFFECTING DIELECTRIC PROPERTIES OF FOODS

The dielectric properties of food play an important role in both RF and MW heating (Piyasena et al., 2003) but these properties are influenced by a variety of factors. The content of moisture is generally a critical factor (Tang, 2005), but the frequency of the applied alternating field, the temperature of the material, and also the density, chemical composition (i.e., fat, protein, carbohydrate, and salt), and structure of the material all have an influence (Piyasena et al., 2003).

In terms of bulk density samples of an air-particle mixture with higher density generally have higher ε' and ε'' values because of less air incorporation within the samples (Nelson & Datta, 2001). In relation to composition, Nelson & Datta (2001) stated that the dielectric properties of materials are dependent on chemical composition and especially on the presence of mobile ions and the permanent dipole moments associated with water.
Limited number of studies has examined the dielectric properties of food and agricultural products at RF frequencies.

Most of additives which have free ions can change the dielectric properties of final product and in turn influence the RF heating process. Within the meat species studied, turkey and chicken breast had the higher values of ε'' than pork, lamb, and beef muscle which was attributed to compositional differences between the products. Dielectric properties of three different beef meat blends (lean, fat, and 50:50 mixture) over a temperature range -18 to $+10°C$ were investigated and it was reported that in the region of thawing ($-3°C$ to $-1°C$), ε' and ε'' values at 27.12 MHz were significantly higher than at other measured temperatures for the three blends.

Composition also significantly influenced the measured dielectric properties at all temperatures used and, thus, the heating patterns during thawing.

14.3.2   OTHER FACTORS AFFECTING RF HEATING

Other properties like specific heat capacity and electrical conductivity of food material will also have an influence on the magnitude of the temperature rise obtained. Since foods are conductive materials, ionic depolarization is observed in the food material when exposed to electric field. Therefore, there is a relationship between electrical conductivity and the dielectric loss factor. Electrical conductivity, σ (in Siemens/meter, S/m) can be formulated as

$$\sigma = 2\pi f \mathcal{E}'' \tag{14.2}$$

where $2\pi f$ can also be expressed as angular velocity (ω) in rad/s (Nelson & Datta 2001; Marra et al. 2009). The electric field distribution in the product can be measured by power absorption of the material. This can be formulated as

$$P = 55.61 + 10^{-14} f E^2 \mathcal{E}'' \tag{14.3}$$

where P = power density (W/m^3), f = the applied frequency (Hz), E = electric field strength (V/m) (Orsat et al., 2004). Penetration depth is defined as the distance that an electromagnetic wave can penetrate perpendicular from the surface to bottom of the material as its power decreases to $1/e$ ($1/2.72$) of its initial power at the surface.

Electrical conductivity (σ) indicates the ability of a material to conduct an electric current. In a dielectric food system, σ is related to ionic depolarization. It contributes to ($\mathcal{E}''$) and in RF ranges can be calculated from the following equation (Piyasena et al., 2003). It is also worth noting that at RF frequencies, $\mathcal{E}''$ in liquid or semi-liquid foods can be estimated ($\mathcal{E}''$) with reasonable accuracy by measuring σ value of the material using an electric conductivity meter (Guan et al., 2004).

14.4 PENETRATION DEPTH

The depth in a material where the energy of a plane wave propagating perpendicular to the surface has decreased to $1/e$ of the surface value approximately 37% of its initial value (Bengtsson & Risman, 1971). Penetration depth (dp) is used as an indicator to select appropriate food thickness to ensure a vertically uniform dielectric heating.

The penetration depth, dp (in m) can be calculated by

$$dp = \frac{C}{2\pi f \sqrt{2} \mathcal{E}'' \sqrt{\sqrt{1 + \tan \delta^2} - 1}} \tag{14.4}$$

where C is the velocity of electromagnetic radiation in vacuum, 3×10^8 m/s. When the dielectric constant and dielectric loss factor is low, the deepest penetration depth is observed (Bengtsson & Risman, 1971). Penetration depth is a crucial dielectric property of a material affecting the uniformity of heat in the RF heating as well as sample size. Larger penetration depth than sample size results in uniform heating with little temperature differential in the final product. In contrast, smaller penetration depth with respect to sample size causes limited heating on the near surface of the product resulting in non-uniform heating. That is very similar to conventional heating (Sumnu, 2001). The penetration depths for various food products at different frequencies in RF range are reported by Wang et al. (2003).

14.4.1 APPLICATIONS OF RF HEATING IN FOOD PROCESSING

Food processing applications for RF heating are less common than for MW heating. The early efforts after World War II employed RF energy for applications, such as the cooking of processed meat products, heating of bread, dehydration, and blanching of vegetables. However, due to the high overall operating costs of RF energy at that stage, commercial installations were few. The next generation of commercial applications for RF energy in the food industry was post-bake drying of cookies and snack foods in the late 1980s (Rice, 1993; Mermelstein, 1998). Later in the 1990s, the area of RF pasteurization was studied with attempts made to improve energy efficiency and solve technical problems, such as run-away heating (Houben et al., 1991; Zhao et al., 2000). This, in turn, has led to recent investigations on RF applicator modifications and dielectric properties of food at RF frequencies (Laycock et al., 2003; Zhang et al., 2004, 2006, 2007; Birla et al., 2005). Studies on the application of RF energy to foods focused on the defrosting of frozen products, which resulted in several commercial production lines in 1960s. RF treated juices were reported to have better bacteriological and organoleptic qualities than the juices treated by conventional thermal methods suggesting potential applications in heat processing for preservation of foods. The application and processing conditions related to quality parameters of foods are summarized in Table 14.1.

RF heating has various commercial applications in the food industry. At present, some developments are being applied in the baking industry. Any new application in the food industry will be in the value-added category of processed foods (Tewari, 2007).

The latter authors applied RF heating to thaw herring and white fish on an industrial scale. The next generation of commercial applications of RF energy was in the bakery industry for final drying of cookies and crackers. RF heating is currently practiced in the baking industry to remove excess moisture from products.

TABLE 14.1 Processing Conditions and Related Quality Aspects of RF Treated Foods.

Processed products	RF Conditions	Quality parameters	Salient findings	References
Beef rolls	50 Ω oven, max power 600 W, 27.12 MHz, 80°C	Hardness, Juiciness, color changes, nutritional characteristics (protein, moisture, and fat content)	Rolls with higher salt content got heated faster because of increased conductivity, hence the final cooked product was found to be juicier, softer, and tender	Tang et al. (2006)
Thawing of lean meat beef pieces	50 Ω oven, max power 0.6 kW, 27.12 MHz	Drip losses	Lower loss of nutrients as compared with conventional methods. Unlike RF thawing where all the samples had similar drip loss, RF tempering resulted in higher drip losses.	Farag et al. (2009)
Pork meat doughs	27 MHz, 100°C, 5 min	Rheological properties	RF treated samples had firmer texture, higher storage and loss moduli compared to samples heated in water bath	van Roon et al. (1994)
Pasteurization of pork meat luncheon rolls	450–550 W power level, 74–80°C circulating water temperature, 25–35 min cooking time	Color, texture, and expressible fluid weights	RF pasteurized rolls had more expressible fluid and exhibited more chewiness and gummy texture	Zhang et al. (2004)
Storage quality of ham	Max power 600 W, max voltage 5 kv, Temp. range 75– 85°C, 27.12 MHz,	Color, texture, cooking, and sensory characteristics	1. RF heating coupled with proper packaging is a useful technique to pasteurize and extend shelf life of ham under refrigerated conditions 2. overall differences in the quality of ham rolls cooked using RF cooking and steam cooking where very minor and a fully cooked like quality can be developed by making some fine adjustments to the RF cooking protocol, such as extending cooking time or heating to high temperatures. Overall RF process is very efficient compared to the steam cooking process for cooking ham meat.	Orsat et al. (2004); Zhang et al. (2006)

TABLE 14.1 *(Continued)*

Processed products	RF Conditions	Quality parameters	Salient findings	References
Poultry (Turkey breast roll)	500 W, 27.12 MHz, 73°C, 40 min	Protein, ash, moisture, fat thiamine, and riboflavin, texture profile, color	Proximate analysis results revealed no significant differences between both RF and steam cooking methods. No differences were observed in most of the parameters derived from texture profile analysis, apart from hardness, which was reported to be higher for RF cooked rolls.	Tang et al. (2005)
Quality of chicken breast meat	74°C, 23.8 min	color	In terms of color quality, RF cooked chicken samples scored higher than water bath cooked ones. It was reported that RF cooking is good, best, quick, and cost-effective alternative for cooking chicken breast.	Kirmaci and Singh (2012)
Shelf stable sterile scrambled eggs	—	Color, texture, *Clostridium sporogenes*	RF sterilization is a better method for producing shelf-stable scrambled eggs.	Luechapattanaporn et al. (2005)
Thawing of fish	6 kW, 36–40 M Hz,	Appearance, flavor, and drip loss	RF thawing of fish is successfully achieved from frozen state (−29°C) to center point temperatures ranging between 0°C and 7°C. Better retention of appearance and flavor as well as reduced drip loss.	Jason and Sanders (1962)
Quality change in fish meal	1.2 kW, 6–14 MHz, 60–90°C	*Salmonella* spp. and *E. coli* 0157:H7, compositional parameters like protein, ash, moisture, fat, FFA, protein digestibility assay, nitrogenous compounds like putrescine, cadaverine, histamine, and tyramine	Protein digestibility assay showed 95% digestibility of fish meal, which was similar to the assay for control samples and confirmed no negative effect of RF treatments. No possible effect of RF heating on nitrogenous composition of fish meal was observed.	Lagunas-Solar et al. (2005)

TABLE 14.1 *(Continued)*

Processed products	RF Conditions	Quality parameters	Salient findings	References
Seafood thawing	—	Drip loss, texture, and other quality parameters	RF thawing is an excellent method when compared with all other methods for thawing including microwave, vacuum, and water heating. RF thawing results in minimum drip loss, which in turn maintain high nutritional value, as well as limiting impact on texture and other quality parameters.	Archer et al. (2008)
Apples (blanching)	—	Color, texture, and sweetness	RF treatment has great scope for meeting the quarantine treatment conditions.	Manzocco et al. (2008)
Quarantine treatment for pest insects in-shell walnuts	0.8 W, 27.12 MHz, 55°C	Oxidative rancidity, PV, fatty acid content, cracking strength, weight change, and sensory analysis during storage	Statistically insignificant ($p > 0.05$) differences were observed in all quality attributes studied between walnuts stored for 10 and 20 d at 35°C and fresh ones. Sensory results also suggested that RF treatment did not affect consumer acceptance of walnuts.	Wang et al. (2002)
Pasteurization of tomato puree	27.12 MHz,	L-ascorbic acid retention and HMF formation and color changes	High retention of L-ascorbic acid and minor formation of HMF were observed after RF heating. Less reduction in redness of RF pasteurized tomato puree.	Felke et al. (2011)
Non-fat dry Milk	75–90°C	Whey protein nitrogen index, solubility, color, *Cronobacter sakazakii, and Salmonella* spp	1–5 log cycle destruction of *Cronobacter sakazakii and Salmonella* spp. Good retention of whey protein nitrogen index and color for treatments done below 80°C, reduction in solubility was observed nutritional values.	Chen et al. (2013)

TABLE 14.1 *(Continued)*

Processed products	RF Conditions	Quality parameters	Salient findings	References
Vegetable powders	27.12 MHz, 6 kW	Dielectric properties of dried vegetable powders and their temperature profile	The relationship between MC, temperature, and dielectric properties of broccoli powder at 13.56 and 27.12 MHz can be described by quadratic models with high correlation coefficients ($R2 > 0.96$). The RF heating rate in samples increased linearly with dielectric loss factor and MC. The information provided in this study is useful to develop an effective RF heating strategy to pasteurize dried vegetable powders.	Ozturk et al. (2016)

RF Postbaking is a proven solution to checking problem in biscuits. Checking arises due to the build-up of stresses in the product piece. These stresses are caused by differential moisture content between the outer surface and the center bone of the biscuit. Conventional bake-ovens utilize indirect heating methods. Often this can present a problem because the material itself is a good insulator and as such, it is difficult to get the heat to penetrate to the center of the product. Therefore, the surface dries out, whilst the core or center of the piece remains high in moisture.

Several studies have been done to investigate the effect of RF heating on meat products. Cooking and/or pasteurization of only a few selected meat products have been investigated. Ryynänen (1995) concluded that RF heating could be favorable for heating of cured whole meat products like hams as the mode of heating is by the depolarization of ions in solution, which are plentiful in a product to which salts have been added. Laycock et al. (2003) evaluated the influence of RF cooking on the quality of ground, comminuted, and muscle meat products after being heated to a center temperature of 72°C. They concluded that RF cooking of processed meat products resulted in decreased cooking time, lower juice losses, acceptable color, water holding capacity, and texture. Another study on post-cooking temperature profiles of meat emulsions (Zhang et al., 2004) suggested that uneven temperature distributions are possible in RF cooked meat products. They found that the temperature differentials were two-fold higher within the RF-cooked sample relative to its steam-cooked counterparts. Recently, a study was done to develop a system for production of cooked meats in casings using RF heating (Brunton et al., 2005). They also compared the quality of RF-heated products with those produced by commercial cooking methods. Zhang et al (2006) developed a method for RF pasteurization of large diameter meat product in casings (i.e., leg and shoulder ham) and compared the sensory attributes of RF pasteurized products with conventionally pasteurized product. It was concluded that RF cooked hams had significantly lower water holding capacities and higher yields as compared to their steam cooked counterparts. The differences in the quality between RF and steam cooked samples were found to be relatively minimal and could be even further reduced by making minor adjustments to the RF cooking procedures.

RF heating may be especially effective for large diameter foods like meat because the low frequency of incident electromagnetic radiation allows greater penetration depths (McKenna et al, 2006). RF heating could be particularly suitable for heat processing of cured whole meat products, such as hams as the principal mode of heating occurs via depolarization of solvated ions (Ryynanen, 1995) which are particularly plentiful in a product to which curing salts have been added. In the past RF heating has been applied to the pasteurization of sausage meat emulsions (Houben et al., 1991) and cured hams (Bengtsson et al., 1970) and has even found commercial application with the development of a method for the continuous production of cooked meats by Tulip International AS Denmark and APV. More recently Laycock et al. (2003) reported that RF heating at 27.12 MHz could serve to reduce cooking times by up to 90% in whole, minced and comminuted beef. However, the eating quality and particularly the texture of some of the products were adversely affected. Whilst the technology is undoubtedly capable of fast and efficient pasteurization of meat products (Laycock et al., 2003; van Roon, Houben et al., 1994; Zhang et al., 2004), aside from the work of Laycock et al. (2003) limited information is available as to the quality of meat products produced by this method. Much of the past work on RF pasteurization of meat products has involved the application of RF to uncased meats (Houben et al., 1991; Laycock et al., 2003; van Roon et al., 1994) which had the advantage of reducing packaging costs for processors but increased the risk of post-process contamination and potentially reduced yield due to juice losses from unpackaged hot products. The principal objectives of this study were to develop a method for RF pasteurization of a large diameter cased meat product and to subsequently compare selected instrumental and sensory quality attributes of these RF pasteurized products with a conventionally pasteurized product.

14.5   EQUIPMENT DESIGN

RF energy is generated by means of an RF generator that produces oscillating fields of electromagnetic energy. The generator is comprised of an oscillator, power supply, and control circuitry (Fig. 14.1).

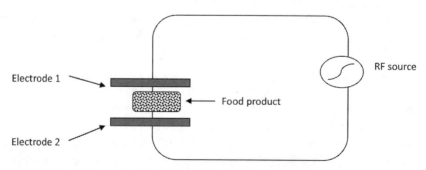

FIGURE 14.1   RF set up for foods (adapted from Koral, 2004).

Heating of poor electrical conductors occurs when a high voltage alternating electric field is applied to a medium sandwiched between two parallel plate electrodes forming a capacitor configuration. The principle on which RF works is such that, during heating, the product to be heated forms a dielectric between two metal capacitor plates, which are alternately charged positively and negatively setting up a high frequency alternating electric field. The design of RF equipment depends on physical characteristics of food including its geometry, shape, electrical conductivity, and dielectric properties like permeability and permittivity, loss tangent and penetration depth. RF heating depends on the dielectric properties of the foods, which is influenced by frequency, temperature, moisture content, and composition (Marra et al., 2009; Piyasena et al., 2003)

Based on electrode configuration RF equipment are classified as

1. "Through-field" applicators with parallel plate RF system and one of the electrodes is grounded which sets up a capacitor to store electric energy. The target material to be heated is placed between, but not touching the parallel electrodes. This is most commonly used electrode configuration for heating thicker materials.
2. "Fringe-field" applicators which consist of a series of bar, rod or narrow plate electrodes which are most suited for heating or drying thin layers (< 10 mm).
3. "Staggered through field" applicators (consisting of rod or tube-shaped electrodes staggered on either side of a belt which is used for heating products of intermediate thickness.

RF sterilization unit with water circulation for meat processing:
A pressurized vessel was developed at Washington State University (Pullman, Washington, USA) for meat processing to provide an overpressure of up to 0.276 MPa gauge (40 psig) that allows foods in large polymeric trays to be heated up to 135°C without bursting (Fig. 14.2). The vessel was constructed with four Ultem polyetherimide (PEI) plastic walls and two parallel aluminum plates as upper and lower lids. Several probe ports permitted the insertion of fiber-optic sensors directly into food packages through sealed thermal wells. In addition, custom-built water conditioning system circulated temperature controlled water of a particular conductivity through the vessel. This system consisted of two exchangers: one used steam for heating, and the other used tap water for cooling. A surge tank was used to help maintain an overpressure with compressed air. The circulating water

temperature could be varied from 20°C to 130°C, an overpressure from 0.136 to 0.204 MPa gauge (20–30 psig).

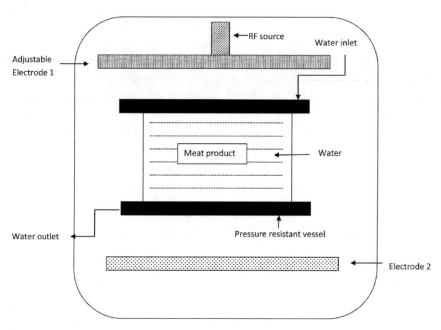

FIGURE 14.2   Water immersion RF systems for meat products (adapted from Wang et al., 2012).

To reduce fringe effect at the interface between the side of the food package and the air in the RF applicators, low-conductivity water was used to immerse the food tray to approximately match the dielectric properties of the food. Together, the water and package present a very flat surface to the imposed field and push the boundary with air away from the food package. This is expected to reduce the effect of variations in package thickness that could lead to non-uniform heating in the food. Temperature controlled water from water conditioning system is required to match the temperature of the heated food to prevent cooling of the package surface. Use of immersion water may compromise energy efficiency in RF pasteurization and sterilization. However, the lost RF energy in the immersion water can be recovered by using it to preheat food trays in industrial applications. It is also possible to minimize the volume of immersion water by lining food packages in continuous systems.

The main purpose of the immersion water was to reduce the fringe effects at the interface between the food package and the air in the RF applicators

(Wang et al., 2003). Based on preliminary results, the electric conductivity of the circulation water was adjusted to 337×10^{-4} S/m at 65°C by mixing tap water with de-ionized water. Computer simulations were also conducted to evaluate the influence of the dielectric properties of each food component on the electric field distribution and heating pattern during RF heating. The measurements indicated small temperature differences in beef meatballs, mozzarella cheese, and sauce when they were properly distributed between layers of noodles. Simulation results suggested that in spite of large differences in electric field intensities in different food components, adequate heat transfer reduced differential heating. Thus, RF heating can be used to process pre-packaged heterogeneous foods and retain product quality.

14.5.1 COMPARATIVE ADVANTAGES OF RF HEATING VERSUS CONVENTIONAL HEATING AND OTHER ELECTRO HEATING METHODS

14.5.1.1 RF VERSUS CONVENTIONAL HEATING METHODS

Conventional food heating methods require heat energy to be generated externally and then transferred to the food material by conduction, convection, or radiation. For products containing particulates, especially when particulates are very large, conventional heating methods require such excessive heat processing that the degradation of outer portion of particulates occurs. In contrast electroheating (e.g., OH, MW, and RF) differs from conventional heating in that heat is generated volumetrically within the material by the passage through, and its interaction with, either alternating electrical current (as in OH) or electromagnetic radiation (formed by the conversion of electrical energy to electromagnetic radiation at MW (300–3000 MHz) or RF (1–300 MHz) frequencies.

14.5.2 RF HEATING VS. OTHER ELECTRO-HEATING METHODS

In OH, the product is placed in direct contact with a pair of electrodes through which generally a low frequency (traditionally 50 or 60 Hz) alternating current is passed into the food product., although higher frequencies (e.g., 10 or 4 kHz) have been shown to further reduce electrochemical reactions. In OH, the product needs to be either unpackaged and in direct contact with the electrodes. In MW heating, special oscillator tubes known as klystrons emit MWs, which are transferred by a waveguide into a metal chamber or cavity where the target material to be heated is placed. Resonant electromagnetic standing wave modes are then established within the cavity although turntable trays and/or stirrers can be used to improve the uniformity of the electromagnetic field within the chamber and around the target material. In contrast, RF energy is generated by a triode valve and is applied to material via a pair of electrodes (Rowley, 2001). RF results in more uniform and simple heating patterns and deeper wave penetration MW results in complex nonuniform standing wave patterns and low power generation levels.

14.6   CONCLUSION

Recent years have shown a substantial number of publications in the area of the quality of RF processed meats but this needs to be replicated across a wider range of food commodities. The availability of such information will help to convince processors of the benefits of this technology. As an alternative to noxious gasses or carbon producing thermal techniques, RF heating is a highly efficient "direct" form of heating, such that no energy is wasted heating large volumes of air or preheating the system itself. The technology is "instant-on, instant-off" using energy only during the treatment process. It is a promising alternative to conventional heating methods and offers advantages, such as reduced cooking times and more uniform heating, maintenance of functionality, consistent log reductions, and simplified process control. It is suitable for large diameter food products, such as meat products resulting in greater penetration depths than MW. This technology is particularly suitable for cured whole meat products as the principal mode of heating occurs via depolarization of solvated ions. Overcooking is avoided while energy is transferred by longer wavelengths. However, the risks of arcing and thermal runaway are the main problems that limit the use of RF heating in the food industry.

14.7   RESEARCH NEEDS AND FUTURE TRENDS

As a potential pasteurization/sterilization technique, more work needs to be published on the effectiveness of RF for inactivating microorganisms and its impact on product quality and shelf life. However, in addition to this, a greater understanding of temperature distribution within products needs to be developed. Conventional heating methods are generally well understood in this context and until similar knowledge is available for RF heating, food processors charged with consumer safety are going to be cautious to change the traditional methods. Designing and scaling-up larger systems for industrial application are essential in order to evaluate the actual capital costs for initial installations and to estimate the energy costs involved in particular treatments, in order to further investigate the efficiency and economics of RF processing. A continuing need for the production of more dielectric property data on foodstuffs and potential packaging is required. This information is the key to improving understanding of temperature distribution but also is important in the design of RF heating systems. Mathematical modeling improves the understanding of RF heating of food and it is essential to the continued development of this technology. More efforts are needed in order to develop computer-aided engineering of processes and plants on an industrial scale.

KEYWORDS

- **radio frequency**
- **food**
- **electroheating**
- **penetration depth**
- **microorganisms**
- **dielectric**

REFERENCES

Archer, C. T.; Activation Domain-dependent Monoubiquitylation of Gal4 Protein is Essential for Promoter Binding in Vivo. *J. Biol. Chem.* **2008**, *283* (18), 12614–12623.

Al-Holy, M.; Ruiter, J.; Lin, M.; Kang, D. H.; Rasco, B. Inactivation of *Listeria innocua* in Nisin-treated Salmon (Oncorhynchus keta) and Sturgeon (Acipenser transmontanus) Caviar Heated by Radio Frequency. *J. Food Prot.* **2004b,** *67* (9), 1848–1854.

Awuah, G. B.; Ramaswamy, H. S.; Economides, A.; Mallikarjuanan, K. "Inactivation of *Escherichia coli* K-12 and *Listeria innocua* in Milk Using Radio Frequency (RF) Heating". *Inn. Food Sci. Emer. Technol.* **2005**, *6* (4), 396–402.

Bengtsson, N.; Risman, P. Dielectric Properties of Foods at 3 GHz as Determined by a Cavity Perturbation Technique. Measurement on Food Materials. *J. Microwave. Power* **1971**, *6* (2), 107–123.

Bengtsson, N. E.; Green, W.; Del Valle, F. R. Radio Frequency Pasteurization of Cured Hams. *J. Food Sci.* **1970,** *35,* 681–687.

Birla, S. L.; Wang, S.; Tang, J.; Fellman, J. K.; Mattinson, D. S.; Lurie, S. Quality of Oranges as Influenced by Potential Radio Frequency Heat Treatments Against Mediterranean Fruit Flies. *Postharvest. Biol. Technol.* **2005,** *38* (1), 66–79.

Birla, S. L.; Wang, S.; Tang, J.; Hallman, G. Improving Heating Uniformity of Fresh Fruit in Radio Frequency Treatments for Pest Control. *Postharvest. Biol. Technol.* **2004,** *33* (2), 205–217.

Brunton, N. P.; Lyng, J. G.; Li, W. Q.; Cronin, D. A.; Morgan, D.; McKenna, B. Effect of Radio-frequency (RF) Heating on the Texture, Colour and Sensory Properties of a Comminuted Pork Meat Product. *Food Res. Intl.* **2005,** *38* (3), 337–344.

Buffler, C. R. Dielectric Properties of Foods and Microwave Materials. In *Microwave Cooking and Processing*; Van Nostrand Reinhold: New York, 1993; pp 46– 69.

Byrne, B.; Lyng, J. G.; Dunne, G.; Bolton, D. J. Radio Frequency Heating of Comminuted Meats–Considerations in Relation to Microbial Challenge Studies. *Food Cont.* **2010,** *21,* 125–131.

Casals, C.; Alandl, I. V.; Picouet, P.; Torrea, R.; Usall, J. Application of Radio Frequency Heating to Control Brown Rot on Peaches and Nectarines. *Postharvest. Biol. Technol.* **2010,** *58,* 218–224.

Chen, C.; Michael, M.; Phebus, R. K.; Thippareddi, H.; Subbiah, J.; Birla, S. L.; Schmidt, K. A. Short Communication: Radio Frequency Dielectric Heating of Non-fat Dry Milk Affects Solubility and Whey Protein Nitrogen Index. *J. Diary. Sci.* **2013,** *96* (3), 1471–1476.

Decareau, R. V. *Food Industry and Trade – Microwave Heating – Industrial Applications;* Academic press, INC.: Orlando, FL, 1985; pp 1–10.

Farag, K. W.; Duggan, E.; Morgan, D. J.; Cronin, D. A.; Lyng, J. G. A Comparison of Conventional and Radio Frequency Defrosting of Lean Beef Meats: Effects on Water Binding Characteristics. *Meat. Sci.* **2009,** *83* (2), 278–284.

Felke, K.; Pfeiffer, T.; Eisner, P.; Schweiggert, V. Radio Frequency Heating. A New Method for Improved Nutritional Quality of Tomato Puree. *Agro Food Ind. Hi Tech.* **2011,** *22* (3), 29–32.

Fung, D. Y. C.; Cunningham, F. E. Effect of Microwaves on Microorganisms in Foods. *Engineering* **1980,** *105* (3), 341–349.

Geveke, D. J.; Brunkhorst, C. Inactivation of *Escherichia Coli* in Apple Juice by Radio Frequency Electric Fields. *J. Food Sci.* **2004,** *69,* 134–138.

Geveke, D. J.; Brunkhorst, C. Radio Frequency Electric Fields Inactivation of *Escherichia Coli* in Apple Cidar. *J. Food Eng.* **2008,** *85,* 215–221.

Geveke, D. J.; Brunkhorst, C.; Fan, X. Radio Frequency Electric Fields Processing of Orange Juice. *Inn. Food Sci. Emerg. Technol.* **2007,** *8,* 549–554.

Guan, D.; Cheng, M.; Wang, Y.; Tang, J. Dielectric Properties of Mashed Potatoes Relevant to Microwave and Radio-frequency Pasteurization and Sterilization Processes. *J. Food Sci.* **2004,** *69* (1), 30–37.

Guo, Q.; Piyasena, P.; Mittal, G. S.; Si, W.; Gong, J. Efficacy of Radio Frequency Cooking in the Reduction of *Escherichia coli* and Shelf Stability of Ground Beef. *Food Microbiol.* **2006,** *23* (2), 112–118.

Houben, J.; Schoenmakers, L.; Van Putten, E.; Van Roon, P.; Krol, B. Radio-frequency Pasteurization of Sausage Emulsions as a Continuous Process. *J. Microwave. Power. Elec. Energy* **1991,** *26* (4), 202–205.

Jason, A. C.; Sanders, H. R. Thawing of Frozen Fish in Moist Air. *Food Technol.* **1962,** *16* (6), 101.

Jones, P. L.; Rowley, A. T. *Industrial Drying of Foods*; Baker; C. G. J., Ed.; Blackie Academic and Professional: London, 1997; Chapter 8.

Kirmaci, B.; Singh, R. K. Quality of Chicken Breast Meat Cooked in a Pilot Scale Radio Frequency Oven. *Inn. Food Sci. Emerg. Technol.* **2012,** *14,* 77–84.

Koral, T Radio Frequency Heating and Post Baking –Amaturing Technology that Can Still Offer Significant Benefits. *Bisc World.* **2004,** *4* (7), 1–5.

Laycock, L.; Piyasena, P.; Mittal, G. S. Radio Frequency Cooking of Ground, Comminuted and Muscle Meat Products. *Meat. Sci.* **2003,** *65* (3), 959–965.

Lagunas-solar, M. C.; Cullor, J. S.; Zeng, N. X.; Truong, T. D.; Essert, T. K.; Smith, W. L.; Pina, C. Disinfection of Dairy and Animal Farm Wastewater with Radiofrequency Power. *J. Dairy. Sci.* 2005, *88,* 4120–4131.

Lee, Toledo, R. T.; Nelson, S. O. The Dielectric Properties of Fresh and Marinated Chicken Breast Meat. Unpublished Data, Personal Communication, 2008.

Luechapattanaporn, K.; Wang, Y.; Wang, J.; Tang, J.; Hallberg, M.; Dunne, C. P. Sterilization of Scrambled Eggs in Military Polymeric Trays by Radio Frequency Energy. *J. Food Sci.* **2005,** *70* (4), 288–294.

Manzocco, L.; Anese, M.; Nicoli, M. C. Radio Frequency Inactivation of Oxidative Food Enzymes in Model Systems and Apple Derivatives. *Food Res. Int.* **2008,** *41* (10), 1044–1049.

Marra, F.; Zhang, L.; Lyng, J. G. Radio Frequency Treatment of Foods: Review of Recent Advances. *J. Food Eng.* **2009,** *91* (4), 497–508.

McKenna, B. M.; Lyng, J.; Brunton, N.; Shirsat, N. Advances in Radio-frequency and Ohmic Heating of Meats. *J. Food Eng.* **2006,** *77* (2), 215–22.

Mermelstein, N. H. Microwave and Radio Frequency Drying. *Food Technol.* **1998,** *52* (11), 84–86.

Nelson, S. O.; Datta, A. K. Dielectric Properties of Food Materials and Electric Field Interactions. In *Handbook of Microwave Technology for Food Applications*; Datta, A. K., Anantheswaran, R. C., Eds.; Marcel Dekker: New York, 2001.

Ohlsson, T. Minimal Processing of Foods with Electric Heating Methods. In *Processing Foods, Quality Optimisation and Process Assessment;* Oliviera, F. A. R., Oliviera, J. C., Hendrickx, M. E., Knorr, D., Gorris, L., Eds.; CRC Press: Boca Raton, FL, 1999; pp 97–105.

Orsat, V.; Bai, L.; Raghavan, G. S. V.; Smith, J. P. Radio-frequency Heating of Ham to Enhance Shelf-life in Vacuum Packaging. *J. Food Process. Eng.* **2004,** *27,* 267–283.

Orsat, V.; Gariepy, Y.; Raghavan, G. S.; Lyew, D. Radio Frequency Treatment for Ready to Eat Fresh Carrots. *Food Res. Intl.* **2001,** *34,* 527–536.

Ozturka, S.; Fanbin, K.; Samir, T.; Rakesh, K. S. Dielectric Properties of Dried Vegetable Powders and Their Temperature Profile during Radio Frequency Heating. *J. Food Eng.* **2016,** *169,* 91–100.

Parker, J.; Reath, M.; Krauss, A.; Campbell, W. Monitoring and Preventing Arc-Induced Wafer Damage in 300mm Manufacturing. *Intergr. Circuit Design Technol..* 2004, 131–134.

Piyasena, P.; Dussault, C.; Koutchma, T.; Ramaswamy, H. S.; Awuah, G. B. Radio Frequency Heating of Foods: Principles, Applications and Related Properties-a Review. *Crit. Rev. Food Sci. Nutr.* **2003,** *43*(6), 587–606.

Rice, J. RF Technology Sharpens Bakery's Competitive Edge. *Food Proc.* **1993,** *6,* 18–24.

Rowley, A. T. Radio Frequency Heating. In *Termal Technologies in Food Processing;* Richardsons, P. Ed.; Woodhead Publishing Ltd.: Cambridge, UK, 2001; pp 163–177.

Ryynanen, S. The Electromagnetic Properties of Food Materials: A Review of the Basic Principles. *J. Food Eng.* **1995,** *26* (4), 409–429.

Samaranayake, C. P.; Sastry, S. K.; Zhang, H. Pulsed Ohmic Heating—a Novel Technique for Minimization of Electrochemical Reactions during Processing. *J. Food Sci.* **2005,** *70* (8), 460–465.

Sona, J.; Mahendran, R. Radio Frequency Heating and its Application in Food Processing: A Review. *Int. J. Cur. Agric. Res.* **2013,** *1,* 042–046.

Sumnu, G. A Review on Microwave Baking of Foods. *Int. J. Food Sci. Technol.* **2001,** *36* (2), 117–127.

Tang, X.; Cronin, D. A.; Brunton, N. P. The Effect of Radio Frequency Heat on Chemical, Physical and Sensory Aspects of Quality in Turkey Breast Rolls. *Food Chem.* **2005,** *93* (1), 1–7.

Tang, X.; Lyng, J. G.; Cronin, D. A.; Durand, C. Radio Frequency Heating of Beef Rolls from Biceps Femoris Muscle. *Meat Sci.* **2006,** *72* (3), 467–474.

Tewari, G. Radio-frequency Heating. In *Advances in Thermal and Non-thermal Food Preservation;* Tewari, G., Juneja, V. K., Eds.; Blackwell Pub.: Hoboken, NJ, 2007.

Tiwari, G.; Wang, S.; Birla, S. L.; Tang, J. Effect of Water-assisted Radio Frequency Heat Treatment on the Quality of "Fuyu" Persimmons. *Biosyst. Eng.* **2008,** *100* (2), 227–234.

Uemura, K. Inactivation if *Bacillus subtilis* Spores in Soybean Milk by Radiofrequency. *J. Food Eng.* **2010,** *100,* 622–626.

van Roon, P. S.; Houben, J. H.; Koolmees, P. A.; van Vliet, T. Mechanical and Microstructural Characteristics of Meat Doughs, Either Heated by a Continuous Process in a Radiofrequency Field or Conventionally in a Water Bath. *Meat Sci.* **1994,** *38* (1), 103–116.

Vela, G. R.; Wu, J. F. Mechanism of Lethal Action of 2,450-MHz Radiation on Microorganisms. *App. Envr. Microbiol.* **1979,** *37* (3), 550–553.

Wang, Y.; Wig, T. D.; Tang, J. Hallberg, L. M. Dielectric Properties of Foods Relevant to RF and Microwave Pasteurization and Sterilization. *J. Food Eng.* **2003,** *57* (3), 257–268.

Wang, S.; Birla, S. L.; Tang, J.; Hansen, J. D. Postharvest Treatment to Control Codling Moth in Fresh Apples Using Water Assisted Radio Frequency Heating. *Postharvest. Biol. Technol.* **2006,** *40* (1), 89–96.

Wang, S.; Tiwari, G.; Jiao, S.; Johnson, J. A.; Tang, J. Developing Postharvest Disinfestation Treatments for Legumes Using Radio Frequency Energy. *Biosyst. Eng.* **2010,** *105* (3), 341–349.

Zhang, L.; Lyng, J. G.; Brunton, N. P. Effect of Radio-frequency Cooking on the Texture, Colour and Sensory Properties of a Large Diameter Comminuted Meat Product. *Meat Sci.* **2004,** *68* (2), 257–268.

Zhang, L.; Lyng, J. G.; Brunton, N. P. Quality of Radio-frequency Heated Pork Leg and Shoulder Ham. *J. Food Eng.* **2006,** *75* (2), 275–287.

Zhang, L.; Lyng, J. G.; Brunton, N. P. The Effect of Fat, Water and Salt on the Thermal and Dielectric Properties of Meat Batter and its Temperature Following Microwave or Radio Frequency Heating. *J. Food Eng.* **2007,** *80* (1), 142–151.

Zhao, Y.; Flugstad, B.; Kolbe, E.; Park, J. E.; Wells, J. H. Using Capacitive (Radio Frequency) Dielectric Heating in Food Processing and Preservation–a Review. *J. Food Process. Eng.* **2000,** *23,* 25–55.

Zhuang, H.; Nelson, S.O.; Trabelsi, S.; Savage, E. M. Dielectric Properties of Uncooked Chicken Breast Muscles from Ten to One Thousand Eight Hundred Megahertz. *Poult Sci.* **2007,** *86,* 2433–2440.

CHAPTER 15

MICROENCAPSULATION OF BIOACTIVE FOOD INGREDIENTS: METHODS, APPLICATIONS, AND CONTROLLED RELEASE MECHANISM—A REVIEW

JEYAKUMARI A.[1], PARVATHY U.[1], ZYNUDHEEN A. A.[2], L. NARASIMHA MURTHY[1], S. VISNUVINAYAGAM[1], and RAVISHANKAR C. N.[2]

[1]*Mumbai Research Centre of ICAR—Central Institute of Fisheries Technology, Vashi 400703, Navi Mumbai, India*

[2]*ICAR—Central Institute of Fisheries Technology, Kochi 682029, India*

Corresponding author. E-mail: jeya131@gmail.com

CONTENTS

Abstract 400
15.1   Introduction 400
15.2   Overview of Microencapsulation Technologies 401
15.3   Microencapsulation Methods 403
15.4   Microencapsulation of Bioactive Ingredients 408
15.5   Characterization of Microcapsules 414
15.6   Controlled Release Mechanism 416
15.7   Conclusion 420
Keywords 420
References 421

ABSTRACT

Microencapsulation is a process of coating of small particles of solid or liquid material (core) with protective coating material (matrix) to produce microcapsules in the micrometer to millimeter range. It is one of the methods of protecting sensitive substances and producing active ingredients with improved properties. Many different active materials like lipids, proteins, vitamins, minerals, enzymes, and flavors have been successfully encapsulated. To produce effective encapsulated products, the choice of coating material and method of microencapsulation process are most important and it also depends on the end use of the product and the processing conditions involved. These microcapsules release their contents at desired rate and time by different release mechanisms, depending on the encapsulated products which provide wide application of food ingredients thereby improving the cost effectiveness for the food manufacturer. This chapter highlights the various microencapsulation methods and its application in the encapsulation of bioactive food ingredients and controlled release mechanisms.

15.1 INTRODUCTION

Consumers' demand for healthy food products is increasing worldwide. Today foods are not intended only to satisfy hunger and provide necessary nutrients for humans. It also intended to prevent nutrition-related diseases and improve physical and mental well-being. In this regard, functional foods play an outstanding role. The foods that enriched with functional components to offer medical and physiological benefits to reduce the risk of chronic diseases beyond their basic nutritional functions are called as functional foods. Bioactives in food are physiologically active components that provide health benefits beyond their nutritional role. These include proteins, lipids, phytosterols, phytochemicals, antioxidants, minerals, vitamins, and probiotic bacteria (Augustin, 2007). These bioactives are very sensitive and their application in food is a great challenge to the industry without affecting their properties. Encapsulation technology has proven to be an excellent tool to protect sensitive food ingredients from degradation that has provided researchers with opportunities to develop foods of novel formulations with improved solutions (Schrooyen, 2001; Pegg & Shahidi, 2007). Microencapsulation is a technique of coating small particles of finely ground solids, drops of liquids, or gaseous components, with protective membranes—microcapsule walls (Calvo et al., 2011). Microcapsules or micron size varies from 2 to 5000 μm. In the food industry, the microencapsulation process can be applied for a variety of reasons, which have been summarized by Desai & Park (2005) as follows as (1) to protect the core material from degradation and to reduce the evaporation rate of the core material to the surrounding environment; (2) to modify the nature of the original material for easier handling; (3) to release the core material slowly over time at the constant rate; (4) to prevent unwanted flavor or taste of the core material; and (5) to separate the components of the mixture that would react one another. Depends on the consumer needs, microencapsulation process has been improved constantly. As a result, it has become an example of a dynamic and technological intensive process method (Boh & Kardos, 2003) characterized by a fast growth of patent in microencapsulation process and its applications, as well as by an increasing number of scientific research articles.

There are separate extensive reviews on microencapsulation techniques used in the food industry. However, there is a need to discuss the different carriers and methods with particular focus on encapsulating bioactive food ingredients. The objective of this chapter is to review the state of the art of microencapsulation technologies in a four perspectives. First, it focuses on theoretical aspects of different types of microencapsulation techniques and criteria required for encapsulating agents. Second, this chapter discusses microencapsulation of different bioactive food ingredients such as omega-3 fatty acids, polyphenols, enzymes, protein hydrolysate and peptides, microorganisms, vitamins, and minerals and its applications. The third section summarizes the characterization of microcapsules and the final part summarizes controlled release mechanisms of microcapsules.

15.2 OVERVIEW OF MICROENCAPSULATION TECHNOLOGIES

The substance that is encapsulated may be called the core material, the active agent, internal phase, or payload phase. The substance that is encapsulating may be called the coating, membrane, shell, carrier material, wall material, external phase, or matrix. Two main types of encapsulates are reservoir type and matrix type (Zuidam & Shimoni, 2010). In reservoir type, the active agents forms a core surrounded by an inert diffusion barrier. It is also called single-core or mono-core or core-shell type. In matrix type, the active agent is dispersed or dissolved in an inert polymer. Coated matrix type is a combination of first two (Fig. 15.1).

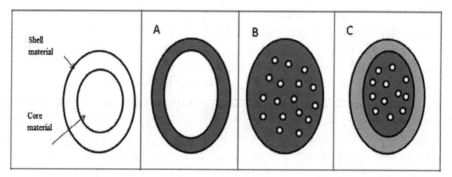

FIGURE 15.1   Morphology of microcapsule (A) reservoir type; (B) matrix type; and (C) coated matrix.

15.2.1   CORE MATERIAL

The core material is the material over which coating has to be applied to serve the specific purpose. Core material may be in form of solids or droplets of liquids and dispersions. The composition of the core material can be varied, as the liquid core can include dispersed and/or dissolved materials. The solid core can be active constituents, stabilizers, diluents, excipients, and release-rate retardants or accelerators (Hammad et al., 2011).

15.2.2   COATING MATERIAL

The coating material should be capable of forming a film that is cohesive with the core material be chemically compatible and nonreactive with the core material and provide the desired coating properties, such as strength, flexibility, impermeability, optical properties, and stability. Coating material should also have the properties such as stabilization of core material, inert toward active ingredients, controlled release under specific conditions, tasteless, non-hygroscopic, no high viscosity, soluble in an aqueous media or solvent, or melting and economical (Hammad et al., 2011).

Coating materials used for microencapsulation of hydrophobic active agent are sugar, starch, glucose syrup, maltodextrin, corn syrup solids, dextran, modified starch, sucrose, dextrin, cyclodextrins, octenyl succinic anhydride (OSA) starch, carboxy methyl cellulose, methylcellulose, ethyl cellulose, nitrocellulose, acetyl cellulose, cellulose acetate-phthalate, cellulose acetate-butyrate-phthalate, gum acacia, gum arabic, agar, sodium alginate, carrageenan, pectin, gluten, casein, gelatin, hemoglobin, peptides, soy, wheat, corn (zein), whey protein concentrate, whey protein isolate, cellulose derivatives, chitosan, etc. (Shahidi & Han, 1993; Robert et al., 2014).

Coating materials used for microencapsulation of hydrophilic active agent are wax, tri-stearic acid, diacylglycerols, monoacylglycerols, oils, hydrogenated fats, hardened oils, phospholipids, fatty acids, plant sterols, sorbitan esters, beeswax, paraffin wax, microcrystalline wax, carnuba wax, shellac, ethyl cellulose, etc. (Shahidi & Han,1993; Robert et al., 2014).

15.3 MICROENCAPSULATION METHODS

The microcapsules are prepared by a variety of methods. The microencapsulation process can be divided in to physical and chemical process. Physical process includes spray drying, spray chilling, rotary disk atomization, fluid bed coating, stationary nozzle coextrusion, centrifugal head coextrusion, submerged nozzle coextrusion, and pan coating. Chemical process includes phase separation, solvent evaporation, solvent extraction, interfacial polymerization, simple, and complex coacervation and in situ polymerization (Zuidam & Heinrich, 2010). Different types on microencapsulation techniques and their properties, merits, and demerits are summarized below.

15.3.1 *PHYSICAL METHODS OF MICROENCAPSULATION*

15.3.1.1 *SPRAY-DRYING*

It is widely used for the encapsulation of food additives, functional ingredients, and flavors. The major process involves dispersion of the substances to be encapsulated in a carrier material followed by atomization and dehydration of the atomized particles. During this process a film is formed at the droplet surface, there by retarding the larger active molecules while the smaller water molecules are evaporated. The particle size varied between 10 and 400 um. Morphology of encapsulates produced by this method are matrix type. Advantages of spray drying are relatively simple method, fast and easy to scale-up, equipment is readily available; cost of spray-drying method is 30–50 times cheaper; both hydrophilic and hydrophobic polymer can be used; ideal for production of sterile materials. Disadvantages of spray drying is that considerable amounts of the material can be lost during the process due to sticking of the microparticles to the wall of the drying chamber; process variables that should be optimized for encapsulation (Adem et al., 2007; Atmane et al., 2006; Zuidam & Shimoni, 2010).

15.3.1.2   SPRAY CHILLING OR SPRAY COOLING

In the spray chilling, the coating material is melted and atomized through a pneumatic nozzle into a vessel generally containing a carbon dioxide ice bath (temperature −50°C) as in a hot-melt fluidized bed followed by droplets adhere on particles and solidify forming a coat film. The particle size varies from 20–200 μm. Morphology of encapsulates produced by this method are matrix type. It is used for encapsulation of aroma compounds to improve heat stability. Advantages of this method are least expensive; active compounds released within a few minutes after being incorporated in the food stuff. A disadvantage of this method is that special handling and storage conditions can be required (Atmane et al., 2006; Zuidam & Shimoni, 2010).

15.3.1.3   FLUID BED COATING

It is widely used in the pharmaceutical and cosmetic industry. Moreover, it is also used in the food industry to encapsulate flavors. The major process involved in fluid bed coating are (1) preparation of coating solution; (2) fluidization of core particles; (3) coating material is sprayed through a nozzle on to the particles and film formation is initiated; and (4) drying. The size of the particle varies from 5 to 5000 μm. Morphology of encapsulates produced by this method are reservoir type. Advantages of fluid bed coating are (1) uniform layer of shell material onto solid particles and (2) high thermal efficiency; lower capital and maintenance costs. Disadvantages of this method include (1) control of air stream and air temperature is a critical factor and (2) to achieve uniform particle, coating droplets must be significantly smaller than core (Zuidam & Shimoni, 2010; Sebastien, 2004).

15.3.1.4   SPINNING DISK AND CENTRIFUGAL CO-EXTRUSION

Spinning disk involves the formation of a suspension of core particles in the coating liquid and the passage of this suspension over a rotating disk under conditions that result in a film of the coating much thinner than the core particle size. Atomization of the mixture at the edge of the disk results in coated core particles. Centrifugal coextrusion is based on a modified double fluid nozzle where the active ingredient is pumped through the inner part of the nozzle while the shell material is pumped through the outer part of the nozzle. At the edge of the nozzle round beads are formed that are constituted of active ingredient in the core and an outer layer of the shell material. The particle size varies from 150 to 8000 μm. Morphology of encapsulates produced by this method are reservoir type. Advantages of this method is that product outputs are comparable or even higher than regular spray drying or spray cooling processes. However, the major problem in this process are difficult to obtain in extremely viscous carbohydrate melts; maintain the emulsion stability; direct observation of the particles during production is more difficult (Zuidam & Shimoni, 2010; Sebastien, 2004).

15.3.1.5 EXTRUSION

In this method, a core material dispersed in a molten carbohydrate passes through a series of dics into a bath of dehydrating liquid. Upon contact with the liquid, the coating material, which forms the encapsulating matrix, hardens, and entraps the core material. Isopropyl alcohol is the most common liquid employed for the dehydration and hardening process. The particle size varies from 200 to 5000 μm. Morphology of encapsulates produced by this method are matrix type. This method is used for encapsulation of flavors. The principal advantage of the extrusion is the stability of flavors against oxidation and also product shelf-life is long (e.g., 5 years for extruded flavor oils). Disadvantages of this method are (1) large particles formed by extrusion and (2) very limited range of shell material is available (Zuidam & Shimoni, 2010; Desai & Park, 2005).

15.3.1.6 FREEZE-DRYING/LYOPHILIZATION

It is one of the most useful processes for drying of thermo sensitive substances. The major steps involved in this process are (1) mixing of core in coating solution; (2) freeze-drying of the mixture; and (3) grinding (option). The particle size varies from 20 to 5000 μm. Morphology of encapsulates produced by this method are matrix type. Advantages of freeze drying includes (1) it is used for encapsulation of heat-sensitive materials and aromas; (2) removal of oxygen and application of low temperature, which helps minimize the product oxidation. Disadvantages of this method are (1) high energy use, the long processing time, and the open porous structure obtained and (2) compared to spray-drying, freeze-drying is up to 30–50 times more expensive (Atmane et al., 2006; Zuidam & Shimoni, 2010).

15.3.2   CHEMICAL METHODS OF MICROENCAPSULATION

15.3.2.1   COACERVATION

In this method, the core material is emulsified in the protein solution, and formation of coacervate wall is initiated by changing either the temperature, pH, or by adding a concentrated salt solution. The resultant microcapsules are isolated by centrifugation or filtration and might be dried by spray drying or fluid bed drying. The particle size varies from 10 to 800 µm. Morphology of encapsulates produced by this method are reservoir type. It is used to encapsulation of polyphenols and water-soluble compounds. Disadvantage of this method is that mass production is difficult due to agglomeration (Fang & Bhandari, 2010; Zuidam & Shimoni, 2010).

15.3.2.2   SUPERCRITICAL FLUIDS TECHNOLOGY

Supercritical fluids have recently been used for encapsulation of heat-sensitive material (enzymes, volatile flavors, etc.). In this method, process concept is very similar to spray drying, but super critical fluids are used for the solubilization/swelling of the coating material and the core material instead of water as in spray drying. When a pressurized supercritical solvent containing the shell material and the active ingredient is released through a small orifice, the abrupt pressure drop causes the desolvation of the shell material and the formation of a coating layer around the active ingredient. Particle size varies from 10 to 400 µm. Morphology of encapsulates produced by this method are matrix type. The main advantage here is no requirements of surfactants, yielding a solvent-free product, and mild process conditions ($T < 30°C$ throughout the process). Disadvantage of this method is that all solutes should be soluble in the supercritical fluid (Zuidam & Shimoni, 2010; Sebastien, 2004).

15.3.2.3   LIPOSOME ENTRAPMENT

Liposome consists of an aqueous phase that is completely surrounded by a phospholipid-based membrane. When phospholipids, such as lecithin, are dispersed in an aqueous phase, the liposomes form spontaneously. One can either have aqueous or lipid soluble material enclosed in the liposome. Microfluidization, ultrasonication, and reverse-phase evaporation technique can be used to produce different varieties of liposomes for specific purposes. The particle size varies from 10 to 1000 µm. Liposomes are mainly studied for clinical application as pharmaceutical drug carriers (Anu Puri et al., 2009) and their use in foods is limited due to its physical and chemical instability. It is mainly used to enhance the ripening of hard cheeses (Anjani et al., 2007). Disadvantage of this method is that high cost, low stability and low encapsulation yield (Pegg & Shahidi, 2007; Zuidam & Shimoni, 2010).

15.3.2.4 COCRYSTALLIZATION

In this method, sucrose is concentrated to the supersaturated state and maintained at a temperature high enough to prevent crystallization. A predetermined amount of core material is then added to the concentrated syrup with vigorous mechanical agitation, thus providing nucleation for the sucrose- ingredient mixture to crystallize. As the syrup reaches the temperature at which transformation and crystallization begin, a substantial amount of heat is emitted. Agitation is continued to promote and extend transformation/ crystallization until the agglomerates are discharged from the vessel. The encapsulated products are then dried to desirable moisture (if necessary). The particle size varies from 2 to 30 µm. Morphology of encapsulates produced by this method appear as cluster like agglomeration. The crystallization process offers several advantages such as improved solubility, homogeneity, hydration, and flowability, core material in a liquid form can be converted into dry powdered form without additional drying. Disadvantage of this method includes the granular product has a lower hygroscopicity, heat sensitive core material may get degrade during process (Pegg & Shahidi, 2007; Zuidam & Shimoni, 2010).

15.3.2.5 INCLUSION COMPLEXATION

In this method, β-cyclodextrin is typically used as the encapsulating medium. β-cyclodextrin is a cyclic glucose oligomer, consisting of seven β-D-glucopyranosyl units linked by α-(1–4) bonds. β-cyclodextrin molecule forms inclusion complexes with compounds that can fit dimensionally into its central cavity. These complexes are formed in a reaction that takes place only in the presence of water. This reaction can be accomplished by stirring or shaking the cyclodextrin and core material to form a complex, which could then be easily filtered and dried. Molecules that are less polar than water (i.e., most flavor substances) and have suitable molecular dimensions to fit inside the cyclodextrin interior can be incorporated into the molecule. The particle size varies from 5 to 15 μm. The main advantage of this method is the unique release characteristics and the protection of unstable and high value specialty flavor chemicals. Disadvantages of this method are (1) limited amount of flavor (9–14%) can be incorporated; (2) cyclodextrin is very expensive; and (3) cyclodextrin can act as an artificial enzyme and it may lead to enhance the rate of hydrolysis which result in undesirable adulteration of the flavor (Pegg & Shahidi, 2007; Zuidam & Shimoni, 2010; Desai & Park, 2005).

15.4 MICROENCAPSULATION OF BIOACTIVE INGREDIENTS

There are numerous methods used for microencapsulation of bioactive ingredients. But no single encapsulation process is adaptable to all core materials or product applications. Microencapsulation methods used for bioactive ingredients such as omega-3 fatty acids, polyphenols, vitamins and minerals, calcium, enzymes, and flavors are discussed below.

15.4.1 ENCAPSULATION OF OMEGA-3 FATTY ACIDS

Omega-3 fatty acids belong to family of polyunsaturated fatty acids (PUFAs). As the name suggests, these have multiple double bonds with the first double bond placed at the third carbon starting from the methyl end. The first member of the omega-3 family is alpha-linolenic acid (ALA, 18:3n-3), which is not synthesized by the human body. However, ALA plays an important role in several physiological functions in human body and hence is recognized as essential in the diet. The other nutritionally important omega-3 fatty acids are the longer chain metabolites of ALA, eicosapentaenoic acid (EPA, 20:5n-3) and docosahexaneoic acid (DHA, 22:6n-3). Major dietary sources of ALA are dark green leafy vegetables, certain nuts, seeds and their oils, flaxseed, hempseed and walnut, canola, perilla, chia, kiwifruit, etc. EPA and DHA are long chain PUFA (LCFUFA) obtained in our diet principally from marine sources. Major dietary sources of EPA and DHA are fatty fish such as herring and mackerel, liver of lean white fish such as cod and halibut, blubber of marine mammals such as whales and seals and algal species (Kris-Etherton *et al.,* 2000).

Due to the highly unsaturated nature of omega-3 fatty acids, these are susceptible to oxidation and readily produce hydroperoxides, off flavors and odors, which are deemed undesirable by consumers. To overcome the above-mentioned problems, the use of microencapsulation technology has been explored by various researchers (Klinkesorn et al., 2005; Jeyakumari et al., 2014; Jeyakumari et al., 2015; Binsi et al., 2017). Omega-3 fatty acids have been microencapsulated using different encapsulation techniques. So far, spray drying, complex coacervation, and extrusion are the most commonly used commercial techniques for microencapsulation of omega-3 fatty acids. Coating material used for microencapsulation of omega-3 fatty acids includes gelatin, maltodextrin, casein, lactose, sodium caseinate, dextrose equivalence, highly branched cyclic dextrin, methylcellulose, hydroxypropyl methylcellulose, n-octenylsuccinate, derivatized starch/glucose syrup or terhalose, sugar beet pectin, gum arabic, corn syrup solids, egg white powder, whey protein isolate, lecithin, whey protein concentrate, modified starches, tapioca starch, waxy maize, lactose, lecithin, chitosan, transglutaminase, zein prolamine (corn protein), soybean soluble polysachharide (SSPS), hydroxypropyl betacyclodextrin (HPBCD), etc. (Pratibha et al., 2014).

15.4.2 ENCAPSULATION OF POLYPHENOLS/FLAVORS

Flavors play an important role in consumer satisfaction and influence further consumption of foods. Food manufacturers are usually concerned about the preservation of aromatic additives, since aroma compounds are not only delicate and volatile, but also very expensive (Atmane et al., 2006). Recently, the market of flavors is focused on using aromatic materials coming from natural sources to replace the use of synthetic flavors gradually (Teixeira et al., 2004). Flavor is indispensable ingredients of food preparations, and their compositions are often highly complex. Commercial food flavors in liquid form are difficult to handle or incorporate into food. Moreover, many flavor components exhibit considerable sensitivity to oxygen, light, and heat. In response to these difficulties, dry flavors have been developed through encapsulation. Encapsulation provides an effective method to protect flavor compounds from evaporation, degradation, and migration from food.

Essential oils (EOs) are volatile, complex mixtures of compounds characterized by a strong odor, and they are formed by aromatic plants as secondary metabolites. Several EOs such as garlic, cinnamon, thyme, oregano, clove, basil, coriander, citrus peel, eucalyptus, ginger, rosemary, and peppermint have been demonstrated various biological properties activities, including antioxidant, anti-inflammatory, antibacterial, and antiviral functions (Bennick, 2002). A large body of preclinical research and epidemiological data suggests that plant polyphenols can slow the progression of certain cancers, reduce the risks of cardiovascular disease, neurodegenerative diseases, diabetes, or osteoporosis suggesting that plant polyphenols might act as potential chemo preventive and anti-cancer agents in humans (Arts & Hollman, 2005; Scalbert et al., 2005). Major polyphenols found in different plant sources are anthocyanidins, catechins, flavanones, isoflavones, flavonols, hydroxybenzoic acids, hydroxycinnamic acids, lignans, tannins (proanthocyanidines), etc. (Fang & Bhandari, 2010). Due to their instability during processing and another unfortunate trait of polypheonls is their potential unpleasant taste, such as astringency which needs to be masked before incorporation into food products (Manach et al., 2004). The utilization of encapsulated polyphenols instead of free compounds can overcome the drawbacks of their instability, alleviate unpleasant tastes or flavors, as well as improve the bioavailability and half-life of the compound in vivo and in vitro. Phenols and phenylpropanoids are classes of compounds very common to EOs. They are sometimes called hemiterpenes. They are found in clove (90%), cassia (80%), basil (75%), cinnamon (73%), oregano (60%), anise (50%), and peppermint (25%).

Polyphenols have been microencapsulated by different methods such as spray drying, coacervation, co-crystallization, freeze drying, molecular inclusion, extrusion, electrostatic extrusion, etc. Coating material used for encapsulation of polyphenols are maltodextrin, gum arabic, chitosan, citrus fruit fiber, colloidal silicon dioxide, sodium caseinate-soy lecithin, skimmed milk powder, whey protein concentrate, gelatin, calcium alginate , chitosan, glucan, ᴋ-carrageenan, sucrose syrup, pullulan, corn syrup solids, glycerin, sodium alginate, cyclodextrin, and modified starch (Fang & Bhandari, 2010; Amr et al., 2015).

15.4.3   ENCAPSULATION OF VITAMINS AND MINERALS

Both lipid-soluble (e.g., vitamin A, β-carotene, vitamins D, E, and K) and water-soluble (e.g., ascorbic acid) vitamins can be encapsulated using various technologies (Wilson & Shah, 2007). The most common reason for encapsulating these ingredients is to extend the shelf-life, either by protecting them against oxidation or by preventing reactions with components in the food system in which they are present. Iron (Fe) is one of the most important elements, and its deficiency affects about one third of the world's population.

The best way to prevent this problem is through the Fe fortification of food. However, the bioavailability of Fe is negatively influenced by interactions with food ingredients such as tannins, phytates, and polyphenols. Moreover, Fe catalyses oxidative processes in fatty acids, vitamins, and amino acids, and consequently alters sensory characteristics and decreases the nutritional value of the food. Microencapsulation can be used to prevent these reactions. The bioavailability of readily water soluble Fe salts such as ferrous sulfate, ferrous lactate is higher than that of poorly water-soluble (e.g., ferrous fumarate) or water-insoluble (e.g., ferrous phosphate) Fe salts. Liposome technology is the method of choice for Fe fortification of fluid food products. Fe bioavailability study on milk enriched with ferrous sulfate encapsulated in a lecithin liposome has been conducted (Boccio et al., 1997; Uicich et al., 1999). Heat treatment and storage for 6 months did not result in decreased Fe bioavailability of the Fe-fortified milk. Bioavailability was similar to absorption of Fe from high-bioavailable $FeSO_4$.

Microencapsulation methods used for vitamins and minerals includes spray drying, spray cooling and spray chilling, liposome entrapment, extrusion, and fluidized bed coating. Coating material used for encapsulation of vitamins and minerals are tripolyphosphate, cross-linked chitosan, starch, β-cyclodextrin, maltodextrin, gum arabic, waxes, fatty acids, water-soluble polymers and water-insoluble monomers, soy lecithin, egg phosphatidylcholine, cholesterol, DL-α-tocopherol, lactose, fructooligosaccharide, polymethacrylate, ethylcellulose, hydrogenated vegetable oil, stearin, gelatin, gum acacia, and β-cyclodextrin (Shabbar Abbas et al., 2012; Wilson & Shah, 2007; Goncalves et al., 2016).

15.4.4 ENCAPSULATION OF CALCIUM

Soya milk contains much less calcium (12 mg/100 g) than cow's milk (120 mg/100 g), which is undesirable from a nutritional point of view. Attempts to fortify soya milk with calcium salts (calcium triphosphate and calcium citrate) have been unsuccessful because this process causes the soybean proteins to coagulate and precipitate. By encapsulating the calcium salt (calcium lactate) in a lecithin liposome it was possible to fortify 100 g soya milk with up to 110 mg calcium, thereby reaching levels equivalent to those in normal cow's milk. The soya milk remained stable at 4°C for at least 1 week. It would be desirable to know how stable a fortified long-shelf-life soya milk would be, and how the calcium bio-availability of soya milk compares with that of cow's milk (Motohiko et al., 2006).

15.4.5   ENCAPSULATION OF ENZYMES

Enzymes are biomacromolecules or in other words complex protein molecules with specific catalytic functions that are produced by all living cells to catalyze the biochemical reactions required for life. Enzymes have some excellent properties (high catalytic activity, selectivity, and specificity). Because of their enormous catalytic power in aqueous solution at normal temperatures and pressures, enzymes are of great commercial and industrial importance. In the microencapsulation method, enzyme is entrapped within a semipermeable membrane so that the activity of enzyme is not affected. But the movement of the substrate to the active site may be restricted by the diffusional limitations especially when large molecules like starch and proteins are used, which can have an adverse effect on the enzyme kinetics (Cisem, 2011).

Microencapsulation methods used for encapsulation of proteolytic and lipolytic enzymes includes complex coacervation, spray drying and liposome entrapment. Coating material used for encapsulation of enzymes are sodium alginate, starch, chitosan, modified chitosan (water soluble), calcium alginate, gum arabic, α-amalase and carrageenan, etc. (Cisem, 2011; Kailasapathy et al. 1998; Anjani et al., 2007).

15.4.6   ENCAPSULATION OF MICROORGANISM

Probiotic bacteria are defined, live microorganisms which, administered in adequate amounts, confer a beneficial physiological effect on the host (humans or animals). These bioactive ingredients have been at the forefront of the development of functional foods, particularly in dairy products (Sanders, 2003).

There are five microencapsulation methods have been applied to probiotics such as spray-coating (fluid bed coating), spray-drying, extrusion, emulsion, and gel particle technologies (which include spray-chilling). Among these spray-coating and gel-particle technologies are most often used for microencapsulation of probiotics (Champagne & Patrick Fustier, 2007). Coating materials used for microencapsulation of microorganisms includes alginate and its combinations, high-amylose corn starch, mixture of xanthan–gellan, carrageenan and its mixtures, gelatin or gelatin and gum, cellulose acetate phthalate, and mixture of chitosan and hexamethylene diisocyanate (Vidhyalakshmi et al., 2009).

15.4.7 ENCAPSULATION OF PROTEIN HYDROLYSATE AND PEPTIDE

Food protein hydrolysates and peptides are considered a category of promising functional food ingredients. However, commercial application of protein hydrolysates and their constituent peptides can be impeded by their low bioavailability, bitter taste, hygroscopicity and likelihood of interacting with the food matrix. Encapsulation as a delivery mechanism can be used to overcome these challenges for improving the bioavailability and organoleptic properties of the peptides. Bioactive peptides are different from other food bioactive compounds such as vitamins or polyphenols in that the chemical species within the protein hydrolysates are highly heterogeneous (McClements, 2014). Bioactive peptides are defined as "food derived components (genuine or generated) that, in addition to their functional value, exert a physiological effect in the body" (Vermeirssen et al., 2007). Consequently, bioactive peptides may need to be isolated from more complex matrices or fractionated prior to encapsulation. Bioactive peptides are primarily encapsulated for the purpose of masking the bitter taste that result from exposure of taste receptors to hydrophobic amino acid residues generated from protein hydrolysis (Erdmann et al., 2008) and also reduction of hygroscopicity. Proteins, polysaccharides, and lipids are the three carrier systems that have been utilized in food peptide encapsulation. The protein and polysaccharide systems mainly aim at masking the bitter taste and reducing the hygroscopicity of protein hydrolysates, whereas the lipid-based systems are intended for use in enhancing the bioavailability and biostability of encapsulated peptides.

Microencapsulation methods used for encapsulation of protein hydrolysate and peptides includes spray drying, coacervation, and liposome entrapment. Coating material used for microencapsulation of protein hydrolysate and peptide includes soy protein isolate, gelatin, whey protein concentrate, alginate, maltodextrin, gum arabic, carboxymethyl gum, pectin, phosphatidyl choline, phosphatidylglycine, lecithin, and stearic acid (Mohan et al., 2015; Yeo et al., 2001).

15.4.8 APPLICATION OF MICROENCAPSULATED BIOACTIVE INGREDIENTS IN FOOD INDUSTRY

Microencapsulation can potentially offer numerous benefits to the materials being encapsulated. Microencapsulated omega-3 fatty acids can be used in a wide assortment of foods. For example, Novomega, omega-3 fatty acids encapsulated product is marketed for use in bakery products. The encapsulation system of the Novomega is specially formulated for long chain n-3 fatty acids, and results in a product that eliminates strong fish oil tastes and odors. Two other fish oil encapsulated powders, Marinol™ omega-3HS, and Marinol DHA HS are marketed in US. Another omega-3 microencapsulated fish oil powder, MEG-3 has been introduced in the Canadian and US markets. These powders have been included in to bakery, milk and beverage markets (Pszczola, 2005). Bakery products are generally used as a source for incorporation of different nutritionally rich ingredients for their diversification (Sudha et al., 2007). Jeyakumari et al. (2016) observed that cookies fortified with fish oil microencapsulates was comparable with neat sample (without fish oil encapsulates). Yep et al. (2002) have shown that bread enriched with microencapsulated tuna oil (MTO) increases DHA and determined the acute and chronic effects of low doses of long chain (LC) n-3 PUFA (100 mg/day) on plasma LC n-3 PUFA levels using a novel delivery form. Agnikumar et al. (2015) reported that cake fortified with microencapsulated fish oil improved the oxidative stability of the product. Further, microencapsulated bioactive ingredients are used in chewing gums, instant desserts, food flavors, instant beverages, prepared dishes, confectionery, teas, instant drink, and extruded snacks (Atmane et al., 2006).

15.5   CHARACTERIZATION OF MICROCAPSULES

Microencapsulated food ingredients can be produced by different microencapsulation techniques using a number of suitable shell materials. The characteristic of the microcapsules are important for both bioavailability and success in the targeted application. The processing parameters and the material characteristics determine the final characteristics of the microcapsules (Zhang et al., 2009). These characteristics generally include determination of encapsulation efficiency (EE), surface oil and payload, particle size, powder flow properties, and shape of microcapsules, stability, and sensory performance.

15.5.1   ENCAPSULATION EFFICIENCY

EE is defined as the ratio of the mass of the core material which is encapsulated in the wall material to the mass of the core material used in the formulation. EE depends on the core-to-wall ratio, the conditions in which encapsulation is carried out, and encapsulation technique or production method utilized.

15.5.2 PAYLOAD

This is the percentage of the active ingredient per gram of the powder. When an encapsulation process achieves a higher load, the production of the microcapsules becomes economical and the process becomes more economically feasible. High payload means that a lower amount of powder is required per serving of the food. Payload is calculated by taking the ratio of mass of encapsulated ingredients to total mass of powder. A Fourier transform infrared (FTIR) spectroscopy based non-destructive method has been recently developed to determine payload (Vongsvivut et al., 2012).

15.5.3 PARTICLE SIZE

The size of microcapsules for food applications should be below 100 μm to avoid impacting the mouth feel of the food product. The distribution of the particle size should also be as narrow as possible in order to maintain product consistency. The size of the microcapsules can be measured using techniques such as laser scattering or particle size imaging using microscopy. High resolution imaging using electron microscopy or confocal laser scanning microscope (CSLM) is useful for studying the detailed morphology of microcapsules. CSLM can be combined with staining techniques to gain better insights into the characteristics and distribution of the hydrophobic core and the hydrophilic shell.

15.5.4 MORPHOLOGY

Different types of microscopes such as scanning electron microscope, high-resolution transmission electron microscopes and atomic force microscope (Wang, 2000; Ruozi et al., 2005) have been used to study the morphological features of microparticles.

15.5.5 DETERMINATION OF POWDER FLOW PROPERTIES

Quality control parameters are bulk density, Carr Index and Hausner Ratio are used to assess powder flow ability (Fitzpatrick & Ahrne, 2005). As known, a dry product with a high bulk density can be stored in small containers compared to a product with low bulk density.

15.5.6 STABILITY

The primary purpose of microencapsulation is to protect the active ingredients against environmental conditions by providing barrier in the form of wall materials. Different types of wall materials offer the stability to different extent of microcapsules. The extent of oxidation can be quantified by measuring the oxygen head space concentration or by accelerated analysis using a rancimat or oxypress (Kulas and Ackman, 2001).

15.5.7 IN VITRO RELEASE STUDIES

In vitro release experiments usually refer to the experiments which are carried out in an environment that resembles the living organism. The main aim of the in vitro release measurements is to understand the mechanism of release action of the active agent from the developed delivery system. In vitro release studies data have a number of potential applications. The data can be used for quality control purposes to ensure the constancy of behavior of a manufactured product (Yang et al., 2005).

15.6 CONTROLLED RELEASE MECHANISM

The quality of microparticles produced by a certain encapsulating method and by the usage of specific conditions is evaluated according to retention of core material and the stability of the system over the time. Afterwards it is also important to evaluate the release systems. Controlled release has been defined as a method by which one or more active agents or ingredients are made available at a desired site and time at a specific rate. They must certify that release of core material occurs at the target site and at the desirable rate and time. The final objectives are the decrease of the loss of target compound during the process and storage, and the optimization of absorption and the increase of effective use, as mentioned previously. The advantages of controlled release are: the active ingredients are released; controlled rates over prolonged periods of time; loss of ingredients, such as vitamins and minerals, during processing and cooking can be avoided or reduced and reactive or incompatible components can be separated (Brannon-Peppas, 1993).

Commonly used methods for controlled release include temperature and moisture release for hydrophilic encapsulants, and thermal release for fat capsules (Reineccius, 1995). The main mechanisms involved in the core release are diffusion, degradation, use of solvent, pH, temperature, swelling or osmotic pressure, and pressure activated release. In practice, a combination of more than one mechanism is used (Desai & Park, 2005).

15.6.1   DIFFUSION-CONTROLLED RELEASE

In diffusion-controlled release, active ingredient is released by diffusion through the polymer (reservoir system) or through the pores pre-existing in the polymer (matrix systems).

Reservoir systems: The principal steps in the release of an active ingredient from a reservoir system are: diffusion of the active agent within the reservoir; dissolution or partitioning of the active agent between the reservoir carrier fluid and the barrier; diffusion through the barrier and partitioning between the barrier and the elution medium (i.e., the surrounding food); transport away from the barrier surface into the food. The release rate from a reservoir system depends on the thickness, the area and the permeability of the barrier (Azevedo et al., 2005).

Matrix systems: The steps involved in the release of an active agent that is dissolved in the matrix are (Azevedo et al., 2005); diffusion of the active agent to the surface of the matrix; partition of the active agent between the matrix and the elution medium (i.e., the surrounding food); transport away from the matrix surface. The release pattern depends on the geometry of the system, the type of carrier material and the loading of the active agent.

15.6.1.1   SWELLING CONTROLLED RELEASE

In swelling-controlled systems, the active agent dissolved or dispersed in a polymeric matrix is unable to diffuse to any significant extent within the matrix. When the polymer matrix is placed in a thermodynamically compatible medium, the polymer swells owing to absorption of fluid from the medium. The active agent in the swollen part of the matrix then diffuses out (Fan & Singh, 1989).

15.6.1.2   RELEASE OF ACTIVE AGENT BY DEGRADATION

Degradation release occurs when enzymes such as proteases and lipases degrade proteins or lipids, respectively (Rosen, 2006). An example is reducing the time required for the ripening of cheddar cheese by 50% compared with the conventional ripening process (Hickey et al., 2007).

15.6.1.3   SOLVENT-ACTIVATED RELEASE

The active ingredient is released when the food material comes in contact with a solvent, resulting in swelling of the microcapsule (analogous to the swelling-controlled systems). For example, microencapsulation of coffee flavors improves the protection from light, heat, and oxidation when in the dry state, but the core is released upon contact with water (Frascareli et al., 2012).

15.6.1.4   pH-CONTROLLED RELEASE

Active ingredient (e.g., an enzyme) is released at a specific pH (this system can be used alone or may be combined with solvent-activated or osmotically controlled release systems). Drugs, vitamins, minerals, essential amino acids, fatty acids, or even whole diets, can be released into the gastro-intestinal tract by enzymatic degradation of digestible microcapsules. Probiotic microorganisms can be microencapsulated to resist the acid pH of the stomach and only be released in the alkaline pH of the intestine (Toldra & Reig, 2011).

15.6.1.5   TEMPERATURE-SENSITIVE RELEASE

In this method, active ingredient is released owing to a change in temperature. There are two different concepts: temperature-sensitive release, reserved for materials that expand or collapse when a critical temperature is reached, and fusion-activated release, which involves melting of the wall material due to temperature increase. In many applications, core materials are released by heat. Temperature indicators for frozen food, aromas for tea, and baking, are based on the effect of melting of the microcapsule wall. Fat-encapsulated cheese flavor used in microwave popcorn, resulting in the uniform distribution of the flavor; the flavor is released when the temperature rises to 57–90°C (Park & Maga, 2006).

15.6.1.6 PRESSURE-ACTIVATED RELEASE

Active ingredient is released when pressure is applied on the walls of the microcapsules (an example would be the release of sweetener and/or flavor in gum when chewed). This principle is applied in pressure-sensitive copying papers (pressure of the pen-ball or typewriter head), multi-component adhesives (activation in a press), deodorants and fungicides for shoes (mechanical pressure caused by walking), polishing pastes (rubbing) and aromas and sweeteners in chewing gums (Wong et al., 2009). Some wall materials and the possible mechanisms for the microcapsules release are listed in Table 15.1.

TABLE 15.1 Wall Materials and Their Potential Release Mechanisms.

Wall materials	Mechanical	Thermal	Dissolution	Chemical
Soluble in water	–	–	–	–
Alginate	×	–	×	–
Carrageenan	×	–	×	–
Caseinate	×	–	×	–
Chitosan	×	–	–	–
Modified cellulose	×	–	×	–
Gelatin	×	–	–	–
Xanthan gum	×	×	–	–
Arabic gum	×	×	–	–
Latex	×	–	×	–
Starch	×	–	×	–
Insoluble in water				
Ethyl cellulose	×	–	–	×
Fatty alcohols	×	×	–	×
Fatty acids	×	×	–	–
Hydrocarbon resin	×	×	–	–
Mono, di, and triacyl glycerol	×	×	–	–
Natural waxes	×	×	–	–
Polyethylene	×	×	–	–

Source: Adapted from Favaro-Trindade et al. (2008).

15.6.2   FACTORS AFFECTING THE RELEASE RATE OF CORE MATERIALS

The main factors affecting the release rates are related to interactions between the wall material and the core. Moreover, wall material properties such as wall thickness, coating layer, density, solubility, plasticizer level, cross-linking pretreatments also affect the release rate of core materials (Shahidi & Han, 1993). Additionally, other factors influence the release, such as the volatility of the core, ratio between the core and wall material, particle size, temperature, pH, and moisture (Roberts & Taylor, 2000).

15.7   CONCLUSION

Microencapsulation is an effective and important tool to protect active agent against oxidation, evaporation, or migration in food. It plays a major role in development of high-quality functional food ingredients with improved physical and functional properties in order to make superior products. To produce effective encapsulated products, the choice of coating material and method of microencapsulation process are most important. Despite the wide range of application of encapsulated products in pharmaceutical and cosmetic industries, microencapsulated product has found a comparatively much smaller market in the food industry. The microencapsulation technology is yet to become a conventional tool for food industry to develop the healthy and novel food products which can be achieved by multidisciplinary-based research approach and consideration of industrial requirements and constraints.

KEYWORDS

- microencapsulation
- microcapsules
- food ingredients
- bioactives
- controlled release

REFERENCES

Adem, G.; Gaelle, R.; Odile, C.; Andree, V.; Remi, S. Application of Spray Drying in Micro-encapsulation of Food Ingredients: An Overview. *Food Res. Int.* **2007**, *40*, 1107–1121.

Agnikumar, S.; Manjusha, L.; Chouksey, M. K.; Tripathi, G.; Venkateshwarlu G. Delivery of Omega-3 Fatty Acids into Cake through Emulsification of Fish Oil Milk and Encapsulation by Spray Drying with Added Polymer. *Dry. Technol.* **2015**, *33* (1), 83–91.

Amr, M. B.; Shabbar, A.; Barkat, A.; Hamid, M.; Mohamed, Y. A.; Ahmed, M.; Li, L. Micro-encapsulation of Oils: A Comprehensive Review of Benefits, Techniques, and Applications. *Comp. Rev. Food Sci. Food Safe.* **2015**, *15*, 143–182. DOI: 10.1111/1541-4337.12179

Anjani, K.; Kailasapathy, K.; Phillips, M. Microencapsulation of Enzymes for Potential Application in Acceleration of Cheese Ripening. *Int. Dairy J.* **2007**, *17*, 79–86.

Anu Puri, Kristin L.; Brandon S.; Jae-Ho, L.; Amichai, Y.; Eli, H.; Robert, B. Lipid-Based Nanoparticles as Pharmaceutical Drug Carriers: From Concepts to Clinic. *Crit. Rev. Ther. Drug Carrier Syst.* **2009**, *26* (6), 523–580.

Arts, I. C. W.; Hollman, P. C. H. Polyphenols and Disease Risk in Epidemiologic Studies. *Am. J. Clin. Nutr.* **2005**, *81*, 317–325.

Atmane, M.; Muriel, J.; Joel, S.; Stephane, D. Flavour Encapsulation and Controlled Release—A Review. *Int. J. Food Sci. Technol.* **2006**, *41*, 1–21.

Augustin, M. A.; Sanguansri, L. Encapsulation of Bioactives. In *Food Materials Science— Principles and Practice;* Aguilera, J. M., Lillford, P. J., Eds.; Springer: New York, NY, 2007; pp 577–601.

Azevedo, H. S.; Reis, R. L. Understanding the Enzymatic Degradation of Biodegradable Polymers and Strategies to Control Their Degradation Rate. In *Biodegradable Systems in Tissue Engineering and Regenerative Medicine;* Reis, R. L., Román, J. S., Eds.; CRC Press: Boca Raton, FL, 2005; 177–201.

Brannon-Peppas, L. Properties and Applications. In *Polymeric Delivery Systems;* EL-Nokaly, M. A., Piatt, D. M., Charpentier, B. A., Eds.; ACS Symposium Series 520. American Chemical Society: Washington, DC, 1993; p 52.

Bennick, A. Interaction of Plant Polyphenols with Salivary Proteins. *Crit. Rev. Oral Biol. Med.* **2002**, *13*, 184–196.

Boccio, J. R.; Zubillaga, M. B.; Caro, R. A.; Gotelli, C. A.; Gotelli, M. J.; Weill, R. A New Product to Fortify Fluid Milk and Dairy Products with High-bioavailable Ferrous Sulfate. *Nutr. Rev.* **1997**, *55*, 240–246.

Binsi, P. K.; Nayak, N.; Sarkar, P. C.; Jeyakumari, A.; Ashraf, M.; Ninan, G.; Ravi Shankar, C. N. Structural and Oxidative Stabilization of Spray Dried Fish Oil Microencapsulates with Gum Arabic and Sage Polyphenols: Characterization and Release Kinetics. *Food Chem.* **2017**, *219*, 158–168.

Boh, B.; Kardos, D. Microcapsule Patents and Products: Innovation and Trend Analysis. Microcapsule Patents and Products, The MML Series, Vol. 6, 2003, pp 47–83.

Calvo, P.; Castano, A. L.; Hernandez, M. T.; Gonzalez-Gomez, D. Effects of Microcapsule Constitution on the Quality of Microencapsulated Walnut Oil. *Eur. J. Lipid Sci. Technol.* **2011**, *113*, 1273–1280.

Champagne, C. P.; Patrick, F. Microencapsulation for the Improved Delivery of Bioactive Compounds into Foods. *Curr. Opin. Biotech.* **2007**, *18*, 184–190.

Cisem, T. Immobilization of Thermophilic Recombinant Esterase Enzyme by Microencapsu-lation in Alginate Chitosan/Cacl₂ Polyelectrolyte Beads. M.Sc. Thesis, 2011.

Desai, K. G. H.; Park, H. J. Recent Developments in Microencapsulation of Food Ingredients. *Dry. Technol.* **2005**, *23,* 1361–1394.

Erdmann, K.; Cheung, B. W. Y.; Schroder, H. The Possible Roles of Food-derived Bioactive Peptides in Reducing the Risk of Cardiovascular Disease. *J. Nutr. Biochem.* **2008**, *19,* 643–654.

Fang, Z.; Bhandari, B. Encapsulation of Polyphenols – A Review. *Trends. Food Sci. Technol.* **2010**, *21,* 510–523.

Fan, L. T.; Singh, S. K. Swelling-Controlled Release. In *Controlled Release: Polymers (Properties and Applications);* Fan, L. T., Singh, S. K., Eds.; Springer-Verlag: Berlin, 1989; Vol. 13, 110–156.

Favaro-Trindade,C. S.; de Pinho, S. C.; Rocha, G. A. Revisao: Microencapsulaçao de Ingredientes Alimenticios. *Br. J. Food Technol.* **2008**, *11* (2), 103–112.

Fitzpatrick, J. J.; Ahrne, L. Food Powder Handling and Processing: Industry Problems, Knowledge Barriers and Research Opportunities. *Chem. Eng. Proc.* **2005**, *44,* 209–214. DOI: 10.1016/j.cep.2004.03.014

Frascareli, E. C.; Silvaa, V. M.; Tonon, R. V.; Hubinger, M. D. Effect of Process Conditions on the Microencapsulation of Coffee Oil by Spray Drying. *Food Bioprod. Process.* **2012**, *90,* 413–424. http://dx.doi.org/10.1016/j.fbp.2011.12.002

Goncalves, A.; Estevinho, B. N.; Rocha, F. Microencapsulation of Vitamin A: A Review. *Trends Food Sci. Technol.* **2016**, *51,* 76–87. DOI: 10.1016/j.tifs.2016.03.001

Hammad, U.; Hemlata, N.; Asif, M. T. M.; Sundara Moorthi, N. Microencapsulation: Process, Techniques and Applications. *Int. J. Res. Pharm. Biomed. Sci.* **2011**, *2* (2), 475–481.

Hickey, D. K.; Kilcawley, K. N.; Beresford, T. P.; Wilkinson, M. G. Lipolysis in Cheddar Cheese Made from Raw, Thermized, and Pasteurized Milks. *J. Dairy Sci.* **2007**, *90* (1), 47–56.

Jeyakumari, A.; Janarthanan, G.; Chouksey, M. K.; Venkateshwarlu, G. Effect of Fish Encapsulates Incorporation on the Physic-chemical and Sensory Properties of Cookies. *J. Food Sci. Technol.* **2016**, *53* (1), 856–863.

Jeyakumari, A.; Kothari, D. C.; Venkateshwarlu, G. Microencapsulation of Fish Oil-milk Based Emulsion by Spray Drying: Impact on Oxidative Stability. *Fish Technol.* **2014**, *51,* 31–37.

Jeyakumari, A.; Kothari, D. C.; Venkateshwarlu, G. Oxidative Stability of Microencapsulated Fish Oil during Refrigerated Storage. *J. Food Process. Preserv.* **2015**, *39,* 1944–1955. DOI:10.1111/jfpp.12433.

Kailasapathy, K.; Lam, S. H.; Hourigan, J. A. Studies on Encapsulating Enzymes to Accelerate Cheese Ripening. *Aus. J. Dairy Tech.* **1998**, *53,* 125.

Klinkesorn, U.; Sophanodora, P.; Chinachoti, P.; McClements D. J.; Decker, E. A. Stability of Spray-Dried Tuna Oil Emulsions Encapsulated with Two-layered Interfacial Membranes. *J. Agric. Food Chem.* **2005**, *53* (21), 8365–8371.

Kris-Etherton, P. M.; Taylor, D. S.; Yu-Poth, S.; Huth, P.; Moriarty, K; Fishell, V.; Hargrove R. L.; Zhao, G.; Etherton, T. D. Polyunsaturated Fatty Acids in the Food Chain in the United States. *Am. J. Clin. Nutr.* **2000**, *71,* 179–188.

Kulas, E.; Ackman, R. G. Different Tocopherols and the Relationship between Two Methods for Determination of Primary Oxidation Products in Fish Oil. *J. Agric. Food Chem.* **2001**, *49,* 1724–1729.

Manach, C.; Scalbert, A.; Morand, C.; Remesy, C.; Jimenez, L.; Polyphenols: Food Sources and Bioavailability. *Am. J. Clin. Nutr.* **2004**, *79,* 727–747.

McClements, D. J.; *Nanoparticle and Microparticle-based Delivery Systems: Encapsulation, Protection and Release of Active Compounds*; CRC Press: Boca Raton, FL, 2014; 572.

Mohan, A.; Subin, R. C. K.; Rajendran; Quan Sophia, H.; Bazinet, L.; Chibuike, C.; Udenigwe. Encapsulation of Food Protein Hydrolysates and Peptides: A Review. *RSC Adv.* **2015,** *5,* 79270–79278.

Motohiko, H.; Hitoshi, T.; Hiroshi, N.; Makoto, K. Calcium Fortification of Soy Milk with Calcium-lecithin Liposome System. *J. Food Sci.* **2006,** *49* (4), 1111–1112.

Park, D.; Maga, J. A. Identification of Key Volatiles Responsible for Odour Quality Differences in Popped Popcorn of Selected Hybrids. *Food Chem.* **2006,** *99* (3), 538–545.

Pegg, R. B.; Shahidi, U. Encapsulation, Stabilization and Controlled Release of Food Ingredients and Bioactives. In *Hand Book of Food Preservation,* 2nd ed.; Rahman, S. M.; Ed.; CRC Press, Taylor & Francis Group: Boca Raton, FL, 2007.

Pratibha, K.; Kim, D.; Colin, J. B.; Benu, A. Microencapsulation of Omega-3 Fatty Acids: A Review of Microencapsulation and Characterization Methods. *J. Fun. Foods.* **2014,** *19,* 868–881. DOI: 10.1016/j.jff.2014.06.029

Pszczola, D. E. Omega-3s Over-come Formulation Problems. *Food Technol.* **2005,** *59,* 48–50.

Reineccius, G. A. Controlled Release Techniques in the Food Industry. In *Encapsulation and Controlled Release of Food Ingredients;* Risch, S. J., Reineccius, G. A., Eds.; ACS Symposium Series 590. American Chemical Society: Washington, DC, 1995; p 8–25.

Roberts, D. D.; Taylor, A. J. Flavor Release. American Chemical Society: Washington, DC, 2000; p 496.

Robert, S.; Ronal, V.; Anikumar, G. G. Introduction to Microencapsulation and Controlled Delivery in Foods. In *Microencapsulation in the Food Industry—A Practical Implementation Guide;* Anikumar, G. G., Niraj, V., Atul, K., Robert, S., Eds.; Elsevier Inc.: San Diego, CA, 2014.

Rosen, R. M. *Delivery System Handbook for Personal Care and Cosmetic Products. Technology, Applications and Formulations.* William Andrew Pub: New York, NY, 2006; p 1095.

Ruozi, B.; Tosi, G.; Forni F.; Fresta, M.; Vandelli, M. A. Atomic Force Microscopy and Photon Correlation Spectroscopy, Two Techniques for Rapid Characterization of Liposomes. *Eur. J. Pharmacol. Sci.* **2005,** *25,* 81–89.

Sanders, M. E. Probiotics: Considerations for Human Health. *Nutr. Rev.* **2003,** *61,* 91–99.

Scalbert, A.; Manach C.; Morand, C.; Remesy, C.; Jimenez, L. Dietary Polyphenols and the Prevention of Diseases. *Crit. Rev. Food Sci. Nutr.* **2005,** *45* (20), 287–306.

Schrooyen, P. M. M.; Meer, R. V. D.; Kruif, C. G. D. Microencapsulation: Its Application in Nutrition. *Proc. Nutr. Soc.* **2001,** *60,* 475–479.

Sebastien, G. Microencapsulation: Industrial Appraisal of Existing Technologies and Trends. *Trend. Food Sci. Technol.* **2004,** *15,* 330–347.

Shabbar, A.; Chang, D. W.; Hayat, K.; Zhang, X. Ascorbic Acid: Microencapsulation Techniques and Trends—A Review. *Food Rev. Int.* **2012,** *28* (4), 343–374. DOI: 10.1080/87559129.2011.635390

Shahidi, F.; Han, X. Q. Encapsulation of Food Ingredients. *Crit. Rev. Food Technol.* **1993,** *33,* 501.

Sudha, M. L.; Vetrimani, R.; Leclavathi, K. Influence of Fibre from Different Cereals on the Rheological Characteristics of Wheat Flour Dough and on Biscuit Quality. *Food Chem.* **2007,** *100* (4), 1365–1370.

Teixeira, M. I.; Andrade, L. R.; Farina, M.; Rocha-Lea, M. H. M. Characterization of Short Chain Fatty Acid Microcapsules Produced by Spray Drying. *Mater. Sci. Eng.* **2004,** *24,* 653–658.

Toldra, F.; Reig, M. Innovations for Healthier Processed Meats. *Trend. Food Sci. Technol.* **2011,** *22* (9), 517–522.

Uicich, R.; Pizarro, F.; Almeida, C.; Diaz, M.; Bocchio, J.; Zubillaga, M.; Carmuega, E.; O'Donnell, A. Bioavailability of Microencapsulated Ferrous Sulfate in Fluid Cow's Milk. Studies in Human Beings. *Nutr. Res.* **1999,** *19,* 893–897.

Vermeirssen, V.; Camp, J. V.; Verstraete, W. Bioavailability of Angiotensin I Converting Enzyme Inhibitory Peptides. *Br. J. Nutr.* **2007,** *92,* 357–366.

Vidhyalakshmi, R.; Bhakyaraj, R.; Subhasree, R. S. Encapsulation "The Future of Probiotics"—A Review. *Adv. Biol. Res.* **2009,** *3* (3–4), 96–103.

Vongsvivut, J.; Heraud, P.; Zhang, W.; Kralovec, J. A.; McNaughton, D.; Barrow, C. J. Quantitative Determination of Fatty Acid Compositions in Micro-encapsulated Fish-oil Supplements Using Fourier Transform Infrared (FTIR) Spectroscopy. *Food Chem.* **2012,** *135,* 603–609.

Wang, Z. L. Transmissions Electron Microscopy of Shape Controlled Nanocrystals and Their Assemblies. *J. Phys. Chem. B.* **2000,** *104,* 1153–1175.

Wilson, N.; Shah, N. P. Microencapsulation of Vitamins—A Review. *Asian Food J.* **2007,** *14* (1), 1–14.

Wong, S. W.; Yu, B.; Curran, P.; Zhou, W. Characterising the Release of Flavor Compounds from Chewing Gum through HS-SPME Analysis and Mathematical Modeling. *Food Chem.* **2009,** *114* (3), 852–858.

Yang, S.; Washington, C. Drug Release from Microparticulate Systems. In *Microencapsulation: Methods and Industrial Applications;* Benita, S., Ed.; CRC Press: Boca Raton, FL, 2005; 183–211.

Yeo, Y.; Namjin, B.; Kinam, P. Microencapsulation Methods for Delivery of Protein Drugs. *Biotechnol. Bioprocess Eng.* **2001,** *6,* 213–230.

Yep, Y.; Li, D.; Mann, N.; Bode, O.; Sinclair, A. Bread Enriched with Microencapsulated Tuna Oil Increases Plasma Docosahexaenoic Acid and Total Omega-3 Fatty Acids in Humans. *Asia Pac. J. Chem. Eng.* **2002,** *11,* 285–291.

Zhang,W.; Yan, C.; May, J.; Barrow, C. J. Whey Protein and Gum Arabic Encapsulated Omega-3 Lipids: The Effect of Material Properties on Coacervation. *Agro Food Ind. Hi-Tech* **2009,** *20,* 18–21.

Zuidam, N. J.; Heinrich J. Encapsulation of Aroma. In *Encapsulation Technologies for Food Active Ingredients and Food Processing;* Zuidam, N. J., Nedovic, V. A., Eds.; Springer: Dordrecht, Netherlands, **2010,** 127–60.

Zuidam Kailasapathy, N. J.; Shimoni E. Overview of Microencapsulates for Use in Food Products or Processes and Methods to Make Them. In *Encapsulation Technologies for Active Food Ingredients and Food Processing;* Zuidam, N. J., Nedović, V. A., Eds.; Springer: New York, NY, **2010.** DOI 10.1007/ 978-1-4419-1008-02

CHAPTER 16

ELECTROSPINNING AS A NOVEL DELIVERY VEHICLE FOR BIOACTIVE COMPOUNDS IN FOOD

RAJAKUMARI RAJENDRAN[1*†], APPARAO GUDIMALLA[1,2*†], RAGHAVENDRA MISHRA[1], SHIVANSHI BAJPAI[3], NANDAKUMAR KALARIKKAL[1,4*], and SABU THOMAS[1,5*]

[1]*International and Inter University Centre for Nanoscience and Nanotechnology, Mahatma Gandhi University, Kottayam, Kerala, India*

[2]*Department of Nanotechnology, Acharya Nagarjuna University, Guntur 22510, Andhra Pradesh, India*

[3]*Yuveraj Dutta PG College, Lakhimpur, India*

[4]*School of Pure and Applied Physics, Mahatma Gandhi University, Kottayam, Kerala, India*

[5]*School of Chemical Sciences, Mahatma Gandhi University, Kottayam, Kerala, India*

[†]**Both authors have equal contribution**

Corresponding authors.
E-mail: sabuthomas@mgu.ac.in, nkkalarikkal@mgu.ac.in

CONTENTS

Abstract 427

16.1 Introduction 427

16.2 Materials for Food Technology 433

16.3 Encapsulating Materials Used in Electrospinning 437

16.4 Nanostructured Electrospun Interlayer of Zein............................441
16.5 3D Food Fabrication 443
16.6 Applications for Food Industries...444
16.7 Conclusion 446
Acknowledgments...448
Keywords...448
References...448

ABSTRACT

Electrospinning is a fiber fabrication technique that makes the polymer solution into fiber/fibers mat at the scale of nano/micrometer. These fibers have high porosity, large surface area and worthy water/air permeability. Due to this reason, electrospun fibers/mat having great attention in the field of food science and delivery systems. Nutraceuticals delivery systems are rapidly growing because these systems contains biopolymers that exhibits fascinating properties like very low toxicity, biocompatibility and biodegradability. These properties makes them as ideal materials to develop nutraceuticals delivery systems with help of electrospinning technique. This chapter explains the delivery systems for nutraceuticals using biopolymers by electrospinning method.

16.1   INTRODUCTION

The interest of novel delivery vehicle for bioactive compounds has focused on the production of new formulations with improved properties, taking much attention to the bioactive compounds from the carrier. Encapsulation has already been commercialized for packaging of active components in food to reduce the interaction of food ingredients one and other during storage or processing. It can control the release of bioactive component and can also enhance the handling process of liquid or gaseous ingredients (Augustin & Sanguansri, 2008; King, 1995).

Many of the previous studies suggested that encapsulation techniques are widely employed in the food industry. The size and shape of the fabricated capsules can vary from nanometers to micro or millimeters range. The morphology and performance of the capsules strongly depend on the material and methods. However, food industry around the world is facing a lot of problems with encapsulation technology, such as degradation of nutraceuticals by light, oxygen, heat, and inefficient delivery of encapsulated bioactive component. Recently, the combination of two or more materials is being used to encapsulate substances. More and more attention has been focused on using those biopolymer complexes of food proteins and polysaccharides as wall materials to encapsulate nutrients (Cooper et al., 2005). Electrospinning is one of the methods to prepare encapsulating materials. Electrospinning is a fiber production method which uses electric force to draw charged threads of polymer solutions or polymer melts up to fiber diameters in the order of some hundred nanometers.

16.1.1 HISTORY OF ELECTROSPINNING

In order to visualize the historical development, the development of electro-spinning way was started, when the behavior of electrostatic and magnetic phenomena on liquid was first proposed by William Gilbert in early 1600 after that many studies were based on the formation of a conical shaped water droplet (Luo et al., 2012; Yurteri et al., 2010). Earlier many studies have explained the influence of surrounding medium such as electrical charge and electrical stress on the stability of liquid droplets (Fong et al., 2007; Rayleigh, 1882; Miao et al., 2010; Zeleny, 1914; Zeleny 1917). Among the methods, the electrospinning technique has been rapidly developed in the last decade for the facile preparation of continuous fibers from submicron down to nanometer diameter. Although there are other methods for preparing one-dimensional fiber with high aspect ratio, such as template synthesis, few methods could match electrospinning in terms of its flexibility, versatility, and ease of fiber production. Electrospinning was invented at the beginning of the last century (1902), but it was not until the mid-1990s that investigators started to realize the huge capability of the procedure of the fiber preparation. The application of this has been widely increased in the past few years, benefitting from the remarkably low cost, flexibility, simplicity, and potential applications of this technique. So far, electrospinning has been used to fabricate an extensive range of nanofibers, such as polymers, metals, ceramics, and composites. In early patent was proposed by Form-hals (1934), this patent was based on the invention of electrospinning on the production of artificial threads and then after he fabricated cellulose acetate based filaments using electric charges and he also published lot of patent on the production of polymer filaments using an electrostatic force (Form-hals, 1934; Taylor, 1964; Taylor, 1969; Chakraborty et al., 2009; Katepalli et al., 2011; Wu & Clark, 2008; Reneker & Yarin, 2008; Huang et al., 2003; Huang et al., 2006; Anton, 1938; Formhals, 1939; Anton, 1940; Subbiah et al., 2005; Bhardwaj & Kundu, 2010; Huang et al., 2003; Yawen et al., 2015; Ko et al., 2003; Zhang, 2012; Liu et al., 2012; Huang et al., 2011; Greiner & Wendorff, 2007; Crespy et al., 2012; Bognitzki et al., 2001; Mieszawska et al., 2007; Li et al., 2004; Augustine et al., 2014; Silke et al., 2002; Fashandi et al., 2012; Zexuan et al., 2011) and also later granted related patents in 1938, 1939, and 1940 (Anton, 1938; Formhals, 1939; Anton, 1940). Form-hals' spinning process consists of a movable thread-collecting device to collect threads in a stretched condition, corresponding that of a spinning drum in conventional spinning (Subbiah et al., 2005).

In middle of the 20th century, Sir Geoffrey Taylor proposed a mathematical model to generate conical shape droplets by the influence of electrical field, later this model known as Taylor cone this model explained the conflict between liquid surface tension and the applied electrical charges, which led to deform the liquid droplet into a cone shape (Taylor, 1964; Taylor, 1969).

However, this technique did not receive much attention until the 1990s, when several research groups (notably that of Reneker) (Reneker, 1996) found that it could be used to generate nanofibers from many organic polymers. Since then, the term "electrospinning" has been simplified and the number of publications about electrospinning has expanded dramatically each year. Today, electrospinning is a well-established technique for generating nanofibers (Li, & Xia, 2004), resultant electrospinning development has regained more attention probably due to a surging interest in nanotechnology. The ultrafine fibers or fibrous formation of various polymers with diameters down to submicron or nanometers can be easily fabricated with this process (Bhardwaj & Kundu, 2010).

The growing realization of the immense potential applications of electrospinning process in multi-disciplinary areas, number of research have been dedicated to the electrospinning process for fiber making from a capital-intensive, large-scale process to a low cost, broadly (Chakraborty et al., 2009; Katepalli et al., 2011; Wu & Clark, 2008). Currently, various types of synthetic and natural polymer-based fibers have been fabricated by electrospinning process (Huang et al., 2003) and the minimum electrospun fiber diameter was found to be less than 1.6 nm (Huang et al., 2006).

16.1.2   PRINCIPLES OF ELECTROSPINNING

Electrospinning is a simple method that utilizes high electrostatic forces to produce fine fibers in the varying diameter range. Electrospinning machine consists four major components: (1) a direct current power supply, (2) a metallic needle with blunt tip, (3) a syringe for containing the electrospun solution, and (4) a grounded conductive collector. A schematic setup for electrospinning machine is shown in Figure 16.1. Considering the electrospinning process, a syringe is filled with a melt-mixture polymer solution and a large voltage (typically kV) is applied between the syringe nozzle and a collector.

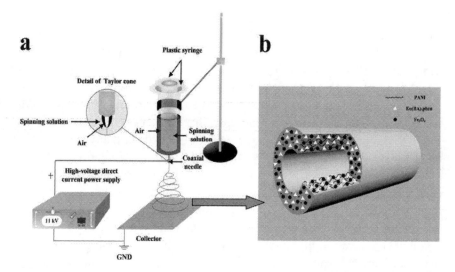

FIGURE 16.1 Schematic illustration of the setup for one-pot coaxial electrospinning process (a) and the as-prepared hollow nanofibers containing europium complex, PANI and Fe_3O_4 NPs (b). (Reprinted with permission from Yawen Liu, Qianli Ma, Xiangting Dong, Wensheng Yu, Jinxian Wang and Guixia Liu. A novel strategy to directly fabricate flexible hollow nanofibers with tunable luminescence–electricity–magnetism trifunctionality using onepot electrospinning (2015). Chem. Phys., **17**, 22977-22984, DOI: 10.1039/c5cp03522h. © 2015, Royal Chemical Society.)

Compared to mechanically or air driven spinning techniques, an electrospinning process, a conical fluid structure called the Taylor cone is formed at the tip of the syringe by the fluid droplet under the applied voltage (Zexuan et al., 2011). During applied demanding voltage, the unpleasant force of the charged polymer alter the surface tension of the solution and a charged jet blow up from the tip of the Taylor cone. If the applied voltage is not high enough, the jet will break up into droplets, an aspect called "Rayleigh instability." If the voltage is sufficiently high, a stable jet will form near the tip of the Taylor cone. Beyond the balanced region, the jet is subject to bending instability (Yarin et al., 2001) that results in the polymer being accumulated on the grounded collector via a whipping motion. As the charged jet accelerates toward regions of reduced potential, the solvent evaporates and the resulting increase in the electrostatic repulsion of the charged polymer causes the fibers to elongate.

Almost any soluble polymer can be subjected to electrospinning if its molecular weight is high. However, the creation of fine nanofibers based on the many processing parameter (such as polymer molecular weight, applied voltage, solution feed percentage, and spinning distance), environmental specifications (such as temperature, humidity, and air velocity in the chamber) and solution properties (such as viscosity, surface tension, and conductivity) (Liu et al., 2012; Huang et al., 2011; Greiner & Wendorff, 2007; Crespy et al., 2012; Bognitzki et al., 2001; Mieszawska et al., 2007; Li et al., 2004). The selecting various polymer blends and tuning the above-mentioned electrospinning process parameters, a wide range of nanofibers can be produced from polymer blends, ceramic precursors, natural polymers, and metal or metal oxides have been spun into an assortment of different fiber morphologies, such as beaded (Huang et al., 2000), ribbon (Fong et al., 2002), porous (Casper et al., 2004), core–shell (McCann et al., 2005), and aligned (Li et al., 2003) fibers. The electrospun membranes have the nano-fibrous structure similar to human tissues and organs; therefore, they can be also used for many biomedical applications (Thavasi et al., 2003) such as tissue engineering (Yoshimoto et al., 2003), medical prostheses (Boland et al., 2004), drug delivery (Katti et al., 2004), and wound healing (Rho et al., 2006). In addition to biomedical applications, they have also been used in other areas, such as sensors (Li et al., 2008), filters, and separation membranes (Gopal et al., 2006), templates for nanotube materials (Bognitzk et al., 2000), protective layer (Schreuder-Gibson et al., 2002), composite materials (Kim et al., 1999), and energy applications (Ramakrishna, et al., 2006). Among the various natural or synthetic polymer-based fibers, the web can be prepared by electrospinning. Recently, nanoparticles/polymer based electrospun fibers exhibit huge potential applications, these composites fiber mat have excellent flexibility, modulus, thermal stability, etc.

An understanding of electrospun fiber formation and modulation of fiber is important in order to fabricate the anticipated nanostructure. Several groups of researchers had explained that the solution concentration, molecular weight, surface tension, conductivity, and viscosity of the polymer solution. Processing parameters and ambient parameters will have an impact on the electrospinning process and the morphology microstructure of the subsequent fibers (Anandhan et al., 2012; Beachley & Wen, 2009; Evcin & Kaya, 2010; Heikkilä & Harlin, 2008). Figure 16.2a Shows the typical SEM image of electrospun PCL membrane, where the fibers have a very smooth surface and constant diameter. However, an increase in a number of nanoparticles causes an increase of the roughness and it is reported, due to increase in viscosity of solution (Augustine et al., 2014).

Many other studies also reported the secondary pore formation on the individual fibers. This might be due to the breath figure formation, which is directly related to atmosphere factors like temperature, viscosity of the solution, nature of the solvent, and humidity (Silke et al., 2002; Fashandi & Karimi, 2012).

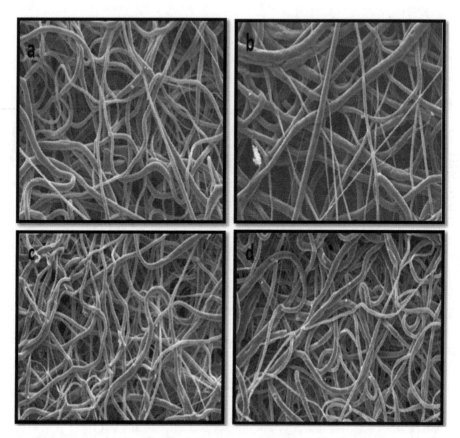

FIGURE 16.2 Scanning electron micrograph of electrospun (a) neat polycaprolactone membrane, (b) polycaprolactone membrane with 0.2 wt% ZnO nanoparticles, (c) polycaprolactone membrane with 0.5 wt% ZnO nanoparticles, and (d) polycaprolactone membrane with 0.8 wt% ZnO nanoparticles (Augustine et al., 2014).

16.1.3 METHODS OF ELECTROSPINNING ENCAPSULATION

Electrospinning as a technology to produce promising and versatile nano-structured food delivery systems has led to great progress due to its simplicity and versatility. Food delivery using polymeric micro/nanostructures is based on the principle that active components can be easily embedded into the polymeric carrier in high loadings without modification of the structure and bioactivity through electrospinning.

The fabricated fiber largely offers various properties (He, et al., 2006; Nair et al., 2004; Verreck et al., 2003; Zong et al., 2002), such as (1) high thermal stability and controlled and fast release kinetics compared to the bulk films with similar thickness, due to the short diffusion passage length; (2) high surface area to volume ratio that enables mass transfer and effective delivery; and (3) efficient release performance of the active component through modulation of the active component/polymer ratio. For encapsulation purposes, various approaches of electrospinning process are available, such as blend electrospinning, co-axial electrospinning, emulsion electrospinning, and surface modification of the electrospun fiber mats. When using electrospinning as an encapsulation technique that comprises the loading approach of the active component would have a direct effect on its deposition. However, the rapid or delayed release performance of the active component from the electrospun fibers will be based on loading approach.

The loading of the effective component can be followed by entrapment and binding approach, it depends on the polymeric carrier and the capable basic properties. In heavy cases, the release of the active component involves diffusion from the polymeric carrier where 100% release of the active component is expected after a certain period of time (Natu, 2010). One of the studies examined the release of a model water-soluble compound from polymeric electrospun fibers (Srikar et al., 2008). This report suggested that the release could happen by desorption of the active component from nonporous in the fibers, or from the outlying surface of the fibers in connection with the release media, where a complete release could not happen.

16.2   MATERIALS FOR FOOD TECHNOLOGY

The polymeric delivery system was first proposed by "Ringsdorf" (1975), in order to achieve depot effects of drugs. After the introduction of polymeric electrospun fibers, it worked as the effective bioactive component delivery system; electrospinning has been referred to a wide range of various natural and synthetic polymers. However, only those that are "generally recognized as safe" (GRAS) are allowed to be used for food science applications. According to US Food and Drug Administration (FDA, 2013), "any substance that is intentionally added to food is a food additive, that is given premarket amend an investigation by FDA, unless the substance is generally recognized as safe, among good experts, as having been reasonably shown to be safe under the conditions of its intended use, or unless they consider the substance is otherwise excluded from the definition of a food additive." Hence, some polymers such as methyl methacrylates that are employed for drug encapsulation might not be widely applicable in food. Only carbohydrates, protein, and lipids biopolymers that are derived from various natural origins are being used as encapsulation wall materials (carrier) for the food industry. Due to performance limitations and technical challenges, these biopolymers have been used for electrospinning process (Kriegel et al., 2008; Schiffman & Schauer, 2008). This limitation and challenges due to their: (1) wide disparity in molecular weight; (2) biovariations due to different resources; (3) high processing cost; (4) high crystallinity and polarity that makes them less soluble in organic solvents; (5) high tendency to form strong hydrogen bonds (high surface tension); and (6) poor mechanical properties and processability of the end products. One of the suggested approaches to overcome these technical issues is to blend biopolymers with synthetic polymers. Synthetic polymers will induce higher molecular entanglement for effective electrospinning process and to improve mechanical properties of the fabricated electrospun fibers (Espíndola-González et al., 2011; Li et al., 2006; Sionkowska, 2011).

The characteristics of biopolymer for encapsulation will also have a vast impact on the encapsulation efficiency and the stability of the encapsulated products. According to an excellent review conducted by Shahidi and Han (1993), an ideal wall material for encapsulation purposes should follow some condition such: (1) good rheological properties at high concentration that is easy to operate during the encapsulation process; (2) good emulsify properties toward the active component and stabilize the emulsion produced; (3) capable of covering and holding the active component within its structure during processing and storage; (4) possibility to complete remove the solvent used during the encapsulation processes through drying or other conditions; (5) capable of protecting the active component against harsh environmental conditions (such as heat, light, and humidity); (6) good solubility in GRAS solvents (e.g., water and ethanol) and inexpensive; and (7) ability to achieve specific capsule solubility and release properties. In addition to that, PEO and polyvinyl alcohol (PVA) biopolymers are being widely used to facilitate the electrospinning process will also be estimated.

Currently, nanotechnology is applied for the characterization, fabrication, and manipulation of biological and nonbiological structures smaller in the range 10–100 nm. These nanometric morphologies have been shown to the unique and novel functional properties (Weiss et al., 2006). Therefore, nanotechnology has been involved in attempts to manufacture the commercial products for much industrial application such as microelectronics, aerospace, and pharmaceutical industries. In the food sector, the applications of nanotechnology are still limited. However, significant scientific advancement and technological achievements in nanotechnology are recent to

enforcement the food industry and associated industries; this affects important aspects of food safety to the molecular synthesis of trendy food products
and ingredients are becoming explore (Huang et al., 2010).

The facts of life that systems mutually structural features on the
nanoscale have physical, chemical, and biological properties substantially
disparate from their macroscopic counterparts is changing the understanding
of biological and physical phenomena in food systems. The concerning
physical, chemical, and biological properties food systems at microscopy
level, the nanotechnology has been helped to understand the biological and
physical phenomena in food systems at nanoscale on the level to twist those
properties and functions in a beneficial way, because foods undergo various
types of postharvest and dealing with convinced modifications that can
change the biological and biochemical functions of the system (Weiss et al.,
2006; Huang et al., 2010; Neethirajan & Jayas, 2011).

16.2.1   NANOTECHNOLOGY: INSPIRATION FROM NATURE

Living organisms are not just a collection of nanoscale objects: Atoms and
molecules are organized in hierarchical structures and dynamic systems
that are the results of millions of years of Mother Nature's experiments. For
example, nerve impulses are generated by nanometer diameter ions of potassium and sodium metal and the size of vital biomolecules such as sugars,
amino acids, and hormones and second example human DNA. The dimension of the most protein and polysaccharide molecules are in nanoscale
dimensions, which indicate that the every living organism on earth exists
because of the interaction, shortage, absorption, location, and presence of
these nanostructures. These nanoscale structures are generated by natural
self-assembly principal. The huge amount of energy and series of optimized processes are always involved to create these types of assembly and
nanoscale structures through the minimization of the overall free energy of a
system to reduce its free pretension, thereby minimizing required activation
energies. Generation of nanoscale structures for food science and technology
is related to depth understanding of thermodynamically driven self-assembly
processes. Areas of review that could prove pleasant in the aside future
include molecular design of protective surface systems (Charpentier, 2005),
surface engineering (Krajewska, 2004), and disparate methods of manufacturing, a well-known as electrospinning (Min et al., 2004) and nanofiltration
(van der Graaf et al., 2005).

16.2.2 POTENTIAL FOOD APPLICATIONS

In food and agricultural systems, nanotechnologies have the potential to impact cover many aspects, such as food safety, disease treatment delivery methods, and new tools for molecular and cellular biology for pathogen detection (Weiss et al., 2006). Nowadays nanotechnology is being used as a tool for achieving further advancements in the food industry such as:

- Increased monitoring of the manufacturing, processing, and shipping of food products through sensors for pathogen and contaminant detection.
- Recent advancement in nanotechnology systems could provide integration of sensing, localization, broadcasting, and remote approach of food products (smart and intelligent systems) and that increased the efficiency, security of food processing, and transportation.
- Devices to uphold historical environmental records of a particular product and tracking of records of a particular product could be also possible.
- Encapsulation and food delivery systems serve as vehicles for upholding, recover, and deliver sensible food ingredients to their specific site of action.

In food engineering, nanotechnology is a new frontier to develop new biosciences and engineering product based on nanosystem. However, applications of nanoscience in food technology are slightly different from these, because food processing is being involved wide variety of raw materials, high biosafety requirements, and well-coordinated technological processes. Four primary areas in food manufacturer may benefit from nanotechnology: evolution of new logical materials, microscale and nanoscale processing, product development and methods, and instrumentation arrangement for improved food security and biosafety. Moreover, the relationship between the morphology of food materials and their total physicochemical properties has been investigated: for example, biopolymers in solutions, gels, and films (Janaswamy & Chandrasekaran, 2005).

Functional nanostructures can be embedded into individual biological molecules which is useful for developing the biosensors that can use natural sugars or proteins as target-recognition groups (Weiss et al., 2006). In sense, there is a vast verity of potential applications of nanotechnology in the food industry; however, adoption of this technology for commercial product may be difficult because they are either costly or infeasible on an industrial scale.

16.2.3 FOOD PRODUCT INNOVATION

Nanotechnology-based food ingredients showed better characteristics such as food flavors and antioxidants (Weiss et al., 2006). The recent advancement in food technology exhibited better functionality of the ingredients in food systems properties along with delivery and controlled protect systems for solubilization of nutraceuticals in foods have been improved (Lawrence & Rees, 2012; Haruyama, 2003). In addition, bioavailability and the ability to disperse the compounds food ingredients such as nanoparticles lycopene and carotenoids are found to be higher than that of their traditionally manufactured counterparts.

16.3 ENCAPSULATING MATERIALS USED IN ELECTROSPINNING

Throughout the world, demand for healthy foods has accrued considerably over the recent years due to growth within the world population associated an increased moralization of healthy lifestyles.

There is a wide verity of sterol lowering practical foods offered in the market, such as cheese, Benecol margarine spreads, which contain extra mesmerized fat-soluble forms of phytosterols or stanols (plant extracts) (Ghorani & Tucker, 2015). Omega-3 is a variety of fatty acids, which occur naturally in foods such as oily fish, a few plants, and seed oils; therefore, these are is a lot of promising substance to be extra to an oversized verity of food merchandise together with marge, milk, fruit juice, and eggs to create functional foods for reduction of vessel risk.

Vitamin D and C also are most vital forms of the substance, which square measure incorporated to fruit juice to lift dietary D (Holick et al., 2011; Rafferty et al., 2007). While a control and target unharness will improve the effectiveness of micronutrients, broadens, and can additionally increase the appliance vary of food ingredients and ensures optimal indefinite quantity with the efficient product (Mozafariet al., 2006). However, the delivery and lifetime of the bioactive compound are forever primarily based on their particle size; for example, the delivery of any bioactive compound to various sites inside the body is directly laid low with the particle size. In some cell lines, only submicron nanoparticles will be absorbed expeditiously, however, not the larger size microparticles (Ezhilarasi et al., 2013). In order to the larger particles, generally unharness encapsulated compounds additional slowly and everywhere longer anticipate periods, meanwhile particle size loss of value introduces large amount bio-adhesive emendation factors, including inflated adhesive force and thence prolonged GI transit time, leading to the next encapsulated compound bioavailability (Chen et al., 2006).

Recently many sorts of strategies square measure used for the assembly of fibrous materials for encapsulation of bioactive compounds, abused substance delivery, as bimolecular sensors and ultra-filtration media, in which electrospinning is best-suited methodology to get nanofibers, such fibrous materials have been studied as potential vehicles for bioactive component and management delivery (Bhushani & Anandharamakrishnan, 2014).

Recently the demand for encapsulation techniques has been increased drastically for the biological preservation of foods, maintain the viability, and tenacity of the probiotic bacteria and bacteriocins overall food process. In this way, proteins and carbohydrates have been utilized because the encapsulating materials for storage (Burgain et al., 2011). Encapsulation of bioactive compounds and probiotic bacteria inside prebiotic substances to build safeguard or perhaps multiplied their survival while passing higher digestive tube and through food process and storage is a region of nice interest for each world and also the food industries (López-Rubio et al., 2012; López-Rubio & Lagaron, 2012).

Micro- and nanoencapsulation have been used for encapsulating the bioactive and sensitive compounds (Ghorani, 2012). This can be performed by numerous methodology like spray drying, freeze drying, emulsification, coacervation, nano-precipitation, and liposome preparation (Bhushani & Anandharamakrishnan, 2014), in which electrospinning is a lot of engaging and appropriate technology to provide encapsulation bioactive compounds while not rupturing the sensitive encapsulated nutrients (due to hot temperature further as toxicity issues related to residual organic agents (López-Rubio et al., 2012; López-Rubio & Lagaron, 2012). Resultant electrospun fibers spun have the potential to serve as a carrier matrix for the look and performance of novel delivery systems for functional product (Ghorani & Tucker, 2015; Bhushani & Anandharamakrishnan, 2014; Ghorani, 2012; Salalha et al., 2006; Heunis et al., 2011).

16.3.1 GELATIN AS A CARRIER

Many of science people team reportable to produce stable and yielded regular fibrils of uniform thickness from gelatin/water through electrospinning method (Zhang et al., 2009; Li et al., 2006). Recently, the large verity of proteins such as bovine serum albumin (BSA), soy protein isolates, and metallic element caseinate has been mixed with gelatin to build the suitable of gelatin as a carrier chemical compound. The proteins/gelatin (1:1) has been used for electrospinning to form spun fiber, obtained spun can have completely different levels of stability. The stability of the spinning process was calculable by observation of the Taylor cone combined with research. If spun samples in a more stable method, it shows homogeneous fibrils thickness.

It also found that a mixture of whey super molecule isolate and gelatin 15 wt% in resolution spun was very stable. Gelatin is a glorious medium for many microorganism growths; hence considerable care acquires to avoid contamination throughout the producing. In order to attenuate the required amount of gelatin within the supermolecule mixture, the various sorts of gelatin are often used in keeping with the top product. Generally, a longer polymer is often spun at a lower concentration, because the overlap concentration of the compound is earned at a lower concentration (Bhardwaj, & Kundu, 2010). Therefore, concentrations of shorter gelatin are used higher than longer gelatin with supermolecule isolate to realize stable spinning method. The longer and shorter gelatin are selected by the bloom numbers, higher boom number are used for longer gelatin, while, for a shorter gelatin, the lower bloom numbers are expected.

Understanding the spin ability of proteins for food application, the precise preparation procedure of the protein solutions is essential because the presence of air bubbles within the sample will hamper the spinning by the formation of mechanical phenomenon cavitations in the presence of an associated acoustic field. In most cases, the electrospinning of globular proteins has been shown unpleasant behavior, due to a less stable spinning process.

Many factors that could play a role to achieve spinning stability, those are a difference in charge densities, impurities, and amount of salts present in the supermolecule solutions. Furthermore, different interactions and compatibility with the carrier compound (gelatin) will have an effect on variations in behavior, just as alternative surface properties. The present of beads in electrospun fibrils square measure enthusiastic about the Taylor cone oscillation, increasing the oscillating Taylor cone will increase the quantity of beads in fibrils the other way around.

16.3.2 PROTEIN-BASED VEHICLES

The electrospinning method is a versatile process for the encapsulation of food ingredients, enzymes, active compounds, and spinning of biopolymers, which connected to the food industries (Ghorani & Tucker, 2015. In this context, many studies explained that the biopolymers and their composites are most tightened materials for active packaging or preservation of nutrient activity for consumption, encapsulation, and delivery systems; this is attributable to unique characteristics of biopolymers like biocompatibility, biodegradability, and antibacterial activity (Bhushani & Anandharamakrishnan, 2014). Resultant, polymer-based, food-grade polymers and biopolymers as bioactive molecule delivery devices have been widely recognized in medical specialty, pharmaceutical sectors, and food industries (Liechty et al., 2010). Recently, macromolecule isolated from the natural substance amaranth (amaranth macromolecule isolate or API) has been used to get ultrathin fibers and it had been found that these fibers morphologies and textures have a lot of applicable for the protein as bioactive encapsulation matrices in food applications (Aceituno-Medina et al., 2013; Neo et al., 2013a; Neo et al., 2013b; Fernandez et al., 2009; Brahatheeswaran et al., 2012; Li et al., 2009; Wongsasulak al., 2014; Yang et al., 2013; Cheng et al., 2013; de Oliveira Mori et al., 2014). However, correct choice as a carrier materials for proteins or active compounds is most crucial things to scale back the adverse result of carrier chemical compound throughout the emotional of bioactive part in physical body, therefore, variety of condition should be followed by the designated chemical compound for electrospinning: it ought to be natural, edible, antitoxic behavior with and while not solvent, and it can be electrospun to create fibers structure while not the necessity to introduce man-made chemical compound as a spinning aid into the mixture (López-Rubio et al., 2012; López-Rubio & Lagaron, 2012).

16.3.3   CARBOHYDRATE BASED

Natural and modified polysaccharides are most favorable vehicles for nano and microencapsulation of active food ingredients and delivery systems, "because of" their biocompatibility, biodegradability, and presence of various types of functionality and high potential to be modified to achieve the required properties (Schiffman & Schauer, 2008). These characteristics of carbohydrates make them flexible carrier's substance for delivery systems and moreover, carbohydrates can easily interact with a wide range of hydrophilic and hydrophobic bioactive food ingredients by their functional groups (Fathi et al., 2014). In order to carbohydrates contain monosaccharide, oligosaccharide, and polysaccharides, they are generally categorized according to their biological origins: higher fabricate foundation (e.g., starch, cellulose, pectin, and guar gum); animal foundation (e.g., chitosan); algal foundation (e.g., alginate and carrageenan), and microbial origin (e.g., xanthan, dextran, and cyclodextrins) (Ghorani & Tucker, 2015).

Chitosan (Desai, 2008a; Desai, 2008b), alginate (Alborzi et al., 2010), starch (Kong & Ziegler, 2012), pullulan (Fabra, et al., 2014), inulin (Jain et al., 2014), and guar gum (Lubambo et al., 2013) are as sufficient for the electro-spinning of outer block materials, similarly, biocompatible integration fibers were electrospun from seed albumin (EA) and polyethylene oxide (PEO) at identified optimum electrospinning conditions (Wongsasulak et al., 2007). To recover the power, conservation, bioavailability, and controlled protect properties of the topic biomolecules, encapsulation has been implemented; it is not solo improved these properties yet, in addition, used the objection-able odors or vogue of the delivered compound (Mascheroni et al., 2013). To outlook, the morphology of encapsulation by electrospinning, a well-known of the raw material investigated that the immobilized fat-soluble vitamin E onto electrospun ester nanofibers mutually a neutral and disk-shaped cross-sectional morphology (Taepaiboon et al., 2007). Another study explained that the perillaldehyde aroma compound immobilized directly in an edible carbohydrate nanofibrous matrix of Pullulan-Cyclodextrin (Mascheroni et al., 2013). This morphology is not only useful for management unleash but improve the stability of sustenance at a lower place lightweight and acidic condition. For example, the B complex (Vitamin B) is vulnerable to degra-dation once exposed to the lightweight and acidic condition. Similarly, the successful application of emulsion electrospinning for the encapsulation of extraordinarily volatile fragrances, namely hydrocarbon in a fibrous matrix has recently been presented (Ghorani & Tucker, 2015; Kayaci & Uyar, 2012).

16.4 NANOSTRUCTURED ELECTROSPUN INTERLAYER OF ZEIN

Proteins and polysaccharides are also widely used in the food system; these materials have excellent properties such as completely biodegradable mate-rials excellent gas and aroma barrier properties at low and intermediate rela-tive humidity (RH) (Miller & Krochta, 1997).

A strategy to prevent the moisture uptake of these hydrophilic materials is to transpire multilayer structures anywhere the hydrophilic hydro colloidal films are sandwiched between hydrophobic biodegradable materials, thus, reducing the plasticization of the inner layer and maintaining its good barrier properties.

Zein is a class of natural prolamin protein, obtained from maize (corn) that has a completely amorphous and hydrophobic character than various proteins consequently of the high content of freezing amino acids, proline, and glutamine, which shows certain specific interesting properties including water resistance, viscosity, and thermal resistance (Cho et al., 2002).

The use of zein has been attracted a growing interest due to their higher tensile strength and lower water vapor permeability properties and potential applications in industrial nanotechnology than other protein-based films (Ghanbarzadeh & Oromiehi, 2008). When zein interlayers have been chemically sandwiched between PHBV polymers it does not suffer plasticization and swelling with very interesting barrier properties (Miguel et al., 1997). Many studies on the combination of proteins or polysaccharides with polyesters reveal that these materials are the promising alternative to diminish the water sensitivity of these hydrocolloids and overall functionalities of thermoplastic bio polyesters, which make them more adequate for the oxygen barrier and food packaging applications.

One study on the behavior of the multilayer films based on plasticized wheat starch sandwiched with different types of biodegradable aliphatic bio polyester's a well known as polylactic acid (PLA), polycaprolactone (PCL), and PHBV reveal that PCL showed medium adhesion values with PLA and PHBV, where PHBV were the least compatible biopolyesters (Martin et al., 2001). While the adhesion between the layers and oxygen barrier properties of multilayer films of PLA containing a zein electrospun fiber interlayer are investigated, the result shows that the influence of oxygen barrier properties (up to 71%) and adhesion between the layers are found through this study.

Recently, production of zein nanofibers via electrospinning is a relatively ideal technology which has been explored. Under the specific conditions, a number of successful electrospinning zein nanofibers experience was reported from acidic and alcoholic solutions (Gopal et al., 2006; Greiner et al., 2006; Torres-Giner et al., 2008). Fibers obtained in this study were in the diameter range from 50 to 6 mm, and typical electrospinning mats are composed of yarns of beads, tubular fibers, cooperative fibers, or ribbon-like fibers, the morphology was strongly dependent on the conditions used for electrospinning and the zein solution properties.

In addition to the pure zein nanofibers, hybrid ultrathin electrospun fibers containing different nature of commercial ceramic nanoparticles have been reported, generating these type of reinforced hybrid nanostructures have been used in coatings, encapsulation, packaging, and various applications (Torres et al., 2010).

Regarding produce fibers through electrospinning process, the characteristics of the fibers was reported to depend on solution properties (polymer concentration, surface tension, electrical conductivity, viscosity, and solvent volatility), environmental situations (humidity, temperature, and air velocity), and development conditions (voltage, flow rate, and spinning distance). As mentioned in many earlier studies, by incorporating the conventional fibers of zein into fabrics, it observed that the permeability to liquids and gasses dropped and increasing their mechanical resistance, wear, and washing staying power are found (Yun et al., 2007).

16.4.1 PREPARATION OF ELECTROSPUN ZEIN INTERLAYER THROUGH ELECTROSPINNING

Preparation and characterization of zein ultrathin fiber mats were released through "Fluidnatek" basic electrospinning equipment by the victimization of 33 wt% of zein in associate in nursing 85 v/v plant product resolutions. The solution was prepared under magnetic stirring at temperature, using a voltage of 12–14 kV and a rate of flow of 0.3 mL/h (Torres-Giner et al., 2008). Before testing, multilayer systems (with and without zein interlayer) were equilibrated in desiccators at 25°C and third RH by victimization silicon dioxide gel, for one week where the analyses were carried out.

16.5 3D FOOD FABRICATION

There is an increasing market prefer for customized food products, practically of which are currently designed and constrained by a specifically trained artisan. The cost for such a limited number of pieces is relatively high. Three-dimensional (3D) food printing, as known as Food Layered Manufacture can be a well-known of strength alternatives to bridge this gap. It aims to perform 3D custom-designed food objects in a layer-by-layer approach, without object-specific tooling, formulate, or human intervention. Thus, these technologies bounce increase field efficiency and reduce manufacturing cost for customized food products fabrication.

16.5.1 ROBOTICS-BASED FOOD MANUFACTURING

The robotic chef is the integrated system to a superbly style professional room. It has had potential applications in both family and industrial environments. According to the feature baking cookies robots can find ingredients; combine them in correct order and place resulting dough in a baking receptacle of kitchen appliance. These robots are well designed to perform everyday manipulation tasks and basic actions like memorizing an object, putting it down or gushing (Kumar, 2012). These robotics-based techniques have been used in traditional food producing industries for production. These cannot solely perform the designed operate and also it will greatly cut back work, save labor cost, and improve the manufacturing efficiency.

The successful implantation of these robotic systems intended to develop 3D food printing techniques at the industrial level. These 3D food-printing techniques have the capability to produce food product in bespoke form, color, flavor, texture, and the even nutritional price at mass level (Van Bommel, 2011). Combining 3DP and digital gastronomy techniques can digitally visualize food manipulation, therefore, making a new area for novel food fabrication at the reasonable value. As a result, a customized food style during a kind of digital 3D model is directly remodeled to a finished output in a bedded arrangement (Levy et al., 2011). Combining 3DP and digital gastronomy techniques cut back digitally portray food beating, appropriately making a trendy area for contemporary food fiction at low-priced value. As a result, a customized food style during a kind of digital 3D model is directly remodeled to a finished product in a bedded struc-ture (Levy et al., 2011). Recently, these 3D printing techniques have been commercialized for property right. Food printer platform basically consists the XYZ 3 Cartesian coordinate system, dispensing/sintering units, and user interface component. During the 3D printing method, the three axes motorized moment and accessories feeding position are tasteful by dealer operated PC system. And food composition boot is deposited/trade basically motive by final cause and protect by layer using the PC technologist. To produce the new dishes instead of merely modifying ancient food fabrication method, at least four essential steps are planned, they are metering, mixing, dispensing, and cooking (heating or cooling) (Zoran & Coelho, 2011). Only the dispensing and preparation functions have been utilized for the present industrial or self-developed food printing platforms. Self-developed plat-forms are consummated the specific would like, such as creating 3D sugar structures, building cheese and chocolate 3D objects from edible ingredients by computer-controlled optical maser machine.

16.6   APPLICATIONS FOR FOOD INDUSTRIES

The development of food technology has been explored by food scientist to the transformation of raw ingredients, by physical or chemical approach into food or of food into at variance forms. In terms to encapsulation techniques has been applied to produce the complex properties from food ingredients, these techniques are permits to the manufacture of food ingredients with new properties that can alter the handling, storage, and transportation of the ingredients and protect the sensitive or active purposeful food parts with inexpensive operations.

Recently, bioactive compound has been used in food packaging due to their antimicrobial activity, minimal health, and ecological issues.

As previously mentioned, electrospinning is a most versatile method to give the encapsulation of bioactive materials for food industries. Thus, product at the metric linear unit (1 mm), submicron or nano (10–100 nm) scale can be obtained by electrospinning (Klaessig et al., 2011). Results from research works reveal the potential of these strategies within the tiny and nanoencapsulation of food bioactive compounds and enzymes and more. These techniques boot conjointly be used within the fabrication of watchful materials for active food packaging applications. Apart from these, a minority explored as a crowning achievement promising relevancy of electrospun fibers are as filtration membranes in food and food manner (Fabra et al., 2013). The optimization of electrospinning parameters for the manufacture of modern biopolymer-based food packaging material with purposeful properties of the encapsulated active compounds has been involved in several studies.

For example, a natural antimicrobial compound, electrospun allyl chemical irritant has been used for wetness triggered unleash of the active compound, this compound was produced by victimization the soy macromolecule isolate (SPI) or polylactic acid (PLA), and b-cyclodextrin. The electrospun fibers exhibit diameters in the range from 200 nm to 2 mm, which will be explored in active packaging applications (Vega-Lugo & Lim, 2009). Similarly, the antimicrobial activity of chitosan was employed to fabricate the zein/chitosan electrospun insoluble fiber mats with biocide properties (Torres-Giner et al., 2009).

These fibers possessing fast unleash of the active compound can be used as safe food contact materials even to be used on dry foods as edible coatings. However, the career challenges associated mutually its nature in packaging systems are the insurance of volatiles from degradation and their mild unleash. When an exemplary bouquet combination perillaldehyde was encapsulated in pullulan and b-cyclodextrin, perceptive nanofibrous membranes were produced. The confidence of the design was furthermore extended in the manner of these membranes as food packaging labels abduction the advice of antimicrobial process of the aroma compound. In the opening of that energy, applications use of electrospun nanofibers inefficient packaging were to can investigate.

Apart from the function of electrospun fibers inobservant and effective packaging materials, they are to potential candidates in the manner in multilayered structures. And the electrospun nanofibers, when used as medium layers in packaging materials, gain their technical, optical, and/or efficient properties and edge completely different multilayered structures (Fabra et al., 2013).

Encapsulations of this compound with zein electrospun fibers permit the work of materials for biopackaging. Also, higher encapsulation composure will be obtained by exploitation concentrically electrospinning plan as compared to uniaxial electrospinning (Pérez-Masiá et al., 2013). Along mutually considering associate encapsulating things, zein nanofibers (100–500 nm) boot additionally be used as strengthening agents in a multilayer structure. The outside studies verify the within realm of possible uses of electrospun fibers in food packaging sector; anyhow, process optimization and scale-up square consider needed for technological production. Further combination of engineering in providing new packaging materials mutually improved technical, barrier and antimicrobial properties together with nanosensors for tracing and watch the presence of food throughout transport and computerized information (Silvestre et al., 2011). Consequently, food packaging applications bounce can be updated already electrospinning approach is formed responsible for providing innovative, sustainable, perceptive, and efficient packaging materials with acceptable automated and optical properties.

16.7 CONCLUSION

In this chapter, a comprehensive discussion is presented on the fundamentals of electrospinning food primarily based wholly polymers, food packing materials, and food technology nanofibers including technique, structure, property characterization, and applications.

Meeting global energy demands is one of the foremost pressing needs people faces among the twenty initial centuries. Nanofibers may play a key role in addressing these problems as a result of their distinctive structures and high area to volume ratios. Among the many nanofiber fabrication ways in which the electrospinning holds significant promise due to its comparatively low price and comparatively high production rate. Many kinds of materials synthesized through electrospinning, from conventional polymers to ceramics, metals, alloys, and composites fibers. Current progress reveals that electrospinning technology plays an important role among the food delivery technologies. Finally, current electrospinning processes commonly involve toxic and corrosive organic solvents in the preparation of precursor solutions. Whereas, in the food sector the food grade and pharma grade solvents are used for the preparation of the precursor solutions. For future research, the fabrication of food composite nanomaterials would possibly be a promising field.

Electrospinning and electrospraying techniques can augment the growth of food technique sector in fields of encapsulation, food packaging, and edible coating. The merits of electrospun fibers and electrosprayed particles with regard to food applications area is classified based on its structural properties. Structural advantages are redeemable size, morphology, highly porous structure, high surface area and intertwined fibrous structure. The potential food-based applications of electrospinning and electrospraying are encapsulation, enzyme immobilization, food packaging, food coating and aids infiltration processes. The encapsulation of food bioactive compounds in electrospun fibers and electrosprayed particles enhances their stability and controlled release properties.

Encapsulation of bioactive compounds and probiotic bacteria among prebiotic substances to defend or enhance their survival whereas passing higher gastrointestinal tract is interest for the food industries. Encapsulation of bioactive compounds in micro or nano-scaled particles is widely used technique. Electrospun nanofibers is recently used as a the delivery system in foods for protecting the nutrients of the bioactive component and delivering it to the target site. The main advantages of electrospun nanofibers for the encapsulation of food bioactives in a sustained and controlled manner. This technique is widely used because of the advantages such as room temperature process, reduces denaturation, efficient encapsulation and enhanced stability of bioactives by the use of food-grade polymers and biopolymers.

We have explained the quality of food-grade electrospinning of gelatin as a carrier compound. The approach for the spinning is applicable for a wide range of various proteins. The behavior of the carrier proteins could be expected from its molecular properties. In order to use these protein fibers for delivering a bioactives, the solubility has to be increased by cross-linking. For the cross-linking several food-grade reagents, such as ferulicacid, other plant phenolics, or the enzyme transglutaminase were used.

In addition to the scientific and technical advances, the regulatory issues (including safety/toxicology and amp; impact of environmental), economics, and consumer acceptance can ultimately dictate its success in food applications. Agricultural producers and food manufacturers might gain competitive position through this technology in the long-term. Consumers would possibly profit from the advances in study that contributes to a competitive and innovative domestic agricultural and food system. This produces new ways to strengthen the safety and organic process of food merchandise.

ACKNOWLEDGMENTS

The authors would like to express their sincere thanks to Department of Science and Technology (DST-Inspire fellowship), New Delhi for their financial support.

KEYWORDS

- **electrospinning**
- **nanofibers**
- **biopolymer**
- **nanotechnology**
- **gelatin**

REFERENCES

Aceituno-Medina, M.; Mendoza, S.; Lagaron, J. M.; López-Rubio, A. *Food Res. Int.* **2013,** *54* (1), 667–674.

Alborzi, S.; Lim, L. T.; Kakuda, Y. *J. Food Sci.* **2010,** *75* (1), C100–C107.

Anandhan, S.; Ponprapakaran, K.; Senthil, T.; George, G. *Int. J. Plastics Technol.* **2012,** *16* (2), 101–116.

Anton, F. U.S. Patent 2,116,942, U.S. Patent and Trademark Office: Washington, DC, 1938.

Anton, F. U.S. Patent 2,187,306, U.S. Patent and Trademark Office: Washington, DC, 1940.

Augustin, M. A.; Sanguansri, L. *Food Materials Science;* Springer: New York, 2008; pp 577–601.

Augustine, R.; Malik, H. N.; Singhal, D. K.; Mukherjee, A.; Malakar, D.; Kalarikkal, N.; Thomas, S. *J. Poly. Res.* **2014,** *21* (3), 1–17.

Beachley, V.; Wen, X. *Mater. Sci. Eng. C.* **2009,** *29* (3), 663–668.

Bhardwaj, N.; Kundu, S. C. Electrospinning: A Fascinating Fiber Fabrication Technique. *Biotechnol. Adv.* **2010,** *28* (3), 325–347. https://www.sciencedirect.com/science/article/pii/S0734975010000066.

Bhushani, J. A.; Anandharamakrishnan, C. *Trends Food Sci. Technol.* **2014,** *38* (1), 21–33.

Bognitzk, M.; Hou, H. Q.; Ishaque, M.; Frese, T.; Hellwig, M.; Schwarte, C.; Schaper, A.; Wendorff, J. H.; Greiner, A. *Adv. Mater.* **2000**, *12,* 637–640.

Bognitzki, M.; Frese, T.; Steinhart, M.; Greiner, A.; Wendorff, J. H.; Schaper, A.; Hellwig, M. *Polym. Eng. Sci.* **2001**, *41* (6), 982–989.

Boland, E. D.; Matthews, J. A.; Pawlowski, K. J.; Simpson, D. G.; Wnek, G. E.; Bowlin, G. L. *Front. Biosci.* **2004**, *19,* 422–1432.

Brahatheeswaran, D.; Mathew, A.; Aswathy, R. G.; Nagaoka, Y.; Venugopal, K.; Yoshida, Y.; Sakthikumar, D. *Biomed. Mater.* **2012**, *7* (4), 045001.

Burgain, J.; Gaiani, C.; Linder, M.; Scher, J. J. *Food Eng.* **2011**, *104*(4), 467–483.

Casper, C. L.; Stephens, J. S.; Tassi, N. G.; Chase, D. B.; Rabolt, J. F. *Macromolecules* **2004**, *37,* 573–578.

Chakraborty, S.; Liao, I. C.; Adler, A.; Leong, K. W. *Adv. Drug Deliv. Rev.* **2009**, *61* (12), 1043–1054.

Charpentier, J. C. *Chem. Eng. J.* **2005**, *107* (1), 3–17.

Chen, L.; Remondetto, G. E.; Subirade, M. *Trends Food Sci. Technol.* **2006**, *17* (5), 272–283.

Cheng, Y. H.; Yang, S. H.; Liu, C. C.; Gefen, A.; Lin, F. H. *Carbohydr. Polym.* **2013**, *92* (2), 1512–1519.

Cho, S. Y.; Park, J. W.; Rhee, C. *LWT Food Sci. Technol.* **2002**, *35* (2), 135–139.

Cooper, C. L.; Dubin, P. L.; Kayitmazer, A. B.; Turksen, S. *Curr. Opin. Colloid Interface Sci.* **2005**, *10* (1), 52–78.

Crespy, D.; Friedemann, K.; Popa, A. M. *Macromol. Rapid Commun.* **2012**, *33* (23), 1978–1995.

de Oliveira Mori, C. L.; dos Passos, N. A.; Oliveira, J. E.; Mattoso, L. H. C.; Mori, F. A.; Carvalho, A. G., Tonoli, G. H. D., et al. *Ind. Crops Prod.* **2014**, *52,* 298–304.

Desai, K.; Kit, K. *Polymer* **2008a**, *49* (19), 4046–4050.

Desai, K.; Kit, K.; Li, J.; Zivanovic, S. *Biomacromolecules* **2008b**, *9* (3), 1000–1006.

Espíndola-González, A.; Martínez-Hernández, A. L.; Fernández-Escobar, F.; Castaño, V. M.; Brostow, W.; Datashvili, T.; Velasco-Santos, C. *Int. J. Mol. Sci.* **2011**, *12* (3), 1908–1920.

Evcin, A.; Kaya, D. A. *Sci. Res. Essays.* **2010**, *5* (23), 3682–3686.

Ezhilarasi, P. N.; Karthik, P.; Chhanwal, N.; Anandharamakrishnan, C. *Food Bioprocess Technol.* **2013**, *6* (3), 628–647.

Fabra, M. J.; Busolo, M. A.; Lopez-Rubio, A.; Lagaron, J. M. *Trends Food Sci. Technol.* **2013**, *31* (1), 79–87.

Fabra, M. J.; López-Rubio, A.; Lagaron, J. M. *Food Hydrocoll.* **2014**, *39,* 77–84.

Fashandi, H.; Karimi, M. *Polymer* **2012**, *53,* 5832–5849.

Fathi, M.; Martin, A.; McClements, D. J. *Trends Food Sci. Technol.* **2014**, *39* (1), 18–39.

Fernandez, A.; Torres-Giner, S.; Lagaron, J. M. *Food Hydrocoll.* **2009**, *23* (5), 1427–1432.

Fong, C. S.; Black, N. D.; Kiefer, P. A.; & Shaw, R. A. *Am. J. Phys.* **2007**, *75* (6), 499–503.

Fong, H.; Liu, W. D.; Wang, C. S.; Vaia, R. A. *Polymer* **2002**, *43,* 775–780.

Formhals, A. Process and Applications for Preparing Artificial Threads. U.S. Patent, 1(975), 504, 1934.

Formhals, A. Method and Apparatus for Spinning. U.S. Patent, 2,160,962, 1939.

Ghanbarzadeh, B.; Oromiehi, A. R. *Int. J. Biol. Macromol.* **2008**, *43* (2), 209–215.

Ghorani, B. Production and Properties of Electrospun Webs for Therapeutic Applications. Doctoral dissertation, University of Leeds, 2012.

Ghorani, B.; Tucker, N. *Food Hydrocoll.* **2015**, *51,* 227–240.

Gopal, R.; Kaur, S.; Ma, Z.; Chan, C.; Ramakrishna, S.; Matsuura, T. *J. Memb. Sci.* **2006,** *281* (1), 581–586.

Greiner, A.; Wendorff, J. H. *Angew. Chem. Int. Ed.* **2007,** *46* (30), 5670–5703.

Greiner, A.; Wendorff, J. H.; Yarin, A. L.; Zussman, E. *Appl. Microbiol. Biotechnol.* **2006,** *71* (4), 387–393.

Haruyama, T. *Adv. Drug Deliv. Rev.* **2003,** *55* (3), 393–401.

He, C. L.; Huang, Z. M.; Han, X. J.; Liu, L.; Zhang, H. S.; Chen, L. S. *J. Macromol. Sci. B.* **2006,** *45* (4), 515–524.

Heikkilä, P.; Harlin, A. *Eur. Polym. J.* **2008,** *44* (10), 3067–3079.

Heunis, T.; Bshena, O.; Klumperman, B.; Dicks, L. *Int. J. Mol. Sci.* **2011,** *12* (4), 2158–2173.

Holick, M. F.; Binkley, N. C.; Bischoff-Ferrari, H. A.; Gordon, C. M.; Hanley, D. A.; Heaney, R. P.; Weaver, C. M. et al. *J. Clin. Endocrinol. Metab.* **2011,** *96* (7), 1911–1930.

Huang, C.; Chen, S.; Lai, C.; Reneker, D. H.; Qiu, H.; Ye, Y.; Hou, H. *Nanotechnology* **2006,** *17* (6), 1558.

Huang, C.; Soenen, S. J.; Rejman, J.; Lucas, B.; Braeckmans, K.; Demeester, J.; De Smedt, S. C. *Chem. Soc. Rev.* **2011,** *40* (5), 2417–2434.

Huang, Q.; Yu, H.; Ru, Q. *J. Food Sci.* **2010,** *75* (1), R50–R57.

Huang, Z. M.; Zhang, Y. Z.; Kotaki, M.; Ramakrishna, S. *Compos. Sci. Technol.* **2003,** *63* (15), 2223–2253.

Huang, Z. M.; Zhang, Y. Z.; Kotaki, M.; Ramakrishna, S. *Compos. Sci. Technol.* **2003,** 63(15), 2223–2253.

Huang, L.; McMillan, R. A.; Apkarian, R. P.; Pourdeyhimi, B.; Conticello, V. P.; Chaikof, E. L. *Macromolecules* **2000,** *33,* 2989–2997.

Jain, A. K.; Sood, V.; Bora, M.; Vasita, R.; Katti, D. S. *Carbohydr. Polym.* **2014,** *112,* 225–234.

Janaswamy, S.; Chandrasekaran, R. *Carbohydr. Polym.* **2005,** *60* (4), 499–505.

Katepalli, H.; Bikshapathi, M.; Sharma, C. S.; Verma, N.; Sharma, A. *Chem. Eng. J.* **2011,** *171* (3), 1194–1200.

Katti, D. S.; Robinson, K. W.; Ko, F. K.; Laurencin, C. T. *J. Biomed. Mater. Res. Part B. Appl. Biomater.* **2004,** *70B,* 286–296.

Kayaci, F.; Uyar, T. *Food Chem.* **2012,** *133* (3), 641–649.

Kim, J. S.; Reneker, D. H. *Polym. Compos.* **1999,** *20,* 124–131.

King, A. H. Encapsulation and Controlled Release of Food Ingredients. In *Encapsulation of Food Ingredients;* Risch, S. J., Reineccius, G. A., Eds.; ACS Symposium Series 590; American Chemical Society: Washington DC, 1995; pp 26–39.

Klaessig, F.; Marrapese, M.; Abe, S. *Nanotechnology Standards*; Springer: New York, 2011; pp 21–52.

Ko, F.; Gogotsi, Y.; Ali, A.; Naguib, N.; Ye, H.; Yang, G. L.; Willis, P. *Adv. Mater.* **2003,** *15* (14), 1161–1165.

Kong, L.; Ziegler, G. R. *Biomacromolecules* **2012,** *13* (8), 2247–2253.

Krajewska, B. *Enzyme Microb. Technol.* **2004,** *35* (2), 126–139.

Kriegel, C.; Kit, K. M.; McClements, D. J.; Weiss, J. *Langmuir* **2008,** *25* (2), 1154–1161.

Kumar, P. A. Towards Socially Intelligent Robots in Human Centered Environment. Doctoral Dissertation, Toulouse, INSA, 2012.

Lawrence, M. J.; Rees, G. D. *Adv. Drug Deliv. Rev.* **2012,** *64,* 175–193.

Levy, G. S.; Angel-Levy, P.; Levy, E. J.; Levy, S. A.; Levy, J. A. U.S. Patent 7,949,616, U.S. Patent and Trademark Office: Washington, DC, 2011.

Li, D.; Babel, A.; Jenekhe, S. A.; Xia, Y. Adv. Mater. **2004,** *16* (22), 2062–2066.

Li, J.; He, A.; Zheng, J.; Han, C. C. *Biomacromolecules* **2006,** *7* (7), 2243–2247.

Li, M.; Mondrinos, M. J.; Chen, X.; Gandhi, M. R.; Ko, F. K.; Lelkes, P. I. *J. Biomed. Mater. Res. A.* **2006,** *79* (4), 963–973.

Li, Y.; Lim, L. T.; Kakuda, Y. *J. Food Sci.* **2009,** *74* (3), C233–C240.

Li, D.; Wang, Y. L.; Xia, Y. N. *Nano Lett.* **2003,** *3,* 1167–1171.

Li, Z. Y.; Zhang, H. N.; Zheng, W.; Wang, W.; Huang, H. M.; Wang, C.; MacDiarmid, A. G.; Wei, Y. *J. Am. Chem. Soc.* **2008,** *130,* 5036–5037.

Li, D.; Xia, Y. N. *Adv. Mater.* **2004,** *16,* 1151–1170.

Liechty, W. B.; Kryscio, D. R.; Slaughter, B. V.; Peppas, N. A. *Annu. Rev. Chem. Biomol. Eng.* **2010,** *1,* 149.

Liu, W.; Thomopoulos, S.; Xia, Y. *Adv. Healthc. Mater.* **2012,** *1* (1), 10–25.

López-Rubio, A.; Lagaron, J. M. *Innov. Food Sci. Emerg. Technol.* **2012,** *13,* 200–206.

López-Rubio, A.; Sanchez, E.; Wilkanowicz, S.; Sanz, Y.; Lagaron, J. M. *Food Hydrocoll.* **2012,** *28* (1), 159–167.

Lubambo, A. F.; de Freitas, R. A.; Sierakowski, M. R.; Lucyszyn, N.; Sassaki, G. L.; Serafim, B. M.; Saul, C. K. *Carbohydr. Polym.* **2013,** *3* (2), 484–491.

Luo, C. J.; Loh, S.; Stride, E.; Edirisinghe, M. *Food Bioprocess Technol.* **2012,** *5* (6), 2285–2300.

Martin, O.; Schwach, E.; Avérous, L.; Couturier, Y. *Starch* **2001,** *53,* 372–380.

Mascheroni, E.; Fuenmayor, C. A.; Cosio, M. S.; Di Silvestro, G.; Piergiovanni, L.; Mannino, S.; Schiraldi, A. *Carbohydr. Polym.* **2013,** *98* (1), 17–25.

McCann, J. T.; Li, D.; Xia, Y. N. *J. Mater. Chem.* **2005,** *15,* 735–738.

Mieszawska, A. J.; Jalilian, R.; Sumanasekera, G. U.; Zamborini, F. P. *Small* **2007,** *3,* 722–756.

Miao, J.; Miyauchi, M.; Simmons, T. J.; Dordick, J. S.; Linhardt, R. J. *J. Nanosci. Nanotechnol.* **2010,** *10* (9), 5507–5519.

Miguel, O.; Fernandez-Berridi, M. J.; Iruin, J. J. *J. Appl. Polym. Sci.* **1997,** *64* (9), 1849–1859.

Miller, K. S.; Krochta, J. M. *Trends Food Sci. Technol.* **1997,** *8* (7), 228–237.

Min, B. M.; Lee, S. W.; Lim, J. N.; You, Y.; Lee, T. S.; Kang, P. H.; Park, W. H. *Polymer* **2004,** *45* (21), 7137–7142.

Mozafari, M. R.; Flanagan, J.; Matia Merino, L.; Awati, A.; Omri, A.; Suntres, Z. E.; Singh, H. *J. Sci. Food Agric.* **2006,** *86* (13), 2038–2045.

Nair, L. S.; Bhattacharyya, S.; Laurencin, C. T. *Expert Opin. Biol. Ther.* **2004,** *4* (5), 659–668.

Natu, M. V.; de Sousa, H. C.; Gil, M. H. *Int. J. Pharm.* **2010,** *397* (1), 50–58.

Neethirajan, S.; Jayas, D. S. *Food Bioprocess Technol.* **2011,** *4* (1), 39–47.

Neo, Y. P.; Ray, S.; Jin, J.; Gizdavic-Nikolaidis, M.; Nieuwoudt, M. K.; Liu, D.; Quek, S. Y. *Food Chem.* **2013a,** *136* (2), 1013–1021.

Neo, Y. P.; Swift, S.; Ray, S.; Gizdavic-Nikolaidis, M.; Jin, J.; Perera, C. O. *Food Chem.* **2013b,** *141* (3), 3192–3200.

Pérez-Masiá, R.; López-Rubio, A.; Lagarón, J. M. *Food Hydrocoll.* **2013,** *30* (1), 182–191.

Rafferty, K.; Walters, G.; Heaney, R. P. *J. Food Sci.* **2007,** *72* (9), R152–R158.

Ramakrishna, S.; Fujihara, K.; Teo, W. E. Yong, T.; Ma, Z. W.; Ramaseshan, R. *Mater. Today* **2006,** *9,* 40–50.

Rayleigh, L. *Lond. Edinb. Dubl. Phil. Mag. J. Sci.* **1882,** *14* (87), 184–186.

Reneker, D. H.; Chun, I. *Nanotechnology* **1996,** *7,* 216–223.

Reneker, D. H.; Yarin, A. L. *Polymer* **2008,** *49* (10), 2387–2425.

Rho, K. S.; Jeong, L.; Lee, G.; Seo, B. M.; Park, Y. J.; Hong, S. D.; Roh, S.; Cho, J. J.; Park, W. H.; Min, B. M. *Biomaterials* **2006,** *27,* 1452–1461.

Ringsdorf, H. Structure and Properties of Pharmacologically Active Polymers. *J. Polym. Sci. Polym. Symp.* **1975,** *51* (1), 135–153.

Salalha, W.; Kuhn, J.; Dror, Y.; Zussman, E. *Nanotechnology* **2006,** *17* (18), 4675.

Schiffman, J. D.; Schauer, C. L. *Polym. Rev.* **2008,** *48* (2), 317–352.

Schreuder-Gibson, H.; Gibson, P.; Senecal, K.; Sennett, M.; Walker, J.; Yeomans, W.; Ziegler, D.; Tsai, P. P. *J. Adv. Mater.* **2002,** *34,* 44–55.

Shahidi, F.; Han, X. Q. *Crit. Rev. Food Sci. Nutr.* **1993,** *33* (6), 501–547.

Silke, M.; Jean, S. S.; Bruce, C. D.; John, F. R. *Macromolecules* **2002,** *35,* 8456–8466.

Silvestre, C.; Duraccio, D.; Cimmino, S. *Prog. Polym. Sci.* **2011,** *36* (12), 1766–1782.

Sionkowska, A. *Prog Polym. Sci.* **2011,** *36* (9), 1254–1276.

Srikar, R.; Yarin, A. L.; Megaridis, C. M.; Bazilevsky, A. V.; Kelley, E. *Langmuir* **2008,** *24* (3), 965–974.

Subbiah, T.; Bhat, G. S.; Tock, R. W.; Parameswaran, S.; Ramkumar, S. S. *J. Appl. Polym. Sci.* **2005,** *96* (2), 557–569.

Taepaiboon, P.; Rungsardthong, U.; Supaphol, P. *Eur. J. Pharm. Biopharm.* **2007,** *67* (2), 387–397.

Taylor, G. *Proc. Math. Phys. Eng. Sci.* **1964,** 280 (1382), 383–397.

Taylor, G. *Proc. Math. Phys. Eng. Sci.* **1969,** *313* (1515), 453–475.

Torres-Giner, S.; Gimenez, E.; Lagaron, J. M. *Food Hydrocoll.* **2008,** *22* (4), 601–614.

Torres Giner, S.; Martinez Abad, A.; Ocio, M. J.; Lagaron, J. M. *J. Food Sci.* **2010,** *75* (6), N69–N79.

Torres-Giner, S.; Ocio, M. J.; Lagaron, J. M. *Carbohydr. Polym.* **2009,** *77* (2), 261–266.

Thavasi, V.; Singh, G.; Ramakrishna, S. *Energy Environ. Sci.* **2008,** *1,* 205–221.

Van Bommel, L. Virtual Pivot Point Control for Running Robots. Doctoral dissertation, TU Delft, Delft University of Technology, 2011.

van der Graaf, S.; Schroen, C. G. P. H.; Boom, R. M. *J. Membr. Sci.* **2005,** *251* (1), 7–15.

Vega-Lugo, A. C.; Lim, L. T. *Food Res. Int.* **2009,** *42* (8), 933–940.

Verreck, G.; Chun, I.; Rosenblatt, J.; Peeters, J.; Van Dijck, A.; Mensch, J.; Brewster, M. E. *J. Control Release* **2003,** *92* (3), 349–360.

Weiss, J.; Takhistov, P.; McClements, D. J. *J. Food Sci.* **2006,** *71* (9), R107–R116.

Wongsasulak, S.; Kit, K. M.; McClements, D. J.; Yoovidhya, T.; Weiss, J. *Polymer* **2007,** *48* (2), 448–457.

Wongsasulak, S.; Pathumban, S.; Yoovidhya, T. *J. Food Eng.* **2014,** *120,* 110–117.

Wu, Y.; Clark, R. L. *J. Biomater. Sci. Polym. Ed.* **2008,** *19* (5), 573–601.

Yang, J. M.; Zha, L. S.; Yu, D. G.; Liu, J. *Colloids Surf. B Biointerfaces* **2013,** *102,* 737–743.

Yarin, A. L.; Koombhongse, S.; Reneker, D. H. *J. Appl. Phys.* **2001,** *89,* 3018–3026.

Yawen, L.; Qianli, M.; Xiangting, D.; Wensheng, Y.; Jinxian, W.; Guixia, L. *Chem. Phys.* **2015,** *17,* 22977–22984. DOI: 10.1039/c5cp03522h

Yoshimoto, H.; Shin, Y. M.; Terai, H. J.; Vacanti, P. *Biomaterials* **2003,** *24,* 2077–2082.

Yun, K. M.; Hogan, C. J.; Matsubayashi, Y.; Kawabe, M.; Iskandar, F.; Okuyama, K. *Chem. Eng. Sci.* **2007,** *62* (17), 4751–4759.

Yurteri, C. U.; Hartman, R. P.; Marijnissen, J. C. *KONA Powder Part. J.* **2010,** *28* (0), 91–115.

Zeleny, J. *Phys. Rev.* **1914,** *3* (2), 69.

Zeleny, J. Instability of Electrified Liquid Surfaces. *Phys. Rev.* **1917,** *10* (1), 1.

Zexuan, D.; Scott Kennedy, J.; Yiquan, W. *J. Power Sources* **2011,** *196,* 4886–4904.

Zhang, C. L.; Lv, K. P.; Cong, H. P.; Yu, S. H. *Small* **2012,** *8* (5), 648–653.

Zhang, S.; Huang, Y.; Yang, X.; Mei, F.; Ma, Q.; Chen, G.; Deng, X. *J. Biomed. Mater. Res. A.* **2009,** *90* (3), 671–679.

Zong, X.; Kim, K.; Fang, D.; Ran, S.; Hsiao, B. S.; Chu, B. *Polymer* **2002,** *43* (16), 4403–4412.

Zoran, A.; Coelho, M. *Leonardo* **2011,** *44* (5), 425–431.

CHAPTER 17

AN UNDERUTILIZED NOVEL XERIC CROP: KAIR (*Capparis decidua*)

YAMINI CHATURVEDI[1*] and RANJANA NAGAR[2]

[1]*Department of H.Sc., Govt. of Rajasthan, Jaipur, Rajasthan, India*

[2]*Department of H.Sc., University of Rajasthan, Jaipur, Rajasthan, India*

Corresponding author. E-mail: yaminichaturvedi@yahoo.co.uk

CONTENTS

Abstract 456
17.1 Introduction 456
17.2 Materials and Methods 457
17.3 Conclusion 467
Keywords 468
References 468

ABSTRACT

Capparis decidua, Kair, a fruit of an arid zone shrub of Rajasthan was analyzed for the concentration of nutrients and antinutrients in fresh and various processed conditions. The crude protein and phosphorus contents were appreciable and fairly high scale of vitamin C was seen in fresh state of Kair. Among processings, fermentation resulted in low Na and K levels with a parallel rise in Fe. The level of antinutrients was also affected markedly following processing. Fermentation was seen to be the best method which improved the nutritive value by decreasing antinutritional factors.

17.1   INTRODUCTION

Kair is the round and green fruit of an arid zone shrub which is eaten extensively in desert region in the form of pickle and vegetable. The fruit belongs to the family *Capparidaceae*. The plants of this family occur in tropical and warm temperate regions of the world and quite a good number of plants are found in xeric conditions. The extreme condition is represented by *Capparis decidua*. The fruit also finds its significant place in Ayurveda (the Indian Science of Life). It has been reported to have antiseptic, digestive purgative, deworming, antiasthmatic, and pyretic properties (Sharma, 1956). It improves blood circulation, reduces edema, tones cardiac efficiency, and removes skin disorders. The bark has also been reported to have a glucoside called gluco-capparin (Bhavamishra, 1969). Some published data also refer to the specific aspect of hypocholesterolemic effect of Kair (Dashorra et al., 1984). Thus, the above-mentioned properties signify the therapeutic importance of the fruit Therapeutic values are always based on the metabolic changes in the pathways for which actual concentration of different nutrients and antinutrients of a particular food material should be known Steinmetz et. al. (1996). This study, therefore, has been undertaken to study the concentration of nutrients and antinutrients of *Capparis decidua* in fresh and various processed conditions.

Before consumption, food materials are processed in various ways which render them more palatable, delicious, and nutritious (Paramajyothi & Mulimani, 1996; Sharma & Kapoor, 1997; Mameesh & Tomar, 1993). Fresh Kair, which is consumed in the form of vegetable and pickle, is also subjected to many traditional processings like:

a.   Blanching
b.   Fermentation.

Changes in the nutrients as well as antinutrients due to the effect of the above mentioned indigenous treatments have also been studied.

17.2   MATERIALS AND METHODS

Fresh Kair was obtained from the vegetable market of Jaipur. The sample was cleaned; fruit stalk was removed and then used for analyses. The fresh fruit was taken as a whole for the estimation of moisture, ash, N content, total fat, crude fiber, Ca, Fe, and antinutrient: oxalate. Ethanolic extracts of the fresh and processed sample were prepared and used for the following estimations.

Nutrients		**Antinutrients**	
A.	Soluble proteins	A.	Polyphenols
B.	Amino acids	B.	Phytic acid
C.	Soluble sugar	C.	Oxalates
D.	Glucose		
E.	Triglycerides		
F.	Phosphorus		
G.	Chloride		
H.	Potassium		
I.	Sodium		

17.2.1

Moisture and ash were determined by AOAC official methods (Raghura-malu et al., 1983). Total nitrogen content was studied by Microkjeldhal method (Hawk et al., 1968). The crude protein content was calculated by multiplying total nitrogen content with the factor 6.25. Soluble proteins were assayed by modified Lowry's procedure (Lowry et al., 1961). Amino acids were estimated by Ninhydrin method (Plummer, 1986).

Total sugars and glucose: Total sugars and glucose were studied by phenol-sulfuric acid method and enzymatic GOD/GPD method (Dubois et al., 1966; Tietz, 1970). Fat and triglycerides were assayed by Fossati's method (McGowan et al., 1983). Crude fiber was determined by the method given in the Manual of Laboratory Techniques (Raghuramalu et al., 1983).

17.2.1.1   MINERALS

Phosphorus was done by Daly's method (Daly, 1972). Calcium was estimated by AOAC official method (Raghuramalu et al., 1983). Iron was studied by Wong's method (Raghuramalu et al., 1983). Chlorine was assayed by mercuric nitrate method (Henry, 1964). Sodium and Potassium were diagnosed by Tietz method (Tietz, 1970).

17.2.1.2   ANTINUTRIENTS

Polyphenols were determined by the modified method of Goldstein and Swain (1963). Phytic acid was diagnosed by Young's method (Young, 1936). Oxalates were determined by the method given in a Manual of Laboratory Techniques (Raghuramalu et al., 1983).

17.2.2   STATISTICAL ANALYSES

Data obtained from the chemical analyses of the samples were triplicate in nature. The mean and standard deviation were calculated and compared the means using two-tailed paired "t" test. The calculated values of "t" were compared with the tabulated value of "t" at 2 degrees of freedom at 5% level (4.303) and 1% level (9.925) of significance. The level of significance was denoted as follows:

0.05	–	Nonsignificant
< 0.5	–	Significant
< 0.01	–	Highly significant

17.2.2.1   FOOD PROCESSING

Blanching and fermentation are the traditional treatments to which Kair are subjected to.

17.2.2.2   BLANCHING

Fresh Kair was kept in boiling water for 3 min, wiped and packed in polythene bags to keep under refrigeration (Palande et al., 1996).

Fermentation: Kair was subjected to soaking for different time periods in salted water. During the summer months in the desert of northern India, when the temperature ranges from 40°C to 48°C, the mixture is kept in earthen pots. Prolonged soaking and heat ferments were the samples. Analyses were done at different time intervals of soaking, namely, 3, 6, 10, and 20 days, respectively.

17.2.3   RESULTS AND DISCUSSION

17.2.3.1   COMPOSITION OF FRESH KAIR

The results of the proximate analyses of the fresh Kair have been shown in Table 17.1A. Moisture content was found to be 69.7%. The crude protein content of the fruit was seen 5.8% which was comparable to the values of fresh and green peas (Gopalan et al., 1996). The flower of *Capparis spinosa* (a Mediterranean shrub) which has close resemblance with Kair was found to have 10% higher moisture whereas protein content was quite comparable, that is (6.8 g%) (Rodrigo et al., 1992). Achinewhu et al. (1995) have determined 3.2 g% crude protein for indigenous wild fruits. This value is about 50% less than the value of Kair.

The ash content of Kair was found to be 1.7 g, to which *Capparis spinosa* is very much similar (1.6 g%) (Rodrigo et al., 1992). Kair was not found to be a good source of fiber (0.98 g%). Though a higher value of fiber was reported for *Capparis spinosa* (5.4 g%). Pramila et al. (1991) also gave the values of crude fiber for some uncommon fruits consumed in hilly area in the range from 1.52 to 3.2/100 g.

Available carbohydrates and the energy value were 20.63 g% and 116 cal./100 g, respectively.

17.2.3.2   MINERALS AND VITAMINS

Table 17.1C shows the mineral and vitamin content of Kair. Fresh Kair showed a high P and Cl content (248.6 and 146.4 mg%), respectively, with low values of Ca (130 mg%) and Fe (1.22 mg%). Where Na level of fresh Kair was found moderate, that is, 54.9 mg%. K content, however, was seen to be high.

TABLE 17.1A Proximate Composition of Fresh and Processed Kair ($n = 3$).

	Fresh	Blanched	3g/100 g	Fermentation (days)		
				6	10	20
Moisture	69.7 ± 0.339	69.83 ± 0.3.05*	70.13 ± 0.0.57*	70.4 ± 0.173	72.03** ± 0.055***	73.09 ± 0.057***
Ash	1.7 ± 0.08	1.7 ± 0.0.1*	1.73 ± 0.01*	1.78 ± 0.05***	1.79 ± 0.00***	1.80 ± 0.05***
Total N	0.928 ± 0.099	0.91 ± 0.009*	0.94 ± 0.005*	0.95 ± 0.00*	0.96 ± 0.005**	1.0 ± 0.00***
Crude protein (N × 6.25)	5.8	5.68	5.8	5.9	6.0	6.25
Crude fat	1.18 ± 0.006	1.19 ± 0.000*	1.19 ± 0.000*	1.190 ± 0.01*	1.17 ± 0.05*	1.18 ± 0.05*
Crude fiber	0.98 ± 0.004	0.98 ± 0.000*	—	—	—	—
Available carbohydrate (by diff.)	20.64	20.62	—	—	—	—
Energy (cal.)	116.3	115.8	—	—	—	—

*Nonsignificant.
**Significant.
***Highly significant.
" ": Not done.

TABLE 17.1B Nutrients in Ethanolic Extract ($n = 3$).

	Fresh	Blanched	3g/100 g	Fermentation (days)		
				6	10	20
Soluble proteins	322.3 ± 1.08	$325.0 \pm 2.6^{*}$	$524.4 \pm 5.0^{***}$	$463.4 \pm 2.0^{***}$	$562.5 \pm 4.06^{***}$	$665.9 \pm 4.5^{***}$
Amino acids	130.2 ± 0.16	$131.4 \pm 1.73^{*}$	$147.4 \pm 1.4^{***}$	$158.8 \pm 1.55^{***}$	$200.9 \pm 1.11^{***}$	$292.5 \pm 2.55^{***}$
Triglycerides	468 ± 1.0	$467.0 \pm 2.0^{*}$	$371.0 \pm 1.0^{***}$	$371.0 \pm 1.0^{***}$	$370.0 \pm 0.00^{***}$	$368.0 \pm 1.0^{***}$
Soluble sugars	3360 ± 1.5	$3350 \pm 2.5^{*}$	$3080 \pm 1.1^{***}$	$2520 \pm 2.5^{***}$	$700 \pm 2.0^{***}$	$470 \pm 1.0^{***}$
Glucose	187.0 ± 1.41	$179.0 \pm 1.57^{***}$	$163.0 \pm 1.52^{***}$	$125.0 \pm 2.25^{***}$	$90.8 \pm 2.15^{***}$	$41.17 \pm 2.40^{***}$

*Non-significant.
***Highly significant.

TABLE 17.1C Minerals and Vitamins ($n = 3$).

	Fresh	Blanched	Fermentation (days)			
			3g/100 g	6	10	20
Phosphorus	248.6 ± 1.14	248.6 ± 1.644*	192.1 ± 1.25***	165.3 ± 1.55***	120.2 ± 1.55***	105.5 ± 1.60***
Calcium	130 ±1.62	130.3 ± 1.57*	126.6 ± 1.36**	115.1 ± 1.66***	101.1 ± 1.35***	89.6 ± 1.37***
Iron	1.22 ± 0.008	1.18 ± 0.032*	1.31 ± 0.01***	1.34 ± 0.02***	1.44 ± 0.02***	1.65 ± 0.035***
Chloride	146.4 ± 2.50	145.5 ± 3.50	140.8 ± 1.27**	135.4 ± 2.55***	117.6 ± 2.43***	101.2 ± 2.40***
Sodium	54.9 ±1.24	54.2 ± 2.20*	49.0 ± 3.04***	37.1 ± 1.86***	5.1 ± 0.01***	7.91 ± 0.02***
Potassium	115.96 ± 2.55	115.9 ± 2.51*	116.0 ± 2.63*	114.2 ± 2.35*	111.2 ± 1.27*	100.5 ± 2.25***
Vitamins						
β-carotene	2.44 ± 0.004	2.42 ± 0.00**	2.39 ± 0.005**	2.39 ± 0.005**	2.39 ± 0.005**	2.39 ± 0.005***
Thiamin	–	–	–	–	–	–
Riboflavin	–	–	–	–	–	–
Niacin	–	–	–	–	–	–
Pyridoxine	–	–	–	–	–	–
Folic acid	–	–	–	–	–	–
Vitamin C	55 ± 3.26	49.1 ± 1.32***	45.1 ± 1.05***	45.0 ± 1.00	42.1 ± 1.32**	40.4 ± 1.58***

*Nonsignificant.
**Significant.
***Highly significant.
" – ": Not detected.

Table 17.1C also indicates the vitamin content of the sample. β carotene was available in moderate amounts, that is, 2.44 mg/100 g. It was seen that a serving of 100 g of fresh Kair could meet the RDI of an adult. Guil et al. (1997) reported carotenoid content of selected 16 wild edible plants which ranged from 4.2 mg to 15.4 mg/100 g.

Similarly, ascorbic acid was also found on fairly high scale (55 mg%). Guil et al. (1997) also studied ascorbic acid of mild edible plants of southeast Spain and found the values within 7–155 mg/100 g.

17.2.3.3   ANTINUTRIENTS

A moderate phytate content (112.8 mg%) was observed in fresh Kair (ID). It was interesting that the polyphenolic content was found much low, that is, 2.3 mg%. However, oxalate content of Kair was 310 mg%, though the toxicity of oxalic acid is quite low. Guil et al. (1997) reported that minimal lethal dose for humans is considered to be about 5 g for an adult. Mosha et al. (1995) assessed selected vegetables and they found traces of oxalates in cabbage and turnip while a high concentration was observed in sweet potato (469 mg%) and peanut greens (407 mg%). Wanasundera and Ravindran (1994) reported oxalate levels in yam tubers in the range of 486–78 mg%, but they further stressed that the analyzed oxalates may not constitute a nutritional concern since 50–75% of the antinutrient was in water-soluble form.

17.2.3.4   EFFECTS OF PROCESSING

Tables 17.1A–D depict the effect of blanching.

Blanching proved influential only for glucose, Na and β-carotene which were reduced significantly after the treatment.

17.2.3.5   FERMENTATION

As described earlier, Kair is treated in a peculiar indigenous way for pickling purposes, that is, fermentation. The findings of this rural technology are summarized in Table 17.1A–D. It is clear from Table 17.1A that prolonged soaking of 20 days resulted in a highly significant ($p < 0.01$) increase in moisture content by 4.8%. Fermentation of Kair showed significant changes in nitrogenous compounds (total nitrogen, crude protein, and amino acids).

TABLE 17.1D Antinutrients ($n = 3$).

	Fresh	Blanched		Fermentation (days)		
			3g/100 g	6	10	20
Polyphenols	2.3 ± 0.016	$2.3 \pm 0.05^{*}$	$0.52 \pm 0.02^{***}$	$0.66 \pm 0.006^{***}$	$0.31 \pm 0.001^{***}$	$0.12 \pm 0.009^{***}$
Phytates	112.8 ± 2.543	$112.4 \pm 2.458^{*}$	$44.3 \pm 1.393^{***}$	$42.8 \pm 1.892^{***}$	$37.7 \pm 1.938^{***}$	$28.9 \pm 2.91^{***}$
Oxalates	310 ± 2.087	$310.0 \pm 2.873^{*}$	$304 \pm 2.011^{**}$	$304 \pm 1.56^{**}$	$295 \pm 2.75^{***}$	$290 \pm 2.39^{***}$

*Non-significant.
**Significant.
***Highly significant.

N content increased to a significant level ($p < 0.01$) by 7% after 20 days of soaking, which resulted in higher value of protein. It was also seen that the difference between and among the days was variable such as the increase in N content on the 10th day was significant one ($p < 0.05$), while on the 20th day it became highly significant ($p < 0.01$). Aremu et al. (1995) also observed an increase in the protein content of cocoa beans following fermentation on the 6th day. Binita and Kheterpaul (1997) studied the effect of probiotic fermentation and found that the treatment markedly improved the in vitro protein digestibility in the fermented mixture which was indigenously developed with barley flour, green gram flour, skimmed milk powder, and tomato pulp. In addition, the fermented Kair also showed a highly significant increase of 5% in the ash content after 20 days.

17.2.3.6   NUTRIENTS IN ETHANOLIC EXTRACT

Table 17.1B shows that soluble proteins were almost doubled themselves following fermentation which was highly significant. Such findings have been reported by other researchers also, such as Sharma and Kapoor (1997) and Pawar and Parlikar (1990) who indicated significant increase in the recovery of soluble proteins on dehulling and soaking of various millets.

A highly significant increase was seen in the amino acid level also which could be attributed to the biochemical changes taking place during prolonged soaking of samples. Oste (1991) reported that digestibility improves with different processings like germination, heating, soaking, and fermentation.

Triglycerides showed a highly significant ($p < 0.01$) fall following soaking where they decreased by 21% in 20 days.

A highly significant reduction of 86% and 78% was observed in case of soluble sugars and glucose, respectively, with fermentation. The maximum fall was seen on the 6th day when the reducing sugars decreased from 87 to 33% and the reduction continued thereafter. Sandhu and Shukla (1996) also reported that reducing sugars fell from 92% to 3% after 20 days. Grewal (1994) studied the effect on available carbohydrate content of soybean and the author found that total soluble sugars, reducing sugars, non-reducing sugars and starch decreased significantly as fermentation progressed, and the extent of reduction of these sugars seemed to be greater at higher temperature. Here, it is noteworthy that Kair is fermented in the extreme summer months when the mercury shows a range of 40–48°C in the arid zone. The possible explanation of the reduction may be due to the metabolism of sugars for energy purposes by anaerobic microflora or alcoholic fermentation (West et al., 1974).

17.2.3.7   MINERALS AND VITAMINS

Levels of minerals are also affected by the treatment in Table 17.1C. A significant ($p < 0.01$) decrease of 57.5%, 31%, and 30.8% was seen in P, Ca, and Cl, respectively, after 20 days of soaking. It can be inferred from Table 17.1C that changes started occurring from 3rd day of fermentation but the rapid changes occurred from the 6th day. Na and K also reduced to highly significant levels ($p < 0.01$) by 85.5 and 13.3%, respectively, with fermentation. Maximum fall was seen on the 10th day, where the fall was observed from 68% to 9.3% for Na and 71–29% for K. The apparent fall in Na level recommends the processed Kair for patients suffering from hypertension and Diabetes mellitus. Dashorra et al. (1984) had seen the effect of Kair on rabbits with hypercholesterolemia and found significant positive results. Neerja and Rajyalakshmi (1996) also found the hypoglycemic effect of processed fenugreek seeds in humans. Traditional fermented foods are medicinal values. The consumption of TEMPE (a traditional fermented food from Indonesia) decreased the cholesterol level which is due to inhibition of hydroxymethylglutaryl co-enzyme reductase by linoleic acid during fermentation (Hermosilla et al., 1993).

Fe content also underwent highly significant ($p < 0.01$) positive changes during prolonged soaking. The Fe mineral started rising from 3rd day which continued till the 20th day. The increase can be attributed to the hydrolysis taking place during fermentation. When the phytase activity increases, thereby it decreases the phytate content and making more and more Fe available in free form. Johnson (1991) reported that fermentation during the production of beer, wine, yoghurt, and African tribal food improved the availability of Zn and Fe with a parallel decrease in phytic acid and saponins.

Two important vitamins, namely β carotene and vitamin C were also affected significantly by fermentation. It decreased the provitamin A content highly after 10 days, though the fall, did not continue thereafter. Vitamin C content also underwent marked losses during soaking. The decrease was most rapid during the first 3 days and then stabilized at a lower-pace. Lyimo et al. (1991) investigated the effects of traditional food processing, storage, and fermentation method on vegetable nutrients and they concluded that storage using fermentation in earthenware pots for 6 weeks resulted in significant losses of vitamin.

17.2.3.8 ANTINUTRIENTS

The level of antinutrients was also affected markedly ($p < 0.01$) following processing. Fermentation brought down the level of polyphenols sharply, that is, 94% after 20 days. Uzogara et al. (1990) found that the concentration of polyphenols in cowpea, calculated as tannic acid, were reduced substantially (69–79%) following treatment with Kanwa salt. Here, it is noteworthy that Kair is soaked in salted water also for a prolonged period of time by the rural folks. Sharma and Kapoor also found reduction in polyphenols of pearl millet due to fermentation. Rao and Deosthale (1982) explained that the loss of polyphenols is expected from leaching of the pigments into the water.

Phytates also dropped by 74% after 20 days. Phytate phosphorus of pearl millet also got reduced from 74% to 60% with soaking (Pawar & Parlikar, 1990). The reduction can be explained by the fact that during soaking phytate, being soluble in aqueous solution, gets destroyed by phytase. Reddy et al. (1982) stated that soaking and fermentation under optimal condition demonstrated to reduce the phytate content of vegetables completely. Sandberg (1991), Bishnoi et al. (1994), and Giami and Wachuku (1997) have also given more or less similar views. Kheterpaul and Chauhan (1991) observed complete elimination of phytic acid of pearl millet low following fermentation for 72 h. Obizoba and Atii (1991) concluded that cooking and fermentation improved the nutrient quality and reduced the antinutritional factors to safe levels much greater than any other processing.

17.3 CONCLUSION

So, finally it can be said that Kair is an underutilized novel xerophytic crop which needs more cultivation on a larger scale for its therapeutic benefits. Moreover, among processing, fermentation was seen to be the best method which improved the nutritive value by decreasing the antinutritional factors remarkably and enhancing the availability of important minerals like P, Fe, etc. Besides, during fermentation 6th day may be mentioned as the key point when maximum biochemical changes were taking place.

With a view to upgrade the rural technology, fermentation can be recommended as a potential method for households. This has high significance for the remote areas of developing countries where people cannot afford commercially available expensive fortified foods.

In addition, knowing the nutritional significance one can commercialize culturally acceptable, traditionally processed foods which may present unidentified novel sources of nutrients.

KEYWORDS

- **Kair**
- **fermentation**
- **processing**
- **nutrients**
- **antinutrients**

REFERENCES

Achinewhu, S. C.; Ogbonna, C. C.; Hart, A. D. Chemical Composition of Indigenous Wild Herbs, Spices, Fruits, Nuts, and Leafy Vegetables Used as Foods. *Plant Food Hum. Nutr.* **1995,** *48* (4), 341–348.

Aremu, C. Y.; Agiang, M. A.; Ayatse, J. O. Nutrient and Antinutrient Profiles of Raw and Fermented Cocoa Beans. *Plant Food Hum. Nutr.* **1995,** *48* (3), 217–223.

Bhavamishra. Bhavprakash Nighantau Tikakar. Chunekar, K. C. *Vatadi Varg* (33). Chaukhambha Vidya Bhawan: Varanasi; 404, 1969.

Binita, R.; Kheterpaul, N. Probiotic Fermentation: Effect of Antinutrients and Digestibility of Starch and Protein of Indigenously Developed Food Mixture. *Nutr. Health* **1997,** *11* (3), 139–147.

Bishnoi, S.; Kehterpaul, N.; Yadav, R. K. Effect of Domestic Processing and Cooking Methods on Phytic Acid and Polyphenol Contents of Pea Cultivars *Pisum sativum. Plant Food Hum. Nutr.* **1994,** *45* (4), 381–388.

Daly, J. A. *Clin. Chem.* **1972,** *18,* 263–265.

Dashorra, M. S.; Sharma, I.; Dixit, V. P. Effect of Kair on Rabbits with Hypercholesterolemia. *J. Ayurveda* **1984,** *29* (2), 43–47.

Dubois, M.; Gilles, K. A.; Hamilton, J. K.; Raber, P. A.; Smith, F. Colorimetric Method for the Determination of Sugars and Related Substances. *Annal. Chem.* **1966,** *98,* 350.

Guil, J. L.; Rodriguez-Garcia, I.; Toriza, E. Nutritional and Toxic Factors in Selected Wild Edible Plants. *Plant Food Hum. Nutr.* **1997,** *51* (2), 99–107.

Goldstein, J. F.; Swain, T. *Phyto Chem.* **1963,** *2,* 37.

Gopalan, C.; Rama Sastri, B. V.; Balasubramanian, S. C. *Nutritive Value of Indian Foods;* ICMR National Institute of Nutrition: Hyderabad, 1996; p 48.

Giami, S. Y.; Wachuku, O. C. Composition and Functional Properties of Unprocessed and Locally Processed Seeds from Three Underutilized Food Sources in Nigeria. *Plant Food Hum. Nutr.* **1997,** *50* (1), 27–36.

Grewal, R. B. Effect of Germination and Indigenous Fermentation on Available Carbohydrate Content of Soybean. *Proc. Nutr. Soc. Ind. XXVII Ann.* **1994,** 63.

Hawk, B. P.; Oser, L. B.; Summerson, H. W. *Hawk's BP Physiological Chemistry;* McGraw Hill Co.: New York, 1968.

Henry, R. J. *Clinical Chemistry. Principles and Techniques.* Harper & Row: New York, 1964; p 122.

Hermosilla, J. A. G.; Jha, H. C.; Egge, H.; Mahumud, M. Isolation and Characterization of Hydroxymethylglutaryl Coenzyme a Reductase Inhibitors from Fermented Soybean Extracts. *J. Clin. Biochem. Nutr.* **1993,** *15,* 163–174.

Johnson, P. E. Effect of Food Processing and Preparation on Mineral Utilization. *Adv. Exp. Med. Biol.* **1991,** *289,* 483–498.

Kheterpaul, N.; Chauhan, B. M. Sequential Fermentation of Pearl Millet by Yeasts and *Lacto bacilli*: Effect on the Antinutrients and In Vitro Digestibility. *Plant Food Hum. Nutr.* **1991,** *41* (4), 321–327.

Lowry, O. H.; Rosebrough, N. J.; Farr, A. L.; Randall, R. J. Protein Measurement with Folin Phenol Reagent. *J. Biol. Chem.* **1961,** *193,* 265–275.

Lyimo, M. H.; Nyagwegwe, S.; Mnkeni, A. P. Investigations on the Effect of Traditional Food Processing, Preservation and Storage Methods on Vegetable Nutrients: A Case Study in Tanzania. *Plant Food Hum. Nutr.* **1991,** *41* (1), 53–57.

Mameesh, M. S.; Tomar, M. Phytate Content of Some Popular Kuwaiti. *Food Cereal Chem.* **1993,** *70,* 502–503.

McGowan, M. W.; Artiss, J. D.; Strandberg, D. R.; Zak, B. A. Peroxidase Coupled Method for the Colorimetric Determination of Serum Triglycerides. *Clin. Chem.* **1983,** *29,* 538.

Mosha, T. C.; Gaga, H. E.; Pace, R. D.; Laswai, H. S.; Mtebe, K. Effect of Blanching on the Content of Antinutritional Factors in Selected Vegetables. *Plant Food Hum. Nutr.* **1995,** *47* (4), 361–367.

Neerja, A.; Rajyalakshmi, P. Hypoglycemic Effect of Processed Fenugreek Seeds in Humans. *J. Food Sci. Technol.* **1996,** *33* (5), 427–430.

Obizoba, I. C.; Atii, J. V. Effect of Soaking, Sprouting, Fermentation and Cooking on Nutrient Composition and Some Antinutritional Factors of Sorghum (Guinesia) Seeds. *Plant Food Hum. Nutr.* **1991,** *41* (3), 203–212.

Oste, R. E. Digestibility of Food Protein. *Adv. Exp. Med. Biol.* **1991,** *289,* 371–378.

Paramajyothi, S.; Mulimani, V. H. Effect of Sprouting, Cooking and Dehulling on Polyhenols of Redgram *Cajanus cajan. J. Food Sci. Technol.* **1996,** *33* (3), 259–260.

Palande, K. B.; Kadlag, R. V.; Kachara, D. P.; Chavan, J. K. Effects of Blanching of Pearl Millet Seeds on Nutritional Composition and Shelf Life of Its Meal. *J. Food Sc. Technol.* **1996,** *33* (2), 153–155.

Pawar, V. D.; Parlikar, G. S. Reducing the Polyphenols and Phytates and Improving the Protein Quality of Pearl Millet by Dehulling and Soaking. *J. Food Sci. Technol.* **1990,** *27* (3), 140–143.

Plummer, D. T. The Quantitative Estimation of Amino Acids Using the Ninhydrin Reaction. In *an Introduction to Practical Biochemistry;* Tata McGraw Hill Publishing Company Ltd.: New Delhi, 1986; p 158.

Pramila, S. S.; Annamma, K.; Raghuwanshi, R. Nutrient Composition of Some Uncommon Foods Consumed by Kumaon and Garhwal Hill Subjects. *J. Food Sci. Technol.* **1991,** *28* (4), 237–238.

Rao, P. V.; Deosthale, Y. C. Tannin Content of Pulses: Variental Differences and Effect of Cooking and Germination. *J. Sci. Food Agric.* **1982,** *33,* 1013–1016.

Raghuramalu, N.; Madhavan Nair, K.; Kalyansundaram, S. *A Manual of Laboratory Techniques;* National Institute of Nutrition: Hyderabad, 1983.

Reddy, N. R.; Sathe, S. K.; Salunke, D. K. Phytates in Legumes and Cereals. *Adv. Food Res.* **1982,** *28,* 1.

Rodrigo, M.; Lazaro, M. J.; Alvarruiz, A.; Ginger, V. Composition of Capers *Capparis spinosa:* Influence of Cultivar, Size, and Harvest Date. *J. Food Sci.* **1992,** *57* (5), 1152–1154.

Sandhu, K. S.; Shukla, F. C. Methods of Pickling of Cucumbers *Cucumis* sativus: A Critical Appraisal. *J. Food Sci. Technol.* **1996,** *33* (6), 455–473.

Sandberg, A. S. The Effect of Food Processing on Phytate Hydrolysis and Availability of Iron and Zinc. *Adv. Exp. Med. Biol.* **1991,** *289,* 499–508.

Sharma, P. V. *Dravyaguna Vijnana Vegetable Drugs III.* Chaukhambha Vidya Bhawan Chowk: Varanasi, 207, 1956; Vol. II, 261.

Sharma, A.; Kapoor, A. C. Effect of Processing on Nutritional Quality of Pearl Millet. *J. Food Sci. Technol.* **1997,** *34* (1), 50–53.

Steinmatz, K. A.; Potter, J. D. Vegetables, Fruit, and Cancer Prevention: A Review. *J. Am. Dietet. Assoc.* **1996,** *96* (10), 1027–1039.

Tietz, N. W. *Fundamentals of Clinical Chemistry;* Elsevier: Amsterdam, Netherlands, 1970.

Uzogara, S. G.; Moston, I. D.; Daniel, J. W. Changes in Some Antinutrients of Cowpeas *Vigna unguiculata.* Processed with "Kanwa" Alkaline Salt. *Plant Food Hum. Nutr.* **1990,** *40* (4), 249–258.

Wanasundera, J. P.; Ravindran, G. Nutritional Assessment of *Yam Dioscorea alata* Tubers. *Plant Food Hum. Nutr.* **1994,** *46* (1), 33–39.

West, E. S.; Todd, W. R.; Mason, H. S.; Brugger, T. T. V. *Text Book of Biochemistry,* 4th ed.; Oxford and Publishing Co. Pvt. Ltd.: Delhi, India, 1974.

Young, L. The Determination of Phytic Acid. *Biochem. J.* **1936,** *30,* 252–257.

CHAPTER 18

VALUE ADDITION OF UNDERUTILIZED CROPS OF INDIA BY EXTRUSION COOKING TECHNOLOGY

DUYI SAMYOR, AMIT BARAN DAS, and SANKAR CHANDRA DEKA*

Department of Food Engineering and Technology, Tezpur University, Napaam, Tezpur 784028, Sonitpur, Assam, India

Corresponding author. E-mail: sankar@tezu.ernet.in

CONTENTS

Abstract 472
18.1   Introduction 472
18.2   Pseudocereals and Millets 473
18.3   Value Addition 477
18.4   Extrusion Cooking Technology 478
18.5   Types of Extrudate Products 484
18.6   Conclusion 486
Keywords 486
References 496

ABSTRACT

India is an agrarian country and produces various types of crops in the different parts of country. Even though these crops, namely, grain amaranth, buckwheat, finger millet, pearl millet, horse gram, red rice, black rice, etc., have commercial potential, still termed as "Underutilized" or "Neglected" or "Orphan" crops. These crops are the powerhouse of essential amino acids, phenolic acid, anthocyanin, vitamins, minerals, etc. Value addition of underutilized crops is growing in the food markets. Incorporation of various ingredients add varieties and enhance textural or nutritional quality of final products in the market. Extrusion cooking, high-temperature short time (HTST) is an important and popular food processing technique. This chapter describes the importance of incorporation of these crops in the development of value-added products using extrusion cooking technology.

18.1　INTRODUCTION

India is an agriculture-dependent or agrarian country. Plant biodiversity indicates the primary source for food, feed, shelter, medicines, etc., and sustainable earth (Pugalenthi et al., 2005; WCMC, 1992). Since time immemorial, our predecessor starts cultivating few hundred of species of available plants in various part of the world and transformed them to crop plants through genetic alternation by conscious and unconscious selection (UNEP, 1995).

However, at present, only 150 plant species are used and commercialized on a global scale significantly. Out of 80,000 usable plants, about 30,000 plants reported edible and approximately 7000 plants were cultivated by mankind in nature and total of only 158 plants are widely used for food (Wilson, 1992). According to Harlan (1992), among these food plants, 30 crops provide 90% of the world's food and only 10 crops supply 75% of the world's food budget. Only three crops, rice, wheat, and maize provide 60% of the world's total food requirement. The studies showed that only 50% of the world's requirements for calories are met by only three crops, that is, rice, wheat, and maize. Apparently, ~700 estimated species can play a vital role in poor people's lively hood and can be a means of commercialization. Although having commercial potential, they can also provide important environmental services, as they are adapted to marginal soil and climatic condition. According to Eyzaguirre et al. (1999), "Underutilized" term represents a category of crops which once grown more extensively, but cultivated in limited areas because of economic, agronomic, or genetic reasons. They are also been termed as "Neglected" or "Orphan" crops since they have received scant attention from research and development and very little scientific data available to them. Padulosi and Hoeschle-Zeledon (2004) defined them as the group of crops with potential economic value but remains "underexploited" or "underdeveloped."

Underutilized species can be a beneficial component of human health and have great potential to combat the lifestyle diseases. Various researchers have reported their investigated data to uncover new information in the utilization of these crops for the benefit of economy of the country. The underutilized crops are species, which can be used traditionally or given less important for food, fibers, fodder, oil, or medicinal properties. Being named as less popular or neglected crops, it has the underexploited potential to contribute to food security, nutrition, health, income generation, and environmental services. In Table 18.1, some of the underutilized crops are illustrated elaborately. There are three major categories of crops included in the discussion, namely, pseudocereals and millets, cereals, and legumes.

Value addition of underutilized crops is growing in the food markets. Incorporation of various ingredients add varieties in the items and enhance their textural or nutritional quality. Extrusion cooking, popularly known as HTST is an important and popular food processing technique classified as a process to produce a fiber-rich product (Gaossong & Vasanthan, 2000). The process is not only popular for developing ready-to-eat (RTE) food items but nutritionally to some extent may improve the bioavailability of bioactive compounds. A complex is formed with protein which later broken down in the human body, yielding antioxidant activity per second. Various fruit and vegetable by-products are also incorporated in the preparation of extrudate to increase the level of bioactive compounds (Brennan et al., 2011).

In this chapter, eight different types of underutilized crops, such as buckwheat, grain amaranth, finger millet, pearl millet, black rice, red rice, sorghum, and horse gram found in India have been elaborately discussed.

18.2 PSEUDOCEREALS AND MILLETS

18.2.1 BUCKWHEAT (Fagopyrum esculentum)

The distribution of *Fagopyrum spp.* has been reported from the different states of Northeast region India, namely Sikkim, Meghalaya, Arunachal Pradesh, Assam, Manipur, and Nagaland (Anonymous, 1987–2001). Joshi (1999) also reported that the practice of buckwheat cultivation was evident at Arunachal Pradesh and Sikkim, respectively. There are certain tribes like Monpas and Sherdukpen from Tawang district of Arunachal Pradesh, who use this grain as a staple food in the olden days. Leh and Kargil districts of Ladakh also reported the presence of buckwheat cultivation (Ahmad & Raj, 2012). Nutritional content of essential amino acids, such as lysine, threonine, and tryptophan was also found in higher amount (Lyman et al., 1956; Bonafaccia et al., 2003). Buckwheat (*Fagopyrum esculentum*) in Table 18.1 is classified as a pseudocereal. In Uttaranchal, local liquors called *pechuwi* and *chhang* are prepared from buckwheat (Joshi & Paroda, 1991a). It has been stated that due to the heat treatment, this grain changes its chemical composition and also functional properties of bioactive compounds. They also related the antioxidant capacity to the flavonoids concentration after hydrothermal treatment. Zielinski et al. (2009) related the antioxidant capacity of buckwheat products to the concentration of flavonoids after hydrothermal treatment.

18.2.2   AMARANTHUS (Amaranthus hypochondriacus)

Amaranthus (*Amaranthus hypochondriacus*) illustrated in Table 18.1 is known popularly as amaranth or pigweed. The tiny seeds of these crops are compared favorably with maize and other true cereals (Sauer, 1967). Comparatively, amaranth grains reported the very high content of proteins and balanced amino acids than conventional cereals grain (Pisarikova et al., 2006). Other usages are the preparation of making candy, bread, biscuits, flakes, cake, pastry, crackers, ice cream, and lysine-rich baby foods (Joshi & Rana, 1991b).

18.2.3   FINGER MILLET (Eleusine coracana)

Finger millet *(Eleusine coracana)* is widely cultivated in Africa and South Asia countries. This grain is estimated to be 10% of the world's 30 million tons of millet produced (Dida et al., 2008). Finger millet is said to be one of the oldest crops in India and Archaya (2009) stated that in Ancient Sanskrit Literature it was referred as "nrttakondaka" means "Dancing grain.' Finger millet (*Eleusine coracana*) locally known as "Ragi" a dark brown seed coat grain is nutritionally popular because of its rich source of calcium. In a household, ragi is used for the preparation of flour, pudding, porridge, and roti (Chaturvedi et al., 2008). In food industries, different items made out of ragi are composite flours (Dendy, 1992), popped grain (Nirmala, 2000) and malting that is, weaning foods (Malleshi & Desikachar, 1986). It has been found out that malting characteristics of finger millet are superior and better than the other millets grains (Pawar et al., 2007). Malting characteristics of finger millet are superior and better than others (Pawar et al., 2007). Amount of vitamin C is enhanced and the amount of synthesized lysine and tryptophan are more (Desai et al., 2010).

18.2.4 PEARL MILLET (Pennisetum americanum)

Pearl millet (*Pennisetum americanum*) showed in Table 18.1, also known as bajra in India, is famous for its high amount of macro- and micro-nutrients content (Sihag et al., 2015). Unfortunately, this crop is always less popular and termed as underutilized crop even though it can be grown at low maintenance cost and one of the staples for below poverty line for economic reasons. It has a drought-resistant property and acts as a principle source of energy, protein, fat, and minerals for poor people living in the drought-prone area (Sade, 2009; Sehgal & Kwatra, 2006; Malik et al., 2002). Pearl millet is a rich source of energy (361 Kcal/100 g) which is highest among other cereal when compared to wheat (346 Kcal/100 g), rice (345 Kcal/100 g), maize (125 Kcal/100 g), and sorghum (349 Kcal/100 g) recorded as per the Nutritive value of Indian foods (NIN, 2003).

TABLE 18.1   List of Underutilized Crops Reported by Various Researcher.

Category	Underutilized crops	Scientific name	References
Pseudocereals and millets	Buckwheat	*Fagopyrum esculentum*	Hore and Rathi (2002)
	Grain amaranth	*Amaranthus hypochondriacus*	Milan-Carrillo et al. (2012)
	Finger millet	*Eleusinecoracana*	Majumdar et al. (2006).
	Pearl millet	*Pennisetum glaucum* L	Sihag et al. (2015)
Cereal	Black rice	*Oryza sativa* L.	Saikia et al. (2012)
	Red rice	*Oryza sativa* L.	Samyor et al. (2015b)
	Sorghum	*Sorghum bicolor* L. Moenc)	Stefoska-Needham et al. (2015)
Legume	Horse gram	*Macrotyloma uniflorum*	Marimuthu and Krishnamoorthi (2013)

18.2.5   BLACK RICE (Oryza sativa L)

Black rice (*Oryza sativa* L) is black scented rice locally known as Cv. Poireton is found in the valley of Manipur, India (Table 18.1). It is with a dark purple color pericarp. Various researchers have found out that the presence of anthocyanins in black rice has a health-promoting food ingredients due to their antioxidant activity (Nam et al., 2006; Philpott et al., 2006) and anticancer (Hyun & Chung, 2004), hypoglycemic, and anti-inflammatory effects (Tsuda et al., 2003) is locally known as Cv. Poireton in Manipur. Traditionally they are healthy and rich in phytochemicals content. North-Eastern region of India is very popular for the four types of germplasm which are abundantly available. They are quite rich in rice aromatic compound, 2-acetyl-1-pyrroline (Bradbury et al., 2005; Buttery et al., 1983).GC-MS analysis of purple rice and black rice bran extract collected from Manipur, India showed the presence of nine compounds (Das et al., 2016).

18.2.6   RED RICE (Oryza sativa L)

In Sanskrit, the word "nivara" means *niv* which implies fattening or nourishing. Between two wild types of red rice species; *Oryza nivara* rice has exceptional medicinal values, namely, to enhance the body elements, removal of toxic metabolites, rejuvenation of the body, blood pressure reduction, and prevention of various skin problems (Ahuja et al., 2007). Red rice (*Oryza sativa* L) listed in Table 18.1, is one of the underutilized crops of Arunachal Pradesh locally known as Umling ame (Samyor et al., 2015a). Since time immemorial, ancient Ayurvedic considered the red rice as a nutritive food and medicinal quality rice. Even today, in the different states of India, namely, Karnataka, Madhya Pradesh, Kerala, Tamil Nadu, Uttar Pradesh, the Western Ghats, and Himachal Pradesh, different varieties of red rice used to treat skin diseases, blood pressure, fever, paralysis, rheumatism, and leucorrhea and for lactation. Nivara rice is a red rice from Kerala which is widely used in Ayurvedic practice to exclude toxins and delay premature aging in the human body (Ahuja et al., 2008). It has a health-promoting potential and substantial antioxidant content which help in reduction of reactive cell-damaging free radicals (Oki et al., 2002). Recently, a study conducted by Samyor et al. (2015b) revealed that the some pigmented rice cultivars from the state of Arunachal Pradesh content high amount of micronutrient content such as Mg, K, Na, etc.

18.2.7   SORGHUM (Sorghum bicolor L. Moench)

Sorghum (*Sorghum bicolor* L. Moench) is a staple food for a large population of Africa, India, and the semi-arid parts of the tropics (FAO, 1997). Because of its gluten-free characteristic, sorghum flour gained popularity and may be a replacement for wheat flour in gluten-free products, such as cakes, biscuits, and other RTE products. Sorghum (*Sorghum bicolor* L.) showed in Table 18.1, an underutilized crop is grown worldwide and ranks fifth in global cereal production, after maize, rice, wheat, and barley (FAO, 2012). This grain is reported to be rich in carbohydrates, fiber, vitamins, minerals, and phytochemicals including tannins, phenolic acids, anthocyanins, phytosterols, and policosanols. The type of phytochemicals in some sorghum varieties have been purported to reduce the risk of certain types of cancer, cardiovascular disease, obesity, and diabetes (Awika et al., 2004).

18.2.8   HORSE GRAM (Macrotyloma uniflorum)

Horse gram (*Macrotyloma uniflorum*) (Table 18.1) is extensively grown in India, mainly for animal feed. It is normally used to feed horses even though it is one of the dishes for low-income community/group in India. Traditionally Ayurveda cuisine considered horse gram with medicinal qualities and prescribed during the illness like jaundice or water retention and as part of a weight loss diet (Aiyer, 1990). However, application of thermal processing improves the production of a new product or ingredient. Earlier many researchers have suggested that the thermal processing can improve the texture, palatability, and inactivation of heat labile toxic compounds and enzyme inhibitors in legumes which were unknown to the farmers previously (Ghorpade et al., 1986).

18.3   VALUE ADDITION

India is well known for its agrarian economy and the farmers are struggling hard to increase their income by adding value to their raw horticultural products.

The concept of farm fresh products or value of farm products has increased in endless ways especially among the urban people by cleaning and cooling, packaging, processing, distributing, cooking, combining, churning, culturing, grinding, hulling, extracting, drying, smoking, handcrafting, spinning, weaving, labeling, or packaging (Richards et al., 1996). In generalized term, value addition is a process of increasing the economic value and consumer appeal of a product produce from its raw form to give more desirable form such as apple pie, jams, jellies, pickles, etc. The primary reason for processing post-harvested crops like fruits and vegetables is to extend the shelf-life beyond the period. In industries, due to the market forces, a greater extent for product differentiation and added value to raw commodities started taking place, namely, healthy foods, improved food processor for their productivity, and technological advancement (Royer, 1995). Some examples of value-added products from the low grade rice from Andhra Pradesh such as burfi, noodles, and extruded snack product and vennaundalu (butter coated balls), palathalikalu (dough rolled into strips, steamed/cooked in milk); rice semolina (instant kheer mix and instant upma mix), and flaked rice (nutritious bar) (Anitha & Rajyalakshmi, 2014).

18.4   EXTRUSION COOKING TECHNOLOGY

In the food industry, extrusion cooking has been continuously developed since its invention (Maskan & Altan, 2011; Bouvier & Campanella, 2014). Extrusion cooking of foods has been practiced for 50 years. Olden days, the extruder was engaged in the production of macaroni and RTE cereal pellets only but now ingredients are transformed into various modified intermediate and finished products, respectively (Camire et al., 1990; Camire, 2003; Singh et al., 2007; Riaz et al., 2009). Brennan et al. (2012) stated that extrusion-cooking technology has two forms-cold (below 70°C) and hot (above 70°C). Extrusion cooking techniques are preferable than other techniques due to its high productivity and significant nutrient retention, owing to the high temperature and short time required (Guy, 2001a). This technology also reported in the reduction of microbial contamination increase the shelf-life of extrudates product which has a water activity of 0.1–0.4 (Bordoloi & Ganguly, 2014).

In the recent year, HTST extrusion cooking has been used successfully in the food industry. Products like nutritious RTE meals such as a protein-rich instant porridge were available in the markets (Pelembe et al., 2002; Singh, et al., 2007). Extruders come in various design depending on the utility. Two broad categories are single screw extruder and twin extruder.

Single screw extruders consist of the only single rotating screw in a metal barrel. A hopper is located at the upper section. After the sample is fed, the rotating action of the screw leads the sample to the transition section. In the transition section, the screw channel becomes shallower. Due to the release of high mechanical energy in this section, a rise in temperature in the material composition occurred. Starch becomes expanded and gelatinized. This pressure pushed through the products through the die opening (Ramachandra & Thejaswini, 2015).

A twin-screw extruder mainly consists of two parallel screw shafts with the same length. The shaft which rotates in the same directions is called co-rotating and one which rotates in opposite directions called counter-rotating. Furthermore, it is subdivided into full, partial or non-intermeshing units on the basis of the relative position of the screws (Riaz, 2000). It is quite popular for agro-processing industries, such as pet food, cereals, snacks, etc. (White, 1991).

18.4.1   SOURCE OF RAW MATERIAL FOR EXTRUSION COOKING

The raw material used for extrusion cooking are natural biopolymers such as cereals or tuber flours, or oilseed legumes and other protein which are extensively used in the processing of cereal-based flours or starchy materials for the development of RTE snack products. Various studies have reported that wheat and corn flours, rice flour, potato, rye, barley, oats, sorghum, cassava, tapioca, buckwheat, pea flour, and also protein-rich materials such as pressed oilseed cake from soya, sunflower, rape, field bean, fava beans, etc., have been extensively used as a source of raw material (Guy, 2001b).

Recently, various studies were performed in order to evaluate the suitability of other promising food ingredients to snacks production, such as different fruits and vegetables, namely, apple, beetroot, carrot, cranberry, blueberry, cactus fruits, etc. The extrudate products developed were, for example, RTE breakfast cereals, expanded snacks, crisp bread, baby foods, pet foods, aquafeeds, etc. (Gibson & Ashby, 1997; Camire et al., 2007; Moussa-Ayoub et al., 2015; Potter, et al., 2013). Pulses like lentil, chickpea, dry, carioca and green beans, etc., have very little scientific data in value addition. Very few studies are available regarding the incorporation of pulse flours to develop snack-type foods which are actually rich in bioactive compounds with acceptable quality (Berrios, 2006; Berrios et al., 2002; Berrios et al., 2010; Simons et al., 2014). Pulses are protein-rich crops, getting popular because of its functional gluten-free characteristics foods. These crops encourage different metabolic functions, including glycemic and cholesterol indices stabilizations, reduction of body lipids accumulation, promotion of intestinal transit, and may act in the prevention of some cancers, osteoporosis, heart disease, or diabetes (Asif et al., 2013). Therefore, pulses as vegetable protein sources can lead to low glycemic index in extrusion formulations.

18.4.2 EFFECT OF EXTRUSION ON VARIOUS BIOACTIVE COMPOUNDS

According to Singh et al. (2007) effects of extrusion cooking technology on nutritional constituent are ambiguous. Advantages are the destruction of anti-nutritional factors, gelatinization of starch, increased soluble dietary fiber, and reduction of lipid oxidation whereas disadvantages are Maillard reactions which reduce the nutritional value of the protein, depending on the raw material types, their composition and process conditions.

18.4.2.1 CARBOHYDRATES

Carbohydrates composed of simple sugars to more complex molecules, like starch and fiber in the food items.

18.4.2.2 SUGARS

Major sugars, namely, fructose, sucrose, and lactose are a source of quick energy. They provide sweetness. During extrusion, sugars undergo various chemical changes. Therefore, proper selection of sugars for extrusion cooking is a very important factor for value-added product development. Camire (2001) stated that weaning food should be highly digestible on the other hand snack for obese adults expected to have a digestible component in food. Extrusion process at 170–210°C and 13% feed moisture while protein-enriched biscuits preparation, it was observed that 2–20% of the sucrose was lost (Camire et al., 1990).

18.4.2.3 STARCH

Starch is a polysaccharide which composed of starch molecules, namely, amylose and amylopectin. During the process of extrusion, these molecules contribute to gel formation and viscosity to the cooked paste (Singh et al., 2007). During extrusion, the formation of amylose–lipid complex is prominent. The type of food items also predicts the extent of amylose–lipid complex. In high-amylose starch food item, monoglycerides and free fatty acids are more likely to form complex than triglycerides. Other conditions like low feed moisture (19%) and barrel temperature (110–140°C) also lead to the massive complex formation between stearic acid and normal corn-starch, with 25% amylose (Bhatnagar & Hanna, 1994).

18.4.2.4 DIETARY FIBER

Extrusion cooking can produce changes in the structural characteristics and physicochemical properties. The main effect reported is a redistribution of insoluble fiber to soluble fiber (Larrea et al., 2010). It may be due to rupture of covalent and non-covalent bonds between carbohydrates and proteins associated to the fiber which further results in smaller molecular fragments, that would be more soluble (Wang et al., 1993).

18.4.2.5 PROTEIN

The effect of high mechanical shear, temperature, and pressure on protein was observed by the various researchers. Extrusion technology increases the protein digestibility and also undergoes changes in their molecular interactions involving covalent cross-linking, non-covalent molecular interaction, and protein– starch and protein–lipids interactions (Day, 2013). Extrusion could also improve the digestibility of proteins (De Pilli et al., 2011). Rathod and Annapure (2016) stated that extrusion processing caused the significant reduction of protein anti-nutritional factors such as phytic acid, tannin, trypsin inhibitors, and most effective process with improving both in vitro protein digestibility and in vitro starch digestibility (IVSD). Furthermore, it was revealed that pulses like lentil have a great potential for extrusion to produce RTE snacks with good acceptance.

18.4.2.6 LIPID

During extrusion cooking, the raw materials undergo different chemical and structural changes which affect structure formation and texture of the extruded products. Lipid complexation with amylose is one of the very important reactions in extrusion cooking. Alvarez et al. (1990) reported that the decrease of lipid oxidation was observed in extrudate meat-based product at the die increased from 71 to 115°C.

Effect of extrusion on various bioactive compounds, namely, antioxidant capacity total phenolic content (TPC), total anthocyanin content, oxygen radical absorbance capacity, ferric reducing antioxidant power, and 2,2'-diphenyl-1-picrylhydrazyl radical-scavenging activity were given in Table 18.2. There are several critical factors, namely, screw speed, temperature, moisture content, etc., influencing these compounds negatively as well as positively during extrusion. Some of the extrudate products showed the decrease in antioxidant activity during higher extrusion temperature. It may be because of decarboxylation due to higher barrel temperature and moisture content which may lead to polymerization of phenols and tannin, as a result, affect palatability, and antioxidant activities. (Repo-Carrasco-Valencia et al., 2009a; Repo-Carrasco-Valencia et al., 2009b; Dlamini et al., 2007) (Table 18.2).

But various researchers has also reported that in extrudate products, the activities of bioactive compounds found to be increased also increases, namely, ferulic acid in cereal extrudate product, total phenolic compounds in sweet potato, and cereal blend with vegetable (Zielinski et al., 2001; Shih et al., 2009; Stojceska et al., 2008). According to El-Hady and Habiba (2003), increase in the certain level of the bioactive compound like phenolic acid in the products is possibly from cell wall matrix.

A study conducted by Tiwari and Cummins (2009), reported that cereal products are found to be a rich source of vitamins. White et al. (2010) stated that increased in ORAC values (16–30%) with an increase in barrel temperature possibly due to the products formed during Maillard reaction. Yilmaz and Toledo (2005) reported that Maillard reaction products obtained from heated histidine and glucose have peroxyl radical scavenging activity. Hence, it can relate strongly to oxygen radical absorbance capacity. Ghumman et al. (2016) reported that extrudate product made of pulses at low feed moisture can be used to develop RTE puddings with high protein, viscosity, and digestibility. Extrusion cooking drastically reduced fatty acid and tocopherol contents of the corn-based extrudates at various contents of amaranth or quinoa ranging from 20 to 50% but just a slight change in effect on total phenolic compounds content and folate were observed. During the extrusion process, it is believed that fatty acids form amylose–lipid complexes while phenolic compounds and folate released from the food matrix (Diaz et al., 2016).

TABLE 18.2 Effect of Extrusion on the Various Underutilized Crops.

	Underutilized crops	Type of extruder	Temperature	Products	Effect of extrusion on bioactive compounds	References
Pseudocereals and millets	Buckwheat	Twin-screw extruder	170°C	Buckwheat flour	Antioxidant capacity (%) of the buckwheat flour decreased after extrusion.	Sensoy et al. (2006)
	Grain amaranth	Single-screw extruder	130°C	Extruded amaranth flours	Antioxidant capacity of extruded amaranth flour and total phenolic content (TPC) increases after the extrusion.	Milan-Carrillo et al. (2012)
	Finger millet	Twin-screw extruder	(80°C, 100°C, 120°C, and 140°C)	Encapsulated powder of juice of red cactus pear	Higher screw speed and temperature result in a lower t_{rm} (mean residence time) of the material in the extruder	Ruiz-Gutiér-rez et al. (2015)
	Pearl millet	Single-screw extruder	110°C	Pearl millet extrudate	In 9 h soaking + 40 h controlled germination + pearling + extrusion cooking showed decreases in total polyphenols and β-carotene in the sample.	Sihag et al. (2015)
Cereal	Black rice	Twin-screw extruder	120°C	Milled rice fraction	In the milled rice fraction, increase in total phenolic content, total anthocyanin content (and total antioxidant capacity were observed.	Ti et al. (2015)
	Red rice	Twin-screw extruder	150°C	Rice extrudates (12% feed moisture)	Total phenolic content, total anthocyanin content and ferric reducing antioxidant power were decreasing but 2,2′-diphenyl-1-picrylhy-drazyl radical-scavenging activity was increasing in the extrudate (feed moisture, 12 %)	Sompong et al. (2011)
	Sorghum	Single-screw (110–150°C)		Sorghum extrudate	Extrudates obtained from three varieties of sorghum, namely, SPV1411, SPV 1595, and M-35-1 reported that loss of tannin (%) was more in M-35-1.	Manisha et al. (2013)
Legumes	Horse gram	Twin-screw extruder	180°C	Rice–horse gram flour blend extrudate	In the blend extrudate, total phenolic content was decreasing after extrusion.	Gat and An-anthanarayan (2015)

18.5   TYPES OF EXTRUDATE PRODUCTS

18.5.1   READY-TO-EAT PRODUCTS

RTE products are defined as "any food for consumption without further heating or processing." Examples were open and pre-wrapped RTE products (Food Standards Agency, 2011). Under this definition, various processed foods can be categorized under RTE products, namely, biscuits, crisps, bread, pies, sandwiches and rolls, dairy products (milk, cheese, and spreads), prepared salads and vegetables, and fruit (Fast, 1999). Cereal RTE segment became quite popular among the RTE market, namely, breakfast cereals, extruded cereal shapes, and cereal biscuits/bars. Originally, these breakfast cereal market started in United States of America by the late eighteen hundreds and the beginning of the twentieth century and hence, the Kellogg brothers of healthy vegetarian foods started relating food to human nutrition (Fast, 1999). Souza and Menezes (2008) reported that the high levels of vegetable protein with semi defatted Brazil nut cake and cassava flour mixtures. The high nuts mixtures containing high levels of protein, fat, and ash and the mixtures with less Brazil nut present higher levels of carbohydrate.

18.5.2   TEXTURIZED VEGETABLE PROTEIN

Texturized vegetable protein is a restructuring of protein molecules (usually soy protein) into a layered, cross-linked mass which is resistant to disruption upon processing. Two classifications of TVP are extrusion-cooked meat extenders and extrusion-cooked meat analogs (Alam et al., 2016). According to Hayashi et al. (1992), extruder barrel temperature is mainly for the texturization of the dehulled whole soybean. In the entire process, the critical factor in protein cross-linking reactions is melt temperature. Trinci (1992) stated that TVP is the production of fibrous and elastic meat structures such as shredded TVP and mycoprotein or milk protein preparations for the human consumption.

18.5.3   PASTA PRODUCTS

Pasta is basically wheat-based products that are formed from wheat dough, water, sometimes eggs are added and no leavening is required. Li et al.

(2014) stated that pasta is a useful carrier which acts as a nutrition enhancer or providing specific physiological functions required in the body. Various studies have conducted on the pasta products to know the effect of extrusion on semolina, hydration level (Carini et al., 2010) and also flaxseed flour concentration on the physical and cooking characteristics of extruded pasta (Manthey et al., 2008). Wojtowicz and Moscicki (2011) reported the production of enriched precooked pasta-like products with the incorporation of wheat bran.

18.5.4 MEAT PRODUCTS

Meat and meat items have been texturized by extrusion to give a better palatability in the final products. In the market, extruded chicken products are available (Hsieh et al., 1991). According to Silva et al. (2010) after using processing technology steps, namely, milling, lyophilization, and defatting, bovine rumen reported having 98% of protein (on a dry solid basis) with digestibility corrected amino acid scores of 100%, respectively. Smithey et al. (1995) stated that ground beef chuck was extruded with nonmeat binders, namely, yellow corn flour, white corn flour, and white corn flour with soy fiber.

18.5.5 CONFECTIONERY PRODUCTS

Twin-screw extruders are used in confectionery products preparation to mix well the ingredients. The extrusion process used in the confectionary industry to control temperatures of the heat sensitive materials and to incorporate additional fat, milk, sugars, nuts, and other ingredients (Best, 1994).

18.5.6 PET FOOD AND ANIMAL FEED PRODUCTS

Extrusion technology is extensively being used for pet/animal products. Early studies have been done on the feed production and the nutritional effects of expansion and short-time extrusion were also reported (Plavnik & Wan, 1995). A various researcher like Kawauchi et al. (2011) and Fischer et al. (2012) reported that fiber supplemented extruded foods are being produced by most pet food companies nowadays to dilute energy density which can promote specific gut and general health benefits.

18.6 CONCLUSION

Extrusion cooking technology gains much popularity especially in the development of starchy food items can be a new ray of hope for the under-utilized crops of India. Millets which is still known as poor man's crop yet to enter consumer's market with its new avatar of new product formulation with increase palatability. The process of value addition not only going to increase the economic value but also consumer appeal of a product produce from its raw form to give more desirable form. Due to increasingly hectic lifestyles in urban society, consumers and double-income households, they prefer breakfast cereals items with other health or nutritional benefits, such as reduced sugar, high fiber, vitamin, or phenolic enriched cereals. Because of its ease of preparation and time-saving nature consumers are opting break-fast cereals products in the market. Consumer demand for new and health beneficial food product is increasing day by day and to meet the demand there is a tremendous effort going on in the food industries from diverse sources. Hence, the effective implementation of the extrusion cooking tech-nology for the development of value-added product from underutilized crop can be a way to combat poverty and boon to the economy of India.

KEYWORDS

- buckwheat
- finger millet
- pseudocereals
- extrusion cooking
- ready-to-eat products

REFERENCES

Alam, M. S.; Kaur, J.; Khaira, H.; Gupta, K. Extrusion and Extruded Products: Changes in Quality Attributes as Affected by Extrusion Process Parameters: A Review. *Crit Rev. Food Sci. Nutr.* **2016**, *56*, 445–473.
Asif, M.; Rooney, L, W.; Ali, R.; Riaz, M, N. Application and Opportunities of Pulses in Food system: A Review. *Crit. Rev. Food Sci. Nutr.* **2013**, *53* (11), 1168–1179.

Anitha, G.; Rajyalakshmi, P. Value Added Products with Popular Low-grade Rice Varieties of Andhra Pradesh. *J. Food Sci. Technol.* **2014,** *51* (12), 3702–3711. DOI: 10.1007/s13197-012-0665-4

Aiyer, Y. N. Horsegram, In *Field Crops of India;* Bangalore Press: Bangalore, India, **1990**; pp 115–117.

Ahmad, F.; Raj, A. Buckwheat: A Legacy on the Verge of Extinction in Ladakh. *Cur. Sci.* **2012,** *103* (1), 1.

Ahuja, U.; Ahuja, S. C.; Thakrar, R.; Singh, R. K. Rice-A Neutraceutical. *Asian Agric. Hist.* **2008,** *2,* 93–108.

Awika, J. M.; Rooney, L. W. Sorghum Phytochemicals and Their Potential Impact on Human Health. *Phyto. Chem.* **2004,** *65,* 1199–1221.

Anonymous. Annual Report, *NBPGR Regional Station,* Barapani, Meghalaya, 1987–2001.

Alvarez, V. B.; Smith, D. M.; Morgan, R. G.; Booren, A. M. Restructuring of Mechanically Deboned Chicken and Nonmeat Binders in a Twin-screw Extruder. *J. Food Sci.* **1990,** *55,* 942–946.

Achaya, K. T. *The Illustrated Food of India A–Z;* Oxford University Press: New Delhi, India, 2009.

Ahuja, U.; Ahuja, S. C.; Chaudhary, N.; Thakrar, R. Red Rices-Past, Present, and Future. *Am. Assoc Heart Failure Nurses* **2007,** *11,* 291–304.

Berrios, J. D. J. *Extrusion Cooking of Legumes: Dry Bean Flours;* Taylor & Francis Group: London, England, 2006.

Berrios, J. D. J.; Camara, M.; Torija, M. E.; Alonso, M. Effect of Extrusion Cooking and Sodium Bicarbonate Addition on the Carbohydrate Composition of Black Bean Flours. *J. Food Proc. Preserv.* **2002,** *26* (2), 113–128.

Berrios, J. D. J.; Morales, P.; Camara, M.; Sanchez Mata, M. C. Carbohydrate Composition of Raw and Extruded Pulse Flours. *Food Res. Int.* **2010,** *40,* 531–536.

Best, E. T. Confectionery Eextrusion. In *The Technology of Extrusion Cooking;* Frame, N. D., Ed.; Blackie Academic & Professional: London, 1994; pp 190–236.

Bonafaccia, G.; Marocchini, M.; Kreft, I. Composition and Technological Properties of the Flour and Bran from Common and Tartary Buckwheat. *Food Chem.* **2003,** *80,* 9–15.

Bordoloi, R.; Ganguly, S. Extrusion Technique in Food Processing and a Review on Its Various Technological Parameters. *Ind. J. Sci. Res. Technol.* **2014,** *2* (1), 1–3.

Bradbury, L. M.; Fitzgerald, T. L.; Henry, R. J.; Jin, Q.; Waters, D. L. The Gene for Fragrance in Rice. *Plant Biotechnol. J.* **2005,** *3,* 363–370.

Brennan, C.; Brennan, M.; Derbyshire, E.; Tiwari, B. Effects of Extrusion on the Polyphenols, Vitamins and Antioxidant Activity of Foods. *Trends Food Sci. Technol.* **2011,** *22,* 570–575.

Brennan, L. M.; Shaw, D. S.; Dishion, T. J.; Wilson, M. Longitudinal Predictors of School-age Academic Achievement: Unique Contributions of Toddler-age Aggression, Oppositionality, Inattention, and Hyperactivity. *J. Abnormal Child Psychol.* **2012,** *40,* 1289–1300.

Buttery, R. G.; Ling, L. C.; Juliano, B. O.; Turnbaugh, J. G. Cooked Rice Aroma and 2-acetyl-1-pyrroline. *J. Agric. Food Chem.* **1983,** *31* (4), 823–826.

Carini, E.; Vittadini, E.; Curti, E.; Antoniazzi, F.; Viazzani, P. Effect of Different Mixers on Physicochemical Properties and Water Status of Extruded and Laminated Fresh Pasta. *Food Chem.* **2010,** *122* (2), 462–469.

Camire, M. E.; Dougherty, M. P.; Briggs, J. L. Functionality of Fruit Powders in Extruded Corn Breakfast Cereal. *Food Chem.* **2007,** *101,* 765–770.

Camire, M. E. Extrusion Cooking. In *The Nutrition Handbook for Food Processors;* Taylor & Francis: Milton Park, Abingdon, 2003; pp 314–330.

Camire, M. E. Extrusion and Nutritional Quality. In *Extrusion Cooking: Technologies and Application;* Guy, R., Ed.; Woodhead Publishing Ltd.: Cambridge, 2001; pp 108–130.

Camire, M. E.; Camire, A. L.; Krumhar, K. Chemical and Nutritional Changes. *Crit. Rev. Food Sci. Nutr.* **1990,** *29,* 35–57.

Chaturvedi, R.; Srivastava, S. Genotype Variations in Physical, Nutritional and Sensory Quality of Popped Grains of Amber and Dark Genotypes of Finger Millet. *J. Food Sci. Technol.* **2008,** *45* (5), 443–446.

Desai, A. D.; Kulkarni, S. S.; Sahoo, A. K.; Ranveer, R. C.; Dandge, P. B. Effect of Supplementation of Malted Ragi Flour on the Nutritional and Sensorial Quality Characteristics of Cake. *Adv. J. Food Sci. Technol.* **2010,** *2,* 67–71.

Diaz, J. M. R. O.; Sundarrajan, L.; Kariluoto, S.; Lampi, A.; Tenitz, S.; Jouppila, K. Effect of Extrusion Cooking on Physical Properties and Chemical Composition of Corn-based Snacks Containing Amaranth and Quinoa: Application of Partial Least Squares Regression. *J. Food Proc. Eng.* **2017,** *40,* 1–15.

Dida, M. M.; Wanyera, N.; Dunn, M. L. H.; Bennetzen, J. L.; Devos, K. M. Population Structure and Diversity in Finger Millet (*Eleusinecoracana*) Germplasm. *Trop. Plant Biol.* 2008, *1* (2), 131–141.

De Pilli, T.; Fiore, A. G.; Giuliani, R.; Derossi, A.; Severini, C. Functional Food Produced by Extrusion: Cooking Technology. In *Food Production: New Research;* Nova Science Publisher: Hauppauge, NY, 2001; pp 1–44.

Dendy, D. A. V. *Composite Flour Past, Present and the Future: A Review with Special Emphasis on the Place of Composite Flour in the Semi-arid Zones. Utilization of Sorghum and Millets;* C. A. B. International: Oxford, 1992; pp 67–73.

Das, A. B.; Goud, V. V.; Das, C. Extraction of Phenolic Compounds and Anthocyanin from Black and Purple Rice Bran (*Oryza sativa L.*) Using Ultrasound: A Comparative Analysis and Phytochemical Profiling. 2016. DOI.org/10.1016/j.indcrop.2016.10.041

Dlamini, N. R.; Taylor, J. R. N.; Rooney, L. W. The Effect of Sorghumtype and Processing on the Antioxidant Properties of African Sorghum-based Foods. *Food Chem.* **2007,** *105* (4), 1412–1419.

Eyzaguirre, P.; Padulos, S.; Hodgkin, T. *IPGRI's Strategy for Neglected and Underutilized Species and the Human Dimension of Agrobiodiversity. Priority Setting for Underutilized and Neglected Plant Species of the Mediterraneanregion;* Report of the IPGRI Conference: Padulosi, S., Ed.; 9–11 February 1998. ICARDA, Aleppo. Syria. International Plant Genetic Resources Institute, Rome, Italy, 1999, pp 1–20.

Food and Agriculture Organization of the United Nations (FAO). Production-Sorghum. 2012. http://faostat.fao.org/site/339/default.aspx (accessed May 22, 2014).

FAO. Production Year Book. Food and Agriculture Organization of the United Nation: Rome, Italy, 1997; Vol. 51, pp 59–79.

Food Standards Agency. *E. coli* O157 Control of Cross-contamination Guidance for Food Business Operators and Enforcement Authorities, 2011. http://www.food.gov.uk/multimedia/pdfs/publication/ ecoliguide0211.pdf (accessed May 28, 2014).

Fast, R. B. *Origins of the US Breakfast Cereal Industry;* Cereal Foods World: St. Paul, MN, 1999; Vol. 44, p 394.

Gaossong, J.; Vasanthan, T. The Effect of Extrusion Cooking on the Primary Structure and Water Solubility of B-glucans from Regular and Waxy Barley. *Cereal Chem.* **2000,** *77,* 396–400.

Gat, Y.; Ananthanarayan, L. Physicochemical, Phytochemical and Nutritional Impact of Fortified Cereal-based Extrudate Snacks Effect of Underutilized Legume Flour Addition and Extrusion Cooking. *Nutrafoods* **2015,** *14,* 141–149. DOI 10.1007/s13749-015-0036-7

Ghumman, A.; Kaur. A.; Singh, N; Singh, B. Effect of Feed Moisture and Extrusion Tempera-ture on Protein Digestibility and Extrusion Behavior of Lentil and Horsegram. *LWT Food Sci. Technol.* **2016,** *70,* 49–357.

Guy, R. *Extrusion Cooking: Technologies and Application;* Woodhead Publishing Ltd.: Cambridge, 2001a; pp 3–5.

Guy, R. *Extrusion Cooking Technologies and Application;* CRC Press: Boca Raton, FL, 2001b; pp 1–200.

Gibson, L. G.; Ashby, M. F. *Cellular Solids: Structure and Properties;* Cambridge University Press: Cambridge, UK, 1997.

Hsieh, E.; Peng, I. C.; Clarke, A. D.; Mulvaney, S. J.; Huff, H. E. Restructuring of Mechani-cally Deboned Turkey by Extrusion Processing Using Cereal Flours as the Binder. *LWT Food Sci. Technol.* **1991,** *24,* 139–144.

Hore, D.; Rathi, R. S. Collection, Cultivation and Characterization of Buckwheat in North-eastern Region of India. *Fagopyrum* **2002,** *19,* 11–15.

Hayashi, N.; Abe, H.; Hayakawa, I.; Fujio, Y. Texturization of Dehulled Whole Soybean with a Twin Screw Extruder and Texture Evaluation. In *Food Processing by Ultra High Pressure Twin-Screw Extrusion;* Hayakawa, A., Ed.; Technomic Publishing Company: Lancaster, PA, 1992; pp 133–146.

Hyun, J. W.; Chung, H. S. Cyanidin and Malvidin from Oryza Sativa Cv. Heugjinjubyeo Mediate Cytotoxicity against Human Monocytic Leukemia Cells by Arrest of G2/M Phase and Induction of Apoptosis. *J. Agric. Food Chem.* **2004,** *52,* 2213–2217.

Joshi, B. D.; Paroda, R. S. *Buckwheat in India:* National Bureau of Plant Genetic Resources (NBPGR)Shimla. *Sci. Monogr.* **1991a,** *2,* 117.

Joshi, B. D.; Rana, R. S. *Grain Amaranth: The Future Food Crop.* National Bureau of Plant Genetic Resources (NBPGR) Shimla. *Sci. Monogr.* **1991b,** *3,* 152.

Joshi, B. D. Status of Buckwheat in India. *Fagopyrum* **1999,** *16,* 7–11.

Kawauchi, I. M.; Sakomura, N. K.; Vasconcellos, R. S.; de-Oliveira, L. D.; Gomes, M. O. S.; Loureiro, B. A.; Carciofi, A. C. Digestibility and Metabolizableenergy of Maize Gluten Feed for Dogs as Measured by Two Different Techniques. *Feed Sci. Technol.* **2011,** *169,* 96–103.

Larrea, C. M. A.; Martínez-Bustos, F.; Yoon, K. C. The Effect of Extruded Orange Pulp on Enzymatic Hydrolysis of Starch and Glucose Retardation Index. *Food Bioprocess Technol.* **2010,** *3,* 684–692.

Lyman, C. M..; Kuliken, K. A.; Hall, F. Essential Amino Acid Content of Farm Feeds. *J. Agric. Food Chem.* **1956,** *4,* 1008–1013.

Li, M.; Zhu, K. X.; Guo, X. N.; Brijs, K.; Zhou, H. M. Natural Additives in Wheat-based Pasta and Noodle Products: Opportunities for Enhanced Nutritional and Functional Proper-ties. *Comprehensive Rev. Food Sci. Food Safety* **2014,** *13,* 347–357.

Manthey, F. A.; Sinha, S.; Wolf-Hall, C. E.; Hall, C. A. Effect of Flaxseed Flour and Pack-aging on Shelf Life of Refrigerated Pasta. *J. Food Process. Preserv.* **2008,** *32* (1), 75–87.

Maskan, M.; Altan, A. *Advances in Food Extrusion Technology;* CRC Press: Boca Raton, FL, 2011.

Moussa-Ayoub, T.; Youssef, K.; El-Samahy, S.; Kroh, L.; Rohn, S. Flavonol Profile of Cactus Fruits (*Opuntiaficus indica*) Enriched Cereal-based Extrudates: Authenticity and Impact of Extrusion. *Food Res. Int.* **2015,** *78,* 442–447. DOI:10.1016/ j.foodres.2015.08.019

Malik, M.; Singh, U.; Dahiya, S. "Nutrient Composition of Pearl Millet as Influenced by Geno-types and Cooking Methods. *J. Food Sci. Technol.* **2002,** *39* (5), 463–468. DOI:10.1016/ S0733-5210(03)00016-X

Milan-Carrillo, J.; Montoya-Rodriguez, A.; Gutierrez-Dorado, R.; Perales-Sanchez, X.; Reyes-Moreno, C. Optimization of Extrusion Process for Producing High Antioxidant Instant Amaranth (*Amaranthus hypochondriacus* L.) Flour Using Response Surface Methodology. *Appl. Math.* **2012,** *3,* 1516–1525.

Marimuthu, K.; Krishnamoorthi, K. Nutrients and Functional Properties of Horse Gram (*Macrotyloma uniflorum*), an Underutilized South Indian Food Legume. *J. Chem. Pharm. Res.* **2013,** *5* (5), 390–394.

Manisha, V. J.; Uday, S. A. Effect of Extrusion Process Parameters and Particle Size of Sorghum Flour on Expanded Snacks Prepared Using Different Varieties of Sorghum (*Sorghum bicolour*L.) Annapure. *J. Agric. Sci. Technol.* **2013,** *3,* 71–85.

Malleshi, N. G.; Desikachar, H. S. E. R. Nutritive Value of Malted Flour. *Qual. Plant Food Hum. Nutr.* **1986,** *36,* 191–196.

Ruiz-Gutiérrez, M. G.; Carlos, A. A.; Armando, Q.; Esther, P.; Teresita, de J.; Ruiz-Anchondo, T.; Báez-González, J. G.; Meléndez-Pizarro, C. O. Effect of Extrusion Cooking on Bioactive Compounds in Encapsulated Red Cactus Pear Powder. *Molecules* **2015,** *20,* 8875–8892. DOI:10.3390/molecules20058875

Nirmala, M.; Rao, S. M. V. S. S. T.; Murlikrishna, G. Carbohydrates and Their Degrading Enzymes from Native and Malted Finger Millet (Ragi, Eleusinecoracana, Indaf-15). *Food Chem.* **2000,** *69,* 175–180.

NIN. *Nutritive Value of Indian Foods;* Gopalan, C., Deosthale., Eds.; National Institute of Nutrition: Hyderabad, 2003.

Nam, S. H.; Choi, S. P.; Kang, M. Y.; Koh, H. J.; Kozukue, N.; Friedman, M. Antioxidative Activities of Bran Extracts from Twenty one Pigmented Rice Cultivars. *Food Chem.* **2006,** *94,* 613–620.

Oki, T.; Masuda, M.; Kobayashi, M.; Nishiba, Y.; Furuta, S.; Suda, I.; Sato, T. Polymeric Procyanidins as Radical-scavenging Components in Red-hulled Rice. *J. Agric. Food Chem.* **2002,** *50,* 7524–7529.

Pugalenthi, M.; Vadivel, V.; Siddhuraju, P. Alternative Food/feed Perspectives of an Underutilized Legume Mucunapruriens Var. Utilis – a Review. *Plants Foods Human Nutr.* **2005,** *60,* 201–218.

Philpott, M.; Gould, K, S.; Lim, C.; Ferguson, L. R. In Situ and in Vitro Antioxidant Activity of Sweet Potato Anthocyanins. *J. Agric. Food Chem.* **2006,** *54,* 1710–1715.

Potter, R.; Stojceska, V.; Plunkett, A. The Use of Fruit Powders in Extruded Snacks Suitable for Children's Diets. *LWT Food Sci. Technol.* **2013,** *51,* 537–544.

Padulosi, S.; Hoeschle-Zeledon, I. Underutilized Plant Species: What are They? *LEISA Magaz.* **2004,** *20* (1), 5–6.

Pawar, P. A.; Dhanvijay, V. P. Weaning Foods: An Overview. *Beverage Food World.* **2007,** *34* (11), 27–33.

Pisarikova, B.; Zraly, Z.; Kracmar, S.; Trckova, M.; Herzig, I. The Use of Amaranth (Genus *Amaranthus* L.) in the Diets for Broiler Chickens. *Vet. Med. Czech.* **2006,** *51,* 399-407.

Pelembe, L. A. M.; Erasmus, C.; Taylor, J. R. N. Development of a Proteinrich Composite Sorghumecowpea Instant Porridge by Extrusion Cooking Process. *LWT Food Sci. Technol.* **2002,** *35,* 120–127.

Plavnik, I.; Wan, D. Nutritional Effects of Expansion and Short Time Extrusion on Feeds for Broilers. *Anim. Feed Sci. Technol.* **1995,** *55,* 247–251.

Richards, K.; Deborah, S. W. *Making it on the Farm: Increasing Sustainability Through Value-added Processing and Marketing*; Southern Sustainable Agriculture Working Group: Fayetteville, AR, 1996; p 40.

Royer, J. Potential for Cooperative Involvement in Vertical Coordination and Value-Added Activities. *Agribusiness* **1995,** *11* (5), 473–481.

Repo-Carrasco-Valencia, R.; de La Cruz, A. A.; Alvarez, J. C. I.; Kallio, H. Chemical and Functional Characterization of Kaiwa (*Chenopodium pallidicaule*) Grain, Extrudate and Bran. *Plant Foods Human Nutr.* **2009a,** *64* (2), 94–101.

Ruiz-Gutiérrez, M. G.; Amaya-Guerra, C. A.; Quintero-Ramos, A.; Pérez-Carrillo, E.; Ruiz-Anchondo, T.; Báez-González, J.; Meléndez-Pizarro, C. O. Effect of Extrusion Cooking on Bioactive Compounds in Encapsulated Red Cactus Pear Powder. *Molecules* **2015,** *20,* 8875–8892. DOI:10.3390/molecules20058875

Riaz, M. N. Introduction to Extruders and Their Principles, In *Extruders in Food Applications;* Riaz, M. N., Ed.; CRC Press: Boca Raton, FL, 2000; pp 5–7.

Riaz, M.; Asif, M.; Ali, R. Stability of Vitamins during Extrusion. *Crit. Rev. Food Sci. Nutr.* **2009,** *49,* 361–368.

Repo-Carrasco-Valencia, R.; Pena, J.; Kallio, H.; Salminen, S. Dietary Fiber and Other Functional Components in Two Varieties of Crude and Extruded Kiwicha (Amaranthuscaudatus). *J. Cereal Sci.* **2009b,** *49* (2), 219–224.

Rathod, R. R.; Annapure, U. S. Effect of Extrusion Process on Antinutritional Factors and Protein and Starch Digestibility of Lentil Splits. *LWT Food Sci. Technol.* **2016,** *66,* 114–123.

Smithey, S. L.; Badding, H. H. E.; Hsieh, E. Processing Parameters and Product Properties of Extruded Beef with Nonmeat Cereal Binders. *LWT Food Sci. Technol.* **1995,** *28,* 386–394.

Saikia, S.; Dutta, H.; Saikia, D.; Mahanta, C. L. Quality Characterisation and Estimation of Phytochemicals Content and Antioxidant Capacity of Aromatic Pigmented and Nonpigmented Rice Varieties. *Food Res. Int.* **2012,** *466,* 334–340.

Silva, A. C. C.; Arêas, E. P. G.; Silva, M. A.; Arêas, J. A. G. Effects of Extrusion on the Emulsifying Properties of Rumen and Soy Protein. *Food Biophys.* **2010,** *5,* 94–102.

Samyor, D.; Deka, S. C.; Das, A. B. Phytochemical and Antioxidant Profile of Pigmented and Nonpigmented Rice Cultivars of Arunachal Pradesh, India. *Int. J. Food Prop.* **2015a,** *19* (5), 1104–1114.

Samyor, D.; Deka, S. C.; Das, A. B. Evaluation of Physical, Thermal, Pasting Characteristics and Mineral Profile of Pigmented and Nonpigmented Rice Cultivars. *J. Food Process. Preserv.* **2015b,** *40,* 174–182. DOI:10.1111/jfpp.12594

Sauer, J. O. The Grain Amaranths and Their Relatives: A Revised Taxonomic Andgeographic Survey. *Ann. Missouri Bot. Gard.* **1967,** *54* (2), 103–107.

Singh, S.; Gamlath, S.; Wakeling, L. Nutritional Aspects of Food Extrusion: A Review. *Int. J. Food Sci. Technol.* **2007,** *42,* 916–929.

Simons, C. W.; Hall, C.; Tulbek, M.; Mendis, M.; Heck, T.; Ogunyemi, S. Acceptability and Characterization of Extruded Pinto, Navy and Black Beans. *J. Sci. Food Agric.* **2014,** *11,* 2287–2291.

Sade, F. O. "Proximate, Antinutritional Factors and Functional Properties of Processed Pearl Millet (*Pennisetumglaucum*)," *J. Food Technol.* **2009,** *7* (3), 92–97.

Sehgal, S. A.; Kwatra, A. Nutritional Evaluation of Pearl Millet Based Sponge Cake. *J. Food Sci. Technol.* **2006,** *43* (3), 312–313.

Sihag, M. K.; Sharma, V.; Arora, S.; Singh, A. K.; Goyal, A.; Lal, D. Effect of Storage Conditions on Senso and Microbial Characteristics of Developed Pearl Millet Based Weaning Food. *Indian J. Dairy Sci.* **2015,** *68,* 463–466.

Stefoska-Needham, S.; Beck, E. J.; Johnson, S. K.; Tapsell, L. C. Sorghum: An Underutilized Cereal Whole Grain with the Potential to Assist in the Prevention of Chronic Disease. *Food Rev. Int.* **2015,** *31* (4), 401–437.

Sompong, R.; Siebenhandl-Ehn, S.; Berghofer, E.; Schoenlechner, R. Extrusion Cooking Properties of White and Coloured Rice Varieties. *Starch Stärke* **2011,** *63,* 55–63.

Shih, M. C.; Kuo, C. C.; Chiang, W. Effects of Drying and Extrusion on Color, Chemical Composition, Antioxidant Activities and Mitogenic Response of Spleen Lymphocytes of Sweet Potatoes. *Food Chem.* **2009,** *117* (1), 114–121.

Stojceska, V.; Ainsworth, P.; Plunkett, A.; Ibanoglu, E.; Ibanoglu, S. Cauliflower by-products as a New Source of Dietary Fibre, Antioxidants and Proteins in Cereal Based Ready-to-eat Expanded. *J. Food Eng.* **2008,** *87* (4), 554–563.

Sensoy, I.; Rosen, R. T.; Ho, C.; Karwe, M. V. Effect of Processing on Buckwheat Phenolics and Antioxidant Activity. *Food Chem.* **2006,** *99,* 388–393.

Tsuda, T.; Horio F.; Uchida, K.; Aoki, H.; Osawa, T. Dietary Cyanidin 3-O-β-D-glucoside-rich Purple Corn Color Prevents Obesity and Ameliorates Hyperglycemia. *J. Nutr.* **2003,** *133,* 2125–2130.

Ti, H.; Zhang, R.; Zhang, M.; Wei, Z.; Chi, J.; Deng, Y.; Zhang, Y. Effect of Extrusion on Phytochemical Profiles in Milled Fractions of Black Rice. *Food Chem.* **2015,** *178,* 186–194.

Trinci, A. P. J. *Mycol. Res.* **1992,** *96,* 1–13.

UNEP. *Global Biodiversity Assessment;* United Nations Environment Programme University Press: Cambridge, UK, 1995.

Wang, W. M.; Klopfenstein, C. F. Effect of Twin Screw Extrusion on the Nutritional Quality of Wheat, Barley, and Oats. *Cereal Chem.* **1993,** *70* (6), 712–715.

Wilson, E. O. *The Diversity of Life;* Penguin: London, UK, 1992.

WCMC. Status of the Earth' Living Resources–World Conservation Monitoring Centre. In *Global Biodiversity;* Chapman and Hall: London, New York, 1992; p 5494.

Wojtowicz, A.; Moscicki, L. Effect of Wheat Bran Addition and Screw Speed on Microstructure and Textural Characteristics of Common Wheat Precooked Pasta-like Products. *Pol. J. Food Nutr. Sci.* **2011,** *61* (2), 101–107.

White, J. L. *Twin Screw Extrusion: Technology and Principles;* Hanser Publishers: New York, 1991.

White, B. L.; Howard, L. R.; Prior, R. L. Polyphenolic Composition and Antioxidant Capacity of Extruded Cranberryaapomace. *J. Agric. Food Chem.* **2010,** *58* (7), 4037–4042.

Yilmaz, Y.; Toledo, R. Antioxidant Activity of Water-soluble Maillard Reaction Products. *Food Chem.* **2005,** *93* (2), 273–278.

Zielinski, H.; Kozlowska, H.; Lewczuk, B. Bioactive Compounds in the Cereal Grains before and after Hydrothermal Processing. *Innov. Food Sci. Emerg. Technol.* **2001,** *2* (3), 159–169.

Zielinski, H.; Michalska, A.; Amigo-Benavent, M.; del Castillo, M. D.; Piskuła, M. K. Changes in Protein Quality and Antioxidant Properties of Buckwheat Seeds and Groats Induced by Roasting. *J. Agric. Food Chem.* **2009,** *57,* 477.

CHAPTER 19

FUNGAL RETTING TECHNOLOGY OF JUTE

AVIJIT DAS*

Division of Quality Evaluation and Improvement, National Institute of Research on Jute and Allied Fibre Technology, Kolkata 700040, India

*E-mail: avijitcrri@gmail.com

CONTENTS

Abstract 494

19.1   Introduction 494

19.2   Socio-Economic Importance of Jute 495

19.3   Anatomy Jute Stem 496

19.4   Composition of Jute Fiber 498

19.5   The Retting Process 499

19.6   Need of Alternate Retting Technologies 502

19.7   Microbes of Jute Retting 503

19.8   Global Climatic Change and Need of Minimum-Water
       Retting Technology 505

19.9   Conclusion 506

Keywords 507

References 507

ABSTRACT

Jute is an important commercial crop and price of its main commercial product, that is, fiber depends on its quality or grade. Retting is one of the most important factors determining quality of jute fiber. Jute is traditionally retted with water which requires large slow-moving water bodies. However, such water bodies are diminishing day by day due to scarce rainfall during the retting period. As a result, farmers have no option but to ret the harvested jute stem in shallow water resulting in poor fiber quality. Moreover, farmers have to go for several charges in a season in the same water bodies which not only deteriorates fiber quality but also causes environmental pollution. Water retting also causes skin infections as farmers have to stand in the polluted water for hours to extract the fiber. In order to overcome these problems, several retting technologies (ribbon retting, accelerated retting, fungal retting, etc.), have been evolved which minimize water requirement to various extent. Among these technologies, fungal retting technology is very promising as it allows jute retting with very little water. Fungal retting technology not only saves water to a greater extent but also eliminates environmental and health hazards but at the same time produces comparable qualities of jute fiber. In spite of being a promising technology, there are issues such as up scaling, cost/benefit ratio, complexity of protocol, etc., which are required to be looked into before farmers can harvest the benefit of this technology.

19.1   INTRODUCTION

Jute is an important vegetable fiber crop next only to cotton in terms of usage, global consumption, production and availability (Bhattacharya, 2012). Every part of jute plant can be used but the main economic product of jute plant is its fiber. Price of the fiber is determined by its quality traits. Quality of jute fiber depends on a number of factors, the most important of which is retting. Retting is a process by which fiber is separated from the underlying hard woody tissues. Jute is conventionally retted through water retting which is an age-old practice in the jute growing areas of India and Bangladesh. This process ideally needs slow moving soft water which is sometimes not available due to scarcity of rains during the retting period. Moreover, due to lack of adequate retting ponds, farmers sometimes have to go for more than one retting in the same water body. This not only deteriorates quality of the fiber but also emits foul smell causing environmental hazards. This has necessitated the development of alternate retting technologies that require minimum water.

19.2   SOCIO-ECONOMIC IMPORTANCE OF JUTE

About 60% of the raw jute in the world is produced in India (Roul, 2009) and it plays an important role in the country's economy. It is viewed that about 4 million farmers, 0.25 million of industrial workers, and 0.5 million traders find gainful employment in various jute sectors (Sen et al., 2006). It generates about 10 million working man-days annually and around 32 lakh farm families seek their livelihood by cultivating jute in the country. Thus, raw jute (jute + mesta together) farming, industry, and trade support livelihood to 14 million people (Das et al., 2006). These apart, the jute industry also contribute to the export earnings to the tune of nearly 12,000 million of rupees yearly (Karmakar et al., 2008).

In India, jute is predominantly cultivated by the marginal (65%) and small (25%) farmers of West Bengal (West Bengal is the major jute growing state sharing about three-fourth of the country's production) contributing about 80% of national jute production (Chapke et al., 2006). Jute crop plays crucial role in the farmers' livelihood by providing monetary as well as non-price benefits. All parts of jute plants are used for benefits. Jute leaves have medicinal values and are consumed after cooking by the people of jute growing areas. Besides, industrial use, jute sticks are traditionally used for fuel, fencing, and support for growing some crop like betel vine. Thus, jute is not only a commercially important crop but also socio-culturally important to the farmers. Jute harvesting takes place at a time when marginal farmers and workers are faced with shortage of their food stocks. The cash derived from sales of jute fiber and the wages received by workers is an important contribution to food security for this vulnerable segment of the population. Improved technologies of jute cultivation enabled farmers to produce 16% more fiber yield and about Rs. 5000 per ha monetary gain. Jute cultivation also generates family employment, source of domestic fuel, and facilitates to keep clean field for cultivation of the next crop (Chapke et al., 2013).

Jute industry, on the other hand, is a source of (1) employment generation and (2) contribution to the national exchequer by way of collection of duties and value-addition. Jute industry provides direct employment to about 2.60 lakh jute mill workers and indirectly generates additional employment of about 1.5 lakh people engaged in R&D work, manufacturing in machinery and spare parts, trade, and other allied. Thus, cultivation of jute and mesta supports livelihood of about 40 lakh farm families spread over West Bengal, Bihar, Orissa, Assam, Andhra Pradesh, Tripura, Meghalaya, and Uttar Pradesh (Pratihar, 2007).

19.3   ANATOMY JUTE STEM

Jute fiber is developed in the outer portion of the secondary phloem, that is, bast of the stem. The fiber strands consist of multi-cellular fiber filaments or strands consisting of variable number of cells cemented together firmly by intercellular substances like pectin, hemicellulose, and lignin.

The basic fiber cell, called the ultimate cell, are long, slender cells tapering at both ends, polygonal in cross-section and have a central hollow known as lumen. The ultimate cells vary in length from 500 to 6500 μm in length and 10–30 μm in diameter (Kundu et al., 1959). The ultimate cells of individual fibers are formed by the alpha-cellulose. The cellulose polymers are closely packed in a highly ordered manner, known as microfibrils, and somewhere those are loosely packed in a less ordered manner known as fringed fibril structure. The close-packed and loosely packed areas are also known as crystalline region and amorphous region, respectively, and their proportion is approximately 2:1. Several microfibrils are cemented together to constitute an elementary fiber. They are either part of compact fiber bundle and loosely located within the cortex tissues. Elementary fibers constitute a fiber bundle and in the phloem, they form rings/pyramidal wedges around the stem, of more or less compact structure. Fiber bundles in each wedge are further arranged in large number (8–24) layers (Sinha et al., 2011). These fiber bundles form definite layers of concentric arcs in succession into a cone. The number of layers in each arc of fiber bundle, their volume, and size can vary, not only among species but also among varieties within a species. The fiber bearing potentiality of jute plants largely depends on the number of fiber trapezoids, size of fiber bundles, and their compactness (Kundu, 1959) (Fig. 19.1).

The cell wall of each ultimate cell is composed of an outer thin primary wall and inner thick secondary wall differing from each other in the molecular architecture. Both these walls of a jute ultimate cell are composed of ultra-fine microfibrils. In the primary wall, the fibrils lie in crisscross manner, whereas, fibrils are arranged almost parallel as right-handed spirals (with an angle of 7–9° with respect to cell axis) in the secondary wall (Roy & Lutfar, 2012) (Fig. 19.2).

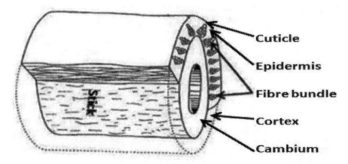

FIGURE 19.1 Schematic diagram of a cross-section of jute stem showing different layers of tissues.

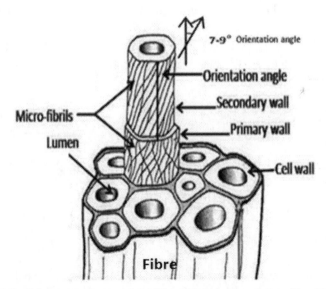

FIGURE 19.2 Schematic representation of the microstructure of a jute fiber (Roy & Lutfar, 2012).

The formation of fiber cells through biosynthesis and their crystallization process are simultaneous. The microscopic observation reveals that crystallization is not uniform throughout the zones of fiber formation—somewhere those are closely packed in a highly ordered manner, known as microfibrils, and somewhere those are loosely packed in a less ordered manner known as fringed fibril structure. The closely packed and loosely packed areas are also known as crystalline region and amorphous region, respectively, and their proportion is approximately 2:1.

19.4   COMPOSITION OF JUTE FIBER

Jute is a lignocellulosic fiber because it is mainly composed of polysac-charides and lignin. It also contains smaller amount of other components, such as fats and waxes, pectin, nitrogenous, coloring and inorganic matters (Sur & Amin, 2010). Lignin incrusts cellulose causing fiber fragility and is collected on the fiber wall as well as in its central part. The cell wall of each plant cell is found to have two divisions—a thin primary wall and a thick secondary wall. The space between the adjacent cells is called middle lamella. Both these walls are composed of ultra-fine micro-fibrils made of cellulose. Cellulose is found in the primary cell wall, whereas, the hemicel-luloses reside mainly in the inter-fibrillar region and lignin in the middle lamella mostly. The cellulose molecule is the basic building unit of plant cell wall and formed by linear polymerization of a series of glucose resi-dues joined to each other by β (1 → 4) glycosidic bond. The repeat unit of cellulose molecule is "cello-biose" consisting of two glucose residues (Table 19.1).

TABLE 19.1   Composition of Jute Fiber.

Constituents (%)	Jute fiber	
	Corchorus capsularis	*Corchorus olitorius*
Alpha cellulose	60.0–63.0	58.0–59.0
Hemicellulose	21.0–24.0	22.0–25.0
Lignin	12.0–13.0	13.0–14.0
Fats and Waxes	0.4–1.0	0.4–0.9
Pectin	0.2–1.5	0.2–0.5
Proteins/nitrogenous matter, etc.	0.80–1.9	0.8–1.6
Ash	0.7–1.2	0.5–1.2

Source: Sur and Amin (2010).

Pectin is a structural polymer contained in the primary cell walls of terrestrial plants. Pectin plays an important role in the fiber as component that binds fibers into bundles and determines the luster and touch of the fiber. These are macromolecular compounds of polygalacturonic acid and agglomerated in the middle lamella. In fiber plants, two fractions of pectin can be found—fraction A which is water soluble and fraction B which is water insoluble. Fibers consist of cells which are bonded to each other by a lamella which consists mostly of pectin B and a small fraction of pectin A.

Fiber bundles are attached to adjacent tissues with a layer of pectin A. Proper removal of pectin substances during processing determines the divisibility and as result fineness of fiber. Pectin A is removed during the retting process by the action of bacterial and fungal pectinases, while pectin B remains with the fiber and determines the compactness of the fiber. Excess removal of pectin and natural waxes will cause the fiber to have unpleasant dry and coarse touch. Complete removal of pectin will result in full disintegration of the fiber bundles into elementary fibers.

A hemicellulose is a heteropolymer, present along with cellulose in almost all plant cell walls. They contain many different sugar monomers, such as xylose, mannose, galactose, rhamnose and arabinose, xylan, glucuronoxylan, arabinoxylan, glucomannan, and xyloglucan. Microfibrils are cross-linked together by hemicelluloses. Lignin is an aromatic heteropolymer of phenylpropanoid units which confers structural rigidity to woody plant tissues and protects them from microbial attack (Higuchi, 1990). It is concentrated mainly in the region of the middle lamella. The amount of lignin in jute is 12–14% depending on the species. Lignin is bound together to the cellulose and hemicelluloses. Lignins assist and strengthen the attachment of hemicelluloses to microfibrils and cause cellulose to be rigid. In the elementary fiber, lignin occurs in the primary wall and outer part of the secondary wall. For processing, lignin is undesirable as touch and flexibility of the fiber become worse. Presence of lignin makes the fiber easily breakable. It also lowers divisibility of the fibers.

Fats and waxes determine soft touch, low friction, and thus ease of moving the fiber. Removal of fats and waxes from the fiber increases friction. Lower inter-fiber friction is desirable for reduced drafting related irregularity but higher inter-fiber friction helps to develop good yarn strength.

19.5   THE RETTING PROCESS

Bast fibers are collected from the phloem or bast surrounding the stem of certain, mainly dicotyledonous plants. In the phloem, bast fibers occur in bundles that are cemented together by pectin and calcium ions. Since the fibers are located in the phloem, they must often be separated from the "woody core" underneath. This process is called retting. Retting of jute is a kind of fermentation process in which the cortical and phloem tissues of the bark of the plants containing free strands are decomposed to separate fiber from non-fibrous woody stem (Asaduzzaman & Abdullah, 1998; Bose, 1969; Ray & Mandal, 1967). Retting is a combination of physical and biochemical processes. The biochemical processes are carried out by enzymes produced by several microorganisms including aerobic and microaerophilic bacteria and fungi. Retting can also be performed by treatment with certain chemicals (for instance high pH and chelating agents) and the process is known as chemical retting. Chemical retting has not become very popular as it cannot be adopted in farmers' filed.

19.5.1 PHYSICAL PROCESS

This is the process of readying jute plants for retting. The harvested jute plants are bundled together and allowed to defoliate by stacking the bundles in a shade. Defoliated jute bundles are steeped in clean and preferably slow moving water with suitable weight to completely immerse the jute bundles. To avoid the discoloration of the fiber seasoned wooden logs, cemented poles, brickbats tied in cement bags can be used as weight and covering materials should preferably be water hyacinth or coconut leaves.

19.5.2 MICROBIAL PROCESS

Steeping of jute plants under water causes water absorption by the plants resulting in swelling and bursting of cuticle at several places releasing soluble constituents, such as sugars, glucosides, nitrogenous compounds, minerals into the surrounding water (Ali, 1958; Ali & Choudhury, 1962a; Ali & Islam, 1965; Ali, 1977) making it nutritionally very rich for the growth of microorganisms present in water as well in the plants (Majumdar & Day, 1977; Mohiuddinn et al., 1978). This water together with the bacteria enters the plant tissues through stomata replacing the intercellular air. The fermentative microorganisms consume the cementing materials namely, the pectins, hemicelluloses, and proteins with release of galacturonic acid and sugar in retting water (Basak et al., 1998). Organic acids, such as butyric acid, acetic acid, lactic acid, etc., have been identified (Ali et al., 1968; Debsharma, 1976). Such a condition favors the growth of mostly microaerophilic bacteria which enter the tissues and decompose the cementing materials mostly pectinous in nature connecting the fiber strands. At this stage, the pH value increases because of the formation of buffering organic compound and the retting of the plant is completed (Ali et al., 1970, 1973; Islam, 2010; Islam & Rahman, 2008). If retting is allowed beyond this stage, microorganisms begin to degrade the fiber cellulose thus weakening the fiber strength.

19.5.3 BIOCHEMISTRY OF RETTING

The process of separation and extraction of fiber from non-fibrous tissues and woody part of the jute stem take place through enzymatic hydrolysis of pectins, gums, and other mucilaginous substances (Dasgupta et al., 1977; Majumdar & Day, 1977). The jute retting microbes should possess high pectinolytic and xylanase (Collins et al., 2005) activities with no or less cellulase (Sarrouh et al., 2012) activity.

The basic skeleton of all pectic substances is poly galacturonic acid. The enzymatic breakdown of pectic substances in the middle lamella is brought about by microbes leading to maceration of plant tissues (Fig. 19.3). Basically, this is catalyzed by the action of two enzymes: Pectin methylesterase (PME), acting on pectins and polygalacturonase (PGase) and acting on pectic acids. PME transforms pectins into pectic acid or polygalacturonic acid by esterification of methoxy group and polygalacturonase causing rapid hydrolysis of the pectic acid into polyglacturonic acid and mono galacturonic acid (Paul, 1977).

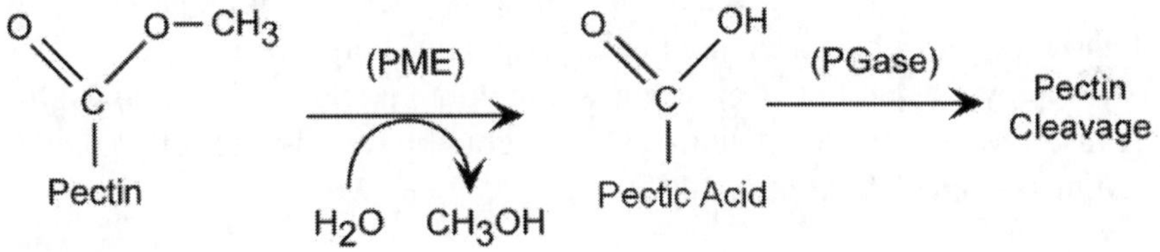

FIGURE 19.3 Degradation of pectin into polyglacturonic acid and monogalacturonic acid.

Another component which entangles the fiber is hemicelluloses. Partial degradation of xylan (major component of jute hemicellulose) is carried out by xylanase (Ahmed & Akhter, 2001; Das et al., 2012) (Fig. 19.4).

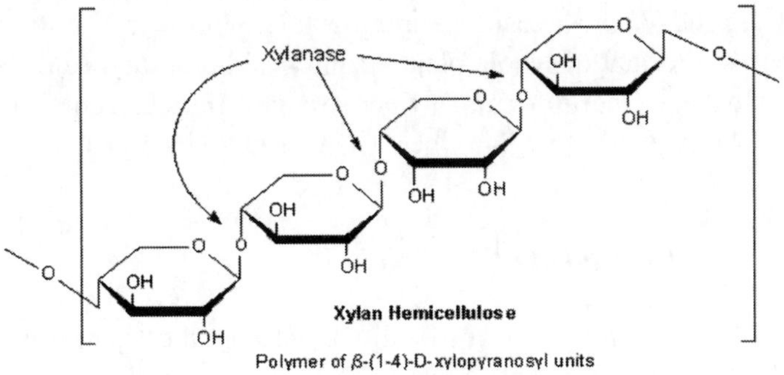

Polymer of β-(1-4)-D-xylopyranosyl units

FIGURE 19.4 Degradation of hemicellulose by xylanase during retting.

19.6   NEED OF ALTERNATE RETTING TECHNOLOGIES

Retting is the most important factor determining the quality of fiber. For conventional water retting of jute which is a general practice in the jute growing areas of the Indian sub-continent large quantity of water is required. About 90% of jute growers ret their harvest in stagnant water using mud, soil, etc., as covering material for proper immersion in the water resulting in poor quality fiber not suitable for high valued diversified products. The condition is further aggravated when jute-farmers ret jute plants in the same water several times in a season due to lack of adequate number of retting ponds. This not only deteriorates fiber quality but also results in health and environmental hazards. To overcome this problem, several alternate retting procedures have been developed.

19.6.1   CHEMICAL RETTING

The cementing materials are removed by boiling with 1.0% sodium hydroxide or 0.5% sulfuric acid at boiling temperature for 6–8 h. By this method, pectic substances, hemicelluloses, and lignins are hydrolyzed and removed by washing in clean water. Ammonium oxalate and sodium sulfate are also found to be very suitable as they donot have adverse effect on fiber qualities (Ahmed & Akhtar, 2001).

19.6.2   RIBBON RETTING

In ribbon retting, at first fiber ribbons are extracted with an extractor or ribboner. Ribbons are then subjected to either conventional water retting or microbial retting. Ribbon retting reduces: (1) the requirement of water from 1:20 to 1:10 substrate liquor ratio; (2) the length of retting time from 14–15 to 7–8 days; and (3) the level of environmental pollution to almost one-fourth in comparison to that of whole plant retting besides assurance of producing better quality jute fiber in terms of fiber strength, fineness, color, lustre, and overall absolutely bark free jute fiber (Alam, 1998; NIRJAFT, 1999).

19.6.3   ACCELERATED RETTING

In this method, a condition is artificially created in the faster growth of the retting microbes resulting in fast retting of jute plants or ribbons. The "*Sonali Sathi*" formulation developed at NIRJAFT, Kolkata is such a technology.

19.6.4   ENZYMATIC RETTING

In this method, instead of using pectinolytic and hemicellulolytic microbes, pectinase and xylanase enzymes are directly used for retting of jute fiber. Enzymatic processes are carried out under controlled conditions based on the type of enzymes. Such enzymes are being commercially produced making retting process faster yielding long quality fibers. Commercial enzyme preparations, namely, Pectinex ultra Sp-L and Flaxzyme consisting predominantly of pectinolytic enzymes were found to ret green jute ribbons within 48 h, producing fairly good quality jute fibers (Majumdar et. al., 1991). In some cases more enzyme formulations are used, for example, Jutex marketed by Indizyme.

19.7   MICROBES OF JUTE RETTING

Various fungi, aerobic and anaerobic bacteria have been reported to be involved in the retting process. The aerobic organisms grow first and consume most of the dissolved oxygen, ultimately creating an environment favorable for the growth of anaerobes. It has been observed that the greater part of decomposition is carried out by anaerobic species. JARI (now known as CRIJAF) scientists isolated species of *Bacillus*, *Clostridium*, *Phoma*, *Aspergillus*, *Mucor*, and *Macrophomina* from retting liquor of jute and studied their pectinolytic and hemicellulolytic activities. They found that only *Clostridium* possesses pectinolytic activity without hemicellulolytic activity (Ray & Mandal, 1967) which is beneficial for retting. Munshi and Chattoo (2008) studied the bacterial population structure of the jute-retting environment. They investigated the presence of microorganisms during the process of jute retting. By constructing two 16S rRNA gene libraries, from two jute-retting locations of West Bengal, they identified phylotypes affiliating to seven bacterial divisions. The bulk of clones came from *Proteobacteria* (~37, 41%) and a comparatively smaller proportion of clones from the divisions—*Firmicutes* (~11, 12%), *Cytophaga–Flexibacter–Bacteroidetes* group (CFB); (~9, 7%), *Verrucomicrobia* (~6, 5%), *Acidobacteria* (~4, 5%), Chlorobiales (~5, 5%), and Actinobacteria (~4, 2%) were identified. Evaluation of the retting waters by whole cell rRNA-targeted fluorescent in situ hybridization, as detected by domain- and group-specific probes, a considerable dominance of the betaproteobacteria (25.9%) along with the CFB group (24.4%) was observed. In addition, 32 bacterial species were isolated on culture media from the two retting environments and identified by 16S rDNA analysis, confirming the presence of phyla, *Proteobacteria* (~47%), *Firmicutes* (~22%), CFB group (~19%), and *Actinobacteria* (~13%) in the retting environment.

Aerobic and anaerobic bacteria were isolated from samples of retted jute stems. These were found to belong to three genera, *Bacillus, Micrococcus,* and *Pseudomonas* with a total of 13 species. Only one new species, *Micrococcus corchorus* and one new variety, namely, *Micrococcus leteus* var. *liquefaciens,* have been reported. Among the aerobes and facultative anaerobes, *Bacillus subtilis* has been found to be most common, while *Bacillus macerans, Bacillus polymyxa, M. corchorus,* and *Pseudomonas aeruginosa,* are the most active retting agents (Ahmad, 2008).

Fungal cultures were efficiently utilized for retting of green ribbon of jute (Haque et al., 1998; Banik et al., 2003). Saprophytic fungus, *Sporotrichum* was found to be capable of retting dry ribbons of jute satisfactorily under laboratory conditions. Another fungus, *Aspergillus* sp. was found to be beneficial in improving the quality of fibers by one or two grades. A number of fungal species, namely *Aspergillus niger, Mucor* sp., *Schizophylum communae, Sporotrichum* sp., *Trichoderma* sp., *Penicillium* sp.1 and sp. 2 isolated from rotten fruits, wood, and manure pits were tested for their retting efficiency on green ribbons of jute (*Corchorus capsularis,* var. CVL-1). Among them, *Sporotrichum* sp., *Schizophylum communae,* and *Trichoderma* sp. retted green ribbons of jute (var. CVL-1) in 7, 9, and 11 days, respectively, while the others did not show any retting efficiency (Haque et al., 2001). Addition of efficient pectinolytic microbial inoculum was found to boost up or improve the ribbon retting process (Haque et al., 1998; Banik et al., 2007). No adverse effects on the fiber bundle strength and fiber yield were observed when *Sporotrichum* sp. was used for retting. According to Pressley Index, fiber strength was found to be 10.82 Ibs/mg and fiber yield was about 2.8 Kg out of 40 Kg green ribbons (Chakravarty et al., 1962).

Scientists at the ICAR-National Institute of Research on Jute and Allied Fibre Technology (NIRJAFT), Kolkata has developed a fungal retting technology which requires little amount water (hence, known as "dry retting"). Here, instead of ribbons, whole jute stems are used. The technology holds promise for the water scarce areas as it allows retting of jute fiber outside water, thus saving water, protecting environment, and eliminating health hazards. Four different pectinolytic fungi—*Aspergillus tamarii, Aspergillus flavus, A. niger,* and *Sporotrichum thermophile*—are used for retting of jute. The fungi are maintained in the laboratory on Potato Dextrose Agar (PDA) medium, and their mass culture is prepared on a mixture of rice-husk and wheat-bran. The slurry prepared from solid-base mass culture by mixing it with water is spread on the defoliated green jute-plants (4–5% of solid base mass culture). The fungus-inoculated plants are wrapped in polythene sheets and are incubated for about 10 days. The fungal growth is monitored in the intervals of 2 days to see for even growth on the plant surface, and if not, plants are over-turned and sprinkled with water. Inoculated fungi grow fast on the surface of the green plant in just moist condition, and feed on the inner gummy and pectin matter facilitating separation of jute fibers and sticks. After retting, fungal growth is removed by wiping with a piece of old gunny bag, and fibers are simply extracted by pulling with hand. The fibers may be washed further in water for brightness and for removing extra adhering gum if any (Banik, 2014). It was possible to produce jute fiber of TD2 grade with this technique (unpublished data from the author). However, the technology needs further refinement to suit the farmers' need as there are certain issues which require further study (Table 19.2).

TABLE 19.2 Fiber and Yarn Characteristics of Fungi Retted Jute Fiber.

Characteristics	Value
Fiber	
Fiber strength	20.5–24.8 g/tex
Fiber fineness	2.8–3.0 tex
Average root content	< 5%
Average fiber grade	TD4
8 lb yarn	
Average tenacity	10.67 cN/tex
Work of rupture	0.86 mJ/tex M
Average hairiness index	11.90
Average breaking load	23.24 N
Um	24.4–27.8%

Source: Banik (2014).

19.8 GLOBAL CLIMATIC CHANGE AND NEED OF MINIMUM-WATER RETTING TECHNOLOGY

Retting and extraction processes have a profound effect on the quality of fiber produced and on the cost of fiber production. Jute retting is still carried out through age-old conventional method requiring presence of adequate water bodies or else an otherwise bumper crop may yield nothing or very poor quality fiber if sufficient water is not available for retting. Retting is such an important step in fiber production that affects the quality of the end products and their competitiveness in the market. Due to global climate change, major jute growing areas in India and Bangladesh are facing acute water scarcity during retting period resulting in poor quality fibers and hence low market value.

Several technologies have been developed which minimize water requirement. Ribbon, fungal, and chemical retting are alternative methods over the conventional water retting. In ribbon, retting barks are removed from jute-plants mechanically or manually in the form of ribbon. The ribbons are coiled and then allowed for retting in water with or without a microbial inoculum.

Ribbon retting has the following advantages over conventional retting: (1) it requires smaller volume of water; (2) it is faster; (3) it produces lesser environmental pollution while; and (4) improving fiber quality (Alam, 1998; NIRJAFT, 1999). Ribbon retting gets completed in a shorter period (due to more surface area) compared to that of stem retting. Ribbon retting is economic and more suitable; however, the ribboning technology presently available is of lower capacity and hence labor intensive. Efforts are being taken by the National Institutes (e.g., NIRJAFT) to develop high capacity ribboner. Chemical retting has not become popular with the farmers because the chemicals are costly and not practicable in farmers' field (Ahmed & Akhter, 2001). Fungal or dry retting can be a better option under water-scarce situation as it works with least amount of water (only for sprinkling and washing). This method is capable of producing high-quality fiber (TD2). However, the method suffers from the following disadvantages: (1) supply and maintenance of the pure fungal culture at the village level and (2) appeared to be a little costlier than conventional retting. The cost can be minimized by using locally available materials and by setting retting on the field itself, thereby eliminating transport cost. Jute researchers must work on these issues so that simpler, cost-effective retting technologies can be developed in the near future.

19.9 CONCLUSION

Jute occupies an important position in the lives of large population of India, Bangladesh, Nepal, Thailand, China, and Myanmar. Realizing its importance in the economy and due to its environment-friendly nature, Food and Agriculture Organization (FAO) declared the year 2009 as the International

Year for Natural Fibers. About 60% of the raw jute in the world is produced in India. It is viewed that about 4 million farmers, 0.25 million industrial workers, and 0.5 million traders find gainful employment in jute sectors (Sen et al., 2006). Besides, the jute based industries also contribute to the export earnings to the tune of nearly 1200 crore of rupees yearly (Karmakar et. al., 2008). Thus, any effort that contributes to higher income from jute will definitely benefit these groups of people. These may include improved production technologies, better post-harvest handling of fiber, and its diversified use. Government and non-Government Organizations must address the issues of high capacity ribboner and develop simpler and cost-effective minimum water retting technologies for water-scarce areas. Current fungal retting technologies hold promise but need further refinement to suit the stakeholders' needs.

KEYWORDS

- jute
- retting
- quality
- fungal
- fiber

REFERENCES

Ahmed, J.; Akhter, F. Jute Retting: An Overview. *J. Biol. Sci.* 2001, *1* (7), 685–688.

Alam, A. In *Retting and Extraction of Jute Problems and Prospects,* Proceedings of International Seminar on Jute and Allied Fibres—Changing Global Scenario, Calcutta, Feb 5–6, 1998; A NIRJAFT Publication: Kolkata, 1998.

Ali, M. M. Aerobic Bacteria Involved in Retting of Jute. *Appl. Microbiol.* **1958,** *6,* 87–89.

Ali, M. M. Aerobic Bacteria Involved in the Retting of Jute. *Appl. Microbiol.* **1977,** *6,* 87–89.

Ali, M. M.; Alam, S. Effect of Different pH Values on Jute Retting. *Bangl. J. Bot.* **1973,** *2,* 19–23.

Ali, M. M.; Choudhury, S. D. Pectic Enzymes in Jute Retting. *Pak. J. Sci. Ind. Res.* **1962a,** *5,* 271–273.

Ali, M. M.; Islam, A. Pectic Enzymes of Penicillium Frequentans Involved in the Retting of Jute. *Pak. J. Sci. Ind. Res.* **1965,** *8,* 47–51.

Ali, M. M.; Sayem, A. Z. M.; Eshaque, A. K. M. Effect of Neutralization of Retting Liquor on the Progress of Retting. *Sci. Ind.* **1970,** *7,* 134–136.

Ali, M. M.; Sayem, A. Z. M.; Alam, S.; Ishaque, M. Relationship of Pectic Enzyme of Macrophomina phaseoli with Stem Rot Disease and Retting of Jute. *Mycopathol. Mycol. Appl.* 1968, *38,* 289–298.

Asaduzzaman, M.; Abdullah, A. B. M. In *Impact of Retting and Other Post Harvest Processes on Quality of Jute Fibre,* Proceedings of International Seminar on Jute and Allied Fibres— Changing Global Scenario, Calcutta, Feb 5–6, 1998; A NIRJAFT Publication: Kolkata, 1998; pp 109–113.

Banik, S. A New Dry Retting Technology for Jute. *ICAR News* **2014,** *20* (2), 2–3.

Banik, S.; Basak, M. K.; Sil, S. C. Effect of Inoculation of Pectinolytic Mixed Bacterial Culture on Improvement of Ribbon Retting of Jute and Kenaf. *Text. Res. J.* **2007,** *33,* 50–51.

Banik, S.; Basak, M. K.; Paul, D.; Nayak, P.; Sardar, D.; Sil, S. C.; Sanpui, B. C.; Ghosh, A. Ribbon Retting of Jute — A Prospective and Eco-Friendly Method for Improvement of Fibre Quality. *Ind. Crops Prod.* **2003,** *17,* 183–190.

Basak, M. K.; Bhaduri, S. K.; Banik, S.; Kundu, S. K.; Sardar, D. In *Some Aspects of Biochemical Changes Associated with Retting of Green Jute Plants,* Proceedings of International Seminar on Jute and Allied Fibres—Changing Global Scenario, Calcutta, Feb 5–6, 1998; A NIRJAFT Publication: Kolkata, 1998; pp 105–108.

Bose, R. G. *The Retting of Jute and Mesta—A Review;* A publication of the Indian Jute Industries Research Association: Calcutta, 1969.

Chakravarty, T.; Bose, R. G.; Basu, S. N. Fungi Growing on Jute Fabrics Deteriorating under Weather Exposure and in Storage. *Appl. Microbiol.* **1962,** *10,* 441–447.

Chapke, R. Role of Jute Cultivation in Farmers' Livelihood. *Indian Res. J. Ext. Edu.* 2013, *13* (1), 1312–135.

Chapke, R.; Biswas, C. R.; Jha, S. K.; Das, S. K. *Technology Evaluation Through Frontline Demonstrations and Its Impact;* CRIJAF Bulletin No. 03: 19. CRIJAF: Barrackpore, India 2006.

Collins, T.; Gerday, C.; Feller, G. Xylanases, Xylanase Families and Extremophilic Xylanases. *FEMS Microbiol Rev.* **2005,** *29,* 3–23.

Das, B.; Chakrabarty, K.; Ghosh, S.; Majumdar, B.; Tripathi, S.; Chakraborty, A. Effect of Efficient Pectinolytic Bacterial Isolates on Retting and Fibre Quality of Jute. *Ind. Crops Prod.* **2012,** *36,* 415–419.

Das, S. K.; Chapke, R. R.; Jha, S. K.; Ghorai, D. *Technology Transfer for Jute-Retrospect and Prospect*; CRIJAF Bulletin No. 10: 34. CRIJAF: Barrackpore, India, 2006.

Dasgupta, P. C.; Sardar, D.; Majumdar, A. K. Chemical Retting of Jute. *Food Farming Agric.* **1976,** *8,* 7–9.

Debsharma, G. D. Biochemical Investigations on Jute Retting. *Indian J. Agric. Sci.* **1976,** *16,* 453–458.

Haque, M. S.; Asaduzzaman, M.; Akhter, F.; Ahmed, Z. Retting of Green Jute Ribbons (Corchorus capsularis var. CVL-1) with Fungal Culture. Online *J. Biol. Sci.* **2001,** *1,* 1012–1014.

Haque, S.; Alamgir, M.; Ahmed, Z. Fungal Efficiency on Retting of Green Ribbon of Jute. J. Asiatic Society, Bangladesh. *Science* **1998,** *24,* 303–307.

Higuchi, T. Lignin biochemistry: Biosynthesis and Biodegradation. *Wood Sci. Technol.* **1990,** *24,* 23–63.

Islam, M. M. In *Jute;* Sultana, N., Ed.; Dynamic Publication: Dhaka, 2010; pp 165.

Islam, M. M.; Rahman, M. In *Hand Book on Agricultural Technologies of Jute, Kenaf and Mesta Crops;* Bangladesh Jute Research Institute: Dhaka, 2008; p 92.

Karmakar, P. G.; Hazra, S. K; Sinha, M. K; Chaudhury, S. K. Breeding for Quantitative Traits and Varietal Development in Jute and Allied Fibres. In *Jute and Allied Fibre Updates: Production and Technology;* NIRJAFT: Kolkata, 2008; pp 67–75.

Kundu, B. C.; Basak, K. C.; Sarkar, P. B. *Jute in India: A Monohgraph;* The Indian Central Jute Committee: Calcutta, 1959; pp 372–380.

Majumdar, A. K.; Day, A. Chemical Constituents of Jute Ribbon and the Materials Removed by Retting. *Food Farm. Agric.* 1977, *9*, 25–26.

Majumdar, S.; Kundu, A. B.; Dey, S.; Ghosh, B. L. Enzymatic Retting of Jute Ribbons. *Int. Biodeterior.* **1991,** *27* (3), 223–235.

Mohiuddin, G.; Chowdhury, M. I.; Kabir, A. K. M. R.; Hasib, S. A. Lignin Content of Jute Cuttings of the White and Tossa Varieties Bangladesh Jute. Bangladesh. *J. Jute Fib. Res.* **1978,** *3*, 27–32.

Munshi, T. K.; Chattoo, B. B. Bacterial Population Structure of the Jute-Retting Environment. *Microbiol Ecol.* **2008,** *56*, 270–82.

NIRJAFT. *Data Book on Jute*; Mitra, B. C., Ed.; (ICAR) Special Publication: Kolkata, India, 1999; p 62.

Paul, N. B. "Modern Concept on Retting and Extraction of Jute" In *Jute Production, Processing and Marketing Technology;* University College of Agriculture, Calcutta Univeristy: Culcutta, 1977; pp 133–140.

Pratihar, J. T. *Jute, the Golden Fibre Present Status and Future of Indian Jute Sector;* Vaanijya, Directorate General of Commercial Intelligence and Statistics: Kolkata, India, 2007; pp 1–17.

Ray, A. K.; Mandal, A. K. Retting and Quality of Jute Fibre. *Jute Bulletin* (Kolkata) **1967,** *July*, 131–139.

Roul, C. *The International Jute Commodity System;* Northern Book Centre: New Delhi, 2009.

Roy, S.; Lutfar, L. B. Bast Fibres: Jute, in the Handbook of Natural Fibres. In *Types, Properties and Factors Affecting Breeding and Cultivation;* Kozlowski, R. M., Ed., Woodhead Publishing Limited: Cambridge, 2012; Vol. 1, pp 24–46.

Sarrouh, B.; Santos, M. T.; Miyoshi, A.; Dias, R.; Azevedo, V. Up-todate Insight on Industrial Enzymes Applications and Global Market. *J. Bioprocess Biotechnol.* 2012, *S4*, 002.

Sen, H. S.; Das, S. K.; Saha, D. Good Beginning has been Made. In *The Hindu Survey of Indian Agriculture;* Kasturi and Sons Ltd: Chennai, 2006; pp 119–125.

Sinha, M. K.; Kar, C. S.; Ramasubramanian, T.; Kundu, A.; Mahapatra, B. S. Corchorus, C. *Wild Crop Relatives: Genomic and Breeding Resources, Industrial Crops;* Kole, Ed.; Springer-Verlag: Heidelberg, 2011; pp 29–61.

Sur, D.; Amin, N. Physics and Chemistry of Jute. In *Jute Basics;* International Jute Study Group: Dhaka, 2010; pp 35–55.

CHAPTER 20

COMPOSTING OF CASHEW APPLE WASTE RESIDUE FOR CULTIVATING PADDY STRAW MUSHROOM

MEGHANA V. DESAI, DAVID G. GOMES, BABU V. VAKIL, and VAIJAYANTI V. RANADE*

Department of Microbiology, Guru Nanak Khalsa College of Arts, Science and Commerce (Affiliated to University of Mumbai), Nathalal Parekh Marg, Matunga, Mumbai 400019, Maharashtra, India

*Corresponding author. E-mail: vaijayantiranade@gmail.com

CONTENTS

Abstract 512
20.1   Introduction 512
20.2   Material and Methods 513
20.3   Conclusion 519
Keywords 519
References 519

ABSTRACT

Cultivation of *Pleurotus sajor caju*, an edible mushroom, utilizing cashew apple waste residue, left post-extraction of juice for feni fermentation was attempted. The bagasse left after the extraction of juice for production of this alcoholic beverage is rich in proteins, minerals, and ascorbic acid. Therefore, it can be utilized as a nutritious alternative substrate for mushroom cultivation post suitable treatments, such as composting. The cultivation of mushroom can be done as a value-added product of cashew apple waste residue, in addition to feni fermentation.

20.1   INTRODUCTION

Cashew apple (*Anacardium occidentale L.*) is one of the important products of the cashew industry as well as the alcoholic beverage industry (Gunjate & Patwardhan, 1995; Ramteke et al., 1984). While cashew nut is an important crop in India, the hard liquor prepared from the apple juice is a common alcoholic beverage (Garruti et al., 2006; Davis, 1999). The soft fibrous fruit juice, rich in nutrients is often employed in the preparation of fermented products such as feni, a distilled liquor (Vankinh et al., 1997).

Mushroom is a great delicacy food worldwide and highly nutritious (Kosaric et al., 1983; Mandeel et al., 2005). *Pleurotus sajor caju,* being a tropical and wide adaptability mushroom, grows easily on organic waste materials (Fulekar, 2005; Looper, 2001). Wide adaptability range of oyster mushroom enables use of cheaper substrates in comparison to button mushroom. Cultivation of edible mushrooms using composted waste materials involve less monetary input and at the same time gives a good yield.

In Goa and Konkan, huge amount of cashew apple residue is being generated after the extraction of juice for feni fermentation. Farmers who carry out such fermentation often consider such residues as waste and are often discarded. This nutritious waste residue can be utilized for soil conditioning and fertility and as a suitable substrate for paddy straw mushroom cultivation. Cultivation of the mushroom can be efficiently carried out in the favorable climatic conditions in these coastal regions. Besides nut, the utilization of cashew apple for feni and the waste for mushroom cultivation gives three fold profitability to the cultivators.

20.2 MATERIAL AND METHODS

20.2.1 PREPARATION OF COMPOST

The composting of cashew apple waste was carried out in wooden boxes (L × B × H cm³) (46 × 30 × 30 cm³) provided with air pockets on four sides for aerobic conditions. Initially, half-inch layer of bagasse was added to wooden box, followed by equal layers of dry leaves, dry grass, and thick fresh cow dung. Finely crushed sun-dried cashew apple waste followed by a thin, half-inch layer of cow dung was added. Finally, moist soil occupied the topmost layer of the compost material. Two more similar layers of these wastes were added, while the last layer was added 12 days post-composting. Wooden boxes were kept moist for ideal composting conditions and heap was mixed every 4 days for maintaining proper aeration. Composting was carried out up to a period of two months untill all the added waste got completely decomposed (Fulekar, 2005).

In this study, some important parameters of compost like temperature, moisture content, specific gravity, pH, alkalinity, electrical conductivity; organic carbon, total nitrogen, and phosphorus were analyzed at regular intervals to check the suitability of compost for oyster cultivation (Murugesan & Rajakumari, 2005).

Temperature was recorded every 4 days, while physical and chemical parameters were analyzed at regular interval of 12 days. The electrical conductivity was determined using a conductivity meter with temperature sensor (Pico⁺, LAB India Instrument Pvt. Ltd, Sr. no: CN05121206). The analysis of the total nitrogen content of the samples was outsourced from Envirolabs, Thane. The sample was analyzed by Macro-Kjeldahl method (APHA) 4500-N-org C using a digestion distillation unit.

20.2.2 PREPARATION OF SPAWN

Pure culture slant of *P. caju* was obtained from Mahatma Phule Krishi Vidyapeeth, Pune for spawn preparation. The mushroom slant culture was transferred on MD6 (Potato Dextrose agar) medium slants and were incubated at 28 ± 2°C in incubator (Remi Instrument Ltd, model. no: CI-12S) till the mycelial growth covered the entire agar surface (7 days) (Morel, 1985; Ghanekar, 2000; Ghanekar, 1999; Chang & Miles, 2004). Two kilograms of good quality, whole wheat grains were boiled in a vessel containing equal amount of water until the grains got swollen without bursting. The grains were allowed to cool after draining the excess water. An amount of 200 g of these grains was autoclaved (Equitron, Sr.No:742IST.JC.146, Medica Instrument Mfg. Co) in 500 mL bottles. Calcium carbonate (6% (w/w)) was added to each bottle and the contents were mixed thoroughly. These bottles were repeatedly sterilized at 25 psi for an hour for 3 days. It was then inoculated with 7 day-old mycelium aseptically after cooling. The bottles were incubated for three weeks at $28 \pm 2°C$ until the substrate became compact and fully colonized with mycelial growth.

20.2.3 CULTIVATION OF PADDY STRAW MUSHROOM

Wheat straw used for cultivation of mushrooms was soaked in water for 6–8 h. After draining the excess water, cashew apple waste compost and wet straw were sterilized separately at 121°C for an hour. Additions of compost substrate and straw after cooling were done in sealable polypropylene bags for three sets of cultivation in the following manner (Donsky, 2000; De Bertoldi, 1985).

1. SET I M—1 kg moist compost substrate + 200 g of grain spawn.
2. SET II M—1 kg moist straw + 200 g of grain spawn.
3. SET III M—500 g moist compost substrate + 500 g moist straw + 200 g of grain spawn.

Multilayered spawning, that is, spawn layer followed by a substrate layer was done. Aeration was ensured by creating air slits in the bag. The spawned bags were kept on a tripod stand in ventilated room at a temperature of $28 \pm 2°C$ (Ghanekar, 1999). The humidity was maintained between 85% and 95% by surrounding it with a constantly watered jute bags. The jute bags were disinfected by treating with concentrated formaldehyde solution before use to prevent contamination by fungal flora. The bags were moistened to allow the mycelial growth during the incubation period. The spawned bags were opened after two and a half weeks. The compact mass of the substrate and mycelium was frequently watered for maintaining the optimal humidity. The wet and dry weight analysis; and biological efficiency were calculated for the fruiting bodies.

20.2.4 RESULTS AND DISCUSSION

During each stage of composting, the organic constituents of the waste undergo transformations which lead to formation of biologically stable components that contribute directly to soil conditioning and fertility. In comparison to fermentation and anaerobiosis, composting is faster, simpler and a safer approach (Looper, 2001; Epstein, 1996). Temperature of compost pile at five different points (four corners and center of the box) was recorded at regular interval of 4 days up to 60 days. Maximum temperature achieved within the pile during this period was 45°C (Table 20.1). The temperature within the pile showed a decreasing trend after 12 days, reaching a constant value of 33°C toward 36th day till end of the process, indicating the growth of mesophilic organisms at this stage (Murugesan & Rajakumari, 2005; De Bertoldi et al., 1985).

The moisture content of the compost was within the lower range of 25–30% during the initial period. This may be due to the increased microbial activity that must have commenced at the initial stages of 4–16 days of composting process, which resulted in increase in temperature of compost pile (Epstein, 1996). The increase in temperature would have resulted in higher water evaporation thus indicating lower moisture content. Another reason may be the loose binding of the particles within the compost pile leading to low water retaining capacity. As the composting process advanced, the compact, clumped or aggregated nature of particles retained more moisture. Moisture content of the final matured compost was 43% as projected in Table 20.1 which is considered safe for addition into soil (De Bertoldi et al., 1985). The specific gravity of compost samples was found to increase with an insignificant factor of 0.1 mg/m^3 within the period of 12–60 days (Table 20.1). The finished compost had an ideal specific gravity of 0.58 g/cc for addition into soil (Murugesan & Rajakumari, 2005; De Bertoldi et al., 1985).

The pH of cashew apple residue utilized during composting was within the range of 3.0–4.0. This might be the reason for initial low pH of the compost. There was a slight increase in pH of the compost from 5.13 to 5.4 during the process. As reported, the final pH of the compost was found to be within the range of 5.0–6.0 as shown in Table 20.1 (Murugesan & Rajakumari, 2005; De Bertoldi et al., 1985). There was a slight increase observed in total alkalinity of the compost samples which was not significant (Table 20.1) (Murugesan & Rajakumari, 2005). pH plays a very crucial role in the availability of plant nutrients. Basic pH increases the availability of phosphorus, manganese, and zinc, while an acidic pH makes available potassium, calcium, nitrogen, copper, and molybdenum. Therefore, ideal pH value will depend on the system to which compost is to be applied. Most of the composts have pH between 5.5 and 8 (De Bertoldi et al., 1985).

Electrical conductivity of the final product was found to be within a safe range of 0–2 dS/m (Table 20.1) indicating that the final product of compost contains sufficient levels of soluble salts to increase fertility of soil (De Bertoldi et al., 1985; Ryan et al., 2001).

In this study, phosphorus content of the compost samples remained constant throughout the composting period. Phosphorus content was within the range of 0.05–0.06 mg/L (Table 20.1). No significant increase was observed in total phosphorus content of compost samples analyzed (De Bertoldi et al., 1985). Phosphorus content of the final compost was very low for its addition in soil for plant growth.

The carbon/nitrogen (C/N) ratios for the compost samples were very high up to 36 days beyond which the ratio reduced significantly (Table 20.1). However, at the end of composting process, the C/N ratio was properly balanced which indicates good microbial activity and stability within the compost pile. The C/N ratio of ready compost was 33, a good indicator of the biological stability of the compost which is optimum for addition into soil (Fulekar, 2005; Murugesan & Rajakumari, 2005; Epstein, 1996).

The water-insoluble matured compost had dark brown color and an earthy odor. The texture of mature compost was loose, homogenous, and coarse granular. The compost volume had reduced to half on completion of composting process.

TABLE 20.1   Variation in Parameters During Composting Process.

Compost characteristics	No. of days				
	12	24	36	48	60
Temperature (°C)	45	36	33	33	32
Moisture content (%)	27.3	40	47	44	43
Specific gravity (g/cc)	0.48	0.51	0.56	0.59	0.58
pH	5.13	5.3	5.3	5.35	5.4
Electrical conductivity (dS/m)	2.23	2.04	1.93	1.93	1.93
Total alkalinity as $CaCO_3$ (mg/L)	11.6	12.2	11.2	11.8	12.5
Amount of alkalinity in sample (mg/g)	0.07	0.1	0.1	0.1	0.1
Total phosphorus content (mg/L)	0.05	0.05	0.06	0.05	0.05
Organic matter (C) (%)	13.7	14.6	16.3	12.06	10.6
Total nitrogen content (N) (%)	0.22	0.21	0.28	0.31	0.32
C/N ratio	62	69	58	38	33

Within 2 days of opening the spawned bag, pinhead fruiting bodies started appearing on all sides of substrate for SET II M and after 3 days for the remaining sets. Mature fruiting bodies of oyster mushroom were harvested at different periods, when the cap started to fold; and had attained a diameter of about 6–8 cm and stem length of 2–3 cm. The fresh wet weight and dry weight of mushrooms were recorded with respect to number of flushes (crops) of individual sets (Table 20.2). The fruiting bodies grown on compact mass of three different substrates (Set I, Set II, and Set III) is shown in Figure 20.1.The harvest yields and biological efficiency (Table 20.3) and the characteristic features of mushrooms cultivated for the three sets (Table 20.4) were evaluated.

TABLE 20.2 Analysis of Number of Flushes Obtained in Three Different Sets.

Set no.	Mushroom weight (g)	1st flush			2nd flush		3rd flush
No. of days		1	2	3	1	2	1
I M	WW	38.77	115.86	20.69	–	–	–
	DW	11.97	27.38	2.82	–	–	–
II M	WW	142.79	102.85	–	–	–	–
	DW	41.59	30	–	–	–	–
III M	WW	16.39	95.86	77.44	29.71	24	24.06
	DW	3.48	28.88	28.43	6.7	5	5.1

WW: wet weight

DW: dry weight

–: no growth

TABLE 20.3 Comparative Harvest Yield and Biological Efficiency Analysis.

Set no.	I M	II M	III M
	Harvest yield (g)		
Wet weight basis	175.32	245.64	267.46
Dry weight basis	42.17	71.59	77.59
	Biological efficiency		
Dry weight of substrate	553.57	310.76	463.59
% biological efficiency	31.67	79.04	57.69

TABLE 20.4   Characteristic Features of Mushrooms for the Three Sets.

Set	Size	Color	Colonization	Number of flushes	Total yield and % biological efficiency
Set I M	Regular or normal	white	Heavily colonized with mushrooms/fruiting bodies well distributed	Single flush	Lower than set I M and II M
Set II M	Large	Grayish white	Heavily colonized with mushrooms/fruiting bodies irregularly distributed	Single flush	Higher that set I M and lower than set II M
Set III M	Normal	white	Heavily colonized with mushrooms/fruiting bodies irregularly distributed	Three flushes	Higher that set I M and set II M

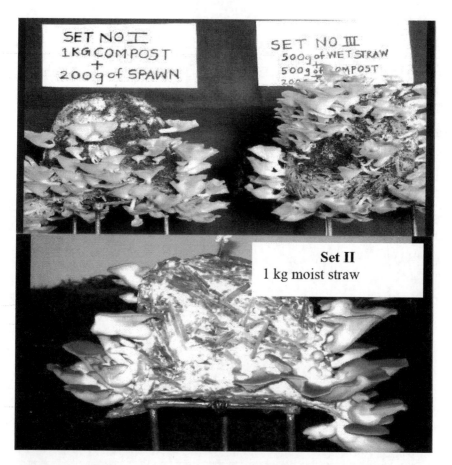

FIGURE 20.1   (See color insert.) Fruiting bodies (sporophores) growing on compact mass of three different sets of substrates of Set I (top left), Set III (top right), and Set II (bottom).

The above data revealed that the yield of mushroom is pronounced in the first two flushes and decreases thereafter. Maximum fruiting bodies were harvested in first flush for SET I M, SET II M, and up to three flushes for SET III M.

On comparison of the three sets, the combination of composted cashew apple waste and straw substrate supported the luxurious growth of *P. sajor-caju* mushroom than composted cashew apple waste alone, where the yield obtained was relatively less (Table 20.3). The performance of the sets was evident by their biological efficiency values, which were highest for composted cashew apple waste and straw (SET III M); and only straw as substrates (SET II M) (Table 20.3).

20.3 CONCLUSION

The study focused on minimizing the wastage of fruit residue and widens the scope for utilization of composted waste as an alternative substrate for production of value-added product like Paddy Straw mushroom. The compost mixed with equal amount of straw has proved to be a good substrate for cultivation of oyster mushrooms. It can be concluded that this cheap fruit waste material is one of the promising substrates for edible mushrooms as highly nutritious food (Vieira & de Andrade, 2016).

KEYWORDS

- **cashew apple waste**
- **feni**
- **oyster mushrooms**
- **composting**

REFERENCES

Chang, S.; Miles, P. G. *Mushrooms: Cultivation, Nutritional Values, Medicinal Effect and Environment*; CRC Press London: New York, Washington, DC, 2004.

Davis K. *Cashew: An ECHO Technical Note*; 1999; pp 1–8, North Fort Myers, Florida.

De Bertoldi, M.; Vallini, G.; Pera, A. *Technological Aspects of Composting Including Modeling and Microbiology in Composting of Agricultural and Other Wastes;* Gasser, J. K. R., Ed.; Elsevier Applied Science Publishers; London, 1985; pp 27.

Donsky, M. A. Growing Mushrooms on Compost; 2000; pp 1–5. https://cdn.preterhuman. net/texts/drugs/shroomery_grow_archive_2003_10_14/shroomery_grow/www.angelfire. com/co/mycosociety/cult3.html

Epstein, E. *The Science of Composting;* CRC Press LLC: 200 N.W. Corporate Blvd., Boca Raton, FL, 1996, Vol. 33431, pp 121–123.

Fulekar, M. H. *Environmental Biotechnology;* Oxford and IBH Publishing Co. Pvt. Ltd: New Delhi, 2005; pp 163–171.

Garruti, D. S.; Franco, M. R. R.; Da Silva, M. A. P.; Janzantti, N. S.; Alves, G. L. Assessment of Aroma Impact Compounds in a Cashew Apple Based Alcoholic Beverage by GC-MS and GC-olfactometry. *Lebensm. Wiss. Technol.* **2006,** *37* (4), 373–378.

Ghanekar, A. *A Handbook of Cultivated Mushrooms;* Shraddha Printers: Ghatkopar, India, 1999; pp 9–64.

Ghanekar, A. *More Facts about Alcoholic Drinks*; Shraddha Printers: Ghatkopar, Mumbai, 2000; pp 27.

Gunjate, R. T.; Patwardhan, M. V. Cashew. In *Handbook Fruit Science and Technology: Production, Composition, Storage and Processing*; Salunkhe, D. K., Kadam, S. S., Eds.; Marcel Dekker, Inc.: New York, 1995; pp: 509–521.

Kosaric, N.; Wieczorek, A.; Cosentino, G. P.; Magee, R. J.; Prenosil, J. E. Ethanol Fermentation. In *Biotechnology;* Rehm, H. J., Reed, G., Dellweg, H., Eds.; Verlag Chemie: Weinheim, Vol. 3, 1983; pp 146–147.

Looper, M. *Whole Animal Composting of Dairy Cattle;* Western Dairy News, Guide D-108, 2001.

Mandeel, Q. A.; Al-Lalith, A. A.; Mohamed, S. A. Cultivation of Oyster Mushrooms (*Pleurotus* species.) on Various Lignocellulosic Wastes, *World J. Microbiol. Biotechnol.* **2005,** *21,* 601–607.

Morel, J. L.; Colin, F.; Germon, J. C.; Godin, P.; Juste, C. Methods for the Evaluation of the Maturity of Municipal Refuse Compost. In *Composting of Agricultural and Other Wastes;* Gasser, J. K. R., Ed.; Elsevier Applied Science Publishers; London, 1985; pp 62.

Murugesan, A. G.; Rajakumari, C. *Environmental Science and Biotechnology Theory and Techniques;* MJP Publishers: Chennai, 2005; pp 243–287.

Ramteke, R. S.; Epieson, W. E.; Singh, N. S.; Chikkaramu, S.; Patwardhan, M. V. Studies on the Preparation of Aaroma Concentrate from Cashew Apple. *J. Food Sci. Technol.* **1984,** *21* (4): 248–249.

Ryan, J.; Estetan, G.; Rashid, A. *Soil and Plant Analysis Laboratory Manual;* International centre for agricultural research in dry areas (ICARDA) and National Agricultural Research Centre (NARC): Alleppo, Syria, 2001; pp 25–49.

Vankinh, L.; Van Do, V.; Phuong, D. D. Chemical Composition of Cashew Apple and Cashew Apple Waste Ensiled with Poultry Litter. *Livestock Res. Rural Dev.* **1997,** *9* (1), 1–3.

Vieira, F. R.; de Andrade, M. C. Optimization of Substrate Preparation for Oyster Mushroom (Pleurotus ostreatus) Cultivation by Studying Different Raw Materials and Substrate Preparation Conditions (composting: phases I and II). *Microbiol. Biotechnol.* **2016,** *32,* 190.

PART IV

Development, Optimization, Characterization, and Applications of Food Products

CHAPTER 21

QUALITY CHARACTERISTICS OF CUSTARD MADE FROM COMPOSITE FLOUR OF GERMINATED FINGER MILLET, RICE, AND SOYBEAN

CHARANJIT S. RIAR* and SURWASE S. BHASKARROA

Department of Food Engineering and Technology, SLIET, Longowal 148106, Sangrur, Punjab, India

Corresponding author. E-mail: charanjitriar@yahoo.com

CONTENTS

Abstract 524
21.1 Introduction 524
21.2 Materials and Methods 526
21.3 Results and Discussions 530
21.4 Summary and Conclusion 537
Keywords 538
References 538

ABSTRACT

Germination along with malting has the ability to improve the nutritional status of food grains, increases the activity of endogenous phytase and α-amylase, formation of polyphenol complexes with proteins, and the gradual degradation of oligosaccharides. Such reductions in phytase may facilitate iron absorption. There could be gastric simulation by enhancing flavor and mouthfeel characteristics as a result of germination followed by malting. Accordingly, this work was carried out to study effect of germination and malting on selected seeds, such as finger millet, rice, soybean composition, and other nutrients with the aim of suitable product development. Germination time (48–60 h) was selected on the basis of seeds sprouting percent (90–95%) and acrospires length (3–5 mm) which according to literature has maximum suitable effect on seed composition and properties. The composite flour was utilized for custard development (product with improved sensory profile). Results indicated that germination had a positive significant effect on nutritional status of raw material used and also helps in improving the sensory profile, such as color, flavor, and mouthfeel characteristics which results due to increase in sensory acceptability of product custard. The product prepared from composite of three flours containing finger millet, rice, and soya bean in the ratio of 20:60:20, respectively, had obtained high score on the basis of descriptive sensory analysis results, particularly for color and flavor characteristics. Having high sensory attributes and nutrients (protein, fiber, minerals, and functional ingredients) rich, easily digestible custard can be developed form combination of germinated legumes and cereals.

21.1   INTRODUCTION

In the germination process, starch, protein, and flatus factors are partially degraded, important for better digestibility. There is also an overall improvement in the flavor profile (Nirmala et al., 2000). Anti-nutrients are chemical substances in food that do not offer nourishment to the body, for example, phytic acid and tannins. The effect of these anti-nutrients in the body depends on the type and the concentration in which it is present in the food material. However, the presence of anti-nutritional factors (tannins and phytates) limits the utilization of the legumes as the main source of protein (Alonso et al., 1998). Germination is one of the most common techniques used to reduce most of the anti-nutritional factors in legumes (Abu-Samaha, 1983) also claimed to improve the nutritive quality of cereals by increasing the contents and availability of essential nutrients and lowering the levels of antinutrients (Osuntogun et al., 1989; Egli et al., 2002).

Millet is one of the oldest foods known to humans and possibly the first cereal grain used for domestic purposes. Finger millet (ragi, *Eleusine coracana*) is an important staple food in the eastern and central Africa as well as some parts of India (Majumder et al., 2006). Its use in making bread mentioned in the Bible. Eleusinian grain is most nutritious among the major cereal grains, its protein content is not only high but of exceptionally good quality. Furthermore, it was reported that it has good amounts of phosphorus, iron, thiamine, riboflavin, and nicotinic acid (Basahy, 1996). Due to the presence of antinutrients in grains, such as tannins and phytates, these micronutrients are less bio-accessible (Ramachandra et al., 1977; Udayasekhara et al., 1988). The malted and fermented ragi flour are extensively used in preparation of weaning food, instant mixes, beverages, and pharmaceutical products (Rao & Muralikrishna, 2001).

Rice has become an attractive ingredient in the extrusion industry due to its bland taste, attractive white color, hypo-allergenicity, and ease of digestion (Kadan et al., 2003). Lang et al. (2007) reported that rice grain contains anti-nutritional factors, which reduce the bioavailability of iron and zinc. Upon germination in brown rice, it frees bound minerals, making them absorbable in our bodies, and also increases taste and tenderness (Kayahara, 2004).

The protein concentration of the soybean is the largest of all legumes. The soybean contains sufficient amount of the indispensable amino acids to satisfy the healthy adult requirements (Ridner, 2006). Khan et al. (2010) indicated that germination of soybean seeds enhances nutritional quality and can be used for variety of medicinal uses.

Custard is a fine-textured food product made from corn starch in which salt; flavoring and coloring agents are added with or without the addition of egg yolk solids, vitamins, and minerals. Custard is primarily consumed either as a breakfast cereal-based food or weaning food in most developing nations of the tropics especially among children (Ihekoronye & Ngoddy, 1985). The fortification of custard with vegetable proteins from oilseeds and legumes has received considerable attention. This is because oilseed and legume proteins are high in lysine; an essential limiting amino acid in most cereals (Enwere, 1998). Jellema et al. (2005) relate the sensory sensation "creamy mouthfeel" in custards to rheological measurements. Sensory evaluation plays an important role in measuring characteristics and acceptability of food products. It can also provide the development technologist with useful information in order to help achieve and control quality, at a level which is particularly acceptable to the consumers (McIlveen & Armstrong, 1996).

Keeping in view the above facts, this study was planned in which the preparation of custard premix was carried out using the composite blend of germinated finger millet, rice, and soybean, followed by product development and its characterization.

21.2 MATERIALS AND METHODS

21.2.1 PROCUREMENT OF RAW MATERIALS

Finger millet (ragi, *E. Coracana*) was purchased from the local market of Parbhani, Maharashtra. Rice (*Oryza sativa*) and soybean *(Glycine max)* were procured from local market of Sangrur, Punjab. All the grains were freshly harvested of having good germination power. After purchasing, these grains were cleaned in unit operation laboratory of the Food Technology Department by using instruments like aspirator, grader, etc., and also by manual inspection.

21.2.2 GERMINATION

Germination of cereals and legumes was done by following different steps like washing, soaking, germination, drying, de-vegetation, grinding, and sieving. Germination of finger millet was carried out as per the method of Mwikya et al. (2000). Germination of paddy was carried out as per the method of Ayernor and Ocloo (2007), whereas soybean was germinated by the procedure as described by Yasmin et al. (2008). Soaked grains were spread on moist cloth to about 2–3 cm thick bed and then covered with another moist cloth. During germination (48–72 h), water was sprinkled occasionally to keep the sprouts moist (Fig. 21.1). The seeds spread for germination were turned and mixed upon once a day. After germination, the sprouts were dried in two steps. Initially, drying was done at 85°C for 15 min and final drying was carried out using cabinet drier at 60–70°C for 4–5 h. Dried sprouts were de-vegetated, that is, the roots and shoots separated from grains after drying by simply rubbing them by hand against gunny cloth and then separated by winnowing. The sprouted grains, after drying and de-vegetation called "green malt" were obtained. Some part of green malt was ground in a flour mill and sieved through 60 mm sieve to make it a homogeneous mixture, stored in airtight container for further analysis.

FIGURE 21.1 Germinated samples of (a) rice, (b) finger millet, and (c) soybean.

21.2.3 CHEMICAL ANALYSIS OF GERMINATED AND UN-GERMINATED FLOURS

Moisture, ash, crude protein, crude fat, and crude fibers were determined by AOAC (1995). The total starch was determined by the method of Chiang and Johnson (1977). Total carbohydrate was quantified by phenol–sulfuric acid method of Dubios et al. (1956) as modified by Wankhade and Tharanthan (1976). Amylose content was determined by the method of Scott et al. (1998). Water absorption and oil binding capacities were determined as per the method of Yamazaki (1953) as modified by Medcalf and Giles (1965). Reducing sugars of germinated and un-germinated flour samples were determined by 3,5-dinitrosalicylic acid (DNS) method of Thimmaiah (1999). Bulk density was determined according to the method of Goula et al. (2004). Color was determined by Lovibond Tintometer protocol having the scale configuration of (aY + 5bR) or (aY + 10bR). The Brookfield rotational viscometer (Model LVT2, Brookfield Engineering Lab, Stoughton, MA, USA) was used to measure the viscosity (consistency). Phytic acid was estimated by the method of Davies and Reid (1979). Total calorific value of the product was determined by the bomb calorimeter as per the instrument protocol. Soluble solid contents of prepared custard were determined as per the method of Gonzalez et al. (2009).

21.2.4   PREPARATION OF CUSTARD POWDER

Green malt was soaked at 25°C for 2–3 h to increase the enzymatic activity. Water was filtered off and the malt dried at 50°C in a cabinet drier followed by grinding in a lab grinder to obtain the flavored malt flour (custard powder).

21.2.5   PREPARATION OF CUSTARD DESSERT SAMPLES

Different combinations of tested malted grains flour used in the preparation of custard powder are shown in Table 21.1. Preparation of the model custard system was performed according to the recipe and protocol of COST Action 921 (Santonico et al., 2008) with minor modifications. Sugar (5%, w/w) was mixed with commercial milk (225 mL) at a temperature of 30°C. The mixture was then placed in a shaking water bath and heated from 30 to 95°C

TABLE 21.1   Formulations of Malted Flour in Custard Samples Preparation.

Samples	Material proportions
A	CS
B	(45:45:10)
C	(60:20:20)
D	(50:30:20)
E	(40:45:15)
F	(40:40:20)
G	(30:50:20)
H	(20:60:20)
I	(20:60:20)

CS: commercial custard; Samples: B–H (germinated flour); Sample: I (un-germinated) (Ratio; finger millet:rice:soybean).

at the rate of 5°C per minute. When the temperature of 60°C was achieved, the custard powder suspension (4%, w/v) prepared in 25 mL of milk separately was added. When the temperature of 95°C was reached, heating was continued at this temperature for further 15 min followed by cooling to 25°C in an ice-water bath with stirring for 15 min The resultant custard was stored at 4°C for 24 h for setting prior to analysis.

21.2.6 DESCRIPTIVE SENSORY EVALUATION

Sensory evaluation was conducted for the custard samples after one-day storage at 4°C temperature. A panel of 17 panelists comprising faculty and staff of Food Engineering and Technology, Department, which falls within the numbers as proposed by Meilgaard et al. (1999) was selected and trained following (ISO, 1993) procedure. Initially, panelists were trained in 2 h sessions prior to evaluation to familiarize with attributes and scaling procedures of custard samples under study. Panelists were screened for sensory evaluation ability (color, physical appearance, flavor, and mouthfeel) as well as their ability to communicate sensory descriptions of products. The selected subjects were trained in analyzing food with different sensory methods and trained in sensory vocabulary by testing samples, followed by creation mutual definitions of the attributes (ISO, 1985, 1991). During panel orientation and selection, six sensory attributes, including appearance (in terms of lumpiness, smoothness, and viscousness) color, flavors, mouth feel (thickness, melting, and creaminess), and after-feel (the oral sensation remaining after swallowing) were rated for all custards as most acceptable sensory attributes of custard. Attribute definition, technique, and scale were provided according to ISO (1985, 1991, 1994) (ANNEXURE I and II). All custard samples were coded with three-digit random numbers and presented to individuals in a tray in individual booths. Orders of serving were completely randomized. Water was provided between samples to cleanse the palate (Yazici & Akgun, 2004; Worrasinchai et al., 2006). The subjects were seated in sensory booths with appropriate ventilation and lighting. During 1–2 h, session subjects were presented with triplicates of each of the three stimuli. First, the custard was smelled and odor attributes were rated. Next, ingested custard was rated on the taste/flavor and mouthfeel attributes. Finally, the custard was swallowed and the two after-feel attributes were rated.

21.2.7 STATISTICAL ANALYSES

The statistical analyses were carried out by one-way analysis of variance (ANOVA) and means were compared using least square difference (LSD) test, with the help of Excel spreadsheets of MS Office 2007 software package and Genstat12 edition.

21.3 RESULTS AND DISCUSSIONS

On the basis of acrospires growth, that is, 3–5 mm in length and degree of germination (about 95%), the temperature and germination time for finger millet, rice, and soybean were 27°C for 72 h, 32°C for 48 h, and 28°C for 48 h, respectively.

21.3.1 ANALYSIS OF FLOURS

Nutritional compositions of ungerminated and germinated finger millet, rice, and soybean flours are shown in Table 21.2 in line. As observed, there was a significant ($p \leq 0.05$) decrease in moisture, total carbohydrates, starch, and amylopectin whereas ash, crude fiber, fat, protein, reducing sugars, and amylose increased significantly ($p \leq 0.05$) after germination in all the three food grain samples analyzed. Azizah and Zainon (1997) demonstrated that crude fiber was decreased in soaked peanut and mung bean, but conversely increased in soaked rice and soybean. In finger millet, phytic acid significantly reduced from 341.21 to 102.42 mg per 100 g of flour, while in rice and soybean phytic acid content was reduced from 981.17 to 213.34 mg and from 35.01 to 25.07 mg, respectively. Kumar et al. (2010) observed that malting of the grains significantly reduced the phytin phosphorus in finger millet. This reduction was accompanied by significant increase in ionizable iron and soluble zinc, indicating the improved availability of these two elements.

21.3.2 PHYSICOCHEMICAL CHARACTERISTICS OF GRAINS/ FLOUR

Physicochemical characteristics of grains are shown in Table 21.3. There was a significant ($p \leq 0.05$) decrease in the bulk density of grains after germination among all the tested samples. This might be due to decrease

TABLE 21.2 Nutritional Composition of Un-germinated and Germinated Flour of Finger Millet, Rice, and Soybean.[*]

Constituents (%) (wb)	UFM	GFM	UR	GR	US	GS
Moisture	12.53_d	11.09_c	8.91_b	7.84_a	8.73_b	8.54_b
Ash	1.70_a	1.95_a	2.45_b	2.95_c	5.01_d	5.14_d
Crude fiber	3.57_a	3.79_a	4.45_b	5.02_c	5.91_d	6.15_d
Fat	1.09_a	1.16_a	2.52_b	2.78_b	19.92_c	22.14_d
Protein	7.51_a	8.62_b	7.81_a	9.51_c	36.43_d	37.54_e
Total carbohydrate	75.67_d	72.16_c	80.35_e	75.15_d	32.51_b	29.42_a
Reducing sugar	1.56_b	8.67_e	0.35_a	4.85_d	0.06_a	2.45c
Starch	65.17_e	42.24_c	70.01_f	52.17_d	11.78_b	5.14_a
Amylose (g/100 g starch)	27.61_d	33.91_e	22.07_b	28.54_d	20.87_a	25.85_c
Amylopectin (g/100 g starch)	72.39b	66.09a	77.93_d	71.46b	79.13_e	74.15_c
Phytic acid (mg/100 g)	341.21_d	102.42_b	981.17_e	213.34_c	35.01a	25.07a

UFM: un-germinated finger millet; GFM: germinated finger millet; UR: un-germinated rice; GR: germinated rice; US: un-germinated soybean; GS: germinated soybean,

[*]Values in triplicate were taken. Values denoted by different small letters subscript row-wise differ significantly ($p < 0.05$).

in endosperm contents due to germination. There was no significant difference between oil binding capacity of germinated and un-germinated finger millet and soybean flour, however, the oil binding capacity of rice was significantly increased from 124.4 to 128.4% after germination. Akaerue and Onwuka (2010) reported that the 24 h sprouting treatment significantly ($p < 0.05$) increased the oil absorption capacity of the mung bean protein isolate whereas a slight reduction was observed in the 36 h sprouted protein isolate.

TABLE 21.3 Physicochemical Characteristics of Un-germinated and Germinated Finger Millet, Rice, and Soybean.[*]

Sample	Bulk density	Oil absorption capacity (%)	Water absorption capacity (%)	Color (% transmittance)			
				R-value	Y-value	B-value	Color reading
UFM	1.021$_d$	106.5$_a$	186.5$_b$	2.2	3.3	2.0	14.3$_e$
GFM	0.703$_b$	108.4$_b$	202.3$_c$	2.0	3.2	2.0	13.2$_d$
UR	0.994$_d$	124.4$_c$	270.5$_d$	1.0	2.0	0.9	7.0$_b$
GR	0.682$_b$	128.4$_e$	276.4$_d$	0.8	1.4	0.5	5.4$_a$
US	0.887$_c$	128.7$_e$	168.4$_a$	1.1	3.6	0.2	9.1$_c$
GS	0.506$_a$	127.2$_d$	167.5$_a$	1.2	3.6	0.2	9.6$_c$

[*]Values in triplicate were taken; UFM: un-germinated finger millet; GFM: germinated finger millet; UR: un-germinated rice; GR: germinated rice; US: un-germinated soybean; GS: germinated soybean. Values denoted by different small letters subscript column-wise differ significantly ($p < 0.05$).

Water binding capacities (WAC) of finger millet and rice flour were significantly ($p < 0.05$) increased from 186.5 to 202% and from 270.5 to 276%, respectively, after germination, while in soybean flour decrease was not significant, that is, from 168.37 to 167.53% after germination. This increase in WAC might have resulted from the hydration of seeds during soaking and sprouting which in turn unfolded the protein, thereby increased its hydrophilic binding sites and exposed them to the aqueous phase. Akaerue and Onwuka (2010) reported that sprouting treatments (24 h) were observed to have significant ($p < 0.05$) increase in the WAC of the protein isolates although there was no significant difference ($p > 0.05$) between the WAC of the 24 and 36 h sprouted protein isolates. Udensi and Okonkwo (2006) reported a higher value of WAC after 24 h germination in Mucuna bean protein isolates. There was no significant difference in color value between germinated and un-germinated flour of grains except for rice grain flour.

21.3.3   CHARACTERIZATION OF CUSTARD

Characteristics of custard samples are shown in Table 21.4. Total soluble solid for all custard samples were found to be in the range of 20–21°B which were significantly higher than the commercial custard sample (14°B). Color of custard changed due to change in malt proportion in the custard powder. The commercial custard was having highest color value. The samples A, D, and H had also shown the higher color values than the other samples. Sample C showed the lowest color values.

TABLE 21.4   Characteristics of Custard Samples Prepared from Commercial and Composite Flour of Germinated Finger Millet, Rice, and Soybean[*].

Samples	Apparent viscosity (μ_a) (mpa. s)	Flow behavior index (n)	Consistency coefficient (k) (mpa. Sn)	T. S. S. (°B)	R-value	Y-value	B-value	Color reading
A	1365_d	0.481_a	3.384_c	14_a	1.7	7.0	0.0	15.5_c
B	840_a	0.751_c	1.626_a	20_b	1.4	2.0	0.4	9.4_b
C	855_a	0.745_c	1.643_a	20_b	1.7	3.0	0.8	11.5_d
D	870_a	0.726_d	1.655_a	21_c	1.5	2.4	0.5	10.5_c
E	940_b	0.705_c	1.768_b	21_c	1.1	1.4	0.7	7.5_a
F	985_b	0.695_c	1.865_c	21_c	1.2	1.5	0.5	7.5_a
G	1025_c	0.665_b	1.914_d	21_c	1.1	2.0	0.5	7.5_a
H	1055_c	0.645_b	1.95_d	21_c	1.1	3.0	1.0	6.9_a
I	1425	0.461_a	3.242_c	13_a	1.7	5.8	0.3	14.8

[*]Values in triplicate were taken; values denoted by different small letters subscript column-wise differ significantly ($p < 0.05$).

Viscosity of all custard samples ranged between 840 and 1365 cP. Viscosity of custard prepared from blended germinated flour of finger millet, rice, and soybean seeds was lower than commercial sample. Mwikya et al. (2000) reported that viscosity of 10% dry matter slurry of the millet decreased within the first 48 h of sprouting. This was caused due to hydrolysis of starch into shorter chain polysaccharides by amylolytic enzymes, as shown by the decrease in starch content and increases in both reducing and non-reducing sugars contents in germinated flours. The viscosity of samples G and H was higher as compared to other samples. Viscosity of samples increased with increase in rice:malt proportion. Starch breakdown proceeds by the combined actions of α-amylase, de-branching enzymes (pullulanase like enzyme), β-amylase, and α-glucosidase in germinated cereal seeds (Zeeman et al., 2007). Sumathi et al. (1995) reported lower viscosities in malted legumes with corresponding increase in amylase activity.

Flow behavior index (*n*) of commercial sample was lower than other samples. Sample H was also having low flow behavior index, 0.645, while sample B had shown highest flow behavior index (0.751). In custard samples, shear stress increased with increase in rate of shear (Fig. 21.2), which indicated that the flow behavior of custard sample was dilatants in nature.

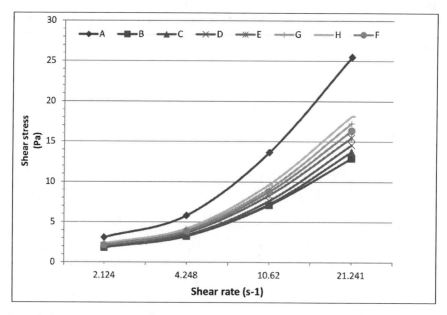

FIGURE 21.2   (See color insert.) Shear stress vs. shear rate relationship of different custard formulations as per Table 21.1.

21.3.4   DESCRIPTIVE SENSORY QUALITY EVALUATION

The descriptive sensory profile data of the custard samples prepared from different custard powder formulations are shown in Table 21.5. All the scores for the coded custard samples were collected and the samples were decoded and the data and comments for custard samples were summarized from sensory evaluation sheets for evaluation. The separate scores and comments obtained for each custard samples and the results thereof are presented as follows.

TABLE 21.5 Descriptive Sensory Score of Custard Samples Prepared from Composite Flour of Germinated Finger Millet, Rice, and Soybean.[*]

Samples	Sensory characteristics					
	Color	Appearance	Aroma	Mouthfeel	Taste	After-feel
A	8.7_a	8.5_a	8.8_a	8.7_a	8.7_a	8.5_a
B	7.1_e	6.7_d	6.8_e	7.5_d	6.6_d	5.9_d
C	6.7_f	6.6_d	7.0_d	6.9_e	6.5_b	6.8_c
D	7.1_e	7.0_c	7.2_d	7.5_d	6.9_e	7.2_b
E	8.0_b	6.9_d	6.5_f	7.9_b	6.6_d	6.8_c
F	7.5_d	8.0_b	8.1_c	7.9_b	7.3_c	8.5_a
G	7.8_c	8.3_a	8.5_b	7.3_c	8.0_b	8.4_a
H	8.5_a	8.4_a	8.5_b	8.7_a	8.7_a	8.5_a
I	8.0_b	7.0_d	6.5_f	6.9_b	6.6_d	6.8_c

[*]Values in triplicate were taken. Values denoted by different small letters subscript column-wise differ significantly ($p < 0.05$).

Sensory scores for the sample H was significantly similar to commercial custard samples which indicated that the sample H was liked most by the panelists. The samples F and G were also having the similar perceptions to that of commercial samples but not for all the sensory properties. This may be due to human perception of looking toward new product or it may be due to influence of color and taste and mouthfeel characteristics. The sensory scores obtained for other samples were, however; significantly lower than that of commercial sample because of increase in finger millet proportion as has been observed in sample C. Also, the decrease in soybean content had a significant negative effect on sensory perceptions of the panelists as has been observed in sample B.

The color value of custard prepared from commercial custard (8.7 values) was more appealing than other custard samples. This may be due to fact that the color of custard prepared from other custard powder samples was darker than the color of custard powder prepared from commercial custard powder except for custard sample H with 60% rice:malt powder having color score of 8.5. The score obtained for commercial custard for aroma was 8.8 while the other custards with 50 and 60% rice were also having comparable aroma scores. Mouthfeel of all the custard samples was described and scored in terms of thickness, melting, and creaminess by the panelists. Custard prepared from custard powder containing 60% rice obtained score of 8.7 which was similar to that of commercial custard sample having similar score of 8.7. The panelists commented that sample H showed higher thickness, showed meltiness, and creaminess when taken in mouth.

Taste of all the custard samples was acceptable by the panelists. The score obtained by commercial custard sample was 8.7. The lowest taste score was obtained by custard prepared from 20% rice malt powder. But the taste of custard prepared from 60% rice malt was highly acceptable by the panelists. Custard prepared from commercial custard powder and custard powder containing 50 and 60% rice malt powder was found to be having good after-feel. Overall acceptability of sample H was found to be highest and comparable to control sample, that is, sample A. Sample G had also shown comparable overall acceptability while sample B had lowest overall acceptability.

21.3.5 NUTRITIONAL ANALYSIS OF FINAL PRODUCT

Table 21.6 shows comparison between the nutritional compositions of finally selected custard prepared from germinated and un-germinated composite flour having proportionate composition as (finger millet (20%), rice (60%), and soybean (20%)). Moisture contents of custard prepared from un-germinated and germinated flour were 28.67 and 30.34%, respectively. Ash and fat contents were not significantly different whereas protein contents increased significantly. Carbohydrate content was significantly reduced from 40.01 to 35.57% in custard prepared from germinated composite flour. The phytic acid contents of custards prepared from un-germinated and germinated seed flour were reduced from 21.14 to 9.18 mg/100 g, respectively. This decrease in anti-nutritional factor was due to metabolic activities including increased enzymatic activity. These results are in agreement with the results obtained in sorghum by Wisal et al. (2005) and in case of soybean by Shipard (2005). Nithya et al. (2006) reported that various processing treatments, such as soaking, sprouting, and roasting could reduce the anti-nutritional factors in pearl millet. The calorific value of custard prepared from germinated flour was also significantly lower than that of the un-germinated flour.

TABLE 21.6 Nutritional Composition of High Sensory Score Custard Sample (Sample-H) of Germinated and that of Un-germinated Composite Flour (Sample-I) Having Similar Composition.[*]

Constituents (%) (wb)	Custard from un-germinated composite flour	Custard from germinated composite flour
Moisture	28.67$_a$	30.34$_b$
Ash	1.34$_a$	1.45$_a$
Fat	3.51$_a$	3.45$_a$
Proteins	7.45$_a$	8.96$_b$
Carbohydrates	40.01$_b$	35.57$_a$
Phytic acid (mg)	21.14$_b$	9.18$_a$
Calorific value (kcal)	221.43$_b$	183.57$_a$

[*]Values in triplicate were taken; values denoted by different small letters subscript row-wise differ significantly ($p < 0.05$).

21.4 SUMMARY AND CONCLUSION

After germination of seeds of finger millet, rice, and soybean, it was observed that germination improved the nutritional composition of seeds. Total carbohydrate and starch contents were significantly reduced during germination while reducing sugar was increased due to starch hydrolyzing enzymes. Bulk density of flour was found to be reduced during germination while color of germinated and un-germinated flour was not significantly different. This study shows that phytic acid was reduced significantly during germination. All custard samples prepared from custard powder containing malted flour of finger millet, rice, and soybean were acceptable as was shown by sensory evaluation. Custard prepared from custard powder (sample H) containing 20% finger millet, 60% rice, and 20% soybean showed higher firmness, consistency, and adhesiveness than custards prepared from custard powder containing low amount of rice content. Custard prepared from germinated flour was digestible because of having low calorific value and reduced phytic acid content.

Thus, it can be concluded that good quality custard can be prepared from non-conventional sources and can be used as a dessert, filling, or sauce can be added to cakes, puddings, pastries, or filled with fruit and served as fruit custard.

KEYWORDS

- **germination**
- **composite flour**
- **custard**
- **descriptive sensory analysis**
- **nutritional analysis**
- **calorific value**

REFERENCES

Abu-Samaha, O. R. Chemical Technological and Nutritional Studies on Lentil. M.Sc. Thesis, Alexandria University, Egypt, 1983.

Akaerue, B. I.; Onwuka, G. I. Evaluation of the Yield, Protein Content and Functional Properties of Mung Bean, Vigna Radiata (L.) Wilczek, Protein Isolates as Affected by Processing. *Pak. J. Nutr.* **2010,** *9,* 728–735.

Alonso, R.; Orue, E.; Marzo, F. Effect of Extrusion and Conventional Processing Methods on Protein and Antinutritional Factor Contents in Pea Seeds. *J. Food Chem.* **1998,** *63,* 505–512.

AOAC. *Official Methods of Analysis of Association of Official Analytical Chemists;* AOAC: Washington, DC, 1995.

Ayernor, G. S.; Ocloo, F. C. K. Physico–chemical Changes and Diastatic Activity Associated with Germinating Paddy Rice (PSB. Rc. 34). *Afr. J. Food Sci.* **2007,** *1,* 037–041.

Azizah, A. H.; Zainon, H. Effect of Processing on Dietary Fiber Contents of Selected Legumes and Cereals. *Malays. J. Nutr.* **1997,** *3* (2), 131–136.

Basahy, A. Y. Nutritional and Chemical Evaluation of Pearl Millet (*Pennisetum typhoides* (Burm f.) Stapf and Hubbard, Poaceae) Grown in the Gizan Area of Saudi Arabia. *Int. J. Food. Sci. Nutr.* **1996,** *47,* 165–169

Chiang, B. Y.; Johnson, J. A. Measurement of Total and Gelatinized Starch by Glucoamylase and O-toluidine. *Cereal Chem.* **1977,** *54,* 429–435.

Davies, N. T.; Reid, H. An Evaluation of Phytate, Zinc, Copper, Iron and Availability from Soy Based Textured Vegetable Protein Meat Substitutes or Meat Extruders. *Br. J. Nutr.* **1979,** *41,* 579.

Dubios, M.; Gilles, K. A.; Hamilton, J. K.; Rebers, P.A.; Smith, F. Colorimetric Method for Determination of Sugars and Related Substances. *Anal. Chem.* **1956,** *28,* 350–356.

Egli, I.; Davidsson, L.; Juillerat, M. A.; Barclay, D.; Hurrell, R. The Influence of Soaking and Germination on the Phytase Activity and Phytic Acid Content of Grains and Seeds Potentially Useful for Complementary Feeding. *J. Food Sci.* **2002,** *67,* 3484–3488.

Enwere, N. J. Foods Plants Origin. Afro-Obis Publications, Ltd.: Nsukka, 1998; pp 74–80.

González-Tomás L.; Bayarri S; Costell E.; Inulin-enriched Dairy Desserts: Physicochemical and Sensory Aspects. *J. Dairy Sci.* **2009,** *92,* 4188–4199.

Goula, A. M.; Adamopoulos, K. G.; Kazakis, N. A. Influence of Spray Drying Conditions on Tomato Powder Properties. *Drying Tech.* **2004,** *22* (5), 1129–1151.

Ihekoronye, A. I.; Ngoddy, P. O. *Integrated Food Science and Technology for the Tropics;* Macmillan Publishers Ltd.: London, Oxford, 1985; pp 262–276.

ISO, 6658. *Sensory Analysis–Methodology–General Guidance,* 1st ed.; International Organization for Standardization: Geneva, 1985; p 14.

ISO, 3972. Sensory Analysis. In *Methodology, Method of Investigating Taste,* 2nd ed.; International Organization for Standardization: Geneva, 1991; p 7.

ISO, 6586-I. *Sensory Analysis: General Guidance for the Selection, Training and Monitoring of Assessors—part I: Selected Assessors;* International Organization for Standardization: Geneva, 1993; p 253.

ISO, 8589. Sensory Analysis. In *General Guidance for the Design of Test Rooms,* 1st ed.; International Organization for Standardization: Geneva, 1994; p 9.

Jellema Renger, H.; Janssen Anke, M.; Terpstra Marjolein, E. J.; De Wijk Rene, A.; Smilde Age, K. Relating the Sensory Sensation "Creamy Mouthfeel" in Custards to Rheological Measurements. *J. Chemom.* **2005,** *19,* 191–200.

Kadan, R. S.; Bryant, R. J.; Pepperman, A. B. Functional Properties of Extruded Rice Flours. *J. Food Sci.* **2003,** *68,* 1669–1672.

Kayahara, H. In *Germinated Brown Rice,* Proceedings of the Department of Sciences of Functional Foods, Shinshu University, Japan, 2004.

Khan, H. N.; Gupta, R.; Farooqi, H.; Habib., A.; Akhtar, P. Biochemical Analysis of *Glycine max* Seeds Under Different Germinating Periods and Densitometric Analysis of Genistein. *J. Phytol.* **2010,** *2,* 83–86.

Kumar, V.; Sinha, A. K.; Makkar, H. P. S.; Becker, K. Dietary Roles of Phytate and Phytase in Human Nutrition: A Review. *Food Chem.* **2010,** *120,* 945–959.

Lang, N. T.; Nguyet, T. A.; Phang, N. V.; Buu, B. C. Breeding For Low Phytic Acid Mutants in Rice (*Oryza sativa* L.). *Omonrice* **2007,** *15,* 29–35.

Majumder, T. K.; Premavalli, K. S.; Bawa, A. S. Effect of Puffing on Calcium and Iron Contents of Ragi Varieties and their Utilization. *J. Food Sci. Tech.* **2006,** *42,* 542–545.

Mcilveen, H.; Armstrong, G. Sensory Analysis and the Food Industry: Can Computers Improve Credibility. *Nutr. Food Sci.* **1996,** *1,* 36–40.

Medcalf, D. G.; Giles, K. A.; Wheat Starches. I. Comparison of Physicochemical Properties. *Cereal Chem.* **1965,** *42,* 558–568.

Meilgaard, M.; Civille, G. V.; Carr, B. T.; Selection and Training of Panel Members. In *Sensory Evaluation Techniques,* 3rd ed.; CRC Press: Boca Raton, FL, 1999; pp 133–59.

Mwikya, S. M.; Camp, J. V.; Yiru, Y.; Huyghebaert, A. Nutrient and Anti-nutrient Changes in Finger Millet (*Eleusine coracan*) during Sprouting. *LWT Food Sci. Technol.* **2000,** *33,* 9–14.

Nirmala, M.; Subba Rao, M. V. S. S. T.; Murlikrishna, G. Carbohydrates and Their Degrading Enzymes from Native and Malted Finger Millet (Ragi, *Eleusine coracana,* Indaf-15). *Food Chem.* **2000,** *69,* 175–180.

Nithya K. S.; Ramachandramurty B.; Krishnamoorthy, V. V. Assessment of Anti Nutritional Factors, Minerals and Enzyme Activities of the Traditional (Co 7) and Hybrid (Cohcu-8) Pearl Millet (*Pennisetum glaucum*) as Influenced by Different Processing Methods. *J. Appl. Sci. Res.* **2006,** *2* (12), 1164–1168.

Osuntogun, B. A.; Adewusi, S. R. A.; Ogundiwin, J. O.; Nwasike, C. C. Effect of Cultivar, Steeping, and Malting on Tannin, Total Polyphenol, and Cyanide Content of Nigerian Sorghum. *Cereal Chem.* **1989,** *66,* 87–89.

Ramachandra G.; Virupaksha T. K.; Shadaksha R. M. Relationship between Tannin Levels and In Vitro Protein Digestibility in Finger Millet (*Eleusina coracana* Gaertn). *J. Agric. Food. Chem.* **1977,** *25,* 1101–1104.

Rao, S. M. V. S. S. T.; Muralikrishna, G. Non-Starch Polysaccharides and Boundphenolic Acids from Native and Malted Finger Millet (Ragi, *Eleusine coracana*, Indaf-15). *Food Chem.* **2001,** *72,* 187–192.

Ridner, E. Soja Propiedades Nutricionales y su Impacto en la Salud. In *Sojapropiedades Nutricionales y su Impacto en la Salud;* Grupo, Q. S. A., Ed.; Sociedad Argentina de Nutrición: Buenos Aires, Argentina, 2006.

Santonico, M.; Pittia, P.; Pennazza, G.; Martinelli, E.; Bernabei, M.; Paolesse, R.; D'amico, A.; Compagnone, D.; Di Natale, C. Study of the Aroma of Artificially Flavoured Custards by Chemical Sensor Array Fingerprinting. *Sens. Actuators B Chem.* **2008,** *133,* 345–351.

Scott, J. M.; Hugh, J. C.; Colin, J. R. A Simple and Rapid Colorimetric Method for the Determination of Amylose in Starch Products. *Starch* **1998,** *50,* 158–163.

Shipard, I.; *How Can I Grow and Use Sprouts as Living Food;*" Stewart Publishing: Toronto, Ontario, 2005.

Sumathi, A.; Malleshi, N. G.; Rao, S. V.; Elaboration of Amylase Activity and Changes in Paste Viscosity of Some Common Indian legumes during Germination. *Plant Foods Human Nutr.* **1995,** *47* (4), 341–347.

Thimmaiah, S. S. Standard Methods of Biochemical Analysis. Blackwell Publishing: New Delhi, India, 1999.

Udayasekhara, R. P.; Deosthale, Y. G. *In Vitro* Availability of Iron and Zinc in white and Coloured Ragi (*Eleusine coracana*): Role of Tannin and Phytate. *Plant Foods Hum. Nutr.* **1988,** *38,* 35–41.

Udensi, E. A.; Okonkwo, A. Effects of Fermentation and Germination on the Physiochemical Properties of *Mucuna cochinchinensis* Protein Isolate. *Afr. J. Biotech.* **2006,** *5,* 896–900.

Wankhade, D. B.; Tharanthan, R. N.; Sesame (*Sesamum indicum*). Carbohydrates. *J. Agri. Food Chem.* **1976,** *21,* 655–659.

Wisal, H.; Samina, M.; Hagir, B.; Elfadil, E.; Abdullahi, H. Effect of Germination, Fermentation and Cooking on Phytic Acid Tannin Contents and HCl-extraactability of Minerals of Sorghum (*sorghum bicholar*) Cultivars, 2005.Worrasinchai, S.; Suphantharika, M.; Pinjai, S.; Jamnong, P. Beta-Glucan Prepared from Spent Brewer's Yeast as a Fat Replacer in Mayonnaise. *Food Hydrocoll.* **2006,** *20,* 68–78.

Yamazaki, W. T. An Alkaline Water Retention Capacity Test for the Evaluation of Cookie Baking Potentialities of Soft Winter Wheat Flours. *Cereal Chem.* **1953,** *30,* 242–246.

Yasmin, A.; Zeb, A.; Khalil, A. W.; Paracha, G. M.; Khattak, A. B. Effect of Processing on Antinutritional Factors of Red Kidney Bean (*Phaseolus vulgaris*) Grains. *Food Bioprocess. Tech.* **2008,** *1* (4), 415–419.

Yazici, F.; Akgun, A. Effect of Some Protein Based Fat Replacers on Physical, Chemical, Textural, and Sensory Properties of Strained Yoghurt. *J. Food. Eng.* **2004,** *62,* 245–254.

CHAPTER 22

KINNOW PEEL–RICE-BASED EXPANDED SNACKS: INVESTIGATING EXTRUDATE CHARACTERISTICS AND OPTIMIZING PROCESS CONDITIONS

HIMANSHU PRABHAKAR*, SHRUTI SHARMA, SAVITA SHARMA, and BALJIT SINGH

Punjab Agricultural University, Ludhiana, Punjab, India

Corresponding author. E-mail: himanshup825@gmail.com

CONTENTS

Abstract 542
22.1 Introduction 542
22.2 Materials and Methods 543
22.3 Conclusion 557
Keywords 557
References 558

ABSTRACT

A study was carried out to develop extruded snacks from kinnow peel using twin-screw intermeshing extruder. Effects of independent variables, namely, moisture (14–18%), screw speed (400–550 rpm), and barrel temperature (120–180°C) on product responses (expansion ratio (ER), bulk density (BD), water absorption and solubility indices, hardness, and overall accept-ability) were evaluated using response surface methodology (RSM) along with central composite rotatable design (CCRD) producing 20 different combinations and for each combination, responses were investigated. An optimization of process variables was attempted for maximum desirability. Multiple regression equations were obtained to describe the effects of each variable on product responses. Results revealed that feed moisture was the most significantly affecting variable. Optimized conditions for preparation of peel-based snacks were 17.7% moisture, 550 rpm screw speed, and 128°C temperature. Sensory evaluation revealed that kinnow peel can be used up to the level of 1% for making extruded snacks.

22.1 INTRODUCTION

Food extrusion is a protean technology which combines a high-temperature short time (HTST) cooking process to form intermediate moisture finished product (Godavarti & Karwe, 1997). Extrusion cooking is at the apogee, claiming to be one of the most popular, and preferred technology in food industries. With changing eating habits and the anytime snacking trend, the global snack foods market is projected to reach about US $31 billion by 2019 (Reuters, 2014). The Indian snacks market is worth around US $3 billion and has an annual growth rate of 15–20% (MOFPI, 2005).

"Kinnow" a hybrid between king and willow mandarins (*Citrus nobilis* Lour × *Citrus deliciosa* Tenora) is one of the important citrus fruit crops in North Indian States (Sharma et al., 2007). India has acquired fourth rank in world (FAO, 2012) contributing 24% of the total citrus produce (Sharma et al., 2007). Among Indian states, Punjab has the highest area under kinnow production (46,000 hectares) and the state registered an output of 988,000 tons of kinnow in 2013–2014 (Gera, 2014). Kinnow peel, along with exhausted pulp contributes to 50–60% of whole kinnow fruit (Brar & Dhillon, 2013). During kinnow processing, the peel is ditched in munic-ipal bins and allowed to rot as there are no recommended provisions for how to handle this so-called unproductive matter. This is where extrusion technology comes in, proffering solution for one of the most conventional complications prevailing in the food industry.

For any extruded product, physical characteristics, such as expansion, density, and hardness are important parameters for the evaluation of consumer acceptability of the final product (Patil et al., 2007). The effects of ingredient properties and processing conditions (such as feed moisture, screw speed, and barrel temperature) on final product quality are also reflected by their influence on process responses or extruder system. The objective of this study was to investigate the process compatibility, optimization of the extrusion conditions, and the effect of feed moisture content, screw speed and barrel temperature on the extrusion behavior and physical properties of kinnow peel-rice blend based snack. The scope of the study was extended to examine the sensory properties of the selected extrudates in terms of overall acceptability.

22.2 MATERIALS AND METHODS

22.2.1 MATERIALS

Kinnow fruit was procured from local market, Ludhiana, dehydrated at 55°C ± 5°C using cabinet drier for 48 h reduced to fine powder and passed through 200 μm sieves for further use in the study. Rice flour, salt, and spices were also procured from local market.

Analyses of proximate composition (namely, moisture content, ash, protein, crude fiber, and fat) of the raw materials and final product were performed according to the approved methods described in AOAC (2000).

22.2.2 FEED PREPARATION AND OPTIMIZING LEVEL OF TREATMENT

Preliminary trials were conducted to decide the levels of dried peel in rice flour, running the experiment with treatment ranging from 1% to 5%. A semi-trained panel of 60 individuals, including students and faculty from the Department of Food Science and Technology, Punjab Agricultural University (India) evaluated the extruded snacks for appearance, color, texture, and overall acceptability on a 9-point hedonic scale from 1 = dislike extremely to 9 = like extremely (Singh et al., 2013). Product having 1% dried peel was chosen and its proximate composition, physical, and functional properties were studied. Experiments were conducted to determine proximate composition of dried peel as well (Table 22.1).

TABLE 22.1 Proximate Chemical Composition of Dried Kinnow Peel ($n = 3$, where n Represents Number of Replicates).

Raw materials	Moisture (%)	Crude protein (%)	Fat (%)	Ash (%)	Crude fiber (%)
Dried peel	3.45 ± 0.12	3.57 ± 0.19	2.09 ± 0.04	3.59 ± 0.01	8.05 ± 0.36

22.2.3 EXTRUDER PROCESS

Co-rotating and intermeshing twin-screw extruder Model BC 21 (Clextral, Firminy, France) were used in the study. The barrel diameter and its length to diameter ratio (L/D) were 2.5 mm and 16:1, respectively. Out of four zones, temperatures of first three barrel zones were maintained at 40, 70, and 100°C, respectively, while the temperature of last zone (compression and die section) was varied according to the experimental design, throughout the study. The diameter of die was 2 mm. The extruder was equipped with a torque indicator showing percentage of torque in proportion to the current drawn by the drive motor. Raw material was metered into the extruder with a single screw volumetric feeder (D. S and M. Modena, Italy). The extruder was thoroughly calibrated with respect to the combinations of feed rate and screw speed to be used. The feed rate was adjusted for optimum filling and the moisture content of feed was varied by injecting water (approximately 50°C) into extruder with water pump while conducting preliminary trials. A variable speed die face cutter with four-bladed knives was used to cut the extrudates (Singh et al., 2013).

22.2.4 PRODUCT RESPONSES

22.2.4.1 SPECIFIC MECHANICAL ENERGY (SME)

Specific mechanical energy (SME) (Wh/kg) was calculated from rated screw speed (682 rpm), motor power rating (8.5 kW), actual screw speed, percent motor torque, and mass flow rate (kg/h) using the following formula (Pansawat et al., 2008).

$$\text{SME}\left(\frac{\text{Wh}}{\text{kg}}\right) = \frac{\text{Actual screw speed}(\text{rpm}) \times \text{Percentmotor torque} \times \text{Motor power rating (Wh)}}{\text{Mass flow rate}\left(\frac{\text{kg}}{\text{h}}\right)}.$$

22.2.4.2 EXPANSION RATIO

The ratio of the diameter of the extrudate and the diameter of the die was used to express the expansion ratio (ER) of the extrudate (Fan et al., 1996). The diameter of the extrudate was determined as the mean of random measurements made with a vernier caliper. The extrudate ER was calculated as

$$\text{ER} = \frac{\text{Diameter of extrudate (mm)}}{\text{Diameter of die (mm)}}.$$

22.2.4.3 BULK DENSITY

The density (g/cc) of the extruded snacks was measured by using a 100 mL graduated cylinder using rapeseed displacement method. The volume of 20 g randomized samples was measured for each test. The ratio of sample weight and the replaced volume in the cylinder was calculated as density (Pan et al., 1998).

$$\text{BD} = \frac{\text{Weight of extrudates(g)}}{\text{Volume of extrudates(cc)}}.$$

22.2.4.4 WATER ABSORPTION INDEX (WAI)

Water absorption index (WAI) of the snacks was determined by method outlined by Anderson et al. (1969). The WAI measures the volume occupied by the granule or starch polymer after swelling in excess of water, that is, it pertains to degree or extent of gelatinization. The ground extrudates were suspended in distilled water at room temperature (34°C) for 30 min, gently stirred during this period, and then centrifuged at 3000 rpm for 10 min. The supernatant liquid was poured carefully into tared evaporating dish. The remaining gel was weighed and WAI was calculated as the grams of gel obtained per gram of solid.

$$\text{WAI} = \frac{\text{Weight of sediments(g)}}{\text{Weight of dry solids(g)}}.$$

22.2.4.5   WATER SOLUBILITY INDEX (WSI)

WSI determines the number of free polysaccharides or polysaccharides released from the granule on addition of excess water. This property is in direct relation to the digestibility of food product. Low molecular weight polysaccharides formed during the process are assimilated into the body promptly. The WSI is the weight of dry solids in the supernatant from the WAI test described above (Anderson et al., 1969) expressed as a percentage of the original weight of the sample.

$$\text{WSI}(\%) = \frac{\text{Weight of dissolved solid in supernatant}}{\text{Weight of dry solids}} \times 100.$$

22.2.4.6   HARDNESS

Textural quality of the snack samples was examined by using a TA-XT2i Texture Analyzer (Stable Microsystems, Surrey, UK). The compression probe (50 mm dia., aluminum cylinder) was used to measure the compression force required for sample breakage which indicates hardness. Testing conditions were 1.0 mm/s pre-test speed, 2.0 mm/s test speed, 10.0 mm/s post-test speed, and 5 mm distance (Bourne, 1978) and readings were taken in triplicates for each sample (CCRD predicted 20 experimental run's sample, and all of the samples were analyzed for abovementioned product responses).

22.2.5   EXPERIMENTAL DESIGN AND DATA ANALYSIS

The CCRD for the three independent variables, that is, moisture (x_1), screw speed (x_2), and barrel temperature (x_3) was performed. The independent variables and variation levels are shown in Table 22.2. The levels of each variable were established on the preliminary trials. The outline of experimental design with the actual level is presented in Table 22.2. The dependent variables were SME, ER, bulk density (BD), WAI, WSI, hardness, and overall acceptability. Response surface methodology (RSM) was used to investigate the effect of extrusion conditions on the product responses. The independent variable levels like feed moisture (12–18%), screw speed (400–550 rpm), and barrel temperature of the last zone (120–180°C) considered for study were selected on the basis of preliminary trials. A rotatable, central composite design (Myers and Montgomery, 2002)) was employed

TABLE 22.2 Effect of Extrusion Condition on Process and Product Responses of Peel Based Rice Extrudates ($n = 3$).

S. No	Extrusion conditions									Responses	
	A: moisture (%)	B:screw speed (rpm)	C: tempera-ture (°C)	SME (Wh/ Kg)	Expansion ratio	Density (g/cm3)	WAI (g/g)	WSI (%)	Hardness (N)	Overall acceptability
1	14 (−1)	400 (−1)	130 (−1)	207.38	3.78	0.270	1.40	70.51	26.04	7.52
2	14 (−1)	400 (−1)	170 (+1)	218.56	3.36	0.23	5.11	49.32	15.74	6.63
3	14 (−1)	550 (+1)	130 (−1)	197.75	3.66	0.26	1.12	72.34	12.38	7.81
4	14 (−1)	550 (+1)	170 (+1)	222.13	3.36	0.21	4.85	34.86	8.47	7.05
5	18 (+1)	400 (−1)	130 (−1)	192.63	3.56	0.21	2.24	58.27	13.85	6.63
6	18 (+1)	400 (−1)	170 (+1)	179.51	3.07	0.19	3.39	49.51	13.97	7.51
7	18 (+1)	550 (+1)	130 (−1)	182.69	3.55	0.21	2.88	65.02	15.67	6.56
8	18 (+1)	550 (+1)	170 (+1)	179.25	3.01	0.17	4.16	32.12	17.26	7.41
9	12.6 (−1.682)	475 (0)	150 (0)	226.25	3.97	0.23	1.10	69.65	10.78	6.43
10	19.4 (+1.682)	475 (0)	150 (0)	230.88	2.89	0.18	5.19	37.61	3.89	6.55
11	16 (0)	349 (−1.682)	150 (0)	186.25	3.71	0.25	2.72	61.83	20.57	7.77
12	16 (0)	601 (+1.682)	150 (0)	181.63	3.62	0.19	3.27	51.92	14.06	7.96
13	16 (0)	475 (0)	161 (−1.682)	205.56	3.61	0.27	2.99	61.44	22.71	7.34
14	16 (0)	475 (0)	203 (+1.682)	161.06	3.13	0.18	3.16	53.93	22.04	7.31
15	16 (0)	475 (0)	150 (0)	209.31	2.97	0.20	1.81	59.61	19.29	7.62
16	16 (0)	475 (0)	150 (0)	210.57	2.98	0.21	1.85	67.92	18.02	7.85
17	16 (0)	475 (0)	150 (0)	209.39	2.98	0.21	1.85	59.69	18.83	7.72
18	16 (0)	475 (0)	150 (0)	209.55	2.97	0.21	1.88	60.33	18.90	7.76
19	16 (0)	475 (0)	150 (0)	210.88	3.11	0.20	1.87	64.45	18.95	7.71
20	16 (0)	475(0)	150 (0)	210.37	3.03	0.22	1.88	66.79	18.95	7.80

to determine the extrusion conditions. The design required 20 experimental runs with eight factorial points, six star corner points, and six center points. Experiments were randomized in order to minimize the systematic bias in observed responses due to extraneous factors. The individual effect of each variable and the effect of interaction in coded levels of variables were determined (Singh et al., 2013).

22.2.6 STATISTICAL ANALYSIS AND OPTIMIZATION

Responses obtained as a result of the proposed experimental design were subjected to regression analysis in order to assess the effects of feed moisture, screw speed, and barrel temperature. Second-order polynomial regression models were established for the dependent variables to fit experimental data for each response using statistical software Design-Expert 9.0.0 (Stat-Ease Inc., Minneapolis, MN).

$$y_i = b_0 + \sum_{i=1}^{a} b_i x_i + \sum_{i=1}^{a} b_{ii} x_i^2 + \sum_{i=1}^{a} \sum_{i=1}^{a} b_{ij} x_i x_j,$$

where x_i (i= 1, 2, 3) are independent variables (moisture, screw speed, and barrel temperature), respectively, and b_0, b_i, b_{ii}, and b_{ij} are coefficient for intercept, linear, quadratic, and interactive effects, respectively. Data were analyzed by multiple regression analysis and statistical significance of the terms was examined by analysis of variance (ANOVA) for each response. The adequacy of regression model was checked by correlation coefficients. The lack-of fit test was used to judge the adequacy of model fit. To aid visualization of variation in responses with respect to processing variables, series of three-dimensional response surfaces plots were drawn. The range of values for SME, ER, BD, WAI, WSI, hardness, and overall acceptability among seven selected samples were used as criteria for numerical optimization. Please compare the SME value with other rice flour based snacks reported. Why do we assess the SME value and where it is useful?

22.2.7 RESULT AND DISCUSSIONS

22.2.7.1 SPECIFIC MECHANICAL ENERGY

The predicted model for SME can be described by the following equation in terms of coded values:

$$\text{SME} = \begin{array}{l} +210.03+1.96x_1 -1.76x_2 -13.66x_3 +6.47x_1^2 -9.31x_2^2 -9.53x_3^2 + \\ 2.86x_1x_2 -0.52x_2x_3 -6.51x_1x_3. \end{array} \quad (22.1)$$

SME is the amount of mechanical energy (work) dissipated as heat inside the material, expressed per unit mass of the material (Godavarti & Karwe, 1997). The measured SME in extrusion cooking of rice flour and kinnow peel blends ranged from 161.06 to 230.88 Wh/kg. An ANOVA was conducted to assess the significant effects of the independent variables on responses and that which of the responses were significantly affected by the varying processing conditions. The coefficient of variation (CV), which indicates the relative dispersion of the experimental points from the predictions of the model, was found to be 2.66% for SME (Table 22.3). A reasonably good coefficient of determination (R^2) and adjusted R^2 value of 0.9983 and 0.9967, respectively, were obtained indicating that the model developed for SME appeared to be appropriate. F value for ER was significant ($p < 0.001$) whereas lack-of fit was not significant ($p > 0.0849$).

TABLE 22.3 ANOVA for the Fit of Experimental Data to RSM.

Regression	Sum of squares						
	SME	Expansion	Bulk density	WAI	WSI	Hardness	Overall acceptability
Adequate precision	96.633	13.933	21.900	135.926	15.997	43.760	23.547
R square	0.9983	0.9603	0.8828	0.9994	0.9551	0.9906	0.9686
Adjusted R square	0.9967	0.7396	0.8608	0.9989	0.9146	0.9822	0.9835
C V (%)	0.50	2.80	5.05	1.57	5.98	4.16	1.24
Lack-of fit	0.0849n.s	0.0559n.s	0.1534n.s	0.0702n.s	0.6614n.s	0.0691n.s	0.3068n.s

n.s, non-significant.

Moisture (x_1) had highly significant positive linear effect while screw speed (x_2), and temperature (x_3) had a highly significant negative linear effect on SME. Increasing feed moisture led to increase in SME. Similar results have been reported by Garber et al. (1998) in corn meal, Kannadhason et al. (2009) and Chevanan et al. (2008) in distiller's dried grains. Increase in moisture content steered the temperature toward declination, resulting in an elevation in SME. However, with increase in screw speed, the SME was observed to decrease. Similar results were observed by Chakraborty and Banerjee (2009) in green gram-rice flour blend and Ojo et al. (2014)

in floating fish extrudate. According to Ojo et al. (2014), increasing screw speed will result in decrease SME only up to a certain point. The relationship between the response and variable beyond this point was discovered to be linear, salt also played crucial role in this occurrence. Hsieh et al. (1993) and Pitts et al. (2014) recorded similar observation that using salt in rice flour led to decrease in SME as screw speed increased. With addition of salt, starch–starch interaction decreased, resulting in increase in melt viscosity, thus decreasing overall SME. A number of authors have reported that with decrease in temperature, SME increases including Dogan and Karwe (2003) in quinoa extrudates, Altan et al. (2008a) and Koksel et al. (2004) in barley based extrudates, Ryu and Ng (2001) in wheat based extrudates and Pathania et al. (2013) in wheat, mung-bean, and rice blend. Increase in temperature allayed the viscosity which ultimately resulted in reduced SME (Chang et al., 1998).

22.2.7.2   EXPANSION RATIO (ER)

The predicted model for ER can be described by the following equation in terms of coded values:

$$\text{ER} = \frac{+3.01 be\,dex_1 - 0.025x_2 - 0.13x_3 + 0.13x_1^2 + 0.21x_2^2 + 0.11x_3^2 +}{8.75x_1x_2 - 6.25x_2x_3 + 0.039x_1x_3.} \tag{22.2}$$

The extruded product usually puffs and changes texture when exiting the die because of the reduction of forces and release of moisture and heat. The measured ER in extrusion cooking of rice flour and kinnow peel blends ranged from 2.89 to 3.97. ANOVA interpreted a highly significant model ($p < 0.0001$) with coefficient of determination (R^2) of 0.9603 and adjusted R^2 value of 0.9245 with CV of 2.80%. The ANOVA for ER of quadratic model (eq 22.2) is given in Table 22.2. F value for ER was significant ($p < 0.001$) whereas lack-of fit was not significant ($p > 0.0589$).

Moisture (x_1), screw speed (x_2), and temperature (x_3) had significant negative linear effect (Fig. 22.1). ER decreased with increase in moisture and similar results have been reported by Kumar et al. (2010) and Ding et al. (2006) in rice based extrudates and Pathania et al. (2013) in wheat, mung bean, and rice blend. With increase in moisture, dough elasticity, and temperature decreased due to reduced friction between dough, screw, and barrel which negatively impacted gelatinization resulting in diminished ER. Similarly, with increase in screw speed, ER decreased. Similar results were observed by Chakraborty and Banerjee (2009) in green gram-rice blend and Ojo et al. (2014) in fish based extrudates. It was observed that ER decreased with increase in temperature and similar results have been reported by Mendoca et al. (2000) in corn bran-corn meal blend, Altan et al. (2008b) in barley-tomato based extrudates and Yagci and Gogus (2008) in rice grit-wheat flour based extrudates. This might have occurred because of defragmentation of starch at higher temperature, that is, temperature beyond gelatinization (Chakraborty & Banerjee, 2009).

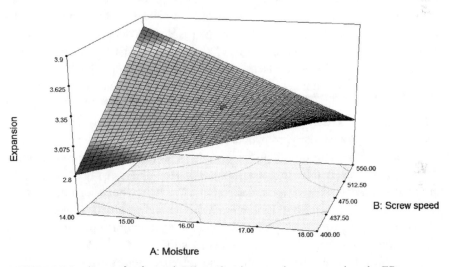

FIGURE 22.1 **(See color insert.)** Effect of moisture and screw speed on the ER.

22.2.7.3 BULK DENSITY (BD)

The predicted model for BD can be described by the following equation in terms of coded values:

$$BD = +0.21be\ desx_1 - 0.011x_2 - 0.025x_3. \tag{22.3}$$

BD is a measure of expansion of extrudate in all directions, unlike ER which considers only the direction perpendicular to extrudate flow. The measured BD in extrusion cooking of rice flour and kinnow peel blends ranged from 0.17 to 0.27 g/cc. ANOVA interpreted a highly significant model ($p < 0.0001$) with coefficient of determination (R^2) of 0.8828 and adjusted R^2 value of 0.8608 with 5.05% CV. F value for BD was significant ($p < 0.001$), whereas lack-of fit was not significant ($p > 0.1534$).

Moisture (x_1), screw speed (x_2), and temperature (x_3) had highly significant negative linear effect. BD decreased with increase in moisture. Similar results have been reported by Asare et al. (2004) in rice-cowpea-groundnut blend, Baik et al. (2004) in rice-chickpea blend and Ding et al. (2006) in rice based extrudates. Increase in moisture content mitigated the viscosity of the product resulting in reduced SME but presence of salt and spices in the formulation mix might have absorbed surplus moisture resulting in increased flow and reduced BD. It has been observed that at higher temperature, increase in moisture content directed declination of BD as also reported by Barrett and Peleg (1992) and Pan et al. (1998). Corresponding to the previous outcome, increasing screw speed resulted in decreased BD. Similar results were observed by Chevanan et al. (2008) in distiller's dried grains, Kumar et al. (2010) in rice extrudates and Hagenimana et al. (2006) in rice flour. BD tended to decrease at greater screw speed because of development of high pressure due to revolving screw inside the barrel. As the product started exiting the die, pressure gradient gave rise to puffed and expanded product with low density. BD decreased with increase in temperature as observed by other authors including Kumar et al. (2010) in rice based extrudates, Altan et al. (2008a) in grape pomace-barley blend, Altan et al. (2008b) in tomato pomace-barley blend, Case et al. (1992) in wheat-corn flour based extrudates, and Mercier and Feillet (1975) in cereal products. Kokesel et al. (2004) explained this occurrence as increase in temperature gave rise to a higher potential energy for the flash-off of super-heated water from extrudates as they left the die. With increase in barrel temperature, extrudates exiting die lost more moisture, and became lightweight puffed product.

22.2.7.4 WATER ABSORPTION INDEX (WAI)

The predicted model for WAI can be described by the following equation in terms of coded values:

$$\text{WAI} = \frac{+1.86 + 1.23x_1 + 0.035x_2 - 13.66x_3 + 0.46x_1^2 + 0.40x_2^2 + 0.43x_3^2 +}{0.019x_1x_2 + 0.24x_2x_3 - 0.63x_1x_3.} \tag{22.4}$$

WAI is basically a measure of capacity of starch granules to hold an excess of water forming a gel thus ensuring degree of gelatinization. The measured WAI in extrusion cooking of rice flour and kinnow peel blends ranged from 1.10 to 5.19 g/g. The ANOVA for WAI of quadratic model (eq 22.4) is given in Table 22.3. Regression model fitted to experimental results had very high coefficient of determination (R^2) and adjusted R^2 are 0.9994 and 0.9989, respectively, with CV of 1.57%. F value for WAI was significant ($p < 0.001$), whereas lack-of fit was not significant ($p > 0.0702$).

Moisture (x_1), screw speed (x_2), and temperature (x_3) had a highly significant positive linear effect on WAI. During extrusion process, it was observed that WAI increased with increase in moisture. Similar results have been reported by Hagenimana et al. (2006) in rice flour, Chang et al. (1998) in jatoba flour-cassava starch blend; Baladran-Quintana et al. (1998) in white pinto bean meal based extrudates, Singh et al. (2013) in potato based snacks, Yagci and Gogus (2008) in rice grit-wheat flour based extrudates and Dogan and Karwe (2003) in quinoa extract. As temperature increased, protein denatured, starch got gelatinized, and swelling of fiber eventuated which, in moderate extrusion conditions, disrupted the molecular structure creating pores leading to penetration of water and thus swelling (Ching et al., 1998). Increasing screw speed resulted in increase in WAI. Similar results were observed by Pathania et al. (2013) in wheat, mung bean, and rice blend. Altan et al. (2008a) reported that with increase in fiber content, increase in screw speed lead to increase in WAI as excess of water was taken up by the fiber. Also, increase in screw speed might have assisted the structural modification of kinnow peel and spice mix present in formulation mix resulting in increased WAI. WAI increased with increasing temperature. Similar results were observed by Hagenimana et al. (2006) and Kumar et al. (2010) in rice based extrudate, Dogan and Karwe (2003) in quinoa extrudate, Pathania et al. (2013) in wheat, mung-bean and rice blend, and Singh et al. (2013) in potato based extrudates. Lee et al. (1999) stated that there was a rapid increase of WAI at lower temperatures because of availability of more undamaged polymer chains as well as hydrophilic groups for binding more water, which resulted in higher values of WAI.

22.2.7.5 WATER SOLUBILITY INDEX (WSI)

The predicted model for WSI can be described by the following equation in terms of coded values:

$$\text{WSI} = \begin{array}{l} +63.18e\,descx_1 - 2.92x_2 - 2.54x_3 - 3.71x_1^2 - 2.56x_2^2 - 2.27x_3^2 \\ -5.05x_1x_2 + 0.25x_3 + 2.13x_1x_3 \end{array} \quad (22.5)$$

WSI determines the amount of free polysaccharide or polysaccharide release from the granule on addition of excess water (Pathania et al., 2013). The measured expansion in extrusion cooking of rice flour and kinnow peel blends ranged from 32.12 to 72.34%. The ANOVA for WSI of quadratic model (eq 22.5) is given in Table 22.3. The coefficient of determination (R^2) and adjusted R^2 for the WSI were 0.9551 and 0.9146, respectively, with CV of 5.98%. F value for ER was significant ($p < 0.001$), whereas lack-of fit was not significant ($p > 0.6614$).

Moisture (x_1), screw speed (x_2), and temperature (x_3) had significant negative linear effect (Fig. 22.2). WSI decreased with increase in moisture. Similar results have been reported by Kumar et al. (2010) and Ding et al. (2006) in rice based extrudates, Pathania et al. (2013) in wheat, mung-bean and rice blend, Onyango et al. (2005) in maize–finger millet blend and Yogci and Gogus (2008) in rice grit-wheat flour based extrudates. Kumar et al. (2010) reported that WSI increased at elevated moisture levels; however, further increase attenuated this response in carrot pomace pulse powder blended with rice flour. Increase in moisture content led to plasticization of extruded melt thus reducing the extent of gelatinization and minor degradation of starch. Increasing screw speed resulted in decrease in WSI. As screw speed elevated, SME got restricted due to presence of salt and spices in formulation, reducing inter-polymer interaction resulting in reduced mechanical shear, retarding synthesis of water-soluble compounds. Correspondingly, WSI decreased with increase in temperature. Similar results were observed by Altan et al. (2008b) in tomato pomace based extruded snacks and Gutkoski and El-Dash (1999) in extruded oats snacks.

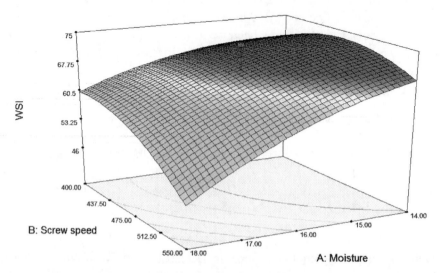

FIGURE 22.2   (See color insert.) Effect of moisture and screw speed on the WSI.

22.2.7.6 HARDNESS

The predicted model for hardness can be described by the following equation in terms of coded values:

$$\text{Hardness} = \begin{array}{l} +18.83ss = x_1 - 1.96x_2 - 0.22x_3 - 4.07x_1^2 - 0.55x_2^2 + 1.24x_3^2 + \\ 0.98x_1x_2 3.26x_2x_3 + 1.99x_1x_3. \end{array} \qquad (22.6)$$

Hardness generally refers to the force required to break the extrudate sample (Altan et al., 2008a, 2008b). The measured hardness in extrusion cooking of rice flour and kinnow pomace blends ranged from 3.89 to 26.04 N. The ANOVA for hardness of quadratic model (eq 22.6) is given in Table 22.3. The values for coefficient of determination (R^2) and adjusted R^2 values obtained were 0.9906 and 0.9822, respectively, with 4.16% of coefficient of variance. F value for SME was significant ($p < 0.001$) whereas lack-of fit was not significant ($p > 0.0691$).

   Moisture (x_1), screw speed (x_2), and temperature (x_3) had a highly significant negative linear effect on hardness. Hardness decreased with increase in feed moisture whereas increasing screw speed resulted in decrease in hardness. Similar results were observed by Altan et al. (2008b) in barley flour and tomato pomace blend, Liu et al. (2000) in oat-corn flour, Wu et al. (2007) in flaxseed-corn meal blend and Ding et al. (2006) in wheat extrudates. High screw speed increased the barrel temperature resulting in higher expansion and decreased hardness. Hardness decreased with increase in temperature. Similar results were observed by Altan et al. (2008b) in barley flour and tomato pomace blend, Sebio and Chang (2000) in yam flour extrudate, Dogan and Karwe (2003) in quinoa extract, Yuliana et al. (2006) in starch extrudates, Yao et al. (2006) in oat flour, Mendonca et al. (2000) in corn meal extrudates, and Keawpeng et al. (2014) in rice extrudates. Increase in temperature resulted in reduced melt viscosity and elevated vapor pressure resulting in bubble growth which acted as a driving force for expansion. As the product expanded, wall thickness decreased resulting in reduced hardness (Ding et al., 2005; Yuliani et al., 2000; Altan et al., 2008b).

22.2.7.7 OVERALL ACCEPTABILITY

The predicted model for overall acceptability can be described by the following equation in terms of coded values:

Overall acceptability $=$

$$+7.75+0.021x_1+0.063x_2-0.070x_3-0.45x_1^2+0.031x_2^2-0.16x_3^2+ \qquad (22.7)$$
$$0.013x_1x_2-0.16x_2x_3+0.42x_1x_3.$$

Overall acceptability score was the mean of different sensory attributes (color, texture, flavor, and taste). The measured overall acceptability of peel based extrudates ranged from 6.43 to 7.96. The ANOVA for overall acceptability of quadratic model (eq 22.7) is given in Table 22.3. The values for coefficient of determination (R^2) and adjusted R^2 values obtained were 0.9835 and 0.9686, respectively, with 1.24% of coefficient of variance. F value for overall acceptability was significant ($p < 0.001$) whereas lack-of fit was not significant ($p > 0.3068$). With increase in moisture and screw speed, sensory scores were recorded to increase indicating a positive effect of these variables on the extrudates. Contrary to that, it was observed that overall acceptability decreased with increase in temperature because of overcooking of extrudates resulting in destruction of certain flavor imparting compounds.

22.2.7.8 OPTIMIZATION

The optimum values for moisture content, temperature and screw speed were observed to be 17.7%, 128°C, and 550 rpm, respectively, with a desirability score of 0.581 out of 1 (Fig. 22.3). The variation between predicted response values and the actual response values was quite reasonable (Table 22.4) as it was recorded to be less than 5%. Extruded snacks were observed to have 3.31% moisture, 8.88% protein, 1.91% crude fiber, 0.22% fat, and 3.25% ash. On subjecting the product to consumer acceptability test ($n = 20$, where n signifies number of participants) the average overall acceptability score was recorded to be 7.50 on a 9-point hedonic scale.

TABLE 22.4 Predicted Responses versus Actual Response.

Values	ER	BD (g/cc)	WAI (g/g)	WSI (%)	Hardness (N)
Predicted	3.34	0.22	3.93	44.36	11.08
Actual	3.32	0.21	4.10	42.27	10.87
Variation (%)	0.60	4.76	4.15	4.94	1.93

Desirability graph

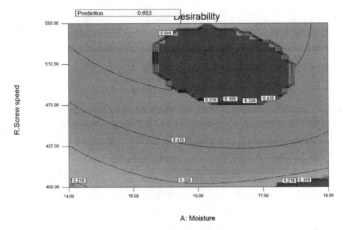

FIGURE 22.3 **(See color insert.)** Desirability function graph.

22.3 CONCLUSION

The results of RSM revealed a significant effect of all three important extrusion variables (feed moisture, screw speed, and barrel temperature) on physical properties of kinnow peel based snacks. Within the experimental range, feed moisture was the most important factor affecting the physical properties of the extrudates. The effect of feed moisture on most of the properties of the extrudates was found to be linear. The findings of this study demonstrate the possibility of using 1% kinnow peel as an ingredient in making extruded products with higher preference levels for parameters of appearance, texture, flavor (taste and odor), and overall acceptability.

KEYWORDS

- **kinnow**
- **peel**
- **extrusion**
- **optimization**
- **RSM**
- **snacks**

REFERENCES

Altan, A.; Mccarthy, K. L.; Medeni, M. Twin Screw Extrusion of Barley Grape Pomace. *J. Food Eng.* **2008a,** *89,* 24–32.

Altan, A.; McCarthy, K. L.; Medeni, M. Evaluation of Snack Foods from Barley–Tomato Pomace Blends by Extrusion Processing. *J. Food Eng.* **2008b,** *84,* 231–242.

Anderson, R. A.; Conway, H. F.; Griffin, E. L. Gelatinization of Corn Grits by Roll and Extrusion Cooking. *Cereal Sci. Today* **1969,** *14,* 4–12.

AOAC. *Official Methods of Analysis,* 17th ed.; Association of Official Analytical Chemists: Washington, DC, 2000.

Asare, E. K.; Sefa-Dedeh, S.; Sakyi-Dawson, E.; Afoakwa, E. O. Application of Response Surface Methodology for Studying the Product Characteristics of Extruded Rice–Cowpea–Groundnut Blends 2000 Corn Bran as a Fiber Source in Expanded Snack. *Int. J. Food Sci. Nutr.* **2004,** *55,* 431–439.

Baik, B. K.; Powers, J.; Nguyen, L. T. Extrusion of Regular and Waxy Barley Flours for Production of Expanded Cereals. *Cereal Chem.* **2004,** *81,* 94–99.

Baladran-Quintana, R. R.; Barbosa-Canovas, G. V.; Zazueta-Morales, J. J.; Anzaldua-Morales, A.; Quintero-Ramos, A. Functional and Nutritional Properties of Extruded Whole Pinto Bean Meal (*Phaseolus vulgaris L.*). *J. Food Sci.* **1998,** *63,* 113–116.

Barrett, A. H.; Peleg, M. Extrudate Cell Structure–Texture Relationships. *J. Food Sci.* **1992,** *57,* 1253–1257.

Bourne, M. C. Texture Profile Analysis. *Food Technol.* **1978,** *33,* 62–66.

Brar, S. K.; Dhillion, G. S. *Biotransformation of Waste Biomass into High Value Biochemicals;* 2013. https://books.google.co.in. (accessedon Nov 10, 2015).

Case, S. E.; Hamann, D. D.; Schwartz, S. J. Effect of Starch Gelatinization on Physical Properties of Extruded Wheat and Corn Based Products. *Cereal Chem.* **1992,** *69,* 401–404.

Chakraborty, P.; Banerjee, S. Optimization of Extrusion Process for Production of Expanded Product from Green Gram and Rice by Response Surface Methodology. *J. Sci. Ind. Res.* **2009,** *68,* 140–148.

Chang, Y. K.; Silvia, M. R.; Gutkoski, L.C.; Sebio, L.; Da Silvia, M. A. A. P. Development of Extruded Snacks Using Jatoba (*Hymenaeastigonocarpa Mart*) Flour and Cassava Starch Blends. *J. Sci. Food Agric.* **1998,** *78,* 59–66.

Chevanan, N.; Rosentrater, K. A.; Muthukumarappan, K. Effect of DDGS, Moisture Content, and Screw Speed on Physical Properties of Extrudates in Single-Screw Extrusion. *Cereal Chem.* **2008,** *85,* 132–139.

Ding, Q. B.; Ainsworth, P.; Plunkett, A.; Tucker, G.; Marson, H. The Effect of Extrusion Conditions on the Functional and Physical Properties of Wheat-based Expanded Snacks. *J. Food Eng.* **2006,** *73,* 142–148.

Dogan, H.; Karwe, M. V. Physicochemical Properties of Quinoa Extrudates. *Food Sci. Technol. Int.* **2003,** *9,* 101–114.

Fan, J.; Mitchell; J. R.; Blanchard, J. M. V. The Effect of Sugars on the Extrusion of Maize Grits: The Role of the Glass Transition in Determining Product Density and Shape. *Int. J. Food Sci. Technol.* **1996,** *31,* 55–65.

FAO. Citrus Fruit Fresh and Processed Statistical Bulletin prepared by Market and Policy Analysis of Raw Materials, Horticulture and Tropical (RAMHOT) Products Team, Trade and Markets Division, FAO: Rome, Italy, 2012.

Garber, B. W.; Hsieh, F.; Huff, H. E. Influence of Particle Size on the Twin-Screw Extrusion of Corn Meal. *Cereal Chem.* **1998,** *74,* 656–661.

Gera Komal, A. Punjab Government Spurs Kinnow Production, Business Standard. http://www.business-standard.com/article/economy-policy/punjab-government-spurs-kinnow-production-114051201209_1.html (accessed Nov 19, 2015).

Godavarti, S.; Karwe, M. V. Determination of Specific Mechanical Energy Distribution on a Twin-Screw Extruder. *J. Agric. Eng. Res.* **1997**, *67*, 277–287.

Gutkoski, L.C.; El-Dash, A. A. Effect of Extrusion Process Variables on Physical and Chemical Properties of Extruded Oat Products. *Plant Food Hum. Nutr.* **1999**, *54*, 315–325.

Hagenimana, A.; Ding, X.; Fang, T. Evaluation of Rice Flour Modified by Extrusion Cooking. *J Cereal Sci.* **2006**, *43*, 38–46.

Hsieh, F.; Grenus, K. M.; Hu, L.; Huff, H. E. Twin-screw Extrusion of Rice Flour with Salt and Sugar. *Cereal Chem.* **1993**, *70*, 493–498.

Kannadhason, S.; Muthukumarappan, K.; Rosentrater, K. A. Effects of Ingredients and Extrusion Parameters on Aquafeeds Containing DDGS and Tapioca Starch. *J. Aquacult. Feed Sci. Nutr.* **2009**, *1*, 6–21.

Keawpeng, I.; Charunuch, C.; Roudaut, G.; Meenune, M. The Optimization of Extrusion Condition of Phatthalung Sungyod Rice Extrudate: A Preliminary Study. *Int. J. Food Res.* **2014**, *21*, 2299–2304.

Koksel, H.; Ryu, G. H.; Basman, A.; Demiralp, H.; Ng, P. K. W. Effects of Extrusion Variables on the Properties of Waxy Hulless Barley Extrudates. *Nahrung* **2004**, *48*, 19–24.

Kumar, N. Development and Characterization of Extruded Product Using Carrot Pomace and Rice Flour. *Int. J. Food Eng.* **2010**, *6*, 3754–3758.

Lee, E. Y.; Ryu, G. H.; Lim, S. T. Effects of Processing Parameters on Physical Properties of Corn Starch Extrudates Expanded Using Supercritical CO_2 Injection. *Cereal Chem.* **1999**, *76* (1), 63–69.

Liu, Y.; Hsieh, F.; Heymann, H.; Huff, H. E. Effect of Process Conditions on the Physical and Sensory Properties of Extruded Oat–Corn Puff. *J. Food Sci.* **2000**, *65*, 1253–1259.

Mendonca, S.; Grossman, M. V. E.; Verhé, R. Corn Bran as a Fibre Source in Expanded Snacks. *LWT Food Sci. Technol.* **2000**, *33*, 2–8.

Mercier, C.; Fillet, P. Modification of Carbohydrate Component of Extrusion Cooking of Cereal Product. *Cereal Chem.* **1975**, *52*, 283–297.

Ministry of Food Processing Industries (MOFPI).Vision 2015. Strategy and Action Plan for Food Processing Industries in India. 2005, *21*, 152–153.

Myers, R. H.; Montgomery, D. C. *Response Surface Methodology;* Wiley: New York, 2002.

Ojo, S. T.; Olukunle, S. T.; Aduewa, T. O.; Ukwenya, A. G. Performance Evaluation of Floating Fish Feed Extruder. *J. Agric. Vet. Sci.***2014**, *7*,103–113.

Onyango, C.; Noetzold, H.; Ziems, A.; Hofmann, T.; Bley, T.; Henle, T. Digestibility and Antinutrient Properties of Acidified and Extruded Maize–Finger Millet Blend in the Production of Uji. *LWT Food Sci. Technol.* **2005**, *38*, 697–707.

Pan, Z.; Zhang, S.; Jane, J. Effects of Extrusion Variables and Chemicals on the Properties of Starch-Based Binders and Processing Conditions. *Cereal Chem.* **1998**, *75*, 541–46.

Pansawat, N.; Jangchud, K.; Jangchud, A.; Wuttijumnong, P.; Saalia, F. K.; Eitenmiller, R. R. Effects of Extrusion Conditions on Secondary Extrusion Variables and Physical Properties of Fish, Rice-Based Snacks. *Lebensm. Wiss. Technol.* **2008**, *41*,632–41.

Pathania, S.; Singh, B.; Sharma, S.; Sharma, V.; Singla, S. Optimization of Extrusion Processing Conditions for Preparation of an Instant Grain Base for Use in Weaning Foods. *Int. J. Eng. Res. Appl.* **2013**, *3*, 1040–1049.

Patil, R. T.; Berrios, J. A. G.; Swansons, B. G. Evaluation of Methods for Expansion Properties of Legume Extrudates. *Appl. Eng. Agric.* **2007**, *23*, 777–783.

Pitts, K. F.; Favaro, J.; Austin, P.; Day, L. Co-effect of Salt and Sugar on Extrusion Processing, Rheology, Structure and Fracture Mechanical Properties of Wheat–Corn Blend. *J. Food Eng.* **2014,** *127*, 58–66.

Reuters. Research and Markets: Global Extruded Snacks Market Outlook 2019—Potato, Corn, Rice, Tapioca, Mixed Grain & Others (Press Release). 2014. Retrieved from http://www.reuters.com/article/research-and-markets-idUSnBw295362a+100+BSW20140929. (accessedon Sept 18 2015).

Ryu, G. H.; Ng, P. K. W. Effects of Selected Process Parameters on Expansion and Mechanical Properties of Wheat Flour and Whole Cornmeal Extrudates. *Starch* **2001,** *53*, 147–154.

Sebio, L.; Chang, Y. K. Effects of Selected Process Parameters in Extrusion of Yam Flour (*Dioscorearotundata*) on Physicochemical Properties of the Extrudates. *Mol. Nutr. Food Res.* **2000,** *44*, 96–101.

Sharma, N.; Kalra, K. L.; Oberoi, H. S.; Bansal. Optimization of Fermentation Parameters for Production of Ethanol from Kinnow Waste and Banana Peels by Simultaneous Saccharification and Fermentation. *Indian J. Microbiol.* **2007,** *47*, 310–316.

Singh, B.; Hussain, Z. S.; Sharma, S. Response Surface Analysis and Process Optimization of Twin Screw Extrusion Cooking of Potato-based Snacks. *J. Food Process. Preserv.* **2013,** *39*, 270–281.

Wu, W.; Huff, H. E.; Hsieh, F. Processing and Properties of Extruded Flaxseed–Corn Puff. *J. Food Process. Preserv.* **2007,** *31*, 211–226.

Yagci, S.; Gogus, F. Development of Extruded Snack from Food By-products: A Response Surface Analysis. *J. Food Process Eng.***2008,** *32*, 565–86.

Yao, N.; Jannink, J.; Alavi, S.; White, P. J. Physical and Sensory Characteristics of Extruded Products Made from Two Oat Lines with Different β-glucan Concentrations. *Cereal Chem.* **2006,** *83*,692–99.

Yuliani, S.; Torley, P. J.; Bhandari, B. Physical and Processing Characteristics of Extrudates Made from Starch and d-limonene Mixtures. *Int. J. Food Sci. Technol.* **2006,** *41*, 83–94

CHAPTER 23

DEVELOPMENT AND PARTIAL CHARACTERIZATION OF BIODEGRADABLE FILM FROM COMPOSITE OF LOTUS RHIZOME STARCH, WHEY PROTEIN CONCENTRATE, AND PSYLLIUM HUSK

SAKSHI SUKHIJA*, SUKHCHARN SINGH, and CHARANJIT S. RIAR

Department of Food Engineering and Technology, Sant Longowal Institute of Engineering and Technology—Deemed University, Longowal 148106, Sangrur, Punjab, India

Corresponding author. E-mail: dietsakshi.2007@gmail.com

CONTENTS

Abstract 562
23.1 Introduction 562
23.2 Materials and Methods 563
23.3 Film Formation 564
23.4 Film Characterization 564
23.5 Results and Discussion 567
23.6 Optimization 571
23.7 Conclusion 572
Keywords 572
References 572

ABSTRACT

The research was carried out for the development of biodegradable film from composite of native lotus rhizome starch (NLS), whey protein concentrate (WPC), and psyllium husk (PH) by casting technique. A central composite rotatable design (CCRD) was applied to analyze the effect of factors, such as NLS (2–3%), WPC (1–2%), PH (0.02–0.04%), and glycerol (GLY, 1.75–2.25%) on the responses (tensile strength (TS), water vapor permeability (WVP), and solubility). A significant effect of independent variables on all the responses was demonstrated by the statistical analysis. The results revealed that WVP, TS, and solubility of biodegradable films were positively affected with increase in concentration of NLS, WPC, and PH, whereas negatively affected by plasticizer (GLY). The optimum concentration with respect to desirable responses as obtained by optimization includes NLS (3.00%), WPC (2.00%), PH (0.04%), and GLY (1.75%) for lower WVP; solubility and higher TS of developed biodegradable film. A high agreement was found between experimental and predicted values.

23.1   INTRODUCTION

Biodegradable polymers originating from renewable sources have spurred interest of researchers because of significant environmental issues caused by petrochemical-based non-biodegradable polymers and depleting petroleum resources. Biodegradable packaging materials developed from natural components including polysaccharide, proteins, and lipids provide an alternative to synthetic polymers. Among polysaccharides, starch is considered to be the most important to develop biodegradable films because of its abundance in nature, capability of forming continuous matrix, and thermoplastic behavior. Several researchers have developed starch-based films using different types of starches like cassava, tapioca, yam, oats, and potato (Mali et al., 2002; Mali et al., 2005; Pelissari et al., 2013; Woggum et al., 2014; Zavareze et al., 2012; Gutierrez et al., 2015). Native lotus rhizome starch (NLS) has not been utilized till date for formation of biodegradable film. However, limited barrier and mechanical properties of starch films need to be improved which may be possible by the addition of protein (whey protein concentrate, WPC) and other novel ingredients like hydrocolloids (psyllium husk, PH) which results in the formation of dense matrix and suitability of developed films for packaging of food products. The varying combination of polysaccharides and proteins may offer possibility to develop suitable composite packaging material to meet consumer expectations and demands (Kristio et al., 2007). Researchers have shown that whey proteins are able to produce flexible and transparent films with excellent resistance to oxygen, aromatic compounds, and oil (Qazanfarzadeh & Kadivar, 2016). The protein network accommodates polysaccharide chains resulting in continuous phase formation in composite films and formation of a dense matrix which leads to improved barrier and mechanical properties (Munoz et al., 2012).

Whey protein is the effluent from cheese making process containing 25–80% protein content and possesses interesting film-forming properties. Psyllium is an annual plant belonging to *Plantago* genus native to India and Iran. The psyllium (*Plantago ovata* Forsk) seed husk is well-known source of psyllium hydrocolloid. The addition of plasticizers in the film forming solution is necessary to reduce the brittleness of film and improve its flexibility by reducing intermolecular forces and increasing the mobility of polymer chain. Plasticizers facilitate polymer chain slipping, decrease network rigidity, and producing a less ordered structure increasing film flexibility (Versino, 2016). The most commonly used plasticizer for film formation is glycerol (GLY). Several studies have been reported for blend of two ingredients for formation of biodegradable and edible films prepared from *Salvia hispanica* + whey protein interaction (Munoz et al., 2012), cassava starch + soy protein concentrate (Chinma et al., 2012), and pea starch + peanut protein isolate (Sun et al., 2013). Only a few authors have manufactured composite films using a three-way ingredient interaction that is, whey protein isolate + gelatin + sodium alginate films (Wang et al., 2010). Therefore, this study aimed at assessing the potential of using NLS, WPC, and PH for the development of composite biodegradable film using central composite rotatable design (CCRD) to determine their physical properties.

23.2 MATERIALS AND METHODS

Fresh lotus rhizomes were procured from certified center for starch extraction. The starch from lotus rhizome was extracted according to the method described by Sukhija et al. (2016). WPC was purchased from Mahaan Proteins Ltd. (New Delhi, India). The protein content of WPC was 70g/100g as estimated by the micro-kjeldhal. PH was purchased from the Sidhpur, Sat-Isabgol Factory (Sidhpur, Gujarat), further ground and sieved through 150 BSS sieve. GLY was purchased from Merck Specialities Pvt. Ltd., Mumbai and was AR grade.

23.3   FILM FORMATION

The ranges for NLS (2–3%), WPC (1–2 %), PH (0.02–0.04%), and GLY (1.75–2.25%) were selected on the basis of preliminary trials. The film-forming solution (150 mL) was prepared in two steps: first, PH (0.04%) was suspended in 50 mL water for 30 min, heated in boiling water bath for 20 min, cooled, and blended with 100 mL suspension of starch (3%), WPC (2%), and GLY (1.75%). The final suspension was blended using a magnetic stirrer. The suspension was then heated at 90°C for 30 min in a water bath to ensure complete gelatinization of starch. The film forming solutions were then uniformly poured on Teflon coated mold (20 × 20 cm) and kept in vacuum oven at 60°C for 30 min for the removal of air bubbles followed by drying at 45°C for 18–20 h. Then, the films were peeled off and stored for conditioning in a desiccator at 25°C and 52% RH (maintained using satu-rated magnesium nitrate ($Mg(NO_3)_2$ solution) for 48 h for further analysis.

23.4   FILM CHARACTERIZATION

23.4.1   MECHANICAL PROPERTIES

Tensile strength (TS) of films was measured according to ASTM D882-12 (ASTM, 2012) with some modifications by using a texture analyzer (Stable Micro Systems, Surrey, UK). The films were initially cut into strips (2 cm × 8 cm) before testing. The initial grip separation and crosshead speed were set at 50 mm and 0.5 mm/s, respectively. TS was calculated by using following equation:

$$TS(MPa) = \frac{Force(N)}{Width(m) \times Thicknesss(m)} \times 100. \tag{23.1}$$

23.4.2   WATER VAPOR PERMEABILITY (WVP)

Water vapor permeability (WVP) is the rate of water vapor transmission (WVT) through a unit area of flat material of unit thickness induced by unit vapor pressure difference between two surfaces, under specified humidity condition of 75% (Souza et al., 2012). WVP of biodegradable films was determined using gravimetric modified cup method ASTM E 96/E96M-15 (ASTM, 2015). Biodegradable films were sealed on the cup mouth with the help of sealant. Anhydrous calcium chloride ($CaCl_2$, 0% RH) was used as desiccant inside the permeation cell and the film to be tested was sealed on the mouth of the cup. The cell was placed inside a desiccator that was maintained at 75% RH using saturated sodium chloride solution to maintain 75% RH gradient across the film. The weight change of the cups with films was recorded at regular time intervals (about 2 h) and the weight change was plotted as function of time. The slope was calculated by linear regression method and the water vapor transmission rate (WVTR) was calculated from the slope of the straight line (g/s) divided by the transfer area (m^2). WVP was calculated using the following equation:

$$WVP = \frac{WVTR}{S(R_1 - R_2)} \times d,$$

(23.2)

where WVP is the water vapor permeability (g.m/Pa.s.m^2), S is the saturation vapor pressure of water (Pa) at the test temperature (30°C), R_1 is the %RH in humidity chamber, R_2 is the % RH in cup and, d is the thickness (m).

23.4.3 SOLUBILITY

Solubility of the film sample was determined according to method described by Romero-Bastida, et al. (2005). The weight of the 20 mm × 20 mm dried films was measured directly and then the films were immersed in 80 mL distilled water at 25°C for 1 h and wobbled slightly periodically, followed by a careful separation of the insoluble film. The separated film samples were first dried in oven (60°C) until constant weight was obtained. The solubility (WS) was computed according to the following equation:

$$Solubility\,(\%) = \frac{(W_i - W_f)}{W_i} \times 100$$

(23.3)

where W_i and W_f is the initial and final dry weight of the sample in g before and after performing test, respectively (g).

23.4.4 EXPERIMENTAL DESIGN

Response surface methodology (RSM) was adopted for the design of experimental combinations based on the multivariate nonlinear model. The optimum levels of the four variables, that is, NLS (%), WPC (%), PH (%), and GLY (%) concentrations were identified regarding three responses: TS, solubility, and WVP of biodegradable films. CCRD of RSM which dictated 30 experimental combinations including six replicates at the center point was used to derive the optimum levels of the independent variables using a four-factor five level (Table 23.1). The order of experiments was fully randomized. This data was analyzed by multiple regressions using the least-squares method. The following second order polynomial equation was used to predict the responses as a function of independent variables:

$$Y = \beta_0 + \beta_1 X_1 + \beta_2 X_2 + \beta_3 X_3 + \beta X_4 + \beta_{11} X_1^2 + \beta_{22} X_2^2 + \beta_{33} X_3^2 + \beta_{44} X_4^2$$
$$+ \beta_{12} X_1 X_2 + \beta_{13} X_1 X_3 + \beta_{14} X_1 X_4 + \beta_{23} X_2 X_3 + \beta_{24} X_2 X_4 + \beta_{34} X_3 X_4 + \varepsilon \quad (23.4)$$

where Y = response variable namely TS (MPa); WVP (g.m/Pa.s.m^2) and solubility (%). The polynomial coefficients were represented by β_0 (constant term), β_1, β_2, β_3, β_4 (linear effects), β_{11}, β_{22}, β_{33}, β_{44} (quadratic effects), and β_{12}, β_{13}, β_{14}, β_{23}, β_{24}, β_{34} (interaction effects). The Design-Expert Software (trial version 9.0.6.2, Stat-Ease Inc., Minneapolis, USA) was used for regression and graphical analysis of the data obtained. The quality of the fitted polynomial models was expressed by the coefficient of determination (R^2), adjusted R^2, and predicted R^2. Analysis of variance (ANOVA) was used to determine the significant effects of variables on the responses.

TABLE 23.1 Experimental Design for Biodegradable Films Runs for Their Corresponding Response Values.

Run	Process variables				Response variables		
	NLS[a] (%)	WPC[a] (%)	PH[a] (%)	GLY[a] (%)	WVP[b] ($\times 10^{-11}$ g.m/ Pa.s. m^2)	TS[b] (MPa)	Solubility (%)
1	2.00	1.00	0.02	1.75	3.54	2.83	24.94
2	3.00	1.00	0.02	1.75	3.83	4.22	19.04
3	2.00	2.00	0.02	1.75	2.72	3.19	21.85
4	3.00	2.00	0.02	1.75	2.69	5.01	16.65
5	2.00	1.00	0.04	1.75	2.92	3.43	21.69
6	3.00	1.00	0.04	1.75	3.38	4.25	18.66
7	2.00	2.00	0.04	1.75	2.43	4.30	18.58
8	3.00	2.00	0.04	1.75	2.17	6.14	16.14
9	2.00	1.00	0.02	2.25	4.09	1.42	27.23
10	3.00	1.00	0.02	2.25	3.98	2.71	26.07
11	2.00	2.00	0.02	2.25	3.51	1.39	28.71
12	3.00	2.00	0.02	2.25	3.49	2.64	26.39
13	2.00	1.00	0.04	2.25	3.89	1.63	27.19

TABLE 23.1 *(Continued)*

Run	Process variables				Response variables		
	NLS[a] (%)	WPC[a] (%)	PH[a] (%)	GLY[a] (%)	WVP[b] ($\times 10^{-11}$ g.m/ Pa.s. m^2)	TS[b] (MPa)	Solubility (%)
14	3.00	1.00	0.04	2.25	3.79	1.46	27.36
15	2.00	2.00	0.04	2.25	3.97	1.42	26.76
16	3.00	2.00	0.04	2.25	3.77	1.68	27.46
17	1.50	1.50	0.03	2.00	3.42	2.59	25.98
18	3.50	1.50	0.03	2.00	3.27	4.69	20.49
19	2.50	0.50	0.03	2.00	4.23	2.04	23.73
20	2.50	2.50	0.03	2.00	2.98	4.21	18.32
21	2.50	1.50	0.01	2.00	3.02	2.08	25.51
22	2.50	1.50	0.05	2.00	2.91	3.67	21.37
23	2.50	1.50	0.03	1.50	2.78	4.73	18.76
24	2.50	1.50	0.03	2.50	4.19	0.69	32.07
25	2.50	1.50	0.03	2.00	2.97	4.29	18.09
26	2.50	1.50	0.03	2.00	3.06	3.91	19.56
27	2.50	1.50	0.03	2.00	2.89	4.45	19.82
28	2.50	1.50	0.03	2.00	2.76	4.07	19.87
29	2.50	1.50	0.03	2.00	2.94	4.41	20.87
30	2.50	1.50	0.03	2.00	2.91	4.23	19.77

[a]NLS, WPC, PH, and GLY are concentration of lotus rhizome starch, whey protein concentrate, psyllium husk, and glycerol, respectively.

[b]WVP and TS are water vapor permeability and tensile strength, respectively.

23.5 RESULTS AND DISCUSSION

23.5.1 TENSILE STRENGTH

TS reflects the durability of films and its ability to enhance the mechanical integrity of food. The ANOVA on the fitted model showed that the concentration of NLS and WPC had highly significant ($p < 0.01$) positive effect on T, whereas plasticizer (GLY) showed significant negative effect ($p < 0.0001$). Furthermore, the mutual interaction between protein-plasticizer and husk-plasticizer showed significant effect ($p < 0.0001$) on TS of films (Fig. 23.1a,b). TS of biopolymer matrices depends on the degree of interaction of components (McHugh et al., 1994). An increase in TS may be expected due to strong interaction between starch and protein that led to the formation of a dense matrix (Munoz et al., 2012). In addition, the hydrocolloid nature of PH also contributed to increase in TS by inducing new links between the starch-protein matrixes (Munoz et al., 2012). Similar studies for increasing in TS with starch-protein and protein-hydrocolloid interactions have been earlier reported (Munoz et al., 2012; Chinma et al., 2012). On the other hand, plasti-cizers modify the properties of biopolymer films by reducing intermolecular forces and promoting formation of weak hydrogen bonds between plasticizer and formed matrix, hence, increasing the mobility of polymer chains causing the mechanical strength to be decreased (Dick et al., 2015).

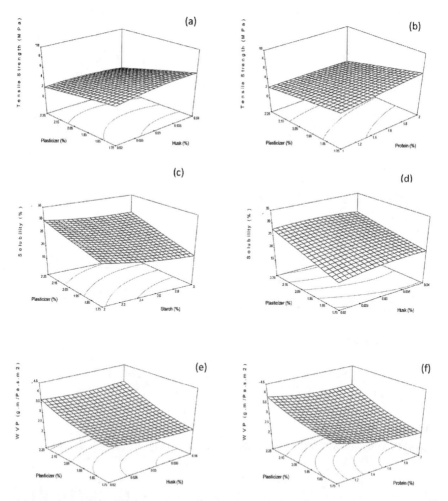

FIGURE 23.1a–f Effect of concentration of starch, protein, husk, and plasticizer on tensile strength, solubility, and water vapor permeability of biodegradable films.

23.5.2 SOLUBILITY

The low solubility of biodegradable film presents its water stability and finds significance in protection of high activity foods or prevention of exudation from frozen or fresh products during processing. Regression coefficient model predicted significant positive effect of starch and protein ($p < 0.0001$); however, PH and plasticizer showered significant negative effect ($p < 0.05$) on solubility of biodegradable films (Table 23.2). Figure 23.1c,d shows the influence of independent variables on the solubility of films. It would be reasonable to speculate that with the increase in concentration of starch and protein solubility of films decreased which might be attributed to their high interaction density and alteration in three-dimensional structure of proteins revealing free —SH groups and hydrophobic side chains during heat treatment (McHugh et al., 1994). However, the increased concentration of PH and GLY presented an increase in solubility of films. An increase in solubility due to PH may be the result of combined effect of branched structure of mucilage present in PH and hydrophilic nature PH and plasticizer resulting in increasing film solubility as more water gets attracted into polymeric matrix creating mobile regions with greater interchain distances by diminishing interactions between biopolymer molecules (Cuq et al., 1997).

TABLE 23.2 Significant Regression Coefficients of Second Order Polynomial Equation for the Response Variables.

Coefficients	Tensile strength (MPa)	Solubility (%)	Water vapor permeability ($\times 10^{-11}$ g.m/Pa.s. m^2)
β_0	4.226**	19.663**	2.921**
β_1	0.529**	−1.256**	−
β_2	0.34**	−0.852**	−0.298**
β_3	0.17*	−0.638**	−0.073*
β_4	−1.129**	3.593**	0.401**
β_{12}	−	−	−
β_{13}	−	0.623*	−
β_{14}	−0.203±	0.872**	−
β_{23}	−	−	0.087*
β_{24}	−0.25*	0.786**	0.166**
β_{34}	−0.303**	0.486*	0.139**

TABLE 23.2   *(Continued)*

Coefficients	Tensile strength (MPa)	Solubility (%)	Water vapor permeability ($\times 10^{-11}$ g.m/Pa.s. m^2)
β_1^2	−0.164[*]	0.916[**]	0.112[**]
β_2^2	−0.293[**]	0.363[**]	0.177[**]
β_3^2	−0.355[**]	0.967[**]	−
β_4^2	−0.397[**]	1.461[**]	0.147[**]
R^2	0.96	0.97	0.97
Adjusted R^2	0.93	0.96	0.94
Pred R^2	0.80	0.90	0.85
Adeq. precision	21.33	28.59	21.15
Lack of fit	0.0590	0.6074	0.2084

[*]Significant at $p < 0.05$.
[**]Significant at $p < 0.0001$.

23.5.3   WATER VAPOR PERMEABILITY

WVP is the most important property of biodegradable films because water is closely related to deteriorative and other reactions (oxidation, browning, vitamin degradation, etc.), texture changes in food in turn affecting stability and quality of food. According to response plot (Fig. 23.1e,f), increase in concentration of starch, protein, and husk positively influenced ($p < 0.0001$) WVP of film. Lower WVP of composite biodegradable films may be attributed to the formation of a continuous phase of polymeric matrix between starch, protein, and husk mucilage as a result of formation of cross-links between starch and protein, whereas, husk mucilage could have acted as a filler resulting the system with lesser free volume and moisture diffusion. Similar studies for decrease in WVP of edible films prepared from *S. hispanica* + whey protein interaction and cassava starch + soy protein concentrate films have been earlier reported (Munoz et al., 2012; Chinma et al., 2012). However, increase in plasticizer concentration adversely affected WVP of films probably due to its penetration into polymeric matrix and diminishing the interaction among molecules, thus increasing free volume and easier diffusion of water molecules resulting in higher WVP of films (Dick et al., 2015; Gontard et al., 1992).

23.6   OPTIMIZATION

In order to optimize the process conditions for film formation, the following considerations were taken: (1) maximization of TS, (2) minimization of WVP, and (3) minimization of solubility. Negligible difference was observed in experimental and predicted values. The solutions analyzed after setting up the optimum concentrations included 3.00% NLS, 2.00% WPC, 0.04% PH, and 1.75% GLY. The desired goals for each independent variable and response were chosen as presented in Table 23.3. Under the suggested optimal conditions, the corresponding values for WVP, TS, and solubility are presented in Table 23.4. The experimental values indicate the suitability of the developed quadratic model and it may be noted that the optimal values are valid within the specified range of independent parameters.

TABLE 23.3   Criteria and Outputs of the Numerical Optimization of the Responses for Biodegradable Films.

Variables	Goal	Lower limit	Upper limit	Lower weight	Upper weight	Importance
NLS[a] (%)	Is in range	2.00	3.00	1	1	3
WPC[a] (%)	Is in range	1.00	2.00	1	1	3
PH[a] (%)	Is in range	0.02	0.04	1	1	3
GLY[a] (%)	Is in range	1.75	2.25	1	1	3
WVP[b] (g.m/Pa.s. m²)	Minimize	2.17	4.23	1	1	3
TS[b] (MPa)	Maximize	0.69	6.14	1	1	3
Solubility (%)	Minimize	16.14	32.07	1	1	3

[a]NLS, WPC, PH, and GLY are concentration of lotus rhizome starch, whey protein concentrate, psyllium husk, and glycerol, respectively .

[b]WVP and TS are water vapor permeability and tensile strength, respectively.

TABLE 23.4   Comparison of the Experimental and Predicted Data for Investigated Properties of Biodegradable Films.

Runs	WVP[b] (×10⁻¹¹ g.m/Pa.s. m²)	TS[b] (MPa)	Solubility (%)
Predicted[a]	2.32	5.91	15.37
Experimental[c]	2.17 ± 0.09	6.06 ± 0.19	15.90 ± 0.23

[a]At optimal point determination by Design-Expert Software.

[b]WVP and TS are water vapor permeability and tensile strength, respectively.

[c]The values are means of triplicate with standard deviation.

23.7 CONCLUSION

The optimization of concentrations of NLS, WPC, PH, and GLY was successfully carried out using CCRD of RSM. High agreement was observed between experimental and predicted values. The optimized biodegradable film possessed maximum TS, minimum solubility, and WVP and the developed film can be further analyzed for desirable properties like light transmittance, elongation (%), transparency, thermal, and morphological characteristics for possible applications.

KEYWORDS

- **biodegradable films**
- **lotus rhizome starch**
- **tensile strength**
- **water vapor permeability**
- **solubility**

REFERENCES

ASTM. Standard Test Method for Tensile Properties of Thin Plastic Sheeting. In *Annual Book of ASTM Standards D882-12;* American Society for Testing and Materials: Philadelphia, 2012.

ASTM Standard Test Methods for Water Vapor Transmission of Materials. In *Annual Book of ASTM Standards E 96/E96M-15;* American Society for Testing Materials: Philadelphia, 2015.

Cuq, B.; Gontard, N.; Aymard, C.; Guilbert, S. Relative Humidity and Temperature Effects on Mechanical and Water Vapor Barrier Properties of Myofibrillar Protein-based Films. *Polym. Gels Netw.* **1997,** 5 (1), 1–15.

Chinma, C. E.; Ariahu, C. C.; Abu, J. O. Development and Characterization of Cassava Starch and Soy Protein Concentrate Based Edible Films. *Int. J. Food Sci. Technol.* **2012,** 47, 383–389.

Dick, M.; Costa, T. M. H.; Gomaa, A.; Subirade, M.; Rios, A. D. O.; Flores, S. H. Edible Film Production from Chia Seed Mucilage: Effect of Glycerol Concentration on Its Physico-chemical and Mechanical Properties. *Carbohydr. Polym.* **2015,** 130, 198–205.

Gutierrez, T. J.; Morales, N. J.; Perez, E.; Tapia, M. S.; Fama, L. Physico-chemical Properties of Edible Films Derived from Native and Phospahted Cush-cush Yam and Cassava Starches. *Food Packag. Shelf Life* **2015,** 3, 1–8.

Gontard, N.; Guilbert, S.; Cuq, J. L. Edible Wheat Gluten Films: Influence of the Main Process Variables on Film Properties Using Response Surface Methodology. *J. Food Sci.* **1992,** *57* (1), 190–195.

Kristio, E.; Biliaderis, C. G.; Zampraka, A. Water Vapor Barrier and Tensile Properties of Composite Caseinate-pllulan Films: Biopolymer Composition Effects and Impact of Beeswax Lamination. *Food Chem.* **2007,** *101,* 753–764.

Munoz, L. A.; Aguilera, J. M.; Rodriguez-Turienzo, L.; Cobos, A.; Diaz, O. Characterization and Microstructure of Films Made from Mucilage of *Salvia hispanica* and Whey Protein Concentrate. *J. Food Eng.* **2012,** *111,* 511–518.

Mali, S.; Grossmann, M. V. E.; Garcia, M. A.; Martino, M. N.; Zaritzky, N. E. Microstructural Characterization of Yam Starch Films. *Carbohydr. Polym.* **2002,** *50,* 379–386.

Mali, S.; Grossmann, M. V. E.; Garcia, M. A.; Martino, M. N.; Zaritzky, N. E. Mechanical and Thermal Properties of Yam Starch Films. *Food Hydrocoll.* **2005,** *19,* 157–164.

McHugh, T. H.; Aujard, J. F.; Krochta, J. M. Plasticized Whey Protein Edible Films: Water Vapor Permeability Properties. *J. Food Sci.* **1994,** *59* (2), 416–419.

Pelissari, F. M.; Andrade-Mahecha, M. M.; Sorbal, P. J. A.; Menegalli, F. C. Comparative Study on the Properties of Flour and Starch Films of Plaintain Bananas (*Musa paradisiaca*). *Food Hydrocoll.* **2013,** *30,* 681–690.

Qazanfarzadeh, Z.; Kadivar, M. Properties of Whey Protein Isolate Nanocomposite Reinforced with Nanocellulose from Oat Husk. *Int. J. Biol. Macromol.* **2016,** *91,* 1134–1140.

Romero-Bastida, C. A.; Bello-Perez, L. A.; Garcia, M. A.; Martino, M. N.; Solorza-Feria, J.; Zaritzky, N. E. Physicochemical and Microstructural Characterization of Films Prepared Thermal and Cold Gelatinization from Non-conventional Sources of Starches. *Carbohydr. Polym.* **2005,** *60,* 235–244.

Sun, Q.; Sun, C.; Xiong, L. Mechanical, Barrier and Morphological Properties of Pea Starch and Peanut Protein Isolate Blend Films. *Carbohydr. Polym.* **2013,** *98,* 630–637.

Sukhija, S.; Singh, S.; Riar, C. S. Isolation of Starches from Different Tubers and Study of Their Physicochemical, Thermal, Rheological and Morphological Characteristics. *Starch Stärke* **2016,** *68,* 160–168.

Souza, A. C.; Benze, R.; Ferrao, E. S.; Ditchfield, C.; Coelho, A. C. V.; Tadini, C. C. Cassava Starch Biodegradable Films: Influence of Glycerol and Clay Nanoparticles Content on Tensile and Barrier Properties and Glass Transition Temperature. *LWT Food Sci. Technol.* **2012,** *46,* 110–117.

Versino, F.; Lopez, O. V.; Garcia, M. A.; Zaritzky, N. E. Starch-based Films and Food Coatings: An Overview. *Starch Starke* **2016,** *68,* 1–12.

Wang, L.; Auty, M. A. E.; Kerry, J. P. Physical Assessment of Composite Biodegradable Films Manufactured Using Whey Protein Isolate, Gelatin and Sodium Alginate. *J. Food Eng.* **2010,** *96,* 199–207.

Woggum, T.; Sirivongpaisal, P.; Wittaya, T. Properties and Characteristics of Dual-modified Rice Starch Based Biodegradable Films. *Int. J. Biol. Macromol.* **2014,** *67,* 490–502.

Zavareze, E. R.; Pinto, V. Z.; Klein, B.; Halal, S. L. M. E.; Elias, M. C.; Harnandez, C. P.; Dias, A. R. G. Development of Oxidized and Heat-moisture Treated Potato Starch Film. *Food Chem.* **2012,** *132,* 344–350.

CHAPTER 24

GREEN LEAF PROTEIN CONCENTRATE AND ITS APPLICATION IN EXTRUDED FOOD PRODUCTS

JAYABRATA SAHA and SANKAR CHANDRA DEKA*

Department of Food Engineering and Technology, Tezpur University, Napaam, Tezpur 784028, Sonitpur, Assam, India

Corresponding author. E-mail: sankar@tezu.ernet.in

CONTENTS

Abstract 576
24.1   General Introduction 576
24.2   Nutritional and Anti-Nutritional Importance of
       Green Leafy Vegetables 577
24.3   Medicinal Importance of Green Leafy Vegetables 579
24.4   Processing Importance of Green Leafy Vegetables 580
24.5   Protein Deficiency and Its Present Scenario 581
24.6   Functional Properties of Leaf Protein Concentrate 582
24.7   Extraction Optimization and Application of Leaf Protein
       Concentrate in Combating the Protein Deficiency 584
24.8   Importance of Cytotoxicity Study 585
24.9   Preparation of Extruded Food Incorporating LPC
       Using Extrusion Technology 586
24.10 Preparation of Pasta (An Extruded Food) by Incorporating
       Leaf Protein Concentrates 587
24.11 Importance of Storage Studies 587
24.12 Conclusion 588
Keywords 588
References 588

ABSTRACT

This chapter describes the general introduction of green leafy vegetables, its nutritional and antinutritional importance, medicinal importance, processing importance, protein deficiency, and its present scenario. Importance of extraction and optimization of leaf protein concentrate (LPC) has been underlined. The functional properties of LPC and its application in the preparation of extruded foods by incorporating it followed by storage studies have been reexamined.

24.1   GENERAL INTRODUCTION

Green leafy vegetables occupy a unique position among all the other vegetables for their flavor, color, and other health benefits. They are considered to be rich in vitamins like ascorbic acids, β-carotenoids, minerals such as zinc and iron, folate, and dietary fiber (Negi & Roy, 2000). There is even a direct correlation with the consumption of green leafy vegetables and fruits with decreased risk of hypertension, cardiovascular diseases, diabetes, stroke, and various forms of cancer (Brandt et al., 2005; Craig, 1999) Although only 10% of calories is replenished by fruits and vegetables, their overall contribution toward health is much high. The importance of leafy vegetables in the developing countries has been discovered in the recent past due to their nutritional and medicinal value. They are also rich sources of proteins, containing about 2–7% and some of them are equal to legumes, soybeans, or whole egg (Aletor et al., 2002). Several workers have advocated the beneficial role of green leafy vegetables and their contribution to health. It has been found that intake of fruits, vegetables, and related nutrients contributed in lowering the risk of Non-Hodgkin's lymphoma among women (Zang et al., 2000). Consumption of dark green leafy vegetables enhanced the serum vitamin A among children (Faber et al., 2002).

Northeast India is well known for its rich biodiversity because the agro-climatic variation supports the growth and the survival of numerous plant species. Majority of the population in this region are tribal, and they depend on forest and wasteland for their food (Das, 1998). People have been using wild plant as food and also medicine since time immemorial (Ramakrishnan, 1998). However, edible green leafy vegetables appear to be underutilized although there is a growing concern on the nutritive values of food. Indian cuisine has a wide range of choice in green leafy vegetables in the daily diet and is widely accepted on account of their low price and can be easily cooked (Gupta & Wagle, 1998). In nature, there are many underutilized green leafy vegetables which have a promising nutritional value that can nourish the ever-increasing human population. Nowadays underutilized green leafy vegetables are gaining popularity as it can increase the per capita availability of foods (Sheela et al., 2004). Results from the work done in the Northeastern region indicate that these vegetables could be further promoted for the better commercialization in order to strengthen the nutritional security in this region (Bhardwaj et al., 2008). Ethnobotanists have advocated the role of green leafy vegetables as food and therapeutics in context to health and response to diseases (Anonymous, 1959).

24.2 NUTRITIONAL AND ANTI-NUTRITIONAL IMPORTANCE OF GREEN LEAFY VEGETABLES

Green leafy vegetables are nutritionally sound as they are the storehouse of moisture, nitrogen, protein, and amino acids like aspartic acid, threonine, serine, glutamic acid, proline, glycine, alanine, cysteine, valine, methionine, isoleucine, leucine, tyrosine, lysine, histidine, and arginine (AOAC, 1990). Reports confirmed the presence of vitamins like vitamin C, thiamin, riboflavin, niacin, protein, fat, sugars like glucose, fructose, sucrose, starch, dietary fiber, organic acids like malic acid, citric acid, oxalic acid, ash, and minerals like sodium, potassium, calcium, iron, magnesium, and zinc contents for 15 Chinese vegetables including aquatica (Candlish, 1983; Candlish et al., 1987) studied the dietary fiber non-cellulose polysaccharide, cellulose, and lignin and starch content in vegetables that are extensively grown and consumed in Southeast Asia including aquatica. The lipid contents like nonpolar lipids, glycolipids, phospholipids, fatty acids, amino acids like aspartame, threonine, serine, glycine, proline, alanine, leucine, tyrosine, histidine, asparagine, and arginine, minerals like calcium, magnesium, iron, zinc, and copper were studied. Micronutrient composition like ash, calcium, iron, zinc, and carotenes like α- and ß-carotenes were determined (Ogle et al., 2001). The dieting patterns of rabbits fed with water spinach were investigated and analyzed as protein supplements for duck feed as basal diet (Ngamsaeng et al., 2004). Ortaliza et al. (1969) estimated the carotene contents and its availability in aquatica and other GLVs. The effects of aquatica on cholesterol metabolism in rats were studied (Chen et al., 1984). Chen et al. (1991) characterized the major carotenoids of water convolvulus (*Ipomoea aquatica*) using various techniques such as open-column, thin layer, high-performance liquid chromatography, and spectrophotometer identified various compounds. Comparing the absorption spectra and retention time with reference standards the compounds identified were ß-carotene, cryptoxanthin, lutein, lutein epoxide, violaxanthin, and neoxanthin. Carotene composition and contents of aquatica and other GLVs were analyzed by AOAC and HPLC methods. Daniel (1989) estimated the polyphenol contents of GLVs and identified quercetin 3'-methyl ether, quercetin 4'-methyl ether and anthocyanins. Flavonoid contents of GLVs like myricetin, quercetin, luteolin, and apigenin contents were determined.

The nutritional value of GLVs was compared with yielding crops such as rice, sugarcane by Munger (1999) where he stated that investment in the crops such as *Ipomoea aquatica* can be cost-effective strategy to supplement caloric and nutritional value of current crop production. Iron, calcium, ß-carotene, ascorbic acid, and oxalic acid content of GLVs consumed by the tribals of Purnia district of Bihar, India were also studied (Rao & Vijay, 2002).The nutrient and anti-nutrient contents of four green leafy vegetables were evaluated by where it was stated that they are the rich sources of ascorbic acid, β-carotene etc. (Kala & Prakash, 2004). In another study conducted by Singh et al. (2001) and Gupta et al. (1989), they stated that green leafy vegetables are rich sources of protein, ascorbic acid, β-carotene, and minerals.

Green leafy vegetables are an important component of the dietary regime of humans because they provide the necessary vitamins and minerals (Fasuyi, 2006). However, they also contain anti-nutrients which reduce the bioavailability of these nutrients (Akindahunsi & Salawu, 2005; Binita & Khetarpaul, 1997). Aletor and Adeogun (1995) reported that some anti-nutrients exhibit protective effects thus making them serve a dual purpose. Oxalate binds to calcium to form calcium oxalate crystals that prevent the absorption and utilization of calcium by the body thereby causing diseases such as rickets and osteomalacia (Ladeji et al., 2004). The calcium crystals may also precipitate around renal tubules causing renal stones. Phytic acid combines with some essential elements such as Fe, Ca, Zn, and P to form insoluble salts called the phytates which are not absorbed by the body, therefore, making these minerals biounavailable. Saponins are naturally oily glycosides occurring in a wide variety of plants. When eaten, they are non-poisonous to warm-blooded animals but are poisonous when injected into the bloodstream (Applebaum et al., 1969). Phytic acid (inositol hexaphosphate) is an organic acid found in plant materials (Heldt, 1997). Phytic acid combines with some essential elements such as iron, calcium, zinc, and phosphorus to form insoluble salts called phytate, which is not absorbed by the body thereby reducing the bioavailability of these elements. Anemia and other mineral deficiency disorders are common in regions where the diet is primarily vegetarian (Erdman, 1979). Tannins have the ability to precipitate certain proteins. They combine with digestive enzymes thereby making them unavailable for digestion (Binita & Khetarpaul, 1997; Abara, 2003).

24.3 MEDICINAL IMPORTANCE OF GREEN LEAFY VEGETABLES

There are 45,000 species of wild plants out of which 9500 species are ethnobotanically important species. Of these, 7500 species are in medicinal use for indigenous health practices. In Ayurvedic system of practice, the extracts of leaves of GLVs like aquatica are administered orally for the disorders like jaundice and nervous debility. This plant is administered against nose bleeding and high blood pressure, opium and arsenic poisoning (Anonymous, 1959; Duke & Ayensu, 1985; Chopra et al., 1956). Dried juice has purgative properties (Anonymous, 1959; Chopra et al., 1956; Nandakarni, 1954). In Assam, the plant is given for nervous and general debility, anthelmintic and also for piles, leucoderma, leprosy, jaundice, and liver complaints (Anonymous, 1959; Chopra et al., 1956; Nandakarni, 1954). The medicinal uses of hitherto report include effect on liver diseases (Nandakarni, 1954; Siddiqui & Husain1992), eye diseases (Jain & Verma, 1981). This plant also possesses anti-oxidant activity and shows moderate anticancer activity against Vero, Hep-2, and A-549 cancer cell lines.

In total, 3900 plant species are used by tribals as food out of which 145 species are root and tubers and 521 species are leafy vegetables (Kamble & Jadav, 2013). Leafy vegetables hold an important place in well-balanced diets. The idea of a well-balanced diet itself has changed in recent years and lesser amounts of red meat and more vegetable and fruits are advised (Ame & Gold, 1996; Lucarini & Canali, 1996; Kratzer & Vohra, 1986). The ethnobotanical reports offer information of the medicinal properties of green leafy vegetables which include details on their anti-diabetic, anti-histaminic, anti-carcinogenic, and antibacterial activities (Kesari et al., 2005; Yamamura et al., 1998; Khanna et al., 2002; Kubo et al., 2004). This health-promoting attributes of green leafy vegetables are linked with their nutritional and non-nutrient content bioactive properties. Green leafy vegetables have long been reported to contribute to the dietary vitamin and mineral intake of local populations (Smith, 1982; Nordeide et al., 1996; WHO, 1990) recommended a minimum daily intake of 400 g fruits and vegetables (WHO, 1990; Rathee et al., 2010) stated the use of *Basella rubra* in habitual headaches and laxatives. Oyewole (2012) and Kachchhava (2006) stated the anti-microbial and anti-inflammatory effect of *Basella alba*. Semwa et al. (2011) stated the anti-bacterial property of *Diplazium esculentum*. Antitumor, antipyretic, antiepileptic, anti-inflammatory, antiulcer, antispasmodic, diuretic, antihypertensive, cholesterol lowering, antioxidant, antidiabetic, hepatoprotective, antibacterial, and antifungal activities of *Moringa oleifera* were stated (Anwar et al., 2007). Hepatoprotective effects and antidiabetic properties of *Brassica juncea* were stated by Walia et al. (2011) and Thakur et al. (2014). Anti-inflammatory effects of *Amaranthus viridis* were stated by Sravan et al. (2011). Antioxidant and anti-inflammatory effects of *Brassica nigra* were stated by Alam et al. (2011). In a study made by Singh et al. (2015) on the medicinal importance of unexploited vegetables under northeastern region they have identified 31 plants having medicinal property which has curative effects on antifungal, anti-bacterial, high blood pressure, menorrhagia, UTI, food poisoning, nose bleeding, sores in tongue, diarrhea, epilepsy, muscular pains, deodorant, disinfectant, etc.

24.4   PROCESSING IMPORTANCE OF GREEN LEAFY VEGETABLES

Candlish (1983) reported that steaming of aquatica and other GLVs had highest carotene content than that of fresh and boiled ones. The presences of cis-isomers of ß-carotene in GLVs were examined before and after traditional Indonesian ways of cooking (Vanderpol et al., 1988). Effect of dehydration on the nutritive value of drumsticks was studied by Joshi and Mehta (2010). Effects of vegetable drying technique on nutrient content were studied by Kirimiri et al. (2010). The yield of carotenoids of GLVs was analyzed by Chen and Han (1990) under different methods of cooking. Mazrizal et al. (1996) reported the retention of vitamin C, iron, and ß-carotene in GLVs under different cooking methods. Total carotenoid and ß-carotene contents during processing like cooking, sun drying, frying, and fermentation were investigated by Speek et al. (1988). Mazrizal et al. (1996) studied the retention of ß-carotene of GLVs when prepared using different cooking methods like boiling, microwave steaming, and frying. In the same line, similar studies were performed by Gupta et al. (2011) where they concluded that dehydration seemed to have a minimal loss of the nutrient and anti-nutrient content of green leafy vegetables. Kailasapathy and Koneshan (1986) studied the loss of ascorbate content in eight leafy vegetables including aquatica due to wilting, from harvest up to a period of 24 h, and under environmental temperatures (24.7–25.8°C) and refrigerated conditions (4.4°C). Study performed on *Vernonia amygdalina* where it was subjected to different conventional food processing techniques, such as soaking in water overnight, blanching, abrasion with and without salt (NaCl) as well as fresh leaf samples and soaking overnight and blanching caused a significant reduction in the nutritional values of bitter leaf than other processing methods studied (Yakubu et al., 2012). In another study performed on the effect of thermal processing on anti-nutrients in common edible green leafy vegetables grown in Ikot Abasi, area of Nigeria it was observed that heating conditions reduced the levels of all the investigated anti-nutrients in vegetables, the greatest reduction in level was observed in oxalates (Udosuro et al., 2013). In another study performed by Patricia et al. (2014), it was observed that long time of cooking higher than 15 min caused negative impact by reducing nutritive value but positive impact by reducing anti-nutrients such as oxalates and phytates.

24.5 PROTEIN DEFICIENCY AND ITS PRESENT SCENARIO

Protein deficiency is one of the major problems in the developing world (Beulajosepin & Mungikar 2013; Govindaraju, 2003). Scientists across the world are facing a grim challenge to shorten this gap of demand and supply of protein deficiency by finding out new resources of plant proteins which are most abundant in the world. Studies of plant proteins as a non-conventional source have been on the increase, due to the new challenges of providing adequate protein for an expanding world population (Zhu & Fu, 2012). A number of vegetable proteins have been tried for incorporation in various food products as functional and nutritional ingredients (Govindaraju, 2003). About 36 million people die every year as a result of hunger on contrary it has been reported that amino acid compositions of leaf protein concentrate (LPC) are as good or as better than that of many common foodstuffs (Gerloff, 1965). Many of the local vegetable materials are underexploited because of inadequate scientific knowledge of their nutritional potentials. Though several works reporting compositional evaluation and functional properties of various types of edible wild plants are in use in the developing countries they abound in literature, and much is still to be done.

The socioeconomic and nutritional status of mostly the rural people of Assam is not satisfactory as compared to the national average. Hence these green leafy vegetables can be used in uplifting the nutritional status of state so that it is at par with that of the country as a whole.

24.6   FUNCTIONAL PROPERTIES OF LEAF PROTEIN CONCENTRATE

The functionality of food protein has been defined as those physical and chemical properties which affect the behavior of proteins in food systems during processing, storage, preparation and consumption (Damodaran, 1997). Many factors influence the functional properties of proteins, including moisture, temperature, pH, enzymes concentrations, reaction time, chemical additives, mechanical processing, ionic strength, and amount, sequence, rate, and time of additives (Kinsella, 1982). It is well known that proteins are of prime importance to health, but they are deficient in diets of most people in the developing countries. Available literature clearly indicates that, apart from lower methionine content, the amino acid profiles of leaf protein from most species compare favorably well with those of soybean, meat, fish, and egg and generally surpass the FAO essential amino acid pattern (Agbede, 2006; Barbeau,1989; Eggum, 1970).

The behavior of functional properties of a protein is affected by both intrinsic and extrinsic factors. The intrinsic factors are shape, size, amino acid composition and sequence, the distribution of net charges, the ratio between hydrophobicity and hydrophilicity, secondary, tertiary, and quaternary structures of the protein as well as the protein's capacity to interact with other components in the food system (Damodaran, 1997). The extrinsic factors that affect the functionality of proteins are pH, temperature, moisture, chemical additives, mechanical processing, enzymes, and ionic strength (Kinsella, 1982).

24.6.1   SOLUBILITY

The solubility of protein is the most important functional property since the protein needs to be soluble in order to be applicable in food systems. Other functional properties like emulsification, foaming, and gelation are dependent on the solubility of proteins (Vaclavik & Christian, 2003). Solubility occurs when equilibrium exists between hydrophilic and hydrophobic interactions. The solubility of a protein is related to the pH, where it is minimal at its isoelectric point, making the environmental pH the most important factor when it comes to the degree of protein solubility. The solubility is also influenced by temperature and ionic strength (Bolontrade et al., 2013). Freezing, heating, drying, and shearing are also factors that have an influence on protein solubility in food systems (Vaclavik & Christian, 2003).

Insoluble proteins are not good for food applications and thus it is important that denaturation caused by factors like heat is controlled so that the protein solubility will not be affected in a negative way (Raikos et al., 2007).

24.6.2 EMULSIONS

Emulsions consist of two liquids that are immiscible, where one of the liquids is dispersed with the other in the form of small droplets. The droplets in an emulsion are called the dispersed (or internal) phase, whereas the surrounding liquid is referred to the continuous (or external) phase (Dickonson & Mclements, 1995). Emulsions can be classified according to the distribution of the oil and the aqueous phase. A system where the oil droplets are dispersed in the aqueous phase is called oil-in-water emulsion (O/W). The examples in food systems include mayonnaise, milk, cream, soups, and sauces. The opposite of an O/W emulsion is water-in-oil (W/O) but there are also water free emulsions and multiple emulsions (O/W/O or W/O/W).

24.6.3 FOAMING

Foam consists of a gas phase, a liquid phase, and a surfactant (e.g., proteins) and the process of whipping or shaking results in the formation of foams. The foods that are made up of foams are whipped toppings, meringues, ice creams, chiffon desserts, and angel cakes (Kinsella, 1982; Yang & Baldwin, 1995). Angel cakes and other baked goods are solid foams. Foams are formed through unfolding and absorption of the protein, at the air–water interface, as well as film formation around the air bubbles. Different proteins have different abilities to form and stabilize foams, and just as in the case of proteins and their different emulsifying properties, this is related to different physical properties of the proteins.

24.6.4 GELATION

Gelation/coagulation is a functional property that is highly important in the food industry. Gels are intermediate between solid and liquid where crosslinking of polymeric molecules makes an intermolecular network within a liquid medium. In food systems, this liquid affects the nature and the magnitude of intermolecular strengths that keep the integrity of the polymeric network (Oakenfull, 1987). Campbell et al. (2003), for example, the making and texturization of foods like cakes, creams, omelets, confectionary, and sauces are dependent on the eggs ability to form gel networks upon heating (Kiosseoglou, 2003). When eggs are used in baked foods, the protein molecules aggregate and form insoluble networks (a gel or coagulum). It is these networks that give the foods like cakes muffins, etc., their height, volume, and stability (American Egg Board, 2013). Studies show that the egg white proteins coagulate at different temperatures when heated, starting at 61.5°C and reach complete coagulation at 73.0°C (Johnson & Zabik, 1981). The proteins of egg white are sometimes referred to as coagulation type proteins (Gossett et al., 1984) and according to Coultate (2009), it is the rapid denaturation of ovalbumin that contributes to the setting of gel when egg white is heated.

24.7 EXTRACTION OPTIMIZATION AND APPLICATION OF LEAF PROTEIN CONCENTRATE IN COMBATING THE PROTEIN DEFICIENCY

One of the major challenges faced by the underdeveloped nation apart from epidemic and endemic disease is malnutrition. Malnutrition affects the socio-economic growth of a nation by stunting a workforce both physically and mentally. The consequence of poverty, poor health, and poor nutrition has multiple effects on the general welfare of the population and it also contributes significantly toward keeping a population in a downward trend of poverty and nutritional insecurity (Fuglie, 2006).

Response surface methodology (RSM) is an effective statistical technique for optimizing complex processes and is one of the commonly used statistical techniques for designing of experiments for food processes and food formulations. RSM has been shown to be an effective tool for optimizing processes by different workers (Zhu & Fu, 2012; Wani et al., 2006; Vatsala et al., 2001; Thakur & Saxena, 2000). Basically, RSM relates product properties by using regression equations that describe interrelations between input parameters and product properties. Colonna et al. (1984) and Onwulata et al. (2000) used RSM to optimize the conditions for an extruder, producing an amaranth-based snack food. Vatsala et al. (2001) used RSM to analyze the effect of corn flour, green gram flour, xanthan, guar gum, arabic gum, and carboxymethyl cellulose (CMC) on the sensory attributes (expansion ratio) of an extruded snack food and found that the responses were affected mostly by changing the levels of corn flour, green gram flour, and guar gum and to a lesser extent by changing the levels of xanthan, arabic gum, and CMC. Onwulata et al. (2000) optimized the ingredients and process conditions for preparing puri using RSM. Some good examples of the appropriate applications of this technique were tried on textured products for the optimization of complex products, properties and many process variables (Sebio & Chang, 2000; Rosell et al., 2001; Latharn, 1997).

The amount of protein that is available from a particular source of food or non-food is called extractable proteins in traditional optimization, the process of protein extraction was achieved by the one-factor-at-a-time approach, which was time-consuming and could ignore the interactions between the variables. In the conventional extraction procedure, many variables such as the extraction time, temperature, and solid–liquid ratio (S/L) significantly influence the extraction efficiency. Ultrasonic-assisted extraction (UAE) has been widely employed in the extraction of valuable compounds from biomass since it has many advantages such as working at (or close to) ambient temperature, a higher efficiency than the conventional extraction methods, and a lower cost (Roldan et al., 2008). Ultrasonic-assisted extraction has been studied in the extraction of biological compounds from different plant materials (Hemwimol et al., 2006; Rodrigues et al., 2008; Velickovic et al., 2008).

24.8   IMPORTANCE OF CYTOTOXICITY STUDY

As it has been already discussed that green leafy vegetables occupy an important position among other vegetables because of its rich nutritional values and significance in food security and are vital for income generation (Akenga, 2005) but despite all these advantages, several studies have reported that some green leafy vegetables are potentially toxic to humans. Hence there is a need to extensively study the wide range of phytochemicals and their toxic behavior in order create an awareness and educate the common people on the possible side effects of these green leafy vegetables which may contain these toxins. The toxicity of these plants are mainly due to various phytochemical groups which includes alkaloids, glycosides, oxalates, resins, essential oils, amino acids, furanocoumarins, polyacetylenes, protein, peptides, coumarins, flavonoids, and glycosides (Pfander, 1984; Concon, 1988; Ezekwe et al., 1999; Palaniswamy et al., 1997; Uiso et al., 1995; FAO, 1992).

24.9 PREPARATION OF EXTRUDED FOOD INCORPORATING LPC USING EXTRUSION TECHNOLOGY

Extrusion is a process in which semi solids are formed through a die. There are different types of extruders which are available in the food market including ram or piston types and screw or worn types (Harper, 1989. It is used to cook food with the act of heating and provide shape to the products. This process results in the formation of the viscous, plastic-like dough from moistened, starchy and proteinaceous foods. The cooking of food results in gelatinization of starch, denaturation of protein, inactivation of raw food enzymes, destruction of naturally occurring toxic substances, diminishing of microbial count in the final product, etc. The hot, plastic extruded product expands rapidly upon discharge due to sudden change (decrease) in pressure. A porous texture is maintained after expansion through cooling and drying. The advantages of food extrusion are high productivity, low cost, versatility, shaping the product, and very less wastage.

Extrusion has been used for large number of food applications due to its unique positive features. It is considered that high temperature and short-time bioreactor is used to transform raw ingredients into a variety of modified intermediate and finished products. The product is treated with heating as well as intense mechanical shearing, compression, and torque which breaks the covalent bonds in biopolymers (Singh et al., 2007). The combined influence of temperature, pressure, shear, and time modifies the functional properties of the food ingredients which in turn helps in preventing the inactivation of the undesirable enzymes that can affect the quality and eliminate several anti-nutritional factors, such as trypsin inhibitors, hemaglutinins, tannins, and phytates (Singh et al., 2007; Cavalho et al., 2004). There are different types of extruders available in the market namely segmented screw/barrel single-screw "wet" extruders, dry extruders, interrupted-flight screw extruders, single screw extruders and twin screw extruders (Riaz, 2000).

During the recent years quite a number of technologies in food processing have emerged which has made an impact on the availability of food products. Food extrusion is one of the latest multidimensional food processing techniques. Great possibilities are offered in food processing field by the use of extrusion technology to modify physicochemical properties of food components. The extruded food, even enhances its biological value, and can be characterized by improving physicochemical properties superior to the original raw material. Extrusion cooking has defined a unique tool in introducing the thermal and mechanical energy to food ingredients, forcing the basic components of the ingredients, such as starch and protein, to undergo chemical and physical changes. Extrusion combines several unit operations including mixing, cooking, kneading, shearing, shaping, and forming so it is a highly versatile unit operation that can be applied to a variety of food processes. Extrusion has been used for years provided the means of producing new and creative foods. One of the major advantages of extrusion cooking is the capability to produce a wide range of finished products with minimum processing time and use of inexpensive raw materials. According to Harper (1989) extrusion has found numerous applications, which includes an array of foods such as nutritious precooked food mixtures for infant feeding and confectionery products like salty and sweet snacks, ready to eat cereals, co-extruded snacks, indirect expanded products, croutons for soups, and salads. It is also used in designing veterinary products such as dry pet foods, fish foods, meat analogs, and materials from defatted high-protein flours.

24.10 PREPARATION OF PASTA (AN EXTRUDED FOOD) BY INCORPORATING LEAF PROTEIN CONCENTRATES

Pasta is a traditional extruded food which has its origins from first century BC (Kruger et al., 1996). It is very popular for its ease of cooking and nutritional qualities. Pasta is traditionally made from semolina including other nontraditional ingredients to create culinary diversity for improving nutritional quality (Toepfer et al., 1972) and is regarded as low glycemic index food (Bjork et al., 2000; Jenkins et al., 1988). Utilization of durum wheat for snack foods has been well identified (Toepfer et al., 1972). As wheat derived staple food; pasta is second to bread in the world consumption (Mariani, 1988). Its worldwide acceptance is attributed to its low cost, ease of preparation, versatility, sensory attributes, and long shelf-life. Pasta is dough which is produced from Durum wheat and is also one of the most popular foods in Mediterranean area, Italy, and other countries of Western Europe. There are two main types of pasta, namely, fresh pasta (non-durum) which is mixed with egg and dried pasta which is made from semolina flour ground from Durum wheat and mixed with water.

24.11 IMPORTANCE OF STORAGE STUDIES

Storage of raw materials of food products is an integral aspect of food industry and is one of the prerequisites for further processing of fresh produce and successful marketing. There is deterioration of quality and loss of valuable nutrients if storage conditions are not adequate. Such deteriorative changes in the food are initiated or accelerated by light such changes as photo-oxidation, destruction of fats, and proteins and inducing changes in the food pigments. Many foods stored at ambient temperature would be subjected to frequent changes in these environmental variations during transportation, distribution, and stocking. Some food products if stored for a prolonged period at a certain combination of ambient condition either intentionally or unintentionally will deteriorate in quality.

24.12   CONCLUSION

It can be concluded that the green leafy vegetables, which are mostly neglected, have a good potential in terms of food value and can serve as an easily accessible food resources. There are plethora of green leafy vegetables whose nutritional and anti-nutritional profiles are yet to be documented and are lying hidden within mother nature. The documentation of these nutritional and anti-nutritional profiles will not only serve as a repository for the future research but will also educate humankind in their daily eating habits. Innovative ways for formulation of recipes will encourage people to consume green leafy vegetable. Further studies on the preservation of these leafy vegetables can be taken up in order to prevent nutritional losses and for availability during lean seasons.

KEYWORDS

- **anti-nutrient**
- **vegetable**
- **hydrophilicity**
- **hydrophobicity**
- **solubility**

REFERENCES

Abara, A. E. Tannin Content of *Dioscorea bulbufera. J. Chem. Soc. Niger* **2003,** *28,* 55–56.

Agbede, J. O. Biochemical and Nutritive Quality of the Seed and Leaf Protein Concentrates form the Seed and Leaf Protein Concentrates from Under-utilized Tree and *Herbaceous legumes*. Ph.D. Thesis, Federal University of Technology, 2006.

Akenga, T. Potential Toxicity of Some Traditional Leafy Vegetables Consumed in Nyang'oma Division, Western Kenya. *Afr. J. Food Agric. Nutr. Dev.* **2005,** *5* (1), 1–13.

Akindahunsi, A. A.; Salawu, S. O. Phytochemical Screening of Nutrients and Antinutrient Composition of Selected Tropical Green Leafy Vegetables. *Afr. J. Biotech.* **2005,** *4* (6), 497–501.

Alam, M. B.; Hossain, M. S; Haque, M. E. Antioxidant and Anti-inflammatory Activities of the Leaf Extract of *Brassica nigra*. *Int. J. Pharm. Sci. Res.* **2011,** *2* (2), 303–310.

Aletor, O. L.; Oshodi, A. A.; Ipinmoroti, K. Chemical Composition of Common Leafy Vegetables and Functional Properties of Their Leaf Protein Concentrates. *Food Chem.* **2002,** *78* (1), 63–68.

Aletor, V. A.; Adeogun, O. A. Nutrient and Antinutrient Components of Some Tropical Leafy Vegetables. *Food Chem.* **1995,** *53* (4), 375–379.

Ame, B.; Gold, L. S. Mom was Right, at Least about Fruits and Vegetables. *J. Chem. Heal Saf.* **1996,** *3,* 17–21.

American Egg Board (Electronic) 2013. http://www.aeb.org/ (accessed April 4, 2013).

Anonymous. *Wealth of India, Raw Materials;* CSIR: New Delhi, 1959; Vol. 5, p 237.

Anwar, F.; Latif, S.; Ashraf, M.; Gilani, A. H. *Moringa oleifera* a Food Plant with Multiple Medicinal Uses. *Phytother. Res.* **2007,** *21* (1), 17–25.

AOAC. *Official Methods of Analysis,* 15th ed.; Association of Official Analytical Chemists: Washington, DC, 1990.

Applebaum, S. W.; Marco, S.; Birk, Y. Saponins as Possible Factors of Resistance of Legume Seeds to the Attack of Insects. *J. Agric. Food Chem.* **1969,** *17* (3), 618–620.

Barbeau, W. E. Nutritional Evaluation of Experimental Weaning Foods Prepared from Green Leaves, Peanut Oil and Legume Flour. *Plant Foods Hum. Nutr.* **1989,** *39* (4), 381–392.

Beulajosepin, E. D.; Mungikar, A. M. Biochemical Analysis of Leaf Protein Concentrate Prepared from Selected Plant Species of Tamil Nadu. *Int. J. Pharm. Res. Dev.* **2013,** 5, 7–14.

Bharadwaj, R.; Singh, R. K.; Sureja, A. K.; Upadhaya, S.; Devi, M.; Singh, A. I. S. H. S. Acta Horticulture 806, International Symposium on Underutilized Plants for Food Security, Nutrition, Income and Sustainable Development, 2008.

Binita, R.; Khetarpaul, N. Probiotic Fermentation: Effect on Antinutrients and Digestability of Starch and Protein of Indigenous Developed Food Mixture. *J. Nutr. Health.* **1997,** *11* (3), 139–147.

Bjorck, I.; Liljeberg, H.; Östman, E. Low Glycaemic Index Foods. *Brit. J. Nutr.* **2000,** *83,* 149–155.

Bolontrade, A. J.; Scilingo, A. A.; Añón, M. C. Amaranth Proteins Foaming Properties: Adsorption Kinetics and Foam Formation—Part 1. *Colloids Surf B. Biointerfaces* **2013,** *105,* 319–327.

Brandt, K.; Lietz, G.; Kobæk-Larsen, M.; Christensen, L. P. Health Promoting Compound in Vegetable and Fruits: A Systematic Approach for Identifying Plant Component with Impact on Human Health. *Trends Food Sci. Technol.* **2005,** *15,* 384–393.

Campbell, L.; Raikos, V.; Euston, S. R. Modification of Functional Properties of Egg-white Proteins. *Food Nahrung.* **2003,** *47* (6), 369–376.

Candlish, J. K.; Gourley, L.; Lee, H. P. Dietary Fiber and Starch Contents of Some Southeast Asian Vegetables. *J. Agric. Food Chem.* **1987,** *35* (3), 319–321.

Candlish, J. K. Tocopherol Content of Some Southeast Asian Foods. *J. Agric. Food Chem.* **1983,** *31* (1), 166–168.

Carvalho, A. S.; Silva, J.; Ho, P.; Teixeira, P.; Malcata, F. X.; Gibbs, P. Relevant Factors for the Preparation of Freeze Dried Lactic Acid Bacteria. *Int. Dairy J.* **2004,** *14* (10), 835–847.

Chen, B. H.; Han, L. H. Effects of Different Cooking Methods on the Yield of Carotenoids in Water Convolvulus (*Ipomoea aquatica*). *J. Food Prot.* **1990,** *53* (12), 1076–1078.

Chen, B. H.; Yang, S. H.; Han, L. H. Characterization of Major Carotenoids in Water Convolvulus (*Ipomoea aquatica*) by Open-column, Thin Layer and High Performance Liquid Chromatography. *J. Chrom.* **1991,** *543,* 147–155.

Chen, M. L.; Chia, D. F.; Run, J. Q. Effect of Dietary Vegetable (*Water convolvulus*) on Cholesterol Metabolism in Rats. *J. Nutr.* **1984,** *114* (3), 503–510.

Chopra, R. N.; Nayar, S. L.; Chopra, I. C. Glossary of Indian Medicinal Plants. CSIR: New Delhi, 1956; p 84.

Colonna, P.; Doublier, I. L.; Melcion, J. P.; De Monredon, F.; Mercier, C. Extrusion Cooking and Drum Drying of Wheat Starch. I. Physical and Macromolecular Modification. *Cereal. Chem.* **1984,** *61* (6), 538–543.

Concon, J. M. Food Toxicology, Part A: *Principles and Concepts;* Marcel Dekker: New York, NY, 1988; p 8.

Coultate, T. P. Food: The Chemistry of Its Components. *Cambridge Royal Soc. Chem.* **2009,** 169–184.

Craig, W. J. Health-promoting Properties of Common Herbs. *Am. J. Clin. Nutr.* **1999,** *70* (3), 491S–499S.

Damodaran, S. Food Proteins: An Overview. In *Food Proteins and Their Applications;* Damodaran, S., Paraf, A., Mercel., Eds.; CRC Press: Boca Raton, FL, 1997.

Daniel, M. Polyphenols of Some Indian Vegetables. *Curr. Sci.* **1989,** *58* (23), 1332–1333.

Das, A. K. The Brahmaputra Changing Ecology. In *The Cultural Dimension of Ecology;* Saraswati, B., Ed.; INGCA and D. K. Print World Pvt. Ltd: New Delhi, 1998. http://ignca. nic.in/cd_07012.htm. Retrieved August 11, 2010.

Dickonson, E.; McClements, D. J. *Advances in Food Colloids;* Chapman and Hall: London, 1995.

Duke, J. A.; Ayensu, E. S. *Medicinal Plants of China;* Reference Publ. Inc.: Algonac, MI, 1985.

Eggum, B. O. Protein Quality of Cassava Leaves. *Br. J. Nutr.* **1970,** *24* (03), 761–768.

Erdman, J. N. Oily Seed Phytates Nutritional Implications. *J. Amer. Oil Chem. Soc.* **1979,** *56* (8), 736–741.

Ezekwe, M. O.; Thomas, R. A.; Membrahtu, T. Nutritive Characterization of Purslane Accessions as Influenced by Planting Date. *Plant Foods Hum. Nutr.* **1999,** *54,* 183–191.

Faber, M. P. M. A.; Venter, S. L.; Dhansay, M. A.; Benade, A. S. Home Gardens Focusing on the Production of Yellow and Darl Green Leafy Vegetables Increase the Serum Retinol Concentrations of 2–5 y Old Children in South Africa. *Am. J. Clin. Nutr.* **2002,** *76* (5), 1048–1054.

Fasuyi, A. O. Nutritional Potentials of Some Tropical Vegetable Meals. Chemical Characterization and Functional Properties. *Afr. J. Biotech.* **2006,** *5* (1), 49–53.

Food and Drug Administration (FAO). *Pathogenic Microorganisms and Natural Toxins Handbook;* Centre for Food Safety and Applied Nutrition: College Park, MD, 1992.

Fuglie, L. J. *The Moringa Tree. A Local Solution to Malnutrition B P 5338;* Church World Service: Dakar, Senegal, 2006; p 3.

Gerloff, E. D. Amino Acid Composition of Leaf Protein Concentrate of Leaf Protein Concentrate. *J. Agric. Food Chem.* **1965,** *13* (2), 139–143.

Gossett, P. W.; Rizvi, S. S.; Baker, R. C. Symposium Gelation in Food Protein Systems, Quantitative Analysis of Gelation in Egg Protein Systems. *Food Tech.* **1984,** *38* (5), 67–96.

Govindaraju, K. *Studies on the Preparation and Characterization of Protein Hydrolysates from Groundnut and Soybean Isolates.* Department of Protein Chemistry and Technology. Central Food Technological Research Institute: Mysore, India, 2003.

Gupta, K.; Barat, G. K.; Wagle, D. S.; Chawla, H. K. Nutrient Contents and Antinutritional Factors in Nonconventional Leafy Vegetables. *Food Chem.* **1989,** *31* (2), 105–116.

Gupta, K.; Wagle, D. S. Nutrition and Anti Nutrition Factors of Green Leafy Vegetables. *J. Agric. Food Chem.* **1998,** *36,* 472–474.

Gupta, S.; Gowri, B. S.; Lakshmi, A. J.; Prakash, J. Retention of Nutrients in Green Leafy Vegetables. *J. Food Sci. Tech.* **2011,** *50* (5), 918–925.

Harper, J. M. Food Extruders and Their Applications. In *Extrusion Cooking;* Mercier, C., Linko, P., Harper, J. M., Eds.; American Association of Cereal Chemists Inc.: St Paul, MN, 1989.

Heldt, H. W. Plant Biochemistry and Molecular Biology. *Photosynthetica* **1997,** *40* (3), 388.

Hemwimol, S.; Pavasant, P.; Shotipruk, A. Ultrasound-assisted Extraction of Anthraquinones from Roots of *Morinda citrifolia. Ultrason. Sonochem.* **2006,** *13* (6), 543–548.

Jain, S. P.; Verma, D. M. *Medicinal Plants in the Folklore of Northeast Haryana;* National Academy Science Letters: India, 1981; Vol. 4(7), pp 269–271.

Jenkins, D. J.; Wolever, T. M.; Jenkins, A. L. Starchy Foods and Glycemic Index. *Diabetes Care* **1988,** *11* (2), 149–159.

Johnson, T. M.; Zabik, M. E. Gelation Properties of Albumen Proteins, Singly and in Combination. *Poultry Sci.* **1981,** *60* (9), 2071–2083.

Joshi, P.; Mehta, D. Effect of Dehydration on the Nutritive Value of Drumstick Leaves. *J. Metabolomics. Syst. Biol.* **2010,** *1* (1), 5–9.

Kachchhava, A. B. *Studies on Anticonvulsant, Analgesic and Anti-inflammatory Activities of Leaf Extracts of Basella Alba;* Dissertation Work Submitted to the Rajiv Gandhi University of Health Science: Bangalore, India, 2006.

Kailasapathy, K.; Koneshan, T. Effect of Wilting on the Ascorbate Content of Selected Fresh Green Leafy Vegetables Consumed in Sri Lanka. *J. Agric. Food Chem.* **1986,** *34* (2), 259–261.

Kala, A.; Prakash, J. Nutrient Composition and Sensory Profiles of Differently Cooked Green Leafy Vegetables. *Int. J. Food Prop.* **2004,** *7* (3), 659–669.

Kamble, V. S.; Jadav, V. D. Traditional Leafy Vegetables: A Future Herbal Medicine. *Int. J. Agric. Food Sci.* **2013,** *3* (2), 56–58.

Kesari, A. N.; Gupta, R. K.; Watal, G. Hypoglycemic Effects of *Murraya koenigii* on Normal and Alloxan-diabetic Rabbits. *J. Ethnopharm.* **2005,** *97* (2), 247–251.

Khanna, A. K.; Rizvi, F.; Chander, R. Lipid Lowering Activity of *Phyllanthus niruri* in Hyperlipemic Rats. *J. Ethnopharm.* **2002,** *82* (1), 19–22.

Kinsella, J. E. *Relationships between Structure and Functional Properties of Food Proteins. Food Proteins;* Fox, P. F., Condon, J. J., Eds.; Applied Science Publishers LTD: England, 1982; pp 51–58.

Kiosseoglou, V. Egg Yolk Protein Gels and Emulsions. *Curr. Opin. Colloid Interface Sci.* **2003,** *8* (4), 365–370.

Kiremire, B. T.; Musinguzi, E.; Kikafunda, J. K.; Lukwago, F. B. Effect of Vegetable Drying Technique on Nutrient Content, a Case Study of South Western Uganda. *Afr. J. Food Agric. Nutr. Dev.* **2010,** *10* (5), 2587–2600.

Kratzer, F.; Vohra, P. The Effect of Diet on Plasma Lipids, Lipoprotein, and Coronary Heart Disease. *J. Amer. Diet. Asso.* **1986,** *88,* 1373–1411.

Kruger, J. E.; Matsuo, R. B.; Dick, J. W. *Pasta and Noodle Technology;* American Association of Cereal Chemists: Eagan, MN, 1996; pp1–12.

Kubo, I.; Fujita, K. I.; Kubo, A.; Nihei, K. I.; Ogura, T. Antibacterial Activity of Coriander
Volatile Compounds Against *Salmonella choleraesuis*. *J. Agric. Food Chem.* **2004,** *52* (11),
3329–3332.

Ladeji, O.; Akin, C. U.; Umaru, H. A. Level of Antinutritional Factors in Vegetables
Commonly Eaten in Nigeria. *Afr. J. Nat. Sci.* **2004,** *7,* 71–73.

Latharn, M. C. *Human Nutrition in the Developing World;* Food and Agricultural Organization of the United Nations: Rome, Italy, 1997; p 522.

Lucarini, S.; Canali, R. *National Research Council, Carcinogens and Anti Carcinogens in the
Human Diet;* National Academy Press: Washington, DC, 1996; pp 221–232.

Masrizal, M. A.; Giraud, D. W.; Driskell, J. A., et al. Retention of Vitamin C, Iron and ß
Carotene in Vegetables Prepared Using Different Cooking Methods. *J. Food Qual.* **1996,**
20 (5), 403–418.

Mariani-Constantini, A. Image and Nutritional Role of Pasta-changing Food Patterns. In
Durum Chemistry and Technology; Lintas, G. C., Ed.; American Association for Clinical
Chemistry: St Paul, MN, 1988.

Munger, H. M. Enhancement of Horticulture Crops for Improved Health. *Hortic. Sci.* **1999,**
34 (7), 1158–1159.

Nandakarni, K. M. *Indian Materia Medica;* Popular Book Depot: Bombay, 1954; Vol. 1, pp
516–518.

Negi, P. S.; Roy, S. K. Effect of Blanching and Drying Methods on β-carotene, Ascorbic
Acid, and Chlorophyll Retention of Leafy Vegetables. *LWT Food Sci. Technol.* **2000,** *33*
(4), 295–298.

Ngamsaeng, A.; Thy, S.; Preston, T. R. *Duckweed Duckweed (Lemna minor) and Water
Spinach Ipomoea aquatica as Protein Supplements for Ducks Fed Broken Rice as the
Basal Diet;* Livestock Research for Rural Development: Cali, Colombia, 2004; Vol. 16,
pp 18–24.

Nordeide, M. B.; Hatløy, A.; Følling, M.; Lied, E.; Oshaug, A. Nutrient Composition and
Nutritional Importance of Green Leaves and Wild Foods in an Agricultural District,
Koutiala, in Southern Mali. *Int. J. Food Sci. Nutr.* **1996,** *46* (6), 455–468.

Oakenfull, D. Gelling Agents. *Crit. Rev. Food Sci. Nutr.* **1987,** *26,* 1–25.

Ogle, M.; Ha, T. A. D.; Generose, M.; Hambraeus, B. Micronutrient Composition and Nutritional Importance of Gathered Vegetables in Vietnam. *Int. J. Food Sci. Nutr.* **2001,** *52* (6),
485–499.

Onwulata, C. I.; Konstance, R. P.; Strange, E. D.; Smith, P. W.; Holsinger, V. H. High-fiber
Snacks Extruded from Triticale and Wheat Formulations. *Cereal. Foods World* **2000,** *45*
(10), 470–47.

Ortaliza, I. C.; Del Rosario, I. F.; Caedo, M. M.; Alcaraz, A. P. The Availability of Carotene
in Some Philippine Vegetables. *Phillipine J. Sci.* **1969,** *98,* 123–131.

Oyewole, O. A. The Antimicrobial Activities of Ethanolic Extracts of *Basella alba* on
Selected Microorganisms. *Sci. J. Microbiol.* **2012,** *1* (5), 113–118.

Palaniswamy, U. R.; McAvoy, R.; Bible, B. Omega-3 Fatty Acid Concentration in *Portulaca
Oleraceae* L. is Altered by the Source of Nitrogen in Hydroponics Solution. *J. Hort. Sci.*
1997, *32,* 462–463.

Pfander, F. A. *Color Atlas of Poisonous Plants: A Handbook for Pharmacists, Doctors, Toxi-
cologists and Biologists;* Wolfe Publishing Limited: London, 1984; pp 10–222.

Raikos, V.; Campbell, L.; Euston, S. R. Rheology and Texture of Hen's Egg Protein Heat-set Gels as Affected by pH and the Addition of Sugar and/or Salt. *Food Hydrocoll.* **2007,** *21* (2), 237–244.

Ramakrishnan, P. S. Ecology and Traditional Wisdom. In *The Cultural Dimension of Ecology;* Saraswati, B., Ed.; New Delhi INGCA and D. K. Print world Pvt.: Delhi, India, 1998. http://ignca.nic.in/cd_07010.htm.

Rao, T. V. R. K.; Vijay, T. Iron, Calcium,β- carotene, Ascorbic Acid and Oxalic Acid Content of Less Common Leafy Vegetables Consumed by the Tribals of Purnia District of Bihar. *J. Food Sci. Technol.* **2002,** *39* (5), 560.

Riaz, M. N. Selecting the Right Extruder. In *Extrusion Cooking Technology and Application;* Wood Head Publishing: Cambridge, England, 2000.

Rodrigues, S.; Pinto, G. A. Optimization of Ultrasound Extraction of Phenolic Compounds from Coconut (*Cocos nucifera*) Shell Powder by Response Surface Methodology. *Ultrason. Sonochem.* **2008,** *15* (1), 95–100.

Roldán-Gutiérrez, J. M.; Ruiz, J. J.; De Castro, M. L. Ultrasound-assisted Dynamic Extraction of Valuable Compounds from Aromatic Plants and Flowers as Compared with Steam Distillation and Superheated Liquid Extraction. *Talanta* **2008,** *75* (5), 1369–1375.

Rosell, C. M.; Rojas, J. A.; De Barber, C. B. Influence of Hydrocolloids on Dough Rheology and Bread Quality. *Food Hydrocoll.* **2001,** *15* (1), 75–81.

Sebio, L.; Chang, Y. K. Effects of Selected Process Parameters in Extrusion of Yam Flour (*Dioscorea rotundata*) on Physicochemical Properties of the Extrudates. *Food Nahrung.* **2000,** *44* (2), 96–101.

Semwa, L. A.; Kaushik, S.; Bhatt, S. P.; Negi, A. Antibacterial Activity of *Diplazium esculentum* (Retz.). *Pharmacog. J.* **2011,** *21* (4), 74–76.

Sheela, K.; Nath, K. G.; Vijayalakshmi, D.; Yankanchi, G. M.; Patil, R. B. Proximate Composition of Underutilized Green Leafy Vegetables in Southern Karnataka. *J. Hum. Ecol.* **2004,** *15* (3), 227–229.

Siddiqui, M. B.; Husain, W. Some Aquatic and Marshy Land Medicinal Plants from Hardoi District of Uttar Pradesh. *Fitoterapia* **1992,** *63,* 245–248.

Singh, G.; Kawatra, A.; Sehgal, S. Nutritional Composition of Selected Green Leafy Vegetables, Herbs, and Carrots. *Plant Foods Hum. Nutr.* **2001,** *56* (4), 359–364.

Singh, S.; Gamlath, S.; Wakeling, L. Nutritional Aspects of Food Extrusion: A Review. *Int. J. Food Sci. Tech.* **2007,** *42* (8), 916–929.

Singh, V.; Shah, K. N.; Rana, D. K. Medicinal Importance of Unexploited Vegetables Under Northeastern Regions of India. *J. Med. Plants Stud.* **2015,** *3,* 33–36.

Smith, I. F. Leafy Vegetables as Sources of Minerals in Southern Nigerian Diets. *Nutr. Rep. Int.* **1982,** *26,* 679–688.

Speek, A. J.; Speek, S. S.; Schreurs, W. H. Total Carotenoids and β- carotenoids of Thai Vegetables and the Effects of Processing. *Food Chem.* **1988,** *27* (4), 245–257.

Sravan, P. M.; Venkateshwarlu, G. K.; Vijaya, B. P.; Suvarna, D.; Dhanalakshmi, C. Effects of Anti-inflammatory Activity of *Amaranthus viridis* Linn. *Ann. Biol. Res.* **2011,** *2* (4), 435–438.

Thakur, A. K.; Chatterjee, S. S.; Kumar, V. Antidepressant-like Effects of *Brassica juncea* Leaves in Diabetic Rodents. *Ind. J. Exp. Biol.* **2014,** *52* (6), 613–622.

Thakur, S.; Saxena, D. C. Formulation of Extruded Snack Food (Gum Based Cereal-pulse Blend): Optimization of Ingredient Levels Using Response Surface Methodology. *LWT Food Sci. Technol.* **2000,** *33* (5), 354–361.

Toepfer, E. W.; Polansky, M. M.; Eheart, J. F.; Slover, H. T.; Morris, E. R.; Hepburn, F. N.; Quackenbush, F. W. Nutrient Composition of Selected Wheat and Wheat Products XI Summary. *Cereal. Chem.* **1972**, *49,* 173–186.

Udousoro, I. I.; Ekop, R. U.; Udo, E. J. Effect of Thermal Processing on Antinutrients in Common Edible Green Leafy Vegetables Grown in Ikot Abasi, Nigeria. *Pak. J. Nutr.* **2013,** *12* (2), 162–167.

Usio, F. C.; Johns, T. Risk Assessment of the Consumption of Pyrrolizidine Alkaloid Containing Indigenous Vegetable *Crotalaria brevidens* (Mitoo). *J. Ecol. Food Nutr.* **1995,** *35,* 111–119.

Vaclavik, V.; Christian, E. *Essentials of Food Science,* 2nd ed.; Kluwer Academic/Plenum Publishers: New York, 2003; p 142.

Van der Pol, F.; Purnomo, S. U.; Van Rosmalen, H. A. Trans–Cis Isomerization of Carotenoids and Its Effect on the Vitamin A Potency of Some Common Indonesian Foods. *Nutr. Rep. Int.* **1988,** *37* (4), 785–793.

Vatsala, C. N.; Saxena, C. D.; Rao, P. H. Optimization of Ingredients and Process Conditions for the Preparation of Puri Using Response Surface Methodology. *Int. J. Food Sci. Tech.* **2001,** *36* (4), 407–414.

Veličković, D. T.; Milenović, D. M.; Ristić, M. S.; Veljković, V. B. Ultrasonic Extraction of Waste Solid Residues from the Salvia Sp. Essential Oil Hydro Distillation. *Biochem. Eng. J.* **2008,** *42* (1), 97–104.

Walia, A.; Malan, R.; Saini, S.; Saini, V.; Gupta, S. Hepatoprotective Effects from the Leaf Extracts of *Brassica juncea* CCl_4 induced Rat Model. *Der Pharm. Sin.* **2011,** *2* (4), 274–285.

Wani, A. A.; Sogi, D. S.; Grover, L.; Saxena, D. C. Effect of Temperature, Alkali Concentration, Mixing Time and Meal/Solvent Ratio on the Extraction of Watermelon Seed Proteins—A Response Surface Approach. *Biosyst. Eng.* **2006,** *94* (1), 67–73.

WHO. *World Health Organization Diet, Nutrition and the Prevention of Chronic Diseases;* Report of a Joint FAO/WHO Expert Consultation, WHO Technical Report Series #916: Geneva,1990.

Yakubu, N.; Amuzat, A. O.; Hamza, R. U. Effect of Processing Methods on the Nutritional Contents of Bitter Leaf (*Vernonia amygdalina*). *Am. J. Food Nutr.* **2012,** *2* (1), 26–30.

Yamamura, S.; Ozawa, K.; Ohtani, K.; Kasai, R.; Yamasaki, K. Antihistaminic Flavones and Aliphatic Glycosides from *Mentha spicata. Phytochem.* **1998,** *48* (1), 131–136.

Yang, S. C.; Baldwin, R. E. Functional Properties of Eggs in Foods. In *Egg Science and Technology,* 4th ed.; Stadelman, W. J., Cotteril, O. J., Eds.; Food Products Press, Haworth Press: Binghamton, NY, 1995; p 405.

Zang, S. M.; Hunter, D. J.; Rosner, B. A.; Giovannucci, E. L.; Colditz, G. A.; Speizer, F. E.; Willett, W. C. Intakes of Fruits, Vegetables and Related Nutrients and Risk of Non-Hodgkins Lymphoma among Women. *Cancer Epi. Biomarkers Prev.* **2000,** *9* (5), 477–485.

Zhu, J.; Qiao, F. Optimization of Ultrasound-assisted Extraction Process of Perilla Seed Meal Proteins. *Food Sci. Biotech.* **2012,** *21,* 1701–1706.

DEVELOPMENT OF SPRAY-DRIED HONEY POWDER WITH VITAMIN C AND ANTIOXIDANT PROPERTIES USING MALTODEXTRIN AS A CARRIER

YOGITA SUHAG[*], GULZAR AHMAD NAYIK, and VIKAS NANDA

Department of Food Engineering and Technology, Sant Longowal Institute of Engineering and Technology—Deemed University, Longowal 148106, Sangrur, Punjab, India

[*]Corresponding author. E-mail: er.yogita18@gmail.com

CONTENTS

Abstract 596
25.1 Introduction 596
25.2 Materials and Methods 597
25.3 Conclusion 606
Keywords 606
References 607

ABSTRACT

This work was aimed at optimizing the formulation of spray-dried nutritionally rich honey powder with the addition of functional compounds, such as aonla (*Emblica officinalis Gaertn*) and basil (*Ocimum sanctum*) extract using response surface methodology (RSM). A Box–Behnken design was carried out with the following independent variables: inlet temperature (160–180°C), maltodextrin (40–50%), feed flow rate (0.08–0.13 mL/s), aonla extract (6–8%), and basil extract (6–8%) with respect to the following product responses namely, bulk density, hygroscopicity, antioxidant activity (AOA), total phenolic content (TPC), and vitamin C content. Statistical analysis illustrated that maltodextrin, aonla extract, and basil extract improved the retention of AOA, TPC, and vitamin C content as these were encapsulated by maltodextrin. The hygroscopicity of honey powder decreased with the increase in inlet temperature and it increased with the increase in feed flow rate. The optimum processing conditions obtained for producing spray-dried honey powder were inlet temperature of 170°C, maltodextrin concentration of 50%, feed flow rate of 0.11 mL/s, aonla extract of 8%, and basil extract of 6%. The experimental and predicted values were found to be higher in agreement.

25.1   INTRODUCTION

Honey is a natural and complex sweetening agent rich in phytochemicals, which is utilized by food processors in large range of food products (Nayik & Nanda, 2015). However, liquid honey creates problems in mass utilization and consumption due to its high viscosity and stickiness. This problem can be overcome by converting liquid honey into powder form. Spray drying proves to be a potential method in conversion of liquids into powder form due to short contact time of droplets and relatively low heat stress makes it suitable for heat-sensitive food components among other drying technologies (Du et al., 2014; Igual et al., 2014). Despite these benefits, the low molecular weight sugars present in honey present limitations like stickiness on chamber wall leading to diminished product yield. In this context, high molecular weight carrier agents like maltodextrin are recommended due to its relatively low cost, neutral aroma and taste, low viscosity at high solids concentrations, and good protection of sensitive components (vitamins and phenolics) against oxidation (Kha et al., 2010). It has been widely reported that maltodextrin improves the drying process and reduce the stickiness and agglomeration problems during storage, hence, improve the product stability. Therefore, it is commonly used for spray-drying processes, such as the drying of acai pulp (Tonon et al., 2008), black mulberry (Fazaeli et al., 2012), chicken meat protein hydrolysate (Kurozawa et al., 2009), cashew apple juice (Oliveira et al., 2009), and sumac extract (Caliskan & Dirim, 2013). Spray-drying process results in significant reduction of the antioxidant activity (AOA) and total phenolic content (TPC) which can be enhanced by the addition of aonla and basil extract (Suhag & Nanda, 2015). Basil (*Ocimum basilicum*) contains hundreds of phytochemicals known as eugenol, carvacrol, eugenol-methyl ether, caryophyllin, ursolic acid, rosmarinic acid, thymol, methyl chavicol, citral, carvacrol, and β-caryophyllene (Prakash & Gupta, 2005). Aonla (*Emblica officinalis*) is the richest source of ascorbic acid (vitamin C) containing chemical substances called leuco-anthocyanidin and polyphenols which retard the oxidation of vitamin C and also contains tannin, polyphenol, pectin, gallic acid, and fiber (Parveen & Khatkar, 2015) (Discuss major properties of aonla and basil extract with reference to this study).

The aim of research work was carried out to optimize the process parameters, namely, inlet air temperature, feed flow rate and concentrations of maltodextrin, aonla extract, and basil extract to develop honey powder with AOA, TPC, and vitamin C retention along with improved bulk density and hygroscopicity.

25.2   MATERIALS AND METHODS

25.2.1   SAMPLE PREPARATION AND SPRAY DRYING

The chemicals Folin–Ciocalteau reagent, gallic acid (GA), sodium carbonate, and maltodextrin (DE-20) were purchased from Loba Chemie Pvt. Ltd., Mumbai; 2,2-diphenyl picrylhydrazyl (DPPH), acetone, and methanol (HPLC grade) purchased from Ranbaxy, New Delhi. Aonla (*E. officinalis*, Neelam variety) was purchased from Punjab Agriculture University (PAU, Ludhiana), and basil leaves (*O. basilicum*, holy basil) purchased from local farmer. Samples of sunflower (*Helianthus annuus*) honey were collected with the help of local beekeepers. The botanical origin of the samples of honey was based on the pollen spectrum (45% and above), which was the ratio of the frequency of each pollen type in honey (Louveaux et al., 1978). The following terms were used for frequency classes: predominant pollen (> 45% of pollen grains counted), secondary pollen (16–45%), important minor pollen (3–15%), and minor pollen (< 3%). Based on the results of preliminary experiments, honey was blended with water in the ratio 1:3.5 to avoid clogging caused by high viscous nature of honey. Maltodextrin was added at three different concentrations, namely, 40%, 45%, and 50% of honey (w/w), 6–8% each of aonla, and basil extract of honey was added in the feed mixture.

A tall type laboratory scale spray dryer (S. M. Scientech, Calcutta, India) with a co-current air-flow was used for spray drying of honey. Two-fluid nozzle atomizer with nozzle tip diameter 0.7 mm and a 1.5 mm diameter nozzle screw cap was used. Three inlet temperatures of 160, 170, and 180°C were set according to design of experiments, while pump rate was adjusted to maintain outlet temperature from 80 to 90°C. After spray drying, powder was collected in an insulated glass bottle connected at the end of cyclone, removed, and then packed in polyethylene bags for further studies.

25.2.2 MOISTURE, pH, PERCENTAGE ACIDITY, AND HMF CONTENT OF AONLA, BASIL, AND HONEY

The samples of honey were analyzed according to the AOAC (2000), methods for moisture, pH, percentage acidity, and hydroxymethylfurfural content, whereas moisture, pH, and percentage acidity of aonla and basil were analyzed by methods as described by Rangana (2000). AOA and TPC of honey, aonla, and basil extract were analyzed by Kamboj et al. (2013) and AOAC (2000). All results are expressed as the average of three replications.

25.2.3 BULK DENSITY

Bulk density (g/mL) was measured by gently pouring the known mass of sample into an empty graduated cylinder, gently tapping 20–25 times, and recording the volume (Goula et al., 2004). All results are expressed as the average of three replications.

25.2.4 HYGROSCOPICITY

An amount of 1 g sample of powder was placed at 25°C in a container with NaCl saturated solution (75% RH). After one week, sample was weighed and hygroscopicity was expressed as g of adsorbed moisture per 100 g dry solids (g/100 g) as the average of three replications (Cai & Corke, 2000).

25.2.5 TOTAL PHENOLIC CONTENT (TPC)

Briefly, 250 mg of sample was mixed with 10 mL of 60% acetone and the mixture was stirred for 30 min at 30°C. Then, 60 μL of supernatant, 300 μL of Folin–Ciocalteau reagent, and 750 μL of 20% sodium carbonate in water were added in 4.75 mL of water. After 30 min absorbance was measured at 760 nm using UV–vis spectrophotometer (Hach DR 6000, Germany) with methanol as the reference. Gallic acid (0–100 mg/L) was used to produce a standard calibration curve. The TPC was expressed in mg of gallic acid equivalents (mg GAE/100 g of spray-dried powder) as the average of three replications (Liu et al., 2008).

25.2.5 DPPH RADICAL SCAVENGING ACTIVITY

An amount of 250 mg of sample was mixed with 10 mL of 60% acetone and the mixture was stirred for 30 min at 30°C. Then, 2 mL of extract was mixed with 2 mL methanolic solution containing 1 mM DPPH. The mixture was shaken vigorously and then left to stand for 30 min in the dark. The absorbance was measured at 517 nm using UV–vis spectrophotometer (Hach DR 6000, Germany). The absorbance of control was obtained by replacing the sample with methanol (Luo et al., 2009). DPPH radical scavenging activity of the sample was calculated as follows and results are expressed as the average of three replications. $(\%) = \dfrac{\text{absorbance of control} - \text{absorbance of sample}}{\text{absorbance of control}} \times 100$.

25.2.6 VITAMIN C CONTENT

Honey powder (1 mg) was treated with 20 mL of 0.4% oxalic acid at room temperature for 5 min and filtered through Whatman No. 4 filter paper. The filtrate (1 mL) mixed with 9 mL 2,6-dichlorophenol-indophenol dye and absorbance was read within 15 min at 520 nm using UV–vis spectrophotometer (Hach DR 6000, Germany) against a blank. Vitamin C was calculated on the basis of the calibration curves of vitamin C and was expressed as mg/100 g of vitamin C as the average of three replications (Egoville et al., 1998).

25.2.7 EXPERIMENT DESIGN

Response surface methodology (RSM) was adopted for the design of experimental combinations based on the multivariate nonlinear model (Montgomery, 2001). This method was applied using Design-Expert version 8.0.7.1 (Statease Inc., Minneapolis, USA) to identify the optimum levels the five variables that is, temperature (°C), feed rate (mL/s), concentration of maltodextrin (%), aonla extract (%), and basil extract (%) regarding five responses: bulk density, hygroscopicity, AOA, TPC, and vitamin C of spray-dried honey powder. A five factor and three level Box–Behnken design consisting of 46 experimental runs including six replicates at center point was employed for this purpose (Tekindal et al., 2012). The order of experiments was fully randomized. This data was analyzed by multiple regressions using the least-squares method. A second-order polynomial equation was used to express the responses as a function of the independent variables which is given below.

$$Y_k = \beta_{k0} + \sum_{i=1}^{n} \beta_{ki} x_i + \sum_{i=1}^{n} \beta_{kii} x_i^2 + \sum_{i=1}^{n-1} \sum_{j=i+1}^{n} \beta_{kij} x_i x_j \tag{25.1}$$

where Y_k = response variable; Y_1 = bulk density (g/mL); Y_2 = hygroscopicity(%); Y_3 = TPC (mg GAE/100 g); Y_4 = AOA (%), and Y_5 = vitamin C (mg/100 g); x_i represent the coded independent variables (x_1 = temperature of inlet air, x_2 = feed flow rate, x_3 = concentration of maltodextrin, x_4 = aonla extract, and x_5 = basil extract); where β_{ko} was the value of the fitted response at the center point of the design, that is, point (0,0,0), β_{ki}, β_{kii}, and β_{kij} were the linear, quadratic, and cross-product regression coefficients, respectively. The test of statistical significance was performed on the total error criteria, with a confidence level of 95%.

The significant terms in the model were found by analysis of variance (ANOVA) for each response. The adequacy of the model was checked by calculating the R^2, adj-R^2, Pred R^2, and Fisher's F-test. The mathematical model is reliable with an R^2 value closer to one. The regression of coefficients was then used to make statistical calculation to generate three-dimensional plots from the regression model. Correlation coefficient between TPC and AOA of spray-dried honey powder was calculated by correlation graphics using software Statistica 7.0.

25.2.8   RESULTS AND DISCUSSION

The values obtained from physicochemical properties of aonla, basil, and honey are shown in Table 25.1. Pollen analysis showed the presence of 64% pollen of *H. annuus*. The moisture content, pH, and percentage acidity of honey, aonla, and basil extract accounted for 18.8%, 82.5%, 82.3%, 3.9%, 3.1%, 4.3%, 1.1%, 2.5%, and 0.7%, respectively. The HMF content of honey was 5.49 mg/100 g, whereas it was not detected in aonla and basil extract. Vitamin C content was not detected in honey whereas vitamin C content of aonla was 449 mg/100 g. The TPC and AOA of aonla and basil were 45.62 mg GAE/100 g, 65.94%, and 48.55 mg GAE/100 g, 69.36 %, respectively. The physico-chemical properties of honey, aonla, and basil extract are given in Table 25.1. The HMF content of honey was observed to be 5.49 mg/100 g whereas it was not detected in aonla and basil extract. Table 25.2 summarizes the significant coefficients of the second-order polynomial equation. For any of the terms in the model, a large regression coefficient and a small p value would indicate a more significant effect on the respective response variables. ANOVA (Table 25.2) showed that the resulting quadratic model adequately represented the experimental data with coefficients of multiple determinations (R^2) of 0.98, 0.99, 0.99, 0.99, and 0.99 for the responses of bulk density, hygroscopicity, TPC, AOA, and vitamin C, respectively. These results suggested that the models used in this study were able to identify the optimum operating conditions of spray drying of honey powder.

TABLE 25.1   Physico-chemical Analyses of Honey, Aonla, and Basil Extract.

Properties	Honey	Aonla extract	Basil extract
Moisture (%)	18.8 ± 1.06	82.5 ± 0.72	82.3 ± 0.89
pH	3.9 ± 0.33	3.1 ± 0.16	4.3 ± 0.24
Percentage acidity	1.1 ± 0.03	2.5 ± 0.17	0.7 ± 0.14
HMF (mg/100 g)	5.49 ± 0.28	ND	ND
Vitamin C (mg/100 g)	ND	449 ± 4.05	7.3 ± 0.35
Total phenolic content (mg GAE/100 g)	38.62 ± 0.93	45.62 ± 1.02	48.55 ± 0.89
Antioxidant activity (% DPPH)	51.89 ± 1.08	65.36 ± 0.63	69.94 ± 1.48

HMF: hydroxymethylfurfural; ND: not detected; DHHP: 2,2-diphenyl picrylhydrazyl; GAE: gallic acid equivalents.

Values are mean of triplicate determinations.

TABLE 25.2 Significant Regression Coefficients of Second-order Polynomial Equation for the Responses.

Coefficients	Bulk density	Hygroscopicity	TPC	AOA	Vitamin C
β_0	0.611**	27.806**	71.6**	57.806**	83.433**
β_1	−0.115**	5.841**	−5.805**	−4.254**	−4.277**
β_2	−0.0542	−1.656**	3.008**	1.302**	1.738**
β_3					
β_4			0.286*		
β_5			0.178*		
β_{12}	−0.0165*	1.42**	0.472**	0.605**	
β_{13}					
β_{14}				−0.455*	
β_1^2		2.239**		−0.514**	
β_2^2			0.840**		0.249**
β_3^2	0.018**		−1.017**	−0.571**	−0.438**
β_4^2	0.021**		−0.734**		−0.232*
β_5^2	0.016**	0.436**	−0.419**	−0.877**	−0.378**
R^2	0.98	0.99	0.99	0.99	0.99
Adjusted R^2	0.97	0.98	0.98	0.98	0.99
Pred R^2	0.95	0.97	0.97	0.96	0.98
Adeq. precision	42.00	50.90	58.58	46.11	68.96
Lack-of fit	0.1528	0.0313	0.0048	0.0049	0.3029

*Significant at $p < 0.05$.

**Significant at $p < 0.0001$.

25.2.8.1 BULK DENSITY

ANOVA predicted that bulk density of honey powder had highly significant ($p < 0.0001$) positive linear effect of inlet temperature (x_1) and maltodextrin (x_3). Increasing inlet temperatures led to lower bulk density, which can be attributed to increased evaporation rate which led to porous or fragmented structure and lower shrinkage of the droplets. Similar result was observed by Julio et al. (2014) and Kha et al. (2010) in the spray drying of blue shark skin protein hydrolysate and gac fruit aril, respectively.

Maltodextrin concentration also resulted in a significant ($p < 0.0001$) decrease in bulk density, probably due to an increase in feed solids concentration which may lead to reduction in particle density due to a rapid surface crust formation at a certain solids concentration attained during the drying process (Fig. 25.1a). On the contrary, Nadeem et al. (2011) indicated significant increase of bulk density of spray-dried soluble sage and avocado powder when the carrier concentration was increased. The authors attributed this behavior to the high degree of agglomeration and to the particle structural collapse, which decreases the powder's volume.

25.2.8.2 HYGROSCOPICITY

The high content of low molecular weight sugars (mainly fructose and glucose) present in honey makes it very hygroscopic. Figure 25.1b presents the contour plots for effect of independent variables on powder hygroscopicity. Maltodextrin concentration was the variable that most affected powder's hygroscopicity. This is due to the fact that maltodextrin is a material with low hygroscopicity and confirms its efficiency as a carrier agent. Similar results were investigated by Moreira et al. (2008) and Tonon et al. (2008) during spray drying of acerola pomace and acai pulp, respectively. Inlet temperature and feed flow rate also influenced powder hygroscopicity. The lowest hygroscopicity values were obtained with increasing flow rates and decreasing temperatures which was related to their greater temperature gradient between the atomized feed and the drying air, resulting in a higher rate of heat transfer for water evaporation, thus producing low-water powders. Igual et al. (2014) explained the fact that the lower degree of hygroscopicity of the spray-dried lulo pulp powders produced at higher temperatures seems to be related to their lower water content.

25.2.8.3 TOTAL PHENOLIC CONTENT

The increase in TPC of honey powder with the increase in honey content might be attributed to the contribution of phenolic compounds of honey, aonla, and basil extract. Response surface plot (Fig. 25.1c) indicated that the effect of inlet temperature had significant negative linear effect on the TPC ($p < 0.0001$). With the increase in inlet air temperature, TPC decreased which can be attributed to high temperature resulting in considerable risk of thermal degradation of bioactive components due to excessive evaporation which could result in fissures and deformations in the wall materials, causing premature release of their contents. A significant ($p < 0.05$) positive influence of addition of basil extract and maltodextrin concentration on TPC response can be clearly seen in the regression model (Table 25.2). Retention of TPC of honey powder was increased with the addition of aonla and basil extract which indicated that TPC was well encapsulated by maltodextrin (Fig. 25.1d). Ramirez et al. (2015) studied the spray-drying encapsulation of model fruit juice (MFJ) and observed that the initial polyphenol content in the MFJ was generally positively affected by an increase in the maltodextrin feed solute concentration and a decrease in the air inlet temperature.

25.2.8.4 DPPH RADICAL SCAVENGING ACTIVITY

The ANOVA on the fitted model showed the significant positive linear effect of maltodextrin concentration (x_3), aonla extract (x_4), and basil extract (x_5) and negative linear effect of inlet temperature (x_1) on the AOA of honey powder ($p < 0.0001$) (Fig. 25.1e,f). As the inlet temperature increased from 160°C to 180°C, polyphenols present in honey, aonla, and basil extract get oxidized resulting in altered molecular structure and degrade thermally leading to lower AOA. Maltodextrin microcapsules promoted a gradual increase in phenolic content and higher AOA of honey powder due to high-DE of maltodextrin (DE-20). Aonla and basil extract ($p < 0.05$) had significant positive linear effect on AOA. The reason might be the addition of aonla and basil extract which is a rich source of natural antioxidants were well encapsulated by maltodextrin. These conclusions are insignificant consensus with the results published by Ho et al. (2015) who worked with spray drying of sim juice observed that increase in temperature from 180°C to 190°C had detrimental effect and positive effect of maltodextrin on the antioxidant retention of spray drying of sim juice. The positive correlation between TPC and AOA was observed to be $r = 0.938$ in honey powder using Pearson's correlation coefficients. Kha et al. (2010) also reported that the strong correlation between total carotenoid and the total AOA was found in spray-dried gac aril powder ($r = 0.94$).

25.2.8.5 VITAMIN C CONTENT

Initially vitamin C content was absent in honey, however, in this study, it was found that vitamin C content of honey powder increased with the incorporation of aonla extract. The response surface plots (figures) showed the significant positive linear effect of aonla extract (x_4) and negative linear effect of inlet temperature (x_1) on the vitamin C content of the honey powder ($p < 0.0001$).

Response surface plot (Fig. 25.1g,h) denoted that increasing inlet air temperature caused a reduction in vitamin C of honey powder, while maltodextrin increased retention of vitamin C because it acts as a wall material to prevent oxidative damage due to oxidation and hydrolysis (Ho et al., 2015).

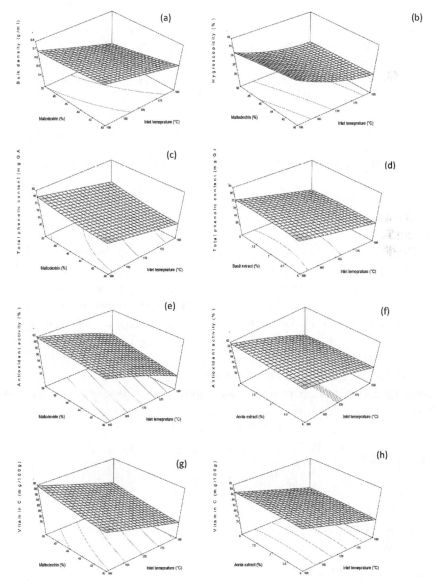

FIGURE 25.1   Effect of concentration of maltodextrin, aonla extract, basil extract, and inlet temperature on bulk density, hygroscopicity, TPC, AOA, and vitamin C content.

25.2.8.6   OPTIMIZATION

The optimization was carried out at selected ranges of inlet temperature, feed flow rate, maltodextrin concentration, aonla extract, and basil extract as 160–180°C, 0.08–0.13 mL/s, 40–50%, 6–8%, and 6–8%, respectively. The desired solution was obtained for the optimum covering criteria with value of 0.725 by applying the desirability function method. The solutions were analyzed after setting up the optimum spray-drying conditions as 170°C for inlet temperature, 0.11 mL/s for feed rate, 50% for maltodextrin concentration, 8% for aonla extract, and 6% for basil extract. The adequacy of the response surface equation was checked by comparing the experimental and predicted values as shown in Table 25.2. Negligible deviations were observed between the experimental and predicted values. The models obtained in this study could be used to optimize the spray-drying process of nutritionally rich honey powder.

25.3   CONCLUSION

The optimization of the spray-drying conditions for the honey powder was successfully executed using the Box–Behnken design of the RSM. The optimum honey powder produced was found nutritionally rich with a high functional value. This could be produced by spray drying by the addition of aonla extract, basil extract, and maltodextrin with low bulk density and hygroscopicity. Strong positive correlation among TPC and AOA of nutritionally rich honey powder developed was also confirmed ($r = 0.938$). The results obtained in this work showed the importance of the simultaneous investigation of processing and feed composition variables during the development of optimized dried honey powder.

KEYWORDS

- honey
- spray drying
- RSM
- antioxidant activity
- vitamin C

REFERENCES

AOAC. Official Methods of Analysis. Sugars and Sugar Products. In *Official Methods of Analysis of AOAC International*; Horwitz, W., Ed.; AOAC: Gaithersburg, MA, 2000; Vol. 2, pp 22–33.

AOAC. *Official Methods of Analysis*, 18th ed.; Association of Official Analytical Chemists: Washington, DC, 2005.

Cai, Y. Z.; Corke, H. Production and Properties of Spray-dried *Amaranthus betacyanin* Pigments. *J. Food Sci.* **2000**, *65*, 1248–1252.

Caliskan, G.; Dirim, S. N. The Effects of the Different Drying Conditions and the Amounts of Maltodextrin Addition during Spray Drying of Sumac Extract. *Food Bioprod. Process.* **2013**, *91*, 539–548.

Du, J.; Ge, Z. Z.; Xu, Z.; Zou, B.; Zhang, Y.; Li, C. M. Comparison of the Efficiency of Five Different Drying Carriers on the Spray Drying of Persimmon Pulp Powders. *Drying Technol.* **2014**, *32*, 1157–1166.

Egoville, M. J.; Sullivan, J. F.; Kozempel, M. F.; Jones, W. J. Ascorbic Acid Determination in Processed Potatoes. *J. Sci. Food Agric.* **1998**, *65*, 91–97.

Fazaeli, M.; Djomeh, Z. E.; Ashtari, A. K.; Omid, M. Effect of Spray Drying Conditions and Feed Composition on the Physical Properties of Black Mulberry Juice Powder. *Food Bioprod. Process.* **2012**, *90*, 667–675.

Goula, A. M.; Adamopoulos, K. G.; Kazakis, N. A. Influence of Spray Drying Conditions on Tomato Powder Properties. *Drying Technol.* **2004**, *22*, 1129–1151.

Ho, L. P.; Pham, A. H.; Le, V. V. M. Effects of Core/wall Ratio and Inlet Temperature on the Retention of Antioxidant Compounds during the Spray Drying of Sim (*Rhodomyrtus tomentosa*) Juice. *J. Food Process. Technol.* **2015**, *39* (6), 2088–2095. DOI:10.1111/jfpp.12452

Igual, M.; Ramires, S.; Mosquera, L. H.; Martinez-Navarrete, N. Optimization of Spray Drying Conditions for Lulo (*Solanum quitoense L.*) Pulp. *Powder Technol.* **2014**, *256*, 233–238.

Julio, C.; Rodriguez, D.; Tonon, R. V.; Hubinger, M. D. Spray Drying of Blue Shark Skin Protein Hydrolysate: Physical, Morphological, and Antioxidant Properties. *Drying Technol.* **2014**, *32*, 1986–1996.

Kamboj, R.; Bera, B. M.; Nanda, V. Evaluation of Physico-chemical Properties, Trace Metal Content and Antioxidant Activity of Indian Honeys. *Int. J. Food Sci. Technol.* **2013**, *48*, 578–587.

Kha, T. C.; Nguyen, M. H.; Roach, P. D. Effect of Spray Drying Conditions on the Physicochemical and Antioxidant Properties of Gac (*Momordica cochinchinensis*) Fruit Aril Powder. *J. Food Eng.* **2010**, *98*, 385–392.

Kurozawa, L. E.; Park, K. J.; Hubinger, M. D. Effect of Maltodextrin and Gum Arabic on Water Sorption and Glass Transition Temperature of Spray Dried Chicken Meat Hydrolysate Protein. *J. Food Eng.* **2009**, *91*, 287–296.

Liu, X.; Cui, C.; Zha, M.; Wang, J.; Luo, W.; Yang, B.; Jiang, Y. Identification of Phenolics in the Fruit of *Emblica* (*Phyllanthus emblica L.*) and Their Antioxidant Activities. *Food Chem.* **2008**, *109*, 909–915.

Louveaux, J.; Maurizio, A.; Vorwohl, G. Methods of Melissopalynology. *Bee World* **1978**, *59*, 139–157.

Luo, W.; Zhao, M.; Yang, B.; Shen, G.; Rao, G. Identification of Bioactive Compounds in *Phyllenthus emblica L.* Fruit and Their Free Radical Scavenging Activities. *Food Chem.* **2009**, *114*, 499–504.

Montgomery, D. C. *Design and Analysis of Experiments;* Wiley M. M. R: New York, 2001; pp 416–419.

Moreira, G. E. G.; Costa, M. G. M.; Souza, A. C. R.; Brito, E. S.; Medeiros, M. F. D.; Azeredo, H. M. C. Physical Properties of Spray Dried Acerola Pomace Extract as Affected by Temperature and Drying Aids. *Food Sci. Technol.* **2008,** *42,* 641–645.

Nadeem, H. S.; Torun, M.; Ozdemir, F. Spray Drying of the Mountain Tea (*Sideritis strica*) Water Extract by Using Different Hydrocolloid Carriers. *LWT Food Sci. Technol.* **2011,** *44,* 1626–1635.

Nayik, G. A.; Nanda, V. Physico-chemical, Enzymatic, Mineral and Colour Characterization of Three Different Varieties of Honeys from Kashmir Valley of India with a Multivariate Approach. *Polish J. Food Sci. Nutr.* **2015,** *65,* 101–108.

Oliveira, M. A.; Maia, G. A.; Figueiredo, R. W.; Souza, A. C. R.; Brito, E. S.; Azeredo H. M. C. Addition of Cashew Tree Gum Tomaltodextrin-based Carriers for Spray Drying of Cashew Apple Juice. *Int. J. Food Sci. Technol.* **2009,** *44,* 641–645.

Parveen, K.; Khatkar, B. S. Physico-chemical Properties and Nutritional Composition of Aonla *(Emblica officinalis)* Varieties. *Int. Food Res. J.* **2015,** *22,* 2358–2363.

Prakash, P.; Gupta, N. Therapeutic Uses of *Ocimum sanctum Linn* (Tulsi) with a Note on Eugenol and Its Pharmacological Actions: A Short Review. *Indian J. Physiol. Pharmacol.* **2005,** *49,* 125–131.

Ramirez, M. J.; Giraldo, G. I.; Orrego, C. E. Modeling and Stability of Polyphenol in Spray-dried and Freeze-dried Fruit Encapsulates. *Powder Technol.* **2015,** *277,* 89–96.

Rangana, S. *Handbook of Analysis and Quality Control for Fruits and Vegetable Products,* 2nd ed.; Tata McGraw-Hill Publishing Company Ltd: Hyderabad, 2000.

Suhag, Y.; Nanda, V. Optimization of Process Parameters to Develop Nutritionally Rich Spray Dried Honey Powder with Vitamin C Content and Antioxidant Properties. *Int. J. Food Sci. Technol.* **2015,** *50,* 1771–1777.

Tekindal, M. A.; Bayrak, H.; Ozkaya, B.; Genc, Y. Box-behnken Experimental Design in Factorial Experiments: The Importance of Bread for Nutrition and Health. *Turk. J. Field Crop* **2012,** *17,* 115–123.

Tonon, R.V.; Brabet, C.; Hubinger, M. D. Influence of Process Conditions on the Physico-chemical Properties of Acai Powder Produced by Spray Drying. *J. Food Eng.* **2008,** *88,* 411–418.

INDEX

A

Acrylamide in food products, 250
  chemistry, 252
    physical parameters of, 253
    structure of, 253
  deep frying, 251
  detection, analytical methods, 263
    GC-MS with bromination, 266–267
    GC-MS without derivatization,
      268–269
    HPLC based methods, 278–280
    LC-based methods, 269, 274, 277–278
  detoxification/mitigation strategies
    additives, impact of, 263
    bakery products, 260
    potato products, 259–260
    raw material, impact of, 260–261
    technology, impact of, 261–263
  formation in food, mechanism of, 254
    almonds, 257–258
    alternative routes, 255–256
    bread, 257
    cereal-based foods, 256–257
    coffee, 258
    Maillard reaction, 255
    potatoes, 256
    processed foods, 255
  processing contaminants, 252
  risk associated
    with exposure and dietary limits, 258
Anacardium occidentale L., 512
Analytical methods
  detection, 263
    GC-MS with bromination, 266–267
    GC-MS without derivatization,
      268–269
    HPLC based methods, 278–280
    LC-based methods, 269, 274, 277–278
Antihypertensive activity
  ACE-I inhibition, 204

ACE-I inhibition, mechanism of action,
    204–205
  fish protein hydrolysates, ACE inhibitory
    activity, 205–206
  hypertension, 204
  rennin-angiotensin system, 204–205
Antioxidant activity (AOA), 596
Aquatic environment, 46
  alginate lyases and agarase, 49–50
  carrageenases, 50
  chitinase and chitosanase, 49
  collagenase, 50
  collagenolytic enzymes, 50
  extremozymes, 50–51
  lipases, 48
  proteases, 47–48

B

Banana plants, 70
Bioactive ingredients
  omega-3 fatty acids, 408–409
  polyphenols/flavors, 409–410
Biodegradable polymers, 562
  film characterization
    experimental design, 565–566
    mechanical properties, 564
    solubility, 565
    water vapor permeability (WVP),
      564–565
  film formation, 564
  materials and methods, 563
  optimization, 571
  results and discussion
    solubility, 569
    tensile strength, 567–568
    water vapor permeability (WVP), 570
Blood–nerve barrier (BNB), 11
Box–Behnken design, 596
Brown bread
  physical properties of
    color attributes, 148–149

loaf weight, 143–145
proofing of dough, 141
sensory analysis, 149–151
texture profile analysis, 145–146
volume of baked product, 143–145
water retention, 143–145
sensory quality of, 133
baked bread loaves, volume, 137
bread crumb, 137–138
brown bread, preparation, 135–136
chemical composition, 137
crumb color, 138
modification, 134
proofing of dough, 137
raw materials, 134
RS AND KF, effect, 139–141
sensory evaluation, 138
starch isolation, 134
statistical analysis, 139
water retention, 137
weight, 137
Bulk density (BD), 551–552

C

Capparis decidua, 446
materials and methods
antinutrients, 458
minerals, 458
moisture and ash, 457
results and discussion
antinutrients, 463, 467
fermentation, 463, 465
fresh kair, composition of, 459
minerals and vitamins, 459, 463, 466
nutrients in ethanolic extract, 465–466
processing, effects of, 463
statistical analyses
blanching, 458
food processing, 458
Cashew apple waste residue for cultivating, composting
material and methods
compost, preparation of, 513
paddy straw mushroom, cultivation of, 514
spawn, preparation of, 513–514
results and discussion, 515–519

Center for Epidemiologic Studies
Depression Scale (CES-D), 7
Central composite rotatable design (CCRD), 542
Coatings
ready-to-eat, improving quality, 299
micro encapsulation, 301–302
natural compounds, use of, 301–302
Collagen, 163–165
basement membrane, 169
conversion of, 170
fibril-forming collagens, 166–169
hydrocolloids, functional properties of, 171, 173
types of, 166
Cross-linking
chemical cross-linking, 182–183
emulsifying capacity, 185
enzymatic cross-linking, 183
foaming properties, 184–185
gel strength, 184
melting point, 184
rheological properties, 186–188
Culinary banana, 108–109
brown bread, physical properties of
color attributes, 148–149
loaf weight, 143–145
proofing of dough, 141
sensory analysis, 149–151
texture profile analysis, 145–146
volume of baked product, 143–145
water retention, 143–145
chemical composition, 114–115
functional properties
freeze-thaw stability, 111–112
paste clarity, 111–112, 117–118
pasting properties, 112, 118
solubility, 111, 116–117
starch swelling power, 111, 116–117
water holding capacity, 111, 116–117
materials and methods
chemical analysis, 110
raw materials, 110
starch isolation, 110
MUSA, 72
banana, 73–75
biochemical properties, ripening stages effect, 75–78

nutritional properties, ripening stages
    effect, 75–78
plantain as value added products,
    78–79
production, 73
peel, chemical properties and bioactive
    compounds, 81–82
cellulose, extraction of, 85–87
utilization as by-products, 83
value added products, 84–85
plantain starch, 79–81
sensory quality of brown bread, 133
baked bread loaves, volume, 137
bread crumb, 137–138
brown bread, preparation, 135–136
chemical composition, 137
crumb color, 138
modification, 134
proofing of dough, 137
raw materials, 134
RS AND KF, effect, 139–141
sensory evaluation, 138
starch isolation, 134
statistical analysis, 139
water retention, 137
weight, 137
structural analysis of
    FTIR spectrum, 112, 120–121
    SEM, morphological analysis, 113, 122
    TGA, thermal stability, 122–123
    thermogravimetric analysis, 113
    X-ray diffraction, 112, 119–120
textural properties, 133
baked bread loaves, volume, 137
bread crumb, 137–138
brown bread, preparation, 135–136
chemical composition, 137
crumb color, 138
proofing of dough, 137
raw materials, 134
RS AND KF, effect, 139–141
sensory evaluation, 138
starch isolation and modification, 134
statistical analysis, 139
water retention, 137
weight, 137
type-II, development of
    autoclaving method, 113

chemical analysis of, 114
cooling method, 113
enzyme debranching method, 113–114
statistical analysis, 114
type-III, development of
chemical analysis, 125–127
enzyme concentration on starch
    debranching, effect, 125
FTIR spectra, 128–129
hydrothermal method, RS production,
    123–124
SEM, 127–128
storage temperature, effect of, 125
thermogravimetric analysis, 129
utilization
bioactive compounds, encapsulation,
    87–91
dehydration of, 91–93
mathematical modeling for thin layer
    drying of, 93–95
maturity stages, 71
value addition
bioactive compounds, encapsulation,
    87–91
dehydration of, 91–93
mathematical modeling for thin layer
    drying of, 93–95
maturity stages, 71
variety of banana, 71
Custard made from composite, quality
    characteristics
descriptive sensory evaluation, 529
dessert samples
    preparation of, 528–529
Eleusinian grain, 525
germinated
    chemical analysis of, 527–528
materials and methods
    raw materials, procurement of, 526
powder
    preparation of, 528
results and discussions
    characterization of, 533–534
    descriptive sensory quality evaluation,
        534–536
    final product, nutritional analysis, 536
    flours, analysis of, 530

grains/flour, physicochemical
    characteristics of, 530–532
rice, 525
seeds
    germination, 526–527
soybean
    protein concentration, 525
statistical analyses, 530
ungerminated flours
    chemical analysis of, 527–528

D

Dense phase carbon dioxide (DPCD), 303
Detection
    analytical methods, 263
        GC-MS with bromination, 266–267
        GC-MS without derivatization,
            268–269
        HPLC based methods, 278–280
        LC-based methods, 269, 274, 277–278
Detoxification/mitigation strategies
    additives, impact of, 263
    bakery products, 260
    potato products, 259–260
    raw material, impact of, 260–261
    technology, impact of, 261–263
Docosahexaenoic acid (DHA), 216
Dorsal root ganglia (DRG), 11
Dorsal root ganglion (DRG), 4

E

Edible polymeric films
    ready-to-eat, improving quality, 299
        micro encapsulation, 301–302
        natural compounds, use of, 301–302
Eicosapentaenoic acid (EPA), 216
Electrolyzed water (EW), 296
Electrospinning as novel delivery vehicle
    encapsulation, 437
        carbohydrate based, 440–441
        gelatin, 438–439
        methods of, 432–433
        protein-based vehicles, 439–440
    food industries
        applications for, 444–446
    history of, 428–429
    materials for food technology, 433–434
        food product innovation, 437

nanotechnology, inspiration from
    nature, 435
    potential food applications, 436
principles of, 429–431
studies, 427
three-dimensional (3D) food printing
    robotics-based food manufacturing,
        443–444
zein, nanostructured electrospun
    interlayer, 441–442
    preparation of, 443
Emerging technologies
    ready-to-eat, improving quality
        high power ultrasound, 311–312
        laser, 308–309
        light-emitting-diodes (LED), 308
        pulsed electric field (PEF), 311
        pulsed light (PL), 309–311
Encapsulation
    electrospinning as novel delivery vehicle,
        437
        carbohydrate based, 440–441
        gelatin, 438–439
        methods of, 432–433
        protein-based vehicles, 439–440
Expansion ratio (ER), 550–551

F

Film characterization
    experimental design, 565–566
    mechanical properties, 564
    solubility, 565
    water vapor permeability (WVP),
        564–565
Fish gelatin
    amino acid composition, 179–180
    cross-linking
        chemical cross-linking, 182–183
        emulsifying capacity, 185
        enzymatic cross-linking, 183
        foaming properties, 184–185
        gel strength, 184
        melting point, 184
        rheological properties, 186–188
    extraction of, 173–176
    gel strength, 178
    issues

gel properties, extraction conditions
effect, 181–182
inferior properties of, 180
melting point, 178–179
physico–chemical properties, 177–178
Fish processing waste protein hydrolysate
antihypertensive activity
ACE-I inhibition, 204
ACE-I inhibition, mechanism of
action, 204–205
fish protein hydrolysates, ACE
inhibitory activity, 205–206
hypertension, 204
rennin-angiotensin system, 204–205
antimicrobials activity, 209–210
antioxidant, 207
anticoagulant activity, 209
fish protein hydrolysates, 208
oxidation mechanism and effect on
food, 206–207
bioactive protein hydrolysates, 201–202
biologically active peptides, 200–201
challenges in utilization of fish waste,
211
hydrolysis
enzymatic, 203
influencing factors, 202–203
mechanism of, 202
opportunities, 211
research gaps, 211
Fish protein hydrolysates (FPH), 61–62
Food irradiation, 303, 320, 329–330
FAO/IAEA/WHO, 322
food borne pathogens, 332
gamma irradiation, 304–305
high dose, applications, 333–334
insect disinfestations, 332
microbial decontamination, 333
shelf-life extension, 332–333
sprout inhibition, 331
ultraviolet radiation (UV-C), 305–307
United States Atomic Energy
Commission (USAEC), 321
Foods preservation
hurdle technology (HT), 352
bakery products, 368–369
dairy product, 366–368
fish products, 362–366

fruit and vegetable products, 355–362
homeostasis, 354
meat, 362–366
metabolic exhaustion, 354
miscellaneous foods, 369–370
multi-target preservation, 355
poultry, 362–366
stress reactions, 355
types, 353–354
Formation in food
mechanism of, 254
almonds, 257–258
alternative routes, 255–256
bread, 257
cereal-based foods, 256–257
coffee, 258
Maillard reaction, 255
potatoes, 256
processed foods, 255
Fourier transforms infrared (FTIR)
spectra, 112
Functional properties
culinary banana resistant starch
freeze-thaw stability, 111–112
paste clarity, 111–112, 117–118
pasting properties, 112, 118
solubility, 111, 116–117
starch swelling power, 111, 116–117
water holding capacity, 111, 116–117
Fungal retting technology
jute, 494
accelerated retting, 502
chemical retting, 502
enzymatic retting, 503

G

Gelatin from cold water fish species
bovine spongiform encephalopathy
(BSE), 161
collagen, 163–165
basement membrane, 169
conversion of, 170
fibril-forming collagens, 166–169
hydrocolloids, functional properties of,
171, 173
types of, 166
gelatin, 162–163
skin of, 161–162

Germination process, 524
Green leaf protein concentrate
    anti-nutritional importance of, 577–579
    cytotoxicity study
        importance of, 585
    extruded food incorporating LPC using
        extrusion technology
        preparation of, 586–587
    medicinal importance, 579–580
    nutritional importance of, 577–579
    pasta (an extruded food) by incorporating
        preparation of, 587
    processing importance of, 580–581
    protein concentrate, functional properties
        application, 584–585
        emulsions, 583
        extraction optimization, 584–585
        foaming, 583
        gelation/coagulation, 583–584
        solubility, 582–583
    protein deficiency, 581
    storage studies
        importance of, 587–588

H

Hasegawa Dementia Scale (HDS-R), 6
Hericium erinaceus, 4
    medicinal properties of
        alcohol-based extracts, 6
        blood–nerve barrier (BNB), 11
        Center for Epidemiologic Studies
            Depression Scale (CES-D), 7
        chemotherapeutic, 5
        dorsal root ganglia (DRG), 11
        double-blind trial, 6
        drug therapy, 10
        Hasegawa Dementia Scale (HDS-R), 6
        Indefinite Complaints Index (ICI), 7
        Kupperman Menopausal Index (KMI),
            7
        methicillin-resistant *Staphylococcus
            aureus* (MRSA), 5
        microsurgical treatments, 10
        mushroom, 7
        nerve injury, 10
        neurotrophic effects, 7
        NG108-15 cells, 8–9
        p38 MAPK in, 11
        Pittsburgh Sleep Quality Index (PSQI),
            7
        protein kinase B (Akt), 11
        rats, CNS, 6
    neuroregenerative potential
        Akt promotes cell, 17
        axon bundles, 20
        axonal reinnervation, 17
        double-labeled immunofluorescence, 17
        hot plate test, 12
        negative control group, 12
        negative control rats, 15
        neurons, 15
        peroneal nerve crush, 14
        rat endothelial cell antigen (RECA-1),
            15
        rat experiment, 11
        thermal hyperalgesia, 14
High-temperature short time (HTST), 467,
    542
Hurdle technology (HT)
    foods preservation, 352
        bakery products, 368–369
        dairy product, 366–368
        fish products, 362–366
        fruit and vegetable products, 355–362
        homeostasis, 354
        meat, 362–366
        metabolic exhaustion, 354
        miscellaneous foods, 369–370
        multi-target preservation, 355
        poultry, 362–366
        stress reactions, 355
        types, 353–354
Hydrolysis
    enzymatic, 203
    influencing factors, 202–203
    mechanism of, 202

I

Indefinite Complaints Index (ICI), 7
Indian livestock industry
    raw meat, 226–227
        export, laws, 238–239
        organized sector of, 227–229
        quality aspect of meat, laws, 236–238
        statistical overview, 235–236

J

Jute, fungal retting technology, 494
  alternate retting technologies
    accelerated retting, 502
    chemical retting, 502
    enzymatic retting, 503
    ribbon retting, 502
  fiber
    composition of, 498–499
  global climatic change, 505–506
  microbes of, 503–505
  minimum water retting
    need of, 505–506
  retting process, 499
    biochemistry of, 501
    microbial process, 500
    physical process, 500
  socio-economic importance, 495–496
  stem
    anatomy of, 496–497

K

Kinnow peel-rice based expanded snacks
  extruder process, 544
  feed preparation, 543–544
  materials and methods, 543
  optimizing level of treatment, 543–544
  product responses
    bulk density, 545
    data analysis, 546–548
    expansion ratio (ER), 545
    experimental design, 546–548
    hardness, 546
    optimization, 548
    specific mechanical energy (SME),
      544–545
    statistical analysis, 548
    water absorption index (WAI), 545
    water solubility index (WSI), 546
  result and discussions
    bulk density (BD), 551–552
    expansion ratio (ER), 550–551
    hardness, 555
    optimization, 556–557
    overall acceptability, 555–556
    specific mechanical energy, 548–550

    water absorption index (WAI),
      552–553
    water solubility index (WSI), 553–554
  unproductive matter, 543
Kupperman Menopausal Index (KMI), 7

L

Lactic acid bacteria (LAB), 28

M

Materials for food technology, 433–434
  food product innovation, 437
  nanotechnology, inspiration from nature,
    435
  potential food applications, 436
Medicinal properties
  Hericium erinaceus
    alcohol-based extracts, 6
    blood–nerve barrier (BNB), 11
    Center for Epidemiologic Studies
      Depression Scale (CES-D), 7
    chemotherapeutic, 5
    dorsal root ganglia (DRG), 11
    double-blind trial, 6
    drug therapy, 10
    Hasegawa Dementia Scale (HDS-R), 6
    Indefinite Complaints Index (ICI), 7
    Kupperman Menopausal Index (KMI),
      7
    methicillin-resistant *Staphylococcus
      aureus* (MRSA), 5
    microsurgical treatments, 10
    mushroom, 7
    nerve injury, 10
    neurotrophic effects, 7
    NG108-15 cells, 8–9
    p38 MAPK in, 11
    Pittsburgh Sleep Quality Index (PSQI),
      7
    protein kinase B (Akt), 11
    rats, CNS, 6
Methicillin-resistant *Staphylococcus aureus*
  (MRSA), 5
Microencapsulation of bioactive food
  ingredients, 400–401
  bioactive ingredients
    applications, 413–414
    omega-3 fatty acids, 408–409

polyphenols/flavors, 409–410
calcium, 411
characterization of
  encapsulation efficiency, 414
  in vitro release studies, 416
  morphology, 415
  particle size, 415
  payload, 415
  powder flow properties, determination,
    415
  stability, 416
chemical methods
  coacervation, 406
  cocrystallization, 407
  inclusion complexation, 407–408
  liposome entrapment, 406–407
  supercritical fluids, 406
controlled release mechanism, 416
  active agent by degradation, release,
    417–418
  PH-controlled release, 418
  pressure-activated release, 419
  solvent-activated release, 418
  swelling controlled release, 417
  temperature-sensitive release, 418
enzymes, 412
microorganism, 412
peptide, 413
physical methods
  centrifugal co-extrusion, 404–405
  extrusion, 405
  fluid bed coating, 404
  freeze-drying/lyophilization, 405
  spinning disk, 404–405
  spray chilling or spray cooling, 404
  spray-drying, 403
protein hydrolysate, 413
rate of core materials
  factors affecting, 420
technologies, 401
  coating material, 402–403
  core material, 402
vitamins and minerals, 410–411
Mitogen-activated protein kinase (MAPK), 4
Modified atmosphere packaging (MAP),
  297–298
MUSA, 72
  banana, 73–75

biochemical properties, ripening stages
    effect, 75–78
nutritional properties, ripening stages
    effect, 75–78
plantain as value added products, 78–79
production, 73

N

Neuroregenerative potential
  Hericium erinaceus
    Akt promotes cell, 17
    axon bundles, 20
    axonal reinnervation, 17
    double-labeled immunofluorescence,
      17
    hot plate test, 12
    negative control group, 12
    negative control rats, 15
    neurons, 15
    peroneal nerve crush, 14
    rat endothelial cell antigen (RECA-1),
      15
    rat experiment, 11
    thermal hyperalgesia, 14

O

Oregano essential oil, effect
  materials and methods
    emulsion and spray drying,
      preparation, 218
    raw materials, 217
  microencapsulates, characterization
    color analysis, 219
    encapsulation efficiency, 218–219
    flow properties, 219
    moisture content, 218
    oxidative stability, determination, 219
    particle morphology, 218
  results and discussion
    microencapsulated fish oil, physical
      properties, 219–221
    oxidative stability of
      microencapsulated fish oil, 221–222

P

Physical methods

bioactive food ingredients, microencapsulation
centrifugal co-extrusion, 404–405
extrusion, 405
fluid bed coating, 404
freeze-drying/lyophilization, 405
spinning disk, 404–405
spray chilling or spray cooling, 404
spray-drying, 403
Pittsburgh Sleep Quality Index (PSQI), 7
Pleurotus sajor caju, 512
Probiotics
facultative aerobes, 27
history and definition, 27
acid, 32
antimicrobial activity, 32–33
Bifidobacterium, 28
bile tolerance, 32
food products, 29
health, effects of, 30
lactic acid bacteria (LAB), 28
Lactobacillus, 28
lactose intolerance, 31
Phoenician era, 29
The Prolongation of Life, 28
human intestinal tract, 26
mechanism
biotechnology and probiotics, 37–38
commercial products, 35–36
food products, 33–34
microorganisms, 33
product considerations, 34–35
Product decontamination
ready-to-eat, improving quality
chemical treatments, 296–297
dense phase carbon dioxide (DPCD), 303
electrolyzed water (EW), 296
modified atmosphere packaging (MAP), 297–298
ozone, 294–295
pressure treatment, 302–303
ultrasound, 295–296
Product responses
kinnow peel-rice
bulk density, 545
data analysis, 546–548
expansion ratio (ER), 545

experimental design, 546–548
hardness, 546
optimization, 548
specific mechanical energy (SME), 544–545
statistical analysis, 548
water absorption index (WAI), 545
water solubility index (WSI), 546
Protein kinase B (Akt), 11

R

Radiation processing
chemical methods, 339
2-alkylcyclobutanones, detection of, 340
biological methods, 340–341
radiation-induced hydrocarbons, detection, 340
dose rate, 320
food irradiation, 320, 329–330
FAO/IAEA/WHO, 322
food borne pathogens, 332
high dose, applications, 333–334
insect disinfestations, 332
microbial decontamination, 333
shelf-life extension, 332–333
sprout inhibition, 331
United States Atomic Energy Commission (USAEC), 321
global population, 319
international status, 341
irradiated foods, wholesomeness
detection of, 337
macronutrients, 336
microbial quality, 335–336
micronutrients, 336
microbial contaminations, 319
national status, 342, 344
physical method
electron spin resonance (ESR), 338
thermoluminescence (TL), 339
process
electron beam, 328–329
gamma irradiator, 324–327
radiation sources, 323
x-ray irradiators, 328–329
Radio frequency (RF), 376
cooks products volumetrically, 378

critical process parameters, mechanisms,
380–381
dielectric properties of foods, factors
affecting, 382
factors affecting, 382–383
electro heat technology, 377
electrostatic discharge, 378
equipment design, 390–392
conventional heating methods, 393
electro heating methods, 393–394
food quality
comparison of, 378
fundamental principles, 379
target material, 380
penetration depth, 383
food processing applications, 384–390
Rat endothelial cell antigen (RECA-1), 15
Raw meat, safety considerations
Indian livestock industry, 226–227
export, laws, 238–239
organized sector of, 227–229
quality aspect of meat, laws, 236–238
statistical overview, 235–236
Indian slaughterhouses for domestic
supply
conditions of, 234–235
unorganized sector, challenges of
agriculture/land availability, 229–230
biological factors, 231–233
breed of animal, 230
chemical factors, 234
consumer preferences, 230–231
economic aspects, 230
enforcement agencies, apathy of, 230
miscellaneous concerns, 231
quality parameters of meat, 239–244
reduced returns, 230
technical barrier, 231
Ready-to-eat, improving quality
coatings, 299
micro encapsulation, 301–302
natural compounds, use of, 301–302
edible polymeric films, 299
micro encapsulation, 301–302
natural compounds, use of, 301–302
emerging technologies
high power ultrasound, 311–312
laser, 308–309

light-emitting-diodes (LED), 308
pulsed electric field (PEF), 311
pulsed light (PL), 309–311
food irradiation, 303
gamma irradiation, 304–305
ultraviolet radiation (UV-C), 305–307
fungal growth, 293
humidity control, 293
International Fresh-Cut Produce
Association (IFPA), 290
polyphenol oxidase (PPO), 292
product decontamination
chemical treatments, 296–297
dense phase carbon dioxide (DPCD),
303
electrolyzed water (EW), 296
modified atmosphere packaging
(MAP), 297–298
ozone, 294–295
pressure treatment, 302–303
ultrasound, 295–296
temperature management, 293
Response surface methodology (RSM), 542,
596

S

Seafood industry
aquatic environment, 46
alginate lyases and agarase, 49–50
carrageenases, 50
chitinase and chitosanase, 49
collagenase, 50
collagenolytic enzymes, 50
extremozymes, 50–51
lipases, 48
proteases, 47–48
carotenoid pigment extraction, 58
caviar and fish roe
production and purification of, 57–58
classification of, 46
fishery products, production, 59
fish sauce, 60
fish silage, 60–61
FPH, 61–62
ripening of fish, 60
flavors
extraction of, 58–59
membranes and organs

enzymatic removal of, 56
pearl essence
    extraction, 56
PUFA-enriched fish oils, 57
recovery and purification
    chromatographic techniques, 52–53
    precipitation method, 52
    solution methods, 53–54
    traditional methods, 51–52
specificity of, 45–46
squid skin, removal of, 55–56
stick water
    viscosity reduction of, 59
Surimi, 54–55
tissue degradation, 55
Specific mechanical energy (SME),
    544–545
Spray-dried honey powder with vitamin C,
    development
    materials and methods
        basil, 598
        bulk density, 598
        DPPH radical scavenging activity, 599
        experiment design, 600
        HMF content of aonla, 598
        honey, 598
        hygroscopicity, 598–599
        moisture, 598
        percentage acidity, 598
        PH, 598
        sample preparation, 597–598
        spray drying, 597–598
        total phenolic content (TPC), 599
        vitamin C content, 599
    Ocimum basilicum, 597
    results and discussion, 601
        bulk density, 602–603
        DPPH radical scavenging activity, 604
        hygroscopicity, 603
        optimization, 606
        total phenolic content (TPC), 603–604
        vitamin C content, 604–605
Structural analysis
    culinary banana resistant starch
        FTIR spectrum, 120–121
        FTIR, spectra, 112
        SEM, morphological analysis, 113, 122
        TGA, thermal stability, 122–123

thermogravimetric analysis, 113
X-ray diffraction, 112, 119–120

T

Textural properties
    culinary banana resistant starch, 133
        baked bread loaves, volume, 137
        bread crumb, 137–138
        brown bread, preparation, 135–136
        chemical composition, 137
        crumb color, 138
        proofing of dough, 137
        raw materials, 134
        RS AND KF, effect, 139–141
        sensory evaluation, 138
        starch isolation and modification, 134
        statistical analysis, 139
        water retention, 137
        weight, 137
Total phenolic content (TPC), 596, 599,
    603–604
Type-II, development of culinary banana
    resistant starch
    modification
        autoclaving method, 113
        chemical analysis of, 114
        cooling method, 113
        enzyme debranching method, 113–114
        statistical analysis, 114
Type-III, development of culinary banana
    resistant starch
    modification
        enzyme concentration on starch
            debranching, effect, 125
        FTIR spectra, 128–129
        hydrothermal method, RS production,
            123–124
    modification
        chemical analysis, 125–
            127
        SEM, 127–128
        storage temperature, effect of, 125
        thermogravimetric analysis, 129

U

Underutilized crops of India, value addition
    extrudate products, types of

animal feed products, 485
meat and meat items, 485
pasta products, 484–485
pet food, 485
ready-to-eat products, 484
texturized vegetable protein, 484
extrusion cooking technology, 478
carbohydrates, 480
dietary fiber, 481
lipid complexation, 481–482
protein, 481
source of raw material, 479–480
starch, 480–481
sugars, 480
millets
Amaranthus hypochondriacus, 474
buckwheat, 473–474
Eleusine coracana, 474–475
Macrotyloma uniflorum, 477
Oryza sativa L., 476
Pennisetum americanum, 475
Sorghum bicolor L. moench, 477
pseudocereals
Amaranthus hypochondriacus, 474
buckwheat, 473–474
Eleusine coracana, 474–475
Macrotyloma uniflorum, 477

Oryza sativa L., 476
Pennisetum americanum, 475
Sorghum bicolor L. moench, 477
value addition, 477–478
United States Atomic Energy Commission
(USAEC), 321
Unorganized sector, challenges
raw meat, safety considerations
agriculture/land availability, 229–230
biological factors, 231–233
breed of animal, 230
chemical factors, 234
consumer preferences, 230–231
economic aspects, 230
enforcement agencies, apathy of, 230
miscellaneous concerns, 231
quality parameters of meat, 239–244
reduced returns, 230
technical barrier, 231
Unproductive matter, 542

W

Water absorption index (WAI), 545,
552–553
Water solubility index (WSI), 546, 553–554
Water vapor permeability (WVP), 564–565